6th International Conference on Advances in Optoelectronics and Micro/Nano-optics (AOM 2017)

Journal of Physics: Conference Series Volume 844

Nanjing, China
23 – 26 April 2017

ISBN: 978-1-5108-4240-3
ISSN: 1742-6588

Printed from e-media with permission by:

Curran Associates, Inc.
57 Morehouse Lane
Red Hook, NY 12571

Some format issues inherent in the e-media version may also appear in this print version.

Printed by Curran Associates, Inc. (2017)

For permission requests, please contact the Institute of Physics
at the address below.

Institute of Physics
Dirac House, Temple Back
Bristol BS1 6BE UK

Phone: 44 1 17 929 7481
Fax: 44 1 17 920 0979

techtracking@iop.org

Additional copies of this publication are available from:

Curran Associates, Inc.
57 Morehouse Lane
Red Hook, NY 12571 USA
Phone: 845-758-0400
Fax: 845-758-2633
Email: curran@proceedings.com
Web: www.proceedings.com

Table of contents

Volume 844

6th conference on Advances in Optoelectronics and Micro/nano-optics
23–26 April 2017, Nanjing, China

Accepted papers received: 10 May 2017
Published online: 7 June 2017

Preface

Optoelectronic Materials and Devices

Micro/nano Optics, Nanophotonics

Plasmonics and Metamaterials

Optical Information Processing

Nano-biophotonics and Life Sciences

Laser Spectroscopy and Microscopy

Micro/nano Manufacturing and Metrology

Fiber-based Micro/nano-optics and Technologies

Optical Super Resolution Microscopy

IOP Conf. Series: Journal of Physics: Conf. Series **844** (2017) 011001 doi:10.1088/1742-6596/844/1/011001

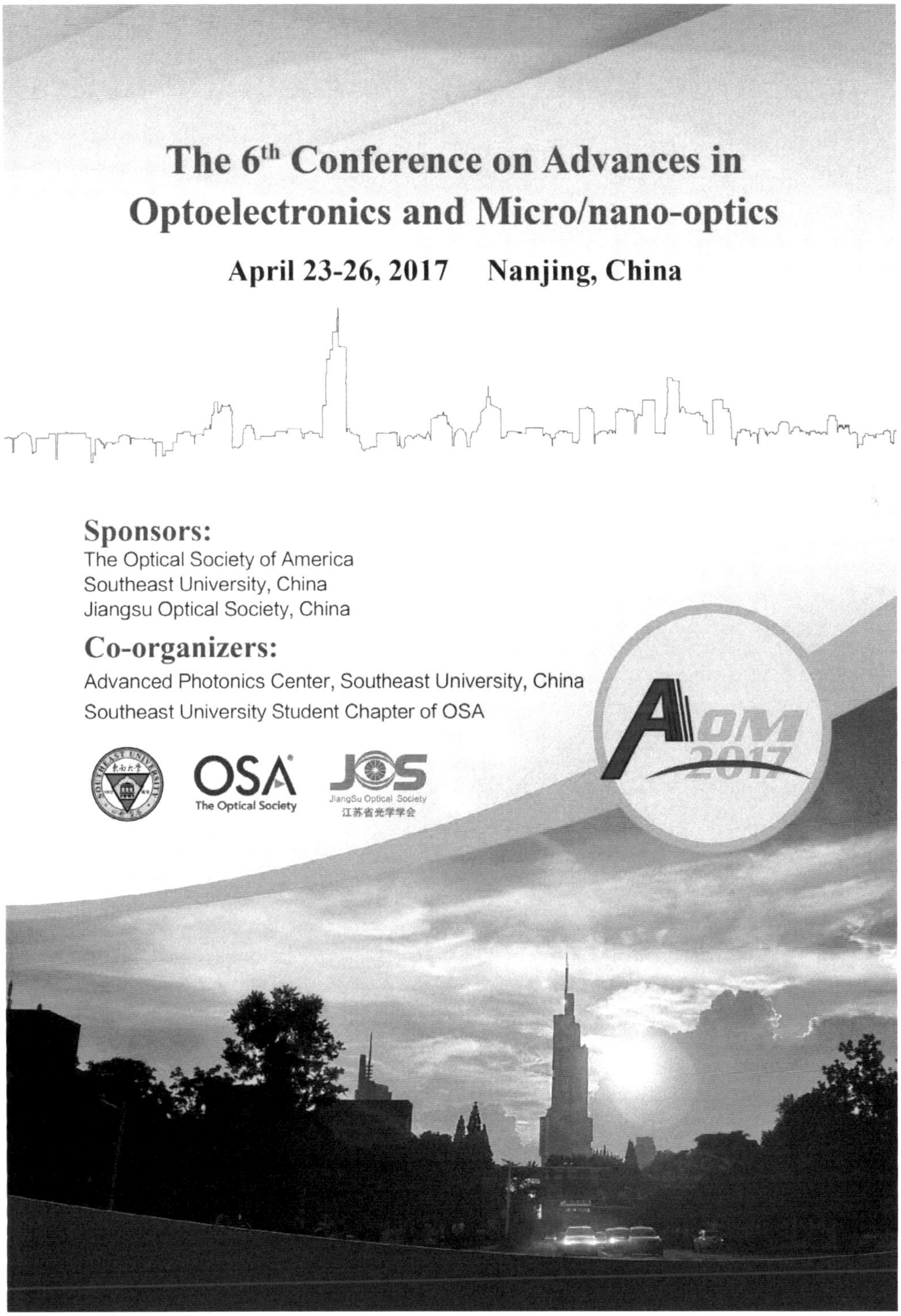

IOP Conf. Series: Journal of Physics: Conf. Series **844** (2017) 011001 doi:10.1088/1742-6596/844/1/011001

The 6th Conference on Advances in Optoelectronics and Micro/nano-optics (AOM 2017)

Nanjing, China
April 23 - 26, 2017

Conference Co-Chairs:
Yiping Cui - Southeast University, China
Xiaocong Yuan - Shenzhen University, China
Shining Zhu - Nanjing University, China

IOP Conf. Series: Journal of Physics: Conf. Series **844** (2017) 011001 doi:10.1088/1742-6596/844/1/011001

WELCOME

Journal of physics: Conference Series is publishing a volume of conference proceedings that contains a selection of papers presented at the 6th Conference on Advances in Optoelectronics and Micro/nano-optics (AOM 2017), which is an OSA topical meeting that started in 2009. AOM 2017, organized by The Optical Society of America, Southeast University, and Jiangsu Optical Society, was successfully held at Nanjing, China from April 23th-26th, 2017. It aims to bring together leading academic scientists, researchers and scholars to exchange and share their experience and research results on all aspects of optoelectronics and micro/nano-optics, and to discuss the practical challenges encountered and the solutions adopted.

Located in Yangtze River Delta area and the center of east China, Nanjing is the capital of Jiangsu province and the second largest city in the east China region, turned out to be an ideal meeting place for domestic and overseas participants of this international conference. The conference program included plenary talks, invited talks, oral and poster contributions. From numerous submissions, 64 of the most promising and IOP-relevant contributions were included in this volume. The submissions present original ideas or results of general significance, supported by clear reasoning, compelling evidence relevant to the research. The authors state clearly the problems and the significance of their research to theory and practice. Being a successful conference, this event gathered more than 300 qualified and high-level researchers and experts, which created a good platform for worldwide researchers and engineers to enjoy the academic communication.

Taking advantage of this opportunity, we would like to thank all participants of this conference, and particularly the authors of all accepted papers for their high quality and fruitful contributions. We do hope that this volume will be beneficial for readers to their future research endeavours and careers. We also gratefully acknowledge tremendous efforts and dedication of many individuals and all the Editorial Board members in IOP Publishing for their support in producing the proceedings of this event.

Sincerely,

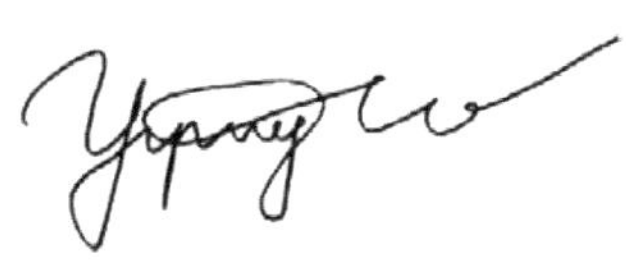

Yiping Cui
Conference Co-Chair

CONFERENCE OFFICIALS

Conference Co-Chairs

Yiping Cui……………………………..……………Southeast University, China
Xiaocong Yuan…………………………………………… Shenzhen University, China
Shining Zhu…………………………………………………Nanjing University, China

Executive Chair

Zhuyuan Wang…………………………………………… Southeast University, China

Technical Program Committee

Guanghui Chen……………...China Electronics Technology Group Corporation, China
Yiping Cui…………………………………………………Southeast University, China
Qihuang Gong……..…………………………………………Peking University, China
Min Gu……………………………...Swinburne University of Technology, Australia
Satoshi Kawata…………………………………………………...Osaka University, Japan
Gong-Ru Lin………………………………National Taiwan University, Taiwan, China
Xu Liu………...Zhejiang University, China
Yanqing Lu…………………………………………………..Nanjing University, China
Ting Mei………………………………..Northwestern Polytechnical University, China
Min Qiu……………………………………………………..Zhejiang University, China
Marek Samoc………………………………Wroclaw University of Technology, Poland
Colin Sheppard…………………………………Italian Institute of Technology, Italy
Charles Surya………………………………Hong Kong Polytechnic University, China
Katarina Svanberg……………………………………………Lund University, Sweden
Sune Svanberg………………………………………………..Lund University, Sweden

IOP Conf. Series: Journal of Physics: Conf. Series **844** (2017) 011001 doi:10.1088/1742-6596/844/1/011001

Jinghua Teng…………………………………………………………….IMRE, Singapore
Dinping Tsai…………………………………………..Academia Sinica, Taiwan, China
Zhuyuan Wang……………………………………………..Southeast University, China
Jingjun Xu……………………………………………………Nankai University, China
Baoli Yao…………………….Xi'an Institute of Optics and Precision Mechanics, China
Xiaocong Yuan……………………………………………Shenzhen University, China
Qiwen Zhan………………………………………………...University of Dayton, USA
Shining Zhu………………………………………………Nanjing University, China

IOP Conf. Series: Journal of Physics: Conf. Series **844** (2017) 011002 doi:10.1088/1742-6596/844/1/011002

Peer review statement

All papers published in this volume of *Journal of Physics: Conference Series* have been peer reviewed through processes administered by the proceedings Editors. Reviews were conducted by expert referees to the professional and scientific standards expected of a proceedings journal published by IOP Publishing.

IOP Conf. Series: Journal of Physics: Conf. Series **844** (2017) 012001 doi:10.1088/1742-6596/844/1/012001

Advanced Optoelectronic Devices based on Si Quantum Dots/Si Nanowires Hetero-structures

J Xu[1,*], Y Y Zhai[1], Y Q Cao[1,2] and K J Chen[1]

[1]National Laboratory of Solid State Microstructures, School of Electronic Science and Engineering and Jiangsu Provincial Key Laboratory of Advanced Photonic and Electronic Materials, Nanjing University, Nanjing210093, China
[2] College of Physics Science and Technology, Yangzhou University, Yangzhou 225002, China
*E-mail:junxu@nju.edu.cn

Abstract. Si quantum dots are currently extensively studied since they can be used to develop many kinds of optoelectronic devices. In this report, we review the fabrication of Si quantum dots (Si QD) /Si nanowires (Si NWs) hetero-structures by deposition of Si QDs/SiO_2 or Si QDs/SiC multilayers on Si NWs arrays. The electroluminescence and photovoltaic devices based on the formed hetero-structures have been prepared and the improved performance is confirmed. It is also found that the surface recombination via the surface defects states on the Si NWs, especially the ones obtained by the long-time etching, may deteriorate the device properties though they exhibit the better anti-reflection characteristics. The possible surface passivation approaches are briefly discussed.

1. Introduction

Nano-crystalline Si quantum dots (Si QDs) have attracted more and more attention in recently years since they can be applied in many kinds of nano-electronic and optoelectronic devices such as next generation of solar cells, non-volatile memories and Si-based light emitting devices [1-3].It is currently an open question to further improve device performance of the Si QDs-based devices. One of the interesting topics is to introduce the novel micro- and/or nano-structures to make the efficient photon management, in which Si-based nanowires structures are one of the potential candidates. For example, it was reported that the enhanced optical absorption and photoluminescence from Si QDs can be achieved by using SiOx nanowires due to the anti-reflection characteristics of the nanostructures [4]. Meanwhile, Si nanowires (Si NWs) fabricated by the metal-assisted chemical etching method were used to get radial p–n junction solar cells, such as crystalline Si NWs cells, a-Si/Si NWs shell/core cells, and organic polymers/Si NWs hybrid cells [5-7]. Si NWs structures can significantly reduce the surface light reflection as well as enhance the incident optical absorption and in turn improve the device performance, which is more suitable for Si-QDs-based optoelectronic devices because the active layer in such devices is usually very thin.

Here, we report the fabrication of hetero-structures containing Si QDs-based multilayers on Si NWs array prepared by the metal-assisted chemical etching technique. The anti-reflection and absorption enhancement behaviors of obtained hetero-structures are demonstrated. The photovoltaic devices and electroluminescence devices based on the Si QD /Si NWs hetero-structures are prepared. Compared with the flat devices, the device performance is obviously improved by using the Si QD /Si NWs hetero-structures. Moreover, it was found that the influence of surface defects states becomes significance with increasing the etching time during the formation of Si NWs as revealed by ESR measurements. The possible passivation techniques are proposed to reduce the surface defects states and in turn to further improve the device properties.

2. Experimental

Si QDs/SiO_2(or SiC) multilayers with various dot sizes were prepared by annealing amorphous Si/SiO_2 (or SiC) stacked structures with controllable amorphous Si sublayer thickness, which were deposited in conventional plasma enhanced chemical vapor deposition (PECVD) system. The detailed preparation parameters and procedures can be found elsewhere [8-9]. The optical bandgaps have been deduced based on the measured optical absorption spectra according to Tauc's plots and we found that the optical bandgap is shifted with the dot size, which provides a flexible way to design the corresponding Si-based photonic and opto-electronic devices.

In order to get Si NWs, the wet electroless etching technique was used to etch the p-type or n-type Cz silicon (100) wafers. The tree-like Si NW arrays were formed due to the selectively anisotropic etching of silicon substrate in aqueous HF solution induced by the Ag nanoparticles. To remove the capped silver, the as-prepared samples were finally dipped into a 30 wt% HNO_3 aqueous solution for 60 s. It is found that the depth of formed Si NW structures is increased by controlling the etching time. We deposited the Si QDs based multilayers on flat and Si NWs arrays to obtain the prototype device structures, especially the photovoltaic devices and light emitting devices.

3. Results and Discussion

The structures of Si NWs were characterized by field emission scanning electron microscopy (SEM, Sigma Zeiss). Figure 1 (a) and (b) show the top-view and cross-sectional SEM images of the Si NW arrays after etching for 8min. The vertically aligned nanowires are clearly identified with high density. The depth of Si NWs is about 450nm while the average diameter is about 50nm after 8min etching. After preparation of Si QDs-based multilayers on the Si NWs array, the top morphologies are still kept due to the good conformal deposition characteristics of Si-based thin films in PECVD system, which was revealed by cross-sectional transmission electron microscopy (not shown in here).

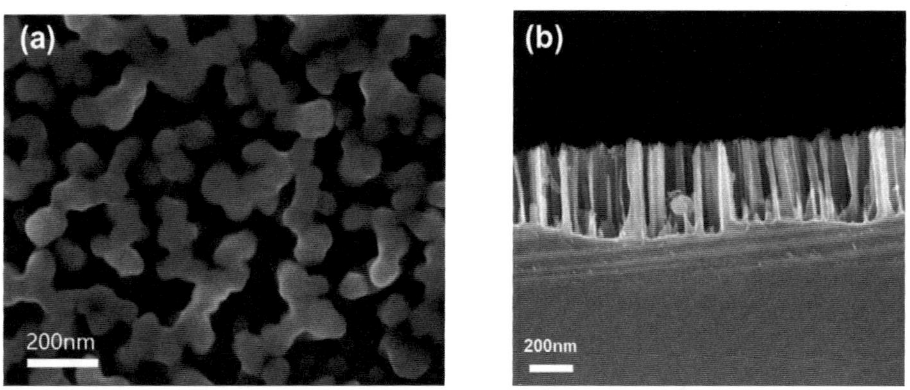

Figure 1. Top (a) and side (b) SEM images of Si NWs after 8 min etching.

A broadband anti-reflection property of Si QDs/Si NWs hetero-structures is found and the reflectance is gradually suppressed with increasing the etching depth of Si NWs. The typical reflection spectra for flat Si substrate, Si NWs array and Si QDs/Si NWs hetero-structures are given in Fig.2. It is found that the reflectance of flat Si substrate is quite high (>30%) in the wavelength range from 200-1100nm. However, the Si NWs array exhibits the reduced reflection in the whole measurement range. In the wavelength range of 200-700nm (UV-visible light range), the reflectance is even less than 5%. After deposition of Si QDs/SiO_2 or Si QDS/SiC multilayers on Si NWs array, the reflection spectrum is slightly changed as shown in Fig.2. The good anti-reflection characteristics of Si NW arrays can be explained in terms of the formation of graded-index layer from the substrate to the top surface which can suppress the reflection obviously. Another important factor is the strong light scatting effect in the Si NWs array, which trap the light due to the increased optical path [10-12]. As a consequence, the optical absorption is significantly enhanced. It is found that the mean absorption weighted by AM 1.5G solar spectrum in the wavelength range from 300 to 1200 nm for Si QDs/SiC multilayers on Si NWs can reach as high as 88.9%, which is enhanced by 46% compared with the flat one [8,13-14].

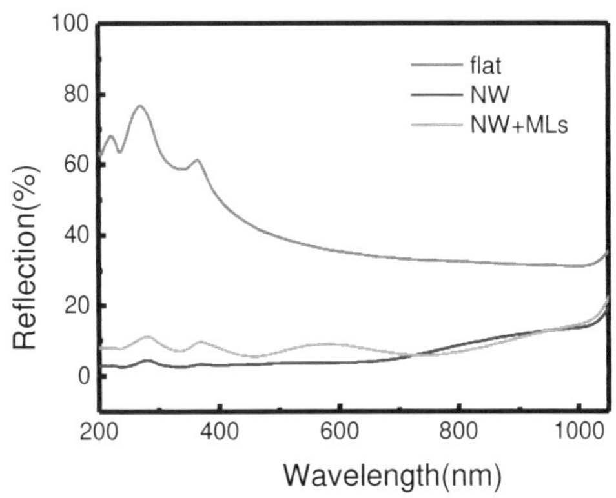

Figure 2. Reflection spectra for flat Si substrate, Si NWs array and Si QDs/Si NWs hetero-structures.

In order to get the device based on Si QDs, the Si QDs (4nm)/SiC (2nm) multilayers and p-type a-Si layer are successively deposited on both flat n-Si wafers and wafers with vertically aligned Si NW arrays to form p-i-n solar cell structures. Figure 3 shows the schematic diagram of device structures. The rectification characteristics are clearly observed, which indicates that the p-i-n structures are well formed by the present approach [8,13]. It is found that the external quantum efficiency (EQE) of Si QDs/Si NWs hetero-structured cell is obviously enhanced over a wide spectral range of 500-1000 nm compared with that of flat reference cell. Meanwhile, the power conversion efficiency based onSi QDs/Si NWs hetero-structures exhibits the best value of 11.3%.

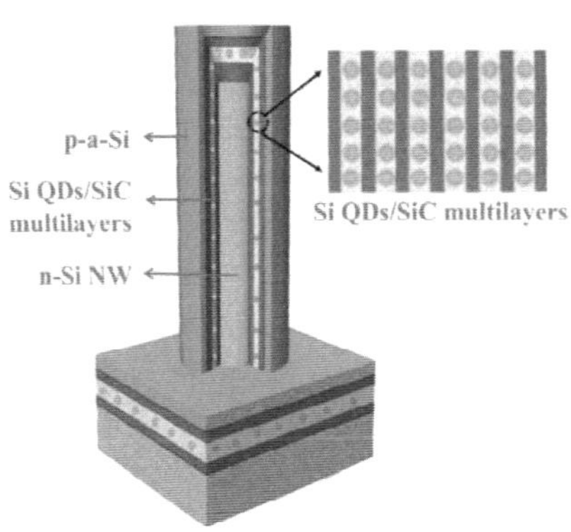

Figure 3. Schematic diagram of photovoltaic device structures containing Si QDs/SiC multilayers deposited on n-Si NWs. The out layer is p-type a-Si film.

IOP Conf. Series: Journal of Physics: Conf. Series **844** (2017) 012001 doi:10.1088/1742-6596/844/1/012001

We also fabricated the light emitting devices containing Si quantum dots/SiO_2 multilayers deposited on Si nanowire arrays. The enhanced electroluminescence has been achieved as shown in Fig. 4. It is found that the electroluminescence intensity is obviously enhanced by using Si QDs/Si NWs hetero-structures compared with that of flat device. At the same injection current (100mA), the integrated electroluminescence intensity is increased by 4-folds and the output power density is in an order of $\mu W/cm^2$. However, it is found that the electroluminescence is first enhanced with increasing the depth of the Si nanowires and then reduced for further increasing the depth though it exhibits the lowest reflectance (~3%). The electron spin resonance (ESR) measurements suggest that the surface defect states may increases after long time etching. Figure 5 is the ESR spectrum of Si NWs sample after 8min etching. It is found that the ESR signal with g value of 2.005 can be observed, which is assigned to Si dangling bonds (P_b- centres) [15]. It is suggested that the surface of Si NWs contain the defects states, which can act as the non-radiative recombination centres to reduce the electroluminescence intensity.

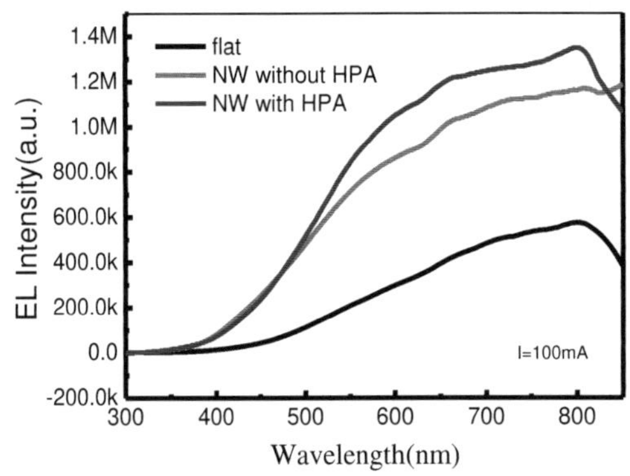

Figure 4. Electroluminescence spectra for flat Si substrate, Si NWs array and Si QDs/Si NWs hetero-structures. The data are collected at the same injection current of 100 mA.

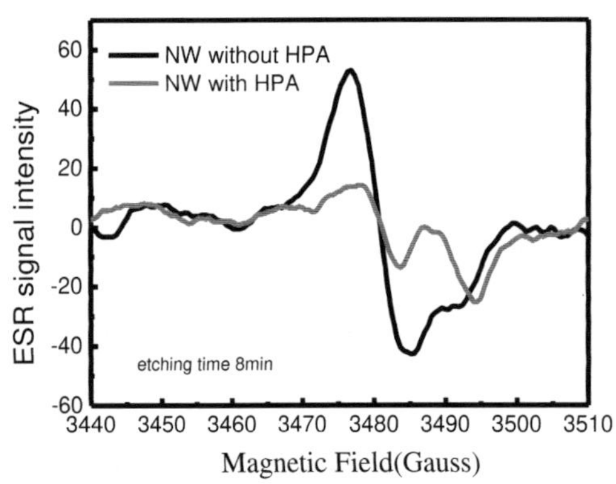

Figure 5. ESR spectra for Si NWs array before and after hydrogen plasma annealing.

In order to reduce the surface states and to further improve the device performance, we proposed the possible passivation techniques such as hydrogen plasma annealing (HPA) treatments or deposition of Al_2O_3 passivation layer. As shown in Fig. 5, the ESR signals related to the dangling bonds are significantly suppressed after HPA treatment. The same passivation effect is also demonstrated by using Al_2O_3 layer. The reduction of surface defects states is indeed improve the device properties. As given in Fig.4, the electroluminescence intensity is further enhanced after HPA treatment and the integrated luminescence intensity is 5 folds stronger than that of flat one and the output power density exceeds 1.0 $\mu W/cm^2$. Meanwhile, we also found that the turn-on voltage of device is reduce to 3.5V and the operation voltage is about 10V, which is compatible with the requirements of Si-based optoelectronic integrated circuits.

4. Conclusions

The Si QDs/Si NWs hetero-structures with good anti-reflection and optical absorption characteristics were fabricated. Based on the hetero-structures, the prepared optoelectronic devices exhibited the enhanced performance. The hydrogen plasma annealing or Al2O3 passivation can significantly suppressed the surface defects states of Si NWs and the device properties can be further improved.

Acknowledgment

This work is supported by the "973 Program" (2013CB632101), "333 project" of Jiangsu Province (BRA2015284) and PAPD..

References

[1] Priolo F, Gregorkiewicz T, Galli M and Krauss T 2014 *Nature Nanotech.* **9** 19
[2] Mu W, Zhang P, Xu J, Sun S, Xu J, Li W, and Chen K 2014 *IEEE J. Sel. Top. Quant. Elec.* **20** 8200106
[3] Ray S, Maikap S, Banerjee W and Das S 2013 *Journal of Physics D: Appl. Phys.* **46** 153001
[4] Samanta A and Das D 2013 *J. Mater. Chem. C* **1** 6623
[5] Peng K, Hu J, Yan Y, Wu Y, Fang H, Xu Y, Lee S and Zhu J 2006 *Adv. Funct. Mater.* **16** 387
[6] Li H, Jia R, Chen C, Xing Z, Ding W, Meng Y, Wu D, Liu X and Ye T 2011 *Appl. Phys. Lett.* **98** 151116
[7] Wang H, Lin T, Hsu C, Tsai M, Huang C, Wei W, Huang M, Chien Y, Yang P, Liu C, Chou L and He J 2013 *ACS Nano* **7** 9325
[8] Cao Y, Xu J, Ge Z, Zhai Y, Li W, Jiang X and Chen K 2015 *J. Mater. Chem. C* **3** 12061
[9] Zhai Y, Cao Y, Lin Z, Qian M, Xu J, Li W, Xu L and Chen K 2016 *IEEE Photonics Journal* **8** 4501708
[10] Yu L Misra S, Wang J, Qian S, Foldyna M, Xu J, Shi Y, Johnson E and Cabarrocas P 2014 *Sci. Rep.* **4** 4357
[11] Wan N, Xu J, Chen G, Gan X, Guo S, Xu L and Chen K 2010 *Acta Mater.* **58** 3068
[12] Li J, Yu H, Li Y, Wang F, Yang M, and Wong S 2011 *Appl. Phys. Lett.* **98** 021905
[13] Cao Y, Ge Z, Jiang X, Xu J, Xu L, Li W, Yu L and Chen K 2016 *Part. & Part. Sys. Charac.* **33** 38
[14] Yu Z, Qian S, Yu L, Misra S, Zhang P, Wang J, Shi Y, Xu L, Xu J and Chen K 2015 *Opt. Expess* **23** 5388
[15] Lu P, Mu W, XU J, Zhang X, Zhang W, Li W, Xu L and Chen K 2016 *Sci. Rep.* **6** 22888

IOP Conf. Series: Journal of Physics: Conf. Series **844** (2017) 012002 doi:10.1088/1742-6596/844/1/012002

InGaAsP/InP Micropillar Cavities for 1.55 μm Quantum-Dot Single Photon Sources

Xiumin Xie[1], Qiang Xu[1], Bizhou Shen[1], Jian Chen[1], Qian Dai[1], Zhu Shi[1], Libo Yu[1], Zhiming M Wang[2] and Hai-Zhi Song[1,2*]

[1]Southwest Institute of Technical Physics, Renminnan Road 4-7, Chengdu 610041, China.

[2]Institute of Fundamental and Frontier Sciences, University of Electronic Science and Technology of China, Jianshebei Road 2-4, Chengdu 610054, China

*Email: hzsong1296@163.com

Abstract Micropillar cavities containing 1.55 μm quantum dots are required for silica-fiber based quantum information processing. The straight way is to construct micropillars consisting of InGaAsP/InP distributed Bragg reflectors. We perform a systematic study mainly on optical properties correlated to the micropillar diameter. As expected, the mode wavelength increases with extending micropillar diameter and tends to saturate at 1.55 μm for the diameter larger than 2.0 μm. Able to be as high as 10^4, the quality factor, however, is almost independent of pillar diameter, which may be the result of the small refractive index contrast. The Purcell factor reaches highest (>130) at pillar diameter of ~0.6 μm, suggesting that these cavities could serve as efficient 1.55 μm single photon sources. The output efficiency increases with extending diameter, and is acceptably good at small diameters. Although not easy in fabrication technique, this cavity provides monolithic scheme for constructing 1.55 μm single photon sources.

1. Introduction

For the purpose of quantum communication over silica-fiber-based networks, InAs/InP quantum dots (QDs) are promising as single photon sources (SPSs) [1]. However, the spontaneous lifetime of InAs/InP QDs is so long (T_1 ~ 1.2 ns) that the demanding GHz emission is not available. A QD SPS also needs to emit highly identical single-photon pulse trains, but the photon indistinguishability of an InAs/InP QD is still bad due to the big gap between the excitonic coherence time (T_2 typically ~100 ps) and the spontaneous lifetime. These two problems can be resolved by introducing InAs/InP QDs into micropillar cavities possessing high Purcell factors F_P. Like the micropillar cavities for InAs/GaAs QDs [2], the straight way to construct an efficient cavity containing InAs/InP QDs might be using a pillar composed of InP-lattice-matched distributed Bragg reflectors (DBRs), but there has been neither experimental nor theoretical trials on this system. In this work, we are going to study and design InGaAsP/InP DBR micropillar cavities for 1.55 μm telecommunication band SPSs.

2. Model and method

The micropillar cavity is a cylinder standing on a semi-infinite InP substrate, as shown in Fig. 1(a). It consists of disk-shaped (in XY plane) periodic InGaAsP/InP pairs coaxially along Z direction on the top and bottom side of an InGaAsP spacer layer. The InGaAsP/InP periodic structures are the DBRs taking the role of reflecting light towards the spacer. The bottom DBR has more pairs of InGaAsP/InP layers than the top DBRs so that there is less useless leakage to the bottom. The thickness of each InP(InGaAsP) layer in DBRs is set as $t_{1(2)} = \lambda_B/(4n_{1(2)})$, where λ_B is the Bragg wavelength around 1.55 μm and $n_{1(2)}$ is the refractive index of InP(InGaAsP). The spacer layer is one wavelength thick in conventional micropillar cavities, or replaced by tapered DBRs in tapered-DBR cavities. Here "taper" means adiabatically deducing the layer thicknesses as the DBR extends towards the cavity center (new spacer) [2]. The layer thicknesses in the tapered DBRs linearly decrease as $t_{1i} = t_1(1-\rho(2i-1))$ for InP and $t_{2i} = t_2(1-2\rho i)$ for InGaAsP, where i is the taper segment number and ρ the tapering slope of layer thickness, in other words, the decreased fraction per tapered layer. In between the tapered DBRs is an InGaAsP spacer layer with thickness $t_0 = t_1(1-2\rho N)$, where N is the total segment number in one tapered DBR. It's worth noting that the spacer here is much thinner than that without tapered DBRs. The InGaAsP layers are lattice-matching to the InP substrate and have an energy gap larger than the photon energy of 1.3 μm wavelength, so that they are extremely transparent for ~1.55 μm photons. A photon emission source, representative of an InAs QD, is inserted in the spacer.

By using finite-difference-time-domain method, the optical properties of this conventional micropillar cavity are simulated with the optical constants of the materials cited or deduced from Ref. [3]. Once a polarized impulse is launched from the photon emission source, the time evolution of the light intensity can be obtained at monitors set in the spacer layer. A Fourier transform gives a spectrum of the electric field intensity, showing some peaks representing the cavity modes. By setting the light source as a narrow-band emission around a mode wavelength λ, we obtain the intensity decay with time t and the steady state distribution, *i.e.* the mode profile. The quality factor Q can be obtained by fitting the exponential light intensity envelope to $\exp(-2\pi ct/Q\lambda)$, where c is the light velocity in vacuum.

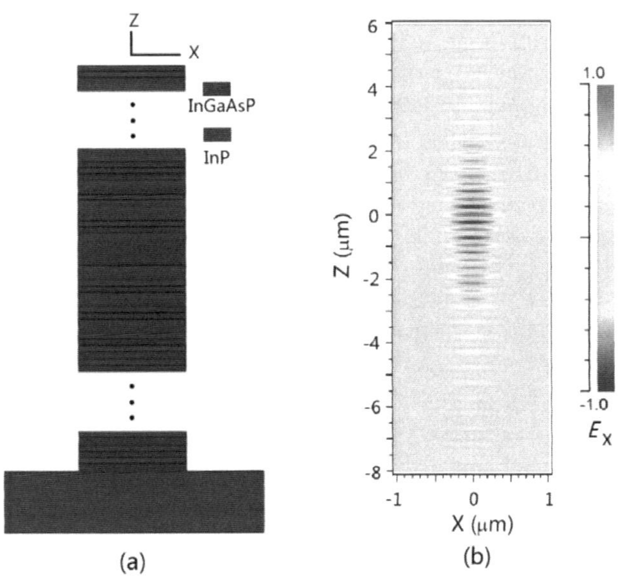

Fig. 1. (a) Schematic cross section of the InGaAsP/InP micropillar cavity without tapered DBRs; (b) Distribution profile of the x-polarized electric field E_x of the fundamental mode.

3. Simulation and Results

The result of the simulation of this InGaAsP/InP micropillar cavity is as follows. At first, there does exist the fundamental mode peaked near 1.55 μm for both conventional and tapered cavities as expected. The x-polarized electric field E_x of the fundamental mode is depicted in Fig. 1(b). It is demonstrated that the profile is highly symmetric and well confined around the spacer layer, which suggests the cavity can be of a good quality.

Massive simulations were performed to obtain the optical characteristics of the fundamental modes with the pillar diameter D varying from 2.5 to 0.4 μm. Figure 2(a) shows the mode wavelength λ as a function of pillar diameter D for conventional and tapered-DBR cavities. At first, λ exhibits a bit blueshift at big D with respect to the designed Bragg wavelength $\lambda_B = 1.55$ μm, and tends to saturate at λ_B for $D > 2$ μm. As D decreases, the blueshift increases at a faster rate. It's noticeable that the blueshift behavior is independent with whether the tapered DBRs are employed or not, and similar to our previous study on conventional Si/SiO_2 micropillar cavities [4] as well as other micropillar cavities with a low index contrast [5], although the total thicknesses of different DBR layers vary a lot (from ~10 to ~30 μm). The λ trend along with D variation can be interpreted by waveguide dispersion that a smaller D brings about more localized geometrical confinement of an optical mode. As a consequence, a smaller D results in a shorter mode wavelength [5].

The quality factor Q as a function of micropillar diameter D with different numbers of DBRs is shown in Fig. 2(b). The Q factor looks more stable than the other micropillars such as GaAs/AlAs [7] and Si/SiO_2-InP [4]. In detail, it tends to saturate at big D and at small D. The saturation for Q happens at D close to 0.6 μm instead of a big value $D > 2$ μm like in Si/SiO_2-InP case [4]. The positive correlation between Q and D derives from the fact that as D decreases, λ strongly deviates from λ_B and the effective incident angle of light on the DBRs increases [6]. A strong Q oscillation with respect to D, which has been demonstrated in previous conventional [2,8] and hybrid [4] micropillar cavities, is hardly observed here. This implies these modes are rather pure, i.e., almost not coupled with the high-order propagating Bloch modes. Physically, this property is related to the small refractive index contrast (~0.2), which makes the cavity more like a waveguide. Anyway, this property benefits to practical applications since one can get good emission linewidth without paying special attention to the cavity size.

A high quality factor Q is the basic and extremely important requirement for an effective cavity SPS. Since the refractive index contrast between InP and InGaAsP is so poor (~0.2), a large number of DBRs would be needed for enough reflectivity to get high Q. It's necessary to examine how Q changes with respect to the numbers of DBR pairs as the thickness of each InP(InGaAsP) layer of the conventional DBR remains at $t_{1(2)} = \lambda_B/(4n_{1(2)})$. Also demonstrated in Fig. 2(b), the quality factor Q increases remarkably as the DBR increases from 16/28 to 40/70 pairs. In detail, on a cavity with 16/28 pairs of top/bottom DBR layers and pillar height ~10 μm, the Q factor is found to be about 250 for $D >$ 0.6 μm. Such a poor quality factor results from the fact that the inadequate DBR pairs can't confine the mode profile efficiently. 30/50 pairs of DBR layers with pillar height ~20 μm lead to Q factor ~1900 for $D > 0.6$ μm. 40/70 pairs of DBR layers with pillar height ~30 μm bring about Q factor up to 6000 for $D > 0.6$ μm. When tapered DBRs are introduced to the spacer, the Q factor enhancement is more significant. For example, when the spacer of the cavity with 30/50 pairs of DBR layers is replaced by tapered DBRs with tapering segment $N = 3$ and tapering slope $\rho = 0.02$, the Q factor reaches ~2200 for $D > 0.8$ μm, about 16% enhancement obtained with merely ~5% pillar height increased. As well, the replacement of tapered DBR for the spacer of the cavity with 40/70 pairs of conventional DBR layers results in ~20% and ~35% Q enhancement with only ~4% and ~8% pillar height increase for $N = 3$, $\rho = 0.02$ and $N = 6$, $\rho = 0.02$, respectively, consistent with Ref. [7]. It is worth stressing that it is possible to get a Q factor as high as 10_4 on a InGaAsP/InP micropillar cavity.

A small mode volume

$$V = \frac{\int \varepsilon(\mathbf{r})\mathbf{E}(\mathbf{r})^2 \mathrm{d}\mathbf{r}}{\varepsilon_{\mathrm{M}} E_{\mathrm{M}}^{\;2}} \tag{1}$$

where ε is the relative dielectric constant, $\mathbf{E}$ the electric field of the light at the position $\mathbf{r}$, and ε_{M} and E_{M} the corresponding values at the point of the maximum light intensity, is always desired for micropillar cavity SPSs since it brings about the capability of separating a single QD resonant with the cavity mode and miniaturizing and integrating the quantum devices. It's thus necessary to check the change of mode volume V against D in the unit of $(\lambda/n)^3$, which is demonstrated in Fig. 2(c), where n is the effective refractive index. It's natural that V monotonically increases with extended D in all cases as expected, and is mainly positively correlated with the micropillar height. The introduction of tapered DBRs to the spacer, on the contrary, helps little on the reduction of V, which exhibits the same behavior as that in previous Si/SiO_2 pillar cavities [9].

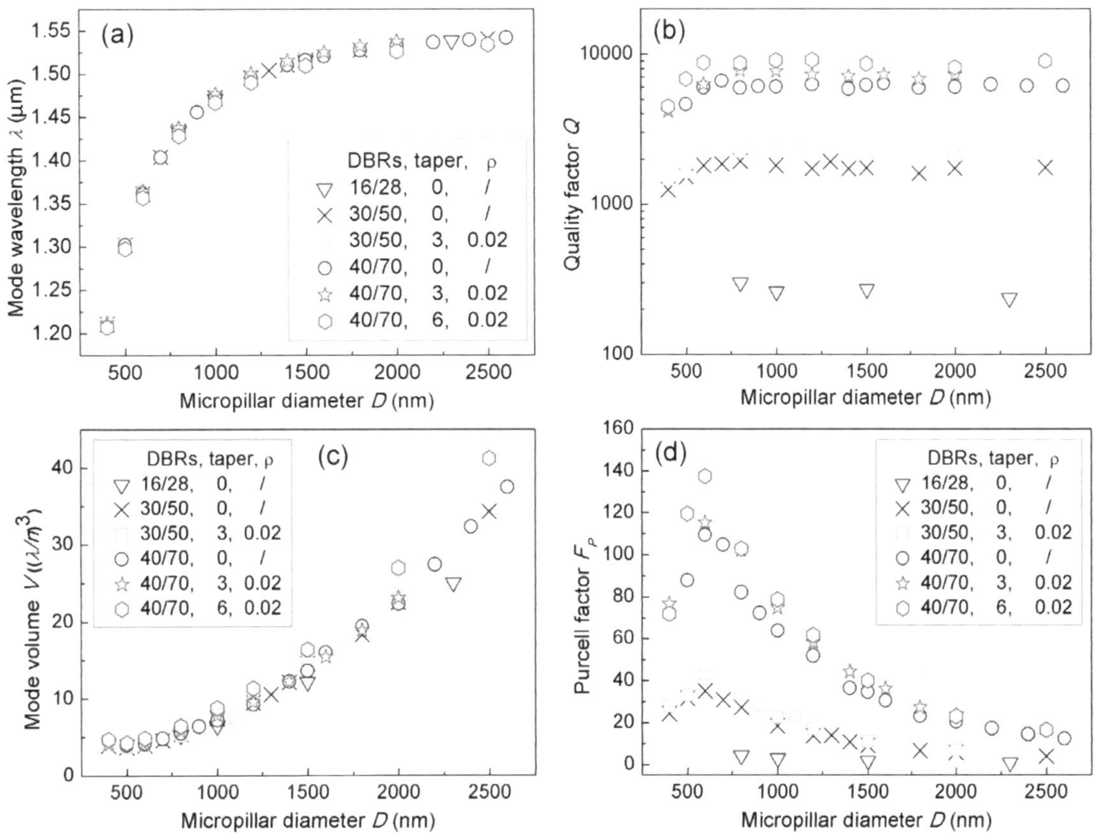

Fig. 2. The mode wavelength (a), Q factor (b), mode volume (c) and Purcell factor (d) as functions of the diameters of InGaAsP/InP micropillar cavities.

One cavity cannot maintain the smallest V and highest Q at the same time. As a tradeoff, it is more meaningful to examine changes of the Purcell factor

$$F_{\mathrm{P}} = \frac{3Q\lambda^3}{4\pi^2 V n^3} \tag{2}$$

with the cavity size, as shown in Fig. 2(d). Besides, high Purcell factors reduce the spontaneous emission lifetimes of the QDs and may thus increase the operation frequency from MHz to GHz and result in highly identical single-photon pulse trains as well. Here F_{P} achieves the peak value at $D = 0.6$ μm for all pillar cavities with or without tapered DBRs. F_{P} decreases rapidly as D deviates away from where the maximum F_{P} locates. Specifically, for the cavity with 16/28 pairs of DBR layers, F_{P} remains below 5 for $D \geq 0.8$ μm. On a cavity with non-tapered 30/50 pairs of DBR layers, there occurs an F_{P} more than 20 as the diameter is below 1 μm, and the maximum F_{P} of ~35 is obtained at $D = 0.6$ μm. F_{P}

becomes smaller at larger diameters because the mode volume is getting larger while the Q factor remains. When the spacer is replaced by 3 segments of tapered DBRs with tapering slope $\rho = 0.02$, F_P goes about 20% higher for all diameters.

More significantly, a cavity with non-tapered 40/70 DBR pairs exhibits F_P from 20 to more than 100. This happens with the pillar diameter being less than 2 μm. Besides, the cavities with 40/70 DBR pairs together with tapered DBRs possess higher F_P than those of non-tapered cavities at the same pillar diameters. More segments of tapered DBRs brings about higher F_P, e.g., 6 segments of tapered DBRs lead to F_P enhancement of ~15% compared to that of 3 segments of tapered DBRs. Although the highest F_P here is high enough for the microcavity to enter a strong coupling regime [9], it's still a practical index that higher F_P is preferred, which corresponds to $D < 1$ μm for micropillar cavities with 30/50 DBRs and $D < 1.8$ μm for cavities with 40/70 DBRs.

Apart from the above parameters, the output efficiency is another figure of merit that affects the quality of the micropillar cavity. Since the QD SPS usually operates under cryogenic temperatures, bad efficiency would give rise to amount of heat generation and thus fierce degradation of the SPS performance. Figure 3 depicts the relationship between the output efficiency ω and the pillar diameter D. Here ω is expressed as

$$\omega = P_{top} / (P_{top} + P_{bottom} + P_{lateral}) \quad (3)$$

where P_{top} represents the direct emergent power flux through the top, P_{bottom} and $P_{lateral}$ are the power flux leakage through the bottom layers and lateral sidewalls, respectively. The output efficiency increases rapidly from a few percent to several tens of percent at extending pillar diameter at small D, and tends to be stable at $D > 2$ μm. In detail, ω tends to saturate at ~73% for cavities with 30/50 pairs of DBRs and ~85% for cavities with 40/70 pairs of DBRs. The bad efficiency at small D is mainly caused by poor lateral confinement of the mode profile. The bottom leakage, however, dominates the power loss at large D. It's noted that the replacement of the spacer by tapered DBRs doesn't make any contribution to ω, unlike those cases to Q and F_P. On the other hand, the micropillar diameter mismatch at highest F_P and best ω suggests that a tradeoff would be made according to practical use.

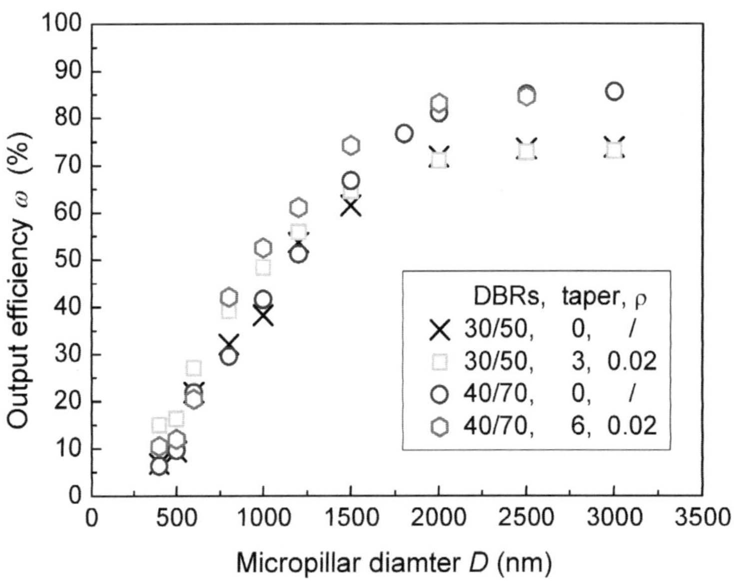

Fig. 3. The output efficiency as a function of the diameters of InGaAsP/InP micropillar cavities.

4. Discussion

With the simulated quality, the above cavity can be considered to be used as a QD SPS. The best single photon generation rate is inversely proportional to the spontaneous lifetime T_1 of the QD excitons. As such, our cavity with $F_P > 10$ could increase the operation frequency from several hundred MHz into GHz band. For photon indistinguishability, there should be a coherence time T_2 comparable or longer than $2T_1$, so a microcavity with $F_P > 2T_1/T_2 \sim 20$ would be required. Obviously, it could be expected that highly indistinguishable 1.55 μm single photons could be produced using the present micropillar cavities by tuning the Bragg wavelength and the tapering slope of tapered DBRs [4]. Furthermore, the best quality also satisfies the strong coupling condition $Q/V^{1/2} > 2200$ [9], so even a coherent SPS at 1.55 μm is also producible by using the present InGaAsP/InP micropillar cavities. Finally, when high output efficiency is considered on account of operation stability, a tradeoff needs to be made since the correlated D for the highest ω deviates a bit from that for the best F_P. For example, output efficiency of ~50% is possible while the Purcell factor remains as high as ~90 with pillar diameter of ~0.8 μm.

The previous 1.55 μm micropillar cavities may be realized by a hybrid fabrication process. Consisting of epitaxially available semiconductor structure, the present cavity provides a monolithic scheme for 1.55 μm QD SPS. Of course, a necessarily 20-30 μm pillar height looks a challenging task, which needs a great effort in fabrication technique.

5. Conclusion

InGaAsP/InP micropillar cavities are studied as candidates for 1.55 μm QD-SPSs. The design condition for efficient SPS is found to be: DBR pairs more than 30/50, tapered DBRs, and pillar diameter less than a specific value, e.g., 1 μm for cavities with 30/50 DBRs and 1.8 μm for cavities with 40/70 DBRs. If the output efficiency is of priority, the pillar diameter can deviate a bit as a tradeoff. Although not technically easy, this monolithic cavity may be producible in the near future.

Reference

[1] K. Takemoto, Y. Nambu, T. Miyazawa, Y. Sakuma, T. Yamamoto, S. Yorozu, and Y. Arakawa, "Quantum key distribution over 120 km using ultrahigh purity single-photon source and superconducting single-photon detectors", Sci. Rep. 5, 14383 (2015).

[2] E. Moreau, I. Robert I, J. M. Gérard, I. Abram, L. Manin, and V. Thierry-Mieg, "Single-mode solid-state single photon source based on isolated quantum dots in pillar microcavities", Appl. Phys. Lett. 79, 2865-2867 (2001).

[3] Palik, E. D., [Handbook of Optical Constants of Solids], Academic Press, New York, 3-1086 (1998).

[4] H. Z. Song, K. Takemoto, T. Miyazawa, M. Takatsu, S. Iwamoto, T. Yamamoto, and Y. Arakawa, "Design of Si/SiO_2 micropillar cavities for Purcell-enhanced single photon emission at 1.55 μm from InAs/InP quantum dots", Opt. Lett. 38, 3241-3244 (2013).

[5] T. Rivera, J. P. Debray, J. M. Gérard, B. Legrand, L. Manin-Ferlazzo, and J. L. Oudar, "Optical losses in plasma-etched AlGaAs microresonators using reflection spectroscopy", Appl. Phys. Lett. 74, 911 (1999).

[6] M. Pelton, J. Vučković, G. S. Solomon, A. Scherer, and Y. Yamamoto, "Three-dimensionally confined modes in micropost microcavities: quality factors and Purcell factors", IEEE J. Quantum Electron. 38, 170 (2002).

[7] M. Lermer, N. Gregersen, F. Dunzer, S. Reitzenstein, S. Höfling, J. Mørk, L. Worschech, M. Kamp, and A. Forchel, "Bloch-wave engineering of quantum dot micropillars for cavity quantum electrodynamics experiments", Phys. Rev. Lett. 108, 057402 (2012).

[8] N. Gregersen, S. Reitzenstein, C. Kistner, M. Strauss, C. Schneider, S. Höfling, L. Worschech, A. Forchel, T. R. Nielsen, J. Mørk, and J. M. Gérard, "Numerical and experimental study of the factor of high-Q micropillar cavities", IEEE J. Quantum Electron. 46, 1470 (2010).

[9] H. Z. Song, K. Takemoto, T. Miyazawa, M. Takatsu, S. Iwamoto, M. Ekawa, T. Yamamoto, and Y. Arakawa, "High quality-factor Si/SiO_2-InP hybrid micropillar cavities with submicrometer diameter for 1.55-μm telecommunication band", Opt. Express. 23, 16264-16272 (2015).

IOP Conf. Series: Journal of Physics: Conf. Series **844** (2017) 012003 doi:10.1088/1742-6596/844/1/012003

Epitaxial growth and characterization of non-polar *a*-plane AlGaN films with MgN/AlGaN insertion layers

Nan Wang[1], Xiong Zhang[1*], Jianguo Zhao[1], Heng Zhang[1], Zili Wu[1], Qian Dai[1], Shuchang Wang[2], Guohua Hu[1], and Yiping Cui[1]

[1]*Advanced Photonics Center, Southeast University, Nanjing 210096, Jiangsu, China*

[2]*College of Physics and Electronic Engineering, Changshu Institute of Technology, Changshu, 215500, Jiangsu, China*

**E-mail: xzhang62@aliyun.com*

Abstract

The MgN/AlGaN insertion layers were applied for the first time in the growth of non-polar *a*-plane AlGaN epi-layers by metal organic chemical vapor deposition technology. The full-width-at-half-maximum value of X-ray rocking curve for the *a*-plane AlGaN epi-layer was decreased by approximately 50.6% and the root-mean-square value of the surface was reduced by 74% by inserting the MgN/AlGaN insertion layers with optimized number of insertion pairs, which revealed that the compressive strain within the *a*-plane AlGaN epi-layers was effectively reduced, leading to significant improvements in crystalline quality and surface morphology, which is very helpful to fabricate high quality AlGaN-based ultraviolet light-emitting-diodes.

Keywords: non-polar *a*-plane AlGaN; MgN/AlGaN insertion layers; crystalline quality; surface morphology.

IOP Conf. Series: Journal of Physics: Conf. Series **844** (2017) 012003 doi:10.1088/1742-6596/844/1/012003

1. Introduction

AlGaN-based ultraviolet light emitting diodes (UV-LEDs) with emission wavelength between 210 and 365 nm are in demand for a wide range of applications, such as sterilization, water/air purification, and medical treatment.[1, 2)] Currently, most of the AlGaN-based UV-LEDs have been fabricated based on polar (0001)-oriented *c*-plane AlGaN epi-layers because there is less difficulty to grow relatively high quality AlGaN epi-layers on polar *c*-plane sapphire substrate than on other orientation sapphire substrates. However, it is well known that the light emission efficiency of the polar *c*-plane AlGaN-based UV-LEDs can be strongly reduced by the quantum confined Stark effect (QCSE) induced by the spontaneous and piezoelectric polarization fields along [0001] *c*-direction.[3)] In contrast, there is no spontaneous polarization-induced built-in electric field in a non-polar AlGaN epi-layer and the QCSE is completely suppressed in non-polar AlGaN-based multiple quantum wells (MQWs).[4)] Hence, high light emission efficiency can be expected for the non-polar AlGaN-based UV-LED. Unfortunately, the hetero-epitaxial growth of non-polar $(11\bar{2}0)$-oriented *a*-plane AlGaN films on semi-polar $(1\bar{1}02)$-oriented *r*-plane sapphire substrates has been proven to be a very challengeable task. Because of the large mismatch both in lattice constant and thermal expansion coefficient between the *r*-plane sapphire substrate and the non-polar AlGaN epi-layer, high density of the misfit dislocations, undulating surface morphology, and strong crystallographic anisotropy are usually generated for the non-polar AlGaN epi-layers during the epitaxial growth process. Therefore, the epitaxial growth of high quality non-polar AlGaN films plays a crucial role in the fabrication of AlGaN-based UV-LEDs.

In order to improve the crystalline quality of the non-polar GaN epi-layers, various kinds of intermediate layers[5-8)] and patterned sapphire substrate (PSS) with different patterns[9-12)] were applied for the epitaxial growth of the non-polar GaN films. However, up to date there were only few reports on the approaches to improve the crystalline quality and the surface morphology for the non-polar AlGaN epi-layers. Even though the first attempt to grow the

IOP Conf. Series: Journal of Physics: Conf. Series **844** (2017) 012003 doi:10.1088/1742-6596/844/1/012003

non-polar *a*-plane AlGaN films under the two-dimensional growth mode was made in 2003,[13] it is still indispensable to further improve the crystalline quality and the surface morphology of non-polar *a*-plane AlGaN epi-layers, especially those with high Al composition (> 0.3) for meeting the needs to fabricate deep UV-LEDs with wavelength less than 320 nm.

In this study, the MgN/AlGaN insertion layers were applied for the first time in the epitaxial growth of non-polar *a*-plane AlGaN films on semi-polar *r*-plane sapphire substrates by metal organic chemical vapor deposition (MOCVD) technology. X-ray diffraction (XRD), transmission electron microscopy (TEM), Raman spectroscopy, photoluminescence (PL) spectroscopy, and atomic force microscopy (AFM) were employed to characterize the structural and optical properties of the grown non-polar *a*-plane AlGaN epi-layers. The characterization results demonstrate that high crystalline quality non-polar *a*-plane AlGaN epi-layers with low dislocations density, smooth surface morphology, low compressive strain, and strong PL intensity have been achieved with the optimized MgN/AlGaN insertion layers.

2. Experimental

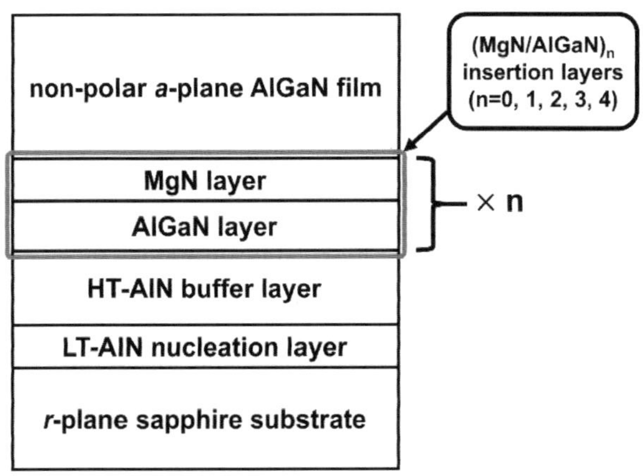

Fig. 1 The schematic layer structures for the non-polar $(11\bar{2}0)$-oriented *a*-plane AlGaN samples grown with various number (n = 0, 1, 2, 3, 4) of pairs of MgN/AlGaN insertion layers.

All the non-polar *a*-plane AlGaN epi-layer samples used in this study were grown on semi-polar *r*-plane sapphire substrates in a low pressure (40 Torr) MOCVD system.

Trimethyl-aluminum (TMAl), trimethyl-gallium (TMGa), bis(cyclopentadienyl)-magnesium (CP_2Mg), and ammonia (NH_3) were used as the precursors for Al, Ga, Mg, and N, respectively, and hydrogen (H_2) was used as the carrier gas. Similar to the epitaxial growth process described in detail in our recently published paper,[14)] for the epitaxial growth of the non-polar *a*-plane AlGaN film without MgN/AlGaN insertion layer which was named as sample A, the semi-polar *r*-plane sapphire substrates were firstly heated up to 1,060 °C in H_2 ambience to remove the surface contamination. A 20 nm-thick low-temperature (LT)-AlN nucleation layer was then deposited on the semi-polar *r*-plane sapphire substrate at a lower temperature of 600 °C. Subsequently, the temperature was raised to 1,100 °C to grow a high-temperature (HT)-AlN buffer layer. Finally, an 800 nm-thick non-polar *a*-plane AlGaN epi-layer was grown on the HT-AlN buffer layer. The epitaxial growth processes and the layer structures for samples B, C, D, and E are basically the same as that for sample A with the exception that 1, 2, 3, and 4 pairs of MgN/AlGaN composite layers were inserted in between the HT-AlN buffer layer and the top AlGaN layer, respectively, as schematically shown in Fig. 1. For the purpose of comparison, the thickness for one pair of the MgN/AlGaN composite layers was fixed at 250 nm throughout this study, and the averaged layer thickness of the MgN film was estimated to be less than 10 nm.

The crystal orientation and the Al composition for the grown non-polar *a*-plane AlGaN epi-layers were determined by means of high resolution (HR)-XRD 2θ-ω scans. In specific, the X-ray rocking curves (XRCs) were measured in two directions that are perpendicular to each other for evaluating the crystalline quality for the non-polar *a*-plane AlGaN epi-layers. The blocking of the threading dislocations (TDs) by the MgN/AlGaN insertion layers was confirmed with TEM measurement. Moreover, the surface morphology for the non-polar *a*-plane AlGaN epi-layers was characterized by AFM measurement. The structural and optical properties for the grown non-polar *a*-plane AlGaN epi-layers were further investigated with Raman and PL spectroscopies conducted at room temperature (RT).

3. Results and discussions

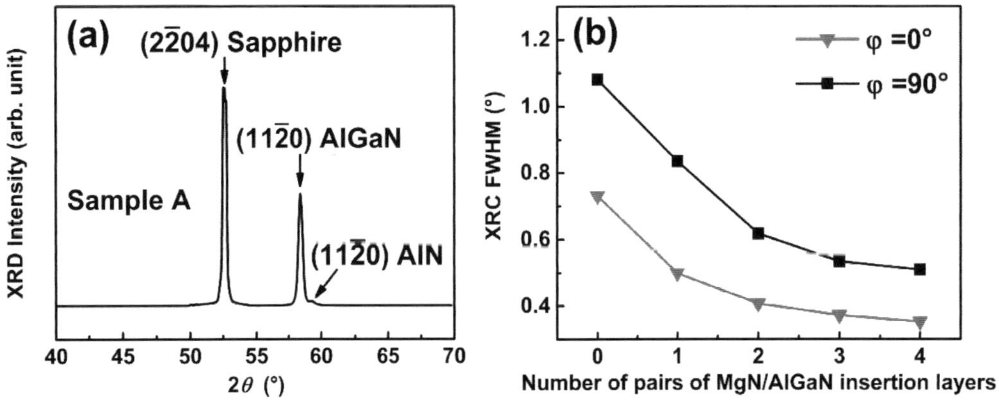

Fig. 2 The HR-XRD 2*θ*-*ω* scanning curve for the non-polar *a*-plane AlGaN epi-layer sample A (a), the XRC FWHM values as a function of the number of pairs of the MgN/AlGaN insertion layers (b).

The HR-XRD 2*θ*-*ω* scanning curve for sample A was shown in Fig. 2(a). It was demonstrated unambiguously that the peaks located at 2*θ*=52.56°, 58.40° and 59.35° were the XRD diffraction peaks from the semi-polar $(2\bar{2}04)$-oriented *r*-plane sapphire substrate, the non-polar $(11\bar{2}0)$-oriented *a*-plane AlGaN and AlN epi-layers, respectively.[15, 16] Extremely similar results were observed for samples B, C, D, and E although they are not shown in Fig. 2(a). Moreover, the Al compositions were determined to be 0.40 for all the five samples A-E by fitting the XRD spectra with an analysis software called "PANalytical X'Pert Epitaxy". Considering the structural anisotropy of the non-polar AlGaN epi-layers, the measurements of the XRCs were performed at azimuth angles of φ= 0° parallel to the [0001] direction and φ= 90° parallel to the $[1\bar{1}00]$ direction, respectively to characterize the crystalline quality of the grown non-polar *a*-plane AlGaN epi-layers. The full width at half maximum (FWHM) values of the XRCs for the five non-polar *a*-plane AlGaN epi-layer samples A-E that have varied number of pairs of the MgN/AlGaN insertion layers were depicted in Fig. 2(b). It was evident that the FWHM values of the XRCs measured along both the [0001] and the $[1\bar{1}00]$ directions decreased remarkably with the increase in the number of pairs of MgN/AlGaN insertion layers, indicating a significant improvement in crystalline quality for the non-polar *a*-plane AlGaN epi-layer samples that were inserted with

MgN/AlGaN composite layers.[17)] In fact, the FWHM values of the XRCs for sample D with three pairs of the MgN/AlGaN insertion layers were found to be 49.1% less along the [0001] direction and 50.6% less along the $[1\bar{1}00]$ direction, respectively than those for sample A without any MgN/AlGaN insertion layers. However, as shown clearly in Fig. 2(b), the FWHM value of the XRC decreased slowly and tended to saturate when the number of pairs of the MgN/AlGaN insertion layers increased further from three for sample D to four for sample E.

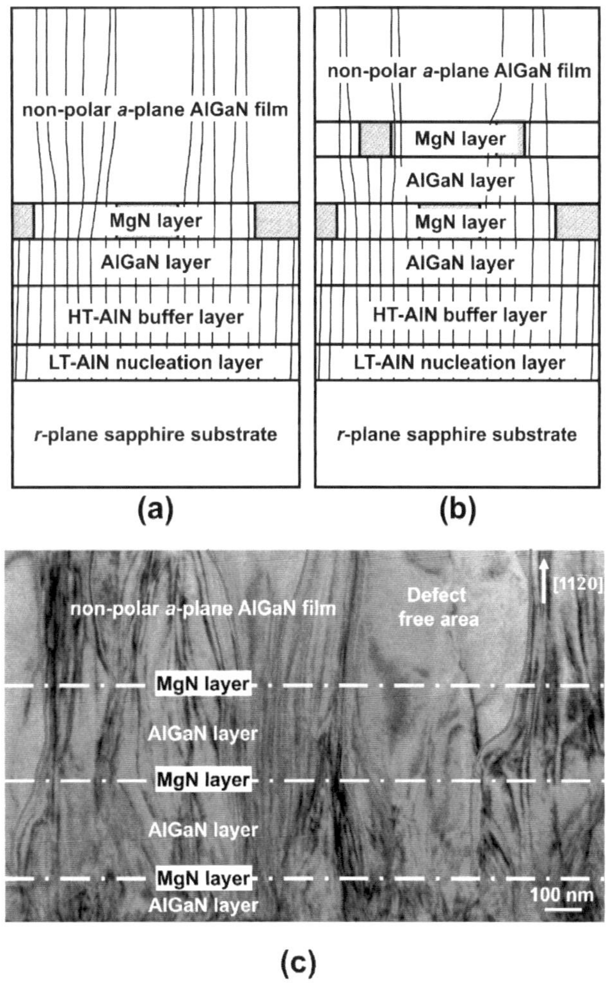

Fig. 3 The growth models of the non-polar $(11\bar{2}0)$-oriented *a*-plane AlGaN epi-layer samples with one pair (a) and two pairs (b) of the MgN/AlGaN insertion layers as well as the bright field image of cross-sectional TEM for sample D (c).

The growth models for the non-polar *a*-plane AlGaN epi-layer samples with one and two pairs of MgN/AlGaN insertion layers as well as the bright field image of the cross-sectional TEM for sample D with three pairs of the MgN/AlGaN insertion layers were demonstrated in Figs. 3(a), 3(b), and 3(c), respectively. It was inferred that the improvement in crystalline quality could be attributed to the blocking of the TDs formed during the growth process of the LT-AlN nucleation layer and at the initial growth stage of the HT-AlN buffer layer by the extra hetero-interfaces associated with the MgN/AlGaN insertion layers.[18,19)] As shown schematically in Fig. 3(a), the TDs in the non-polar *a*-plane AlGaN epi-layer sample with one pair of MgN/AlGaN insertion layers are partially blocked and the defect-free areas with less TDs are created. Moreover, as illustrated in Fig. 3(b), the TDs are further blocked and the defect-free areas are increased markedly in the non-polar *a*-plane AlGaN epi-layer sample with two pairs of MgN/AlGaN insertion layers. On the other hand, as demonstrated clearly by the TEM measurement results shown in Fig. 3(c), the TDs in sample D were indeed effectively blocked or merged with the introduction of the MgN/AlGaN insertion layers, reaching the surface in the form of bundles. As a result, large defect-free areas were observed in the TEM image shown in Fig. 3(c) for sample D, indicating a relatively good coincidence with the models shown in Figs. 3(a) and 3(b).

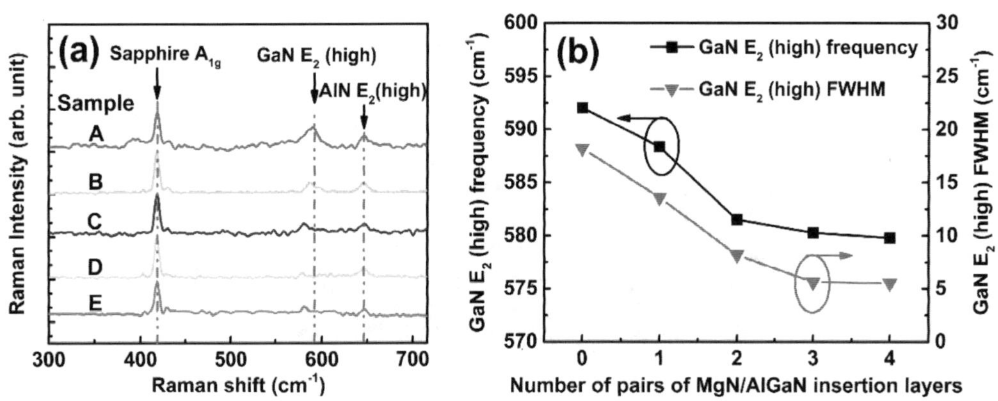

Fig. 4 The RT Raman spectra for the five non-polar *a*-plane AlGaN epi-layer samples A-E measured under the backscattering configuration of $x(-,-)\bar{x}$ (a), the phonon frequency and the FWHM value of GaN-like E_2 (high) mode-related transition for the non-polar *a*-plane

AlGaN epi-layer samples A-E as a function of the number of pairs of the MgN/AlGaN insertion layers (b).

Figure 4(a) shows the RT Raman spectra for the non-polar *a*-plane AlGaN epi-layer samples A-E measured under the backscattering configuration of $x(-,-)\bar{x}$ with particular attentions paid on the behavior of GaN-like E_2 (high) mode. It was well known that under such scattering configurations, E_2 phonon transition related to the strain- and crystalline quality was allowed.[20)] Furthermore, it was found that E_2 (high) modes for the non-polar AlGaN samples included the features of both GaN-like and AlN-like modes.[21)] As shown in Fig. 4(a), two peaks located at 418 cm^{-1} and 646.1 cm^{-1}, respectively, appeared in the Raman spectra for all the samples, corresponding to the A_{1g} mode for the semi-polar $(1\bar{1}02)$-oriented *r*-plane sapphire substrates and the AlN-like E_2 (high) mode for the non-polar $(11\bar{2}0)$-oriented *a*-plane AlGaN epi-layers, respectively. The phonon frequencies and the FWHM values of GaN-like E_2 (high) mode for samples A-E were plotted in Fig. 4(b) to analyze the effects of the MgN/AlGaN insertion layers on the characteristics of GaN-like E_2(high) mode. It is apparent that as the number of the inserted MgN/AlGaN pairs was increased from 0 to 4, the GaN-like E_2 (high) phonon peak was red-shifted monotonously from 592.0 to 579.8 cm^{-1}. Since the GaN-like E_2 (high) phonon peak for the unstrained $Al_{0.40}Ga_{0.60}N$ film was known to be located at 579.8 cm^{-1}, it is reasonable to treat this value as the critical point value between compressive strain and tensile strain.[21)] In other words, lower or higher values than the critical point value implies the existence of a tensile strain or a compressive strain within the epitaxial layers, and the frequency shift from the critical point is proportional to the magnitude of the strain[22, 23)]. Therefore, it could be concluded that samples A-E were all suffering from a compressive strain. This compressive strain, however, decreased monotonously from sample A to sample E, indicating an effective relaxation of the compressive strain within the non-polar *a*-plane AlGaN epi-layers with the increase in the number of pairs of MgN/AlGaN insertion layers from 0 to 4.[14, 24)] Moreover, as shown in Fig. 4(b), the FWHM values of the GaN-like E_2 (high) mode-related peaks decreased evidently from sample A to sample E. Because the FWHM value reflects the crystalline quality, the significant reduction in the FWHM value indicates an improvement in crystalline quality of the grown non-polar *a*-plane AlGaN epi-layers with the insertion of the MgN/AlGaN composite layers.[25)] As mentioned above, the reduction in the dislocation density could be ascribed to the MgN/AlGaN insertion layers-induced enhancement in the opportunity for the dislocations to annihilate, resulting in the improvement in crystalline quality.[26)] In particular, as shown in Fig. 4(a), the GaN-like E_2 (high) peak for sample E was located at 579.8 cm^{-1} which was consistent with the critical point value of 579.8 cm^{-1}. This fact implies that the compressive strain within sample E is nearly completely compensated, or sample E has the lowest dislocation density among the five non-polar *a*-plane AlGaN epi-layer samples A-E. In addition, it was found that both the phonon frequency and the FWHM value of the GaN-like E_2 (high) mode decreased slowly and tended to saturate as the number of pairs of the MgN/AlGaN insertion layers increased further from three for sample D to four for sample E.

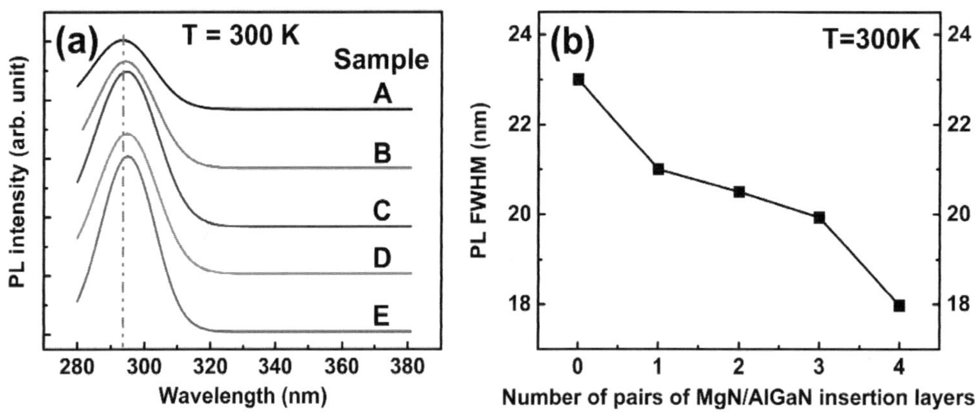

Fig. 5 The PL spectra for the non-polar *a*-plane AlGaN epi-layer samples A-E (a), the FWHM values of the PL emission peaks as a function of the number of pairs of the MgN/AlGaN insertion layers (b).

The PL spectra and the FWHM values of the near band edge (NBE) transition peaks for the grown non-polar *a*-plane AlGaN epi-layer samples as a function of the number of pairs of the MgN/AlGaN insertion layers are plotted in Fig. 5. As shown in Fig. 5(a), the NBE transition peaks for samples A-E were slightly red-shifted from 293.5 to 295.0 nm as the number of the inserted MgN/AlGaN pairs was increased from 0 to 4, indicating clearly the effective relaxation of the compressive strain within the non-polar *a*-plane AlGaN epi-layers.[27)] It was demonstrated clearly in Fig 5(b) that the FWHM value of the PL emission peak decreased monotonously with increasing the number of pairs of the MgN/AlGaN insertion layers from 0 to 4. This result indicates once again that significant improvements in both optical property and crystalline quality could be achieved for the non-polar *a*-plane AlGaN epi-layer samples that were inserted with the MgN/AlGaN composite layers.[28)]

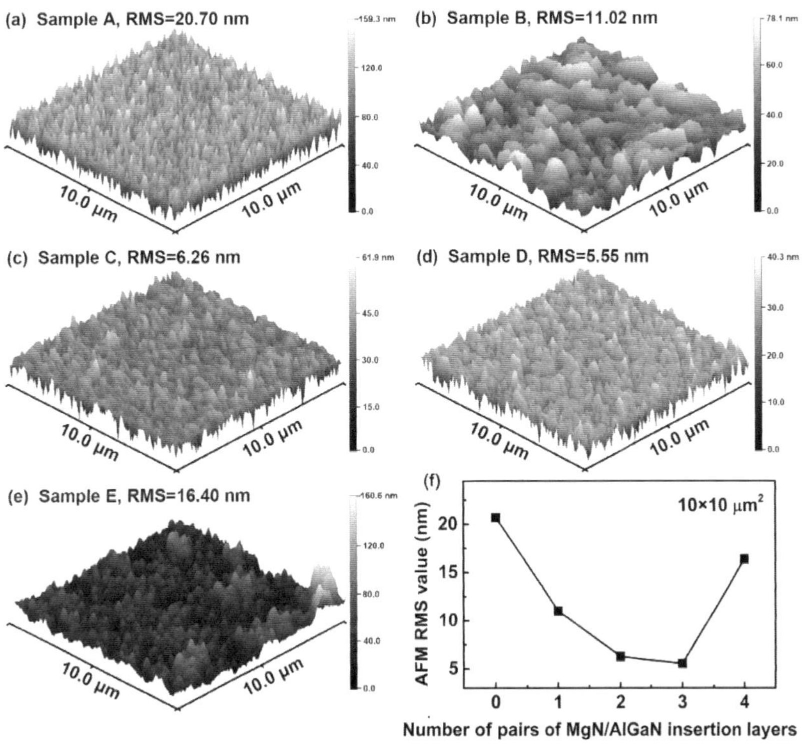

Fig. 6 The AFM images for the non-polar *a*-plane AlGaN epi-layer samples A-E with a detection area of 10×10 μm^2 (a-e), the RMS values as a function of the number of pairs of the MgN/AlGaN insertion layers (f).

The three-dimensional (3D) view AFM images and the root mean square (RMS) values for samples A-E are shown in Fig. 6. The RMS values obtained from the AFM measurements for samples A-D decreased gradually as the number of pairs of the MgN/AlGaN insertion layers increased from 0 for sample A to 3 for sample D, whereas the RMS value increased abruptly as the number of pairs of the MgN/AlGaN insertion layers further increased to 4 for sample E, as shown clearly in Fig. 6(f). This phenomenon demonstrates clearly that the surface morphology for the non-polar *a*-plane AlGaN samples could definitely be improved by inserting the MgN/AlGaN composite layers with an appropriate number of pairs of the MgN/AlGaN insertion layers. Generally, the amelioration in surface morphology could be ascribed to the improvement in both the crystalline quality and the relaxation of the residual compressive strain within the non-polar *a*-plane AlGaN epi-layers.[29)] In fact, the RMS value for sample D with three pairs of the MgN/AlGaN insertion layers was 74% less than that for sample A without any MgN/AlGaN insertion layers, showing the smoothest surface morphology among all the samples. Nevertheless, the RMS value for sample E with four pairs of MgN/AlGaN insertion layers was much larger than that of sample D, which could be explained in terms of such a theory that the growth mode tends to transform from two-dimensional (2D) to 3D growth mode when the thickness of the insertion layers is increased too much, resulting in a degradation in surface morphology.[30)] On the other hand, the surface morphology for the AlGaN epi-layers could also be degraded with the increase in Mg incorporation into the non-polar *a*-plane AlGaN epi-layers.[31)]

4. Conclusions

The MgN/AlGaN insertion layers were applied for the first time in the epitaxial growth of the non-polar *a*-plane AlGaN films on semi-polar *r*-plane sapphire substrates by MOCVD. The FWHM value of the XRC for the non-polar *a*-plane AlGaN epi-layer sample with three pairs of the MgN/AlGaN insertion layers was reduced to approximately a half of that for the sample without any MgN/AlGaN insertion layers. The smallest RMS value obtained from the AFM measurement was also achieved with the non-polar *a*-plane AlGaN epi-layer sample with three pairs of the MgN/AlGaN insertion layers, which was found to be 74% less than that for the sample without any MgN/AlGaN insertion layers. Furthermore, the characterization results of Raman spectroscopy showed that the in-plane compressive strain within the non-polar AlGaN epi-layer samples could be compensated effectively by inserting the MgN/AlGaN composite layers with optimized number of insertion pairs. These results revealed that the defects and misfit dislocations within the grown non-polar *a*-plane AlGaN epi-layers could be significantly reduced with the insertion of the MgN/AlGaN composite layers. The remarkable improvements in optical property, crystalline quality, and surface morphology achieved in this study are very helpful to fabricate high quality non-polar AlGaN-based UV-LEDs.

Acknowledgements

This work was supported by the Key Research and Development Project of Science and Technology Department of Jiangsu Province, People's Republic of China (Grant No. BE2015159). The authors would like to thank Mr. Lang Li for his help with AFM measurements.

References

1) M.S. Shur and R. Gaska: IEEE Trans. Electron Devices **57** [1](2010)12.
2) C.-Y. Cho, J.-J. Kim, S.-J. Lee, S.-H. Hong, K.J. Lee, S.-Y. Yim and S.-J. Park: Appl. Phys. Express **5** [12](2012)122103.
3) T. Deguchi, K. Sekiguchi, A. Nakamura, T. Sota, R. Matsuo, S. Chichibu and S. Nakamura: Jpn. J. Appl. Phys. 2. **38** [8B](1999)L914.
4) P. Waltereit, O. Brandt, A. Trampert, H.T. Grahn, J. Menniger, M. Ramsteiner, M. Reiche and K.H. Ploog: Nature. **406** [6798](2000)865.
5) C.F. Johnston, M.J. Kappers, M.A. Moram, J.L. Hollander and C.J. Humphreys: J. Cryst. Growth **311** [12](2009)3295.
6) A. Chakraborty, K.C. Kim, F. Wu, J.S. Speck, S.P. DenBaars and U.K. Mishra: Appl. Phys. Lett. **89** [4](2006)041903.
7) J.L. Hollander, M.J. Kappers, C. McAleese and C.J. Humphreys: Appl. Phys. Lett. **92** [10](2008)101104.
8) M.J. Kappers, J.L. Hollander, C.F. Johnston, C. McAleese, D.V. Sridhara Rao, A.M. Sanchez, C.J. Humphreys, T.J. Badcock and P. Dawson: J. Cryst. Growth **310** [23](2008)4983.
9) Y. Li, S. You, M. Zhu, L. Zhao, W. Hou, T. Detchprohm, Y. Taniguchi, N. Tamura, S. Tanaka and C. Wetzel: Appl. Phys. Lett. **98** [15](2011)151102.
10) G. Yoo, H. Park, H. Lim, S. Lee, O. Nam, Y. Moon, C. Lim, B. Kong and H. Cho: Jpn. J. Appl. Phys. **50** [4](2011)042103.
11) J.J. Chen, Y.K. Su, C.L. Lin, S.M. Chen, W.L. Li and C.C. Kao: IEEE Photonics Technol. Lett. **20** [13](2008)1193.
12) M. Kim, T. Fujita, S. Fukahori, T. Inazu, C. Pernot, Y. Nagasawa, A. Hirano, M. Ippommatsu, M. Iwaya, T. Takeuchi, S. Kamiyama, M. Yamaguchi, Y. Honda, H. Amano and I. Akasaki: Appl. Phys. Express **4** [9](2011)092102.
13) S. Iyer, D.J. Smith, A. Bhattacharyya, K. Ludwig and T.D. Moustakas: MRS Proc. 743 (2011)133.
14) H. Yang, X. Zhang, S. Wang, Y. Wang, H. Luan, Q. Dai, Z. Wu, J. Zhao and Y. Cui: Superlattices Microstruct. **96** (2016)1.
15) B.A. Haskell, S. Nakamura, S.P. DenBaars and J.S. Speck: Phys Status Solidi B. **244** [8](2007)2847.
16) M.R. Laskar, T. Ganguli, A.A. Rahman, A.P. Shah, M.R. Gokhale and A. Bhattacharya: Phys. Status Solidi: Rapid Res. Lett. **4** [7](2010)163.
17) M.R. Laskar, T. Ganguli, N. Hatui, A.A. Rahman, M.R. Gokhale and A. Bhattacharya: J. Cryst. Growth **315** [1](2011)208.
18) C.F. Johnston, M.J. Kappers, J.S. Barnard and C.J. Humphreys: Phys. Status Solidi (c). **5** [6](2008)1786.
19) Y. Yang, P. Xiang, M. Liu, W. Chen, Z. He, X. Han, Y. Ni, F. Yang, Y. Yao, Z. Wu, Y. Liu and B. Zhang: J. Cryst. Growth **376** (2013)23.
20) M. Kuball: Surf. Interface Anal. **31** [10](2001)987.

21) V.Y. Davydov, I.N. Goncharuk, A.N. Smirnov, A.E. Nikolaev, W.V. Lundin, A.S. Usikov, A.A. Klochikhin, J. Aderhold, J. Graul, O. Semchinova and H. Harima: Phys. Rev. B. **65** [12](2002).
22) A.F. Wilson, A. Wakejima and T. Egawa: Appl. Phys. Express **6** [8](2013)086504.
23) X.-H. Liu, J.-C. Zhang, X.-J. Su, J. Huang, S.-N. Zheng, Y.-Y. Hu, B.-B. Ye, J.-J. Zhao, J.-F. Wang, J.-P. Zhang and K. Xu: Appl. Phys. Express **9** [4](2016)045501.
24) Q. Dai, X. Zhang, J. Zhao, H. Luan, Z. Liang and Y. Cui: Mater. Sci. Semicond. Process. **58** (2017)30.
25) T. Jiang, S.R. Xu, J.C. Zhang, Y. Xie and Y. Hao: Sci Rep. **6** (2016)19955.
26) Y. Li, S. Chen, M. Kong, S. Li, W. Tian, S. Sun, Z. Wu, Y. Fang, J. Dai and C. Chen: J. Appl. Phys. **115** [4](2014)043503.
27) K.-H. Lee, S.H. Park, J.H. Kim, N.H. Kim, M.H. Kim, H. Na and E. Yoon: Thin Solid Films. 518 [22](2010)6365.
28) H.M. Huang, S.C. Ling, J.R. Chen, T.S. Ko, J.C. Li, T.C. Lu, H.C. Kuo and S.C. Wang: J. Cryst. Growth **312** [6](2010)869.
29) P. Tao, H. Liang, X. Xia, Y. Chen, C. Yang, J. Liu, Z. Zhu, Y. Liu, R. Shen, Y. Luo, Y. Zhang and G. Du: Sci. Semicond. Process. **41** (2016)291.
30) K. Tang, S. Huang, S. Gu, S. Zhu, J. Ye, Z. Xu and Y. Zheng: Appl. Surf. Sci. **388** (2016)557.
31) C.-R. Lee, J.-Y. Leem, S.-K. Noh, S.-E. Park, J.-I. Lee, C.-S. Kim, S.-J. Son and K.-Y. Leem: J. Cryst. Growth **193** (1998)300.

Significant light absorption improvement in perovskite/CIGS tandem solar cells with dielectric nanocone structures

Huahua Wang, Boyuan Cai*, and Xiaocong Yuan

Nanophotonics Research Center, Shenzhen University & Key Laboratory of Optoelectronic Devices and Systems of Ministry of Education and Guangdong Province, College of Optoelectronic Engineering, Shenzhen University, Shenzhen, China 518060, E-mail: caiboyuan@126.com

Abstract Here, we propose a novel perovskite/CIGS tandem solar cell geometry with tailored dielectric nanocone structure incorporated on the top surface for light manipulation. Absorption enhancement as high as 15.39% has been achieved both in the top and bottom subcells, leading to a 14.29% thickness reduction of the bottom subcell.

1. Introduction

Recent research into the field of thin-film solar cells has found an attractive technology that utilizes perovskite material as the solar absorber for high efficiency photovoltaic devices [1]. Since the first application of perovskite material in solar cells in 2009, the efficiency has soared from 3.8% to over 20% [2]. However, due to the bandgap limitation of the material, a perovskite-based single junction device can only response to the solar spectrum less than 900 nm for the electron-hole pair generations. Therefore, to obtain even higher efficiency photovoltaic devices, multi-junction devices with appropriate bandgap materials need to be combined with the perovskite material to harvest solar spectrum longer than 900 nm, and advanced light trapping method is deeply necessary. In this work, we utilize high bandgap perovskite material and low bandgap CIGS material to design a tandem solar cell model for harvesting more solar spectrum in the simulation. The detailed layers applied in our model are presented in Fig. 1, based on the individual single junction solar cell device design respectively. To trap light more efficiently in the designed device, ITO dielectric nanocone structures are placed on top of the solar cells, acting light scattering and anti-reflection elements [3]. Through the optimization of the thickness of each subcell for current matching and the nanocone parameters (height and the angle of the cone) for further absorption enhancement, a short current density (J_{SC}) as high as 18.49 mA/cm^2 is achieved, which is much higher compared to the 16.09 mA/cm^2 in flat cells without the nanocone structure, leading to a 15.39% J_{SC} enhancement.

2. Methods and Results

The perovskite/CIGS tandem solar cell integrated with dielectric nanocones on the top surface is illustrated in Fig. 1. Due to that the device performance is determined by the lowest J_{SC} value of each subcell, the appropriate thicknesses of perovskite and CIGS layers for J_{SC} matching need to be investigated first. For the flat solar cell case without nanocones, we choose perovskite absorber for the top subcell (ITO: 80 nm, TiO_2: 50 nm, perovskite: thickness to be optimized, Spiro: 30 nm) and the CIGS absorber for the bottom subcell (ITO: 80 nm, ZnO: 80 nm, CdS: 50 nm, CIGS: thickness to be optimized, MO: 100 nm) considering their bandgap. We utilize the Lumerical FDTD software to build up and simulate the model. To figure out the condition of the J_{SC} matching between each subcell, the thicknesses of perovskite layer are selected to be 100 nm, 150 nm, 200 nm and 250 nm while the thickness of the CIGS layer is swept from 500 to 2500 nm. Through the calculation of the J_{SC} values of each subcell as functions of the thicknesses of the subcells, it can be found that when the thickness of the CIGS layer is 1750 nm and the thickness of the perovskite layer is 150 nm, the J_{SC} value (16.09 mA/cm^2) matches between the top cell and the bottom cell.

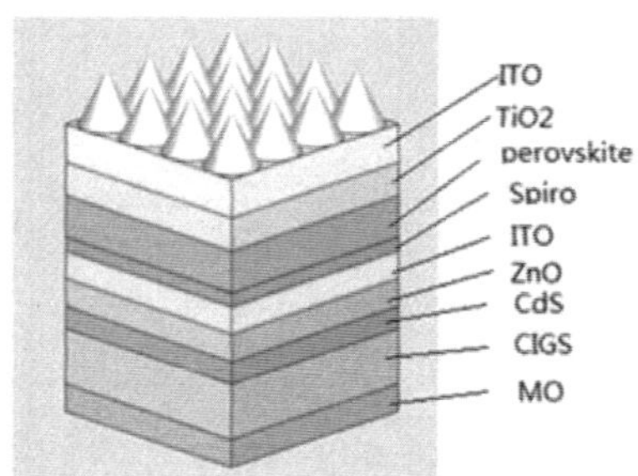

Fig 1. The perovskite/CIGS tandem solar cell structure with nanocones on the top surface.

To further enhance the flat tandem solar cell absorption, we place ITO nanocones on the top surface of the tandem solar cell with the thickness of perovskite layer 150 nm and the thickness of CIGS layer 1750 nm as shown in Fig. 1. For the top surface integrated nanocones, different aspect ratios (determined by the cone height and the cone angle) are optimized for the highest device absorption enhancement [4]. Fig. 2 presents the J_{SC} enhancement mapping in the bottom subcell as a function of the cone height and the cone angle (θ). The J_{SC} value with nanocones is normalized to that of the bottom subcell (16.09 mA/cm^2) in the reference flat tandem solar cell without nanocones. In Fig. 2 (a), it's obvious that the highest enhancement of 18.64% in the bottom subcell can be achieved with the optimized parameters of 500 nm in height and 70° in cone angel for nanocones, leading to an increased J_{SC} value of 19.09 mA/cm^2.

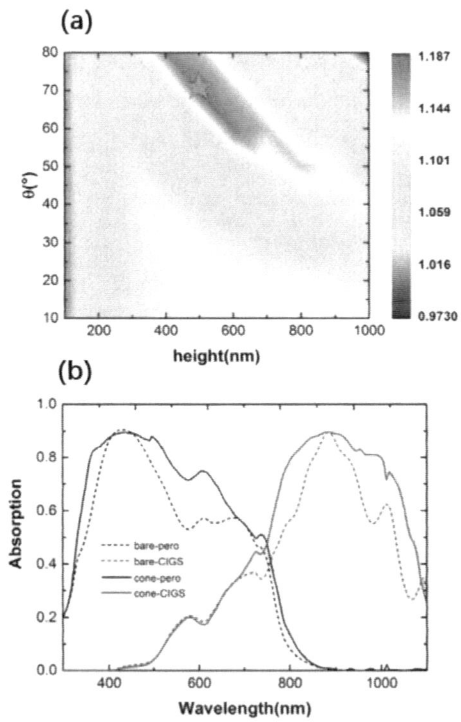

Fig 2. (a) J_{SC} enhancement of the bottom subcell as a function of the height and the cone angel of nanocones. (b) Absorption comparison of the perovskite top subcell and CIGS subcell with/without the optimized cone parameters with cone height of 500 nm and the cone angel of 70°.

Nevertheless, in terms of top cell, the J_{SC} value is 18.48 mA/cm^2 increased about 15.37% under the same parameters. This can be ascribed to the scattering and anti-reflection effects of the nanocones, which increase the light path inside the bottom CIGS subcell and top perovskite subcell respectively. As shown in Fig. 2 (b), the absorption in the bottom subcell can be improved significantly in the wavelength range from 800 nm to 1100 nm due to the scattering effect, while for the perovskite subcell, the absorption in the wavelength from 300 nm to 700 nm is mainly improved. Under this circumstance, it's promising to reduce the thickness of the CIGS layer for cost savings without compromising the overall device performance compared to the flat solar cell device. Through the simulation of nanocones incorporated solar cell with thinner CIGS layer, we can find that the J_{SC} value of 18.49 mA/cm^2 can be obtained with a CIGS layer thickness of only 1500 nm. The matched J_{SC} value can be increased by 15.39% compared to the planar solar cell with a reduced bottom subcell thickness of 14.29%.

3. Conclusion

In summary, a perovskite/CIGS tandem solar cell geometry incorporated with nanocones is systematically investigated for achieving high efficiency perovskite-based thin film solar cells. Through optimizing the nanocone parameters, J_{SC} enhancement as high as 18.64% in bottom subcell can be obtained and the matched J_{SC} value of 18.49 mA/cm^2 is achieved with a reduction of CIGS layer thickness from 1750 nm to 1500 nm. This approach theoretically opens up a pathway for designing perovskite-based low cost high efficiency tandem solar cells.

Acknowledgment

Boyuan Cai thanks The National Natural Science Fund (61605064), The Guangdong Province Natural Science Fund (2016A03031008).

Reference

[1] Martin A. Green and Anita Ho-Baillie, "The emergence of perovskite solar cells," Nature Photon. 8(7), 506–514 (2014).

[2] C. W. Chen and S. Y. Hsiao, "Optical properties of organometal halid perovskite thin films and general device structure design rules for perovskite single and tandem solar cells," J. Mater. Chem. A. 3(17), 9152–9159 (2015).

[3] K. X. Wang and Z. Yu, "Absorption enhancement in ultrathin crystalline silicon solar cells with antireflection and light-trapping nanocone gratings," Nano Lett. 12(3), 1616–1619 (2012).

[4] M. M. Tavakoli and K. T. Tsui, "Highly efficient flexible perovskite solar cells with antireflection and self-Cleaning nanostructures," ACS Nano. 9(10), 10287–10295(2015).

Pattern characteristics and resolution of GaN sample through Parameter of Indirect Microscopic Imaging

N.P. YADAV[1] XUEFENG LIU[1], WEIZE WANG[1], KALEEM ULLAH[1], BIN XU[1]

[1]School of electronic and optical engineering, Nanjing university of science and technology, 200 Xiaolingwei, Nanjing, 210094, China

nagendra1nagendra@gmail.com

Abstract: With smaller DOF (Depth of Focus) achieving focus at the center of the photoresist also became harder. Currently it is big issues to achieve, the super resolution by optical microscope because due to the diffraction limit. In this letter, the pattern characteristics and edge resolution of GaN sample has been studied through an indirect microscope. In this new discovery, we would like to see whether the etched part is fully etched without SiNx in the sample. Some Electronics microscope such as SEM, AFM have already developed for sub nano resolution. But due to time taken, more expensive ,it is not easy to find every place. In this way, we proposed a new imaging method (PIMI), which is differ from all non-optical and traditional optical imaging system. We found that image resolution of GaN samples under measurement is near about 80 nm. All the measurements have been done by PIMI system.

1. Introduction:

Gallium nitride (GaN) is a binary III/V direct bandgap semiconductor commonly used in light-emitting diodes since the 1990s. The compound is a very hard material that has a Wurtzite crystal structure. Its wide band gap of 3.4 eV affords it special properties for applications in optoelectronic, high-power and high-frequency devices. For example, GaN is the substrate which makes violet (405 nm) laser diodes possible, without use of nonlinear optical frequency-doubling. In 1870s, Abbe discovered that when an object is observed by optical devices such as microscopes or telescopes, the features smaller than half of the wavelength of light are not resolvable because of diffraction phenomena [1]. Due to diffraction limit, resolution achievable by using conventional optical microscope is intrinsically limited to approximately two hundred nanometers in visible light. Realizing microscopes with higher resolution have been active research area. At 1930s, microscopes using electrons such as scanning electron microscope (SEM) [2] and transmission electron microscope(TEM) [3] were invented. Those two microscopes have much higher resolutions of tens of nanometers compared to the typical optical microscope due to the shorter de Broglie wavelength of electrons. We have already reported the application of (PIMI) system under the sample of graphene layers[4].It is well known that the resolution and evaluation of the image quality technique is important for imaging system assessment such as radar[5],Atomic force microscope CT[4].etc. We have review our study in image resolution with different methods [8-10].In the present paper, resolution of pattern GaN sample has been studies with the help of parameter of indirect microscopy, which is better as compare to reported before. The details sections are described below.

2. Theory:

With the help of geometric optical coordination system, polarization orientations illustrated and the Jones model [11-14], one can find that during polarization sweeping, the pixel intensity (Ii) depending on the vector addition of two wave fronts of unlike polarization states from the inherent structural anisotropy of the samples under measurement (SUM), varies with input split angle α. This hope can be described by Eq. (1) in a basic form, where Idp is the average intensities over all polarization status transmitted through or reflected from each field points of the SUM, so its value least affected by scatterings near field change of polarization status, $a_0 = \frac{1}{2}$, $a_1 = \cos 2\phi \sin \delta$, $a_2 = \sin 2\phi \sin \delta$ and sin δ is the sine of phase difference between two orthogonal polarization states, φ is the polarization angle of the slow axis. Therefore the ratio between pixel intensity (Ii) and average intensity (Idp) is given in below equation-1.

$$\frac{I_i}{I_{dp}} = a_0 + a_1 \sin 2\alpha + a_2 \cos 2\alpha \tag{1}$$

The field intensities at individual pixel of the CCD matrix were recorded while wide the polarization which is achieved by rotating a linear polarization disc in a manage component. The intensity difference with parameters, such as the split phase and angles were fit to Eq. (1). From the derived equation, the values of the parameters such as Idp, sinδ and φ can be extracted. From the above polarization parameters, using the relationship between Jones's and Mueller's model, the Stokes parameters, S0, S1, S2 and S3 can be further calculated with Eq. (2).

$$\begin{pmatrix} S_0 \\ S_1 \\ S_2 \\ S_3 \end{pmatrix} = \begin{pmatrix} E_x E_x + E_y E_y \\ E_x E_x - E_y E_y \\ E_x E^*_y - E_y E^*_x \\ i(E_x E^*_y - E_y E^*_x) \end{pmatrix} = \begin{pmatrix} I_{dp}(1+\sin\delta) \\ I_{dp}(1+\sin\delta)\cos 2\phi \\ \sqrt{2}\, I_{dp}(1+\sin\delta)\sqrt{\sin 2\phi}\cos\delta \\ \sqrt{2}\, I_{dp}(1+\sin\delta)\sqrt{\sin 2\phi}\sin\delta \end{pmatrix} \tag{2}$$

Where E_x and E_y are the X and Y optical field components, S_0, S_1, S_2, S_3 are stokes parameters along with the depolarization intensity, I_{dp}, phase difference, sinδ, and polarization angle, φ can be retained to each pixel coordinates in the sensor space to form individual parameter maps, or indirect images as we call them, with the contrasts not built by the optical intensity but the spatial polarization parameters related to near field characteristics.

3. Parameter of Indirect Microscopy (PIMI) Setup for Measurement

Based on the above theory [4] we have designed an experimental setup (PIMI) which is shown in fig.1.

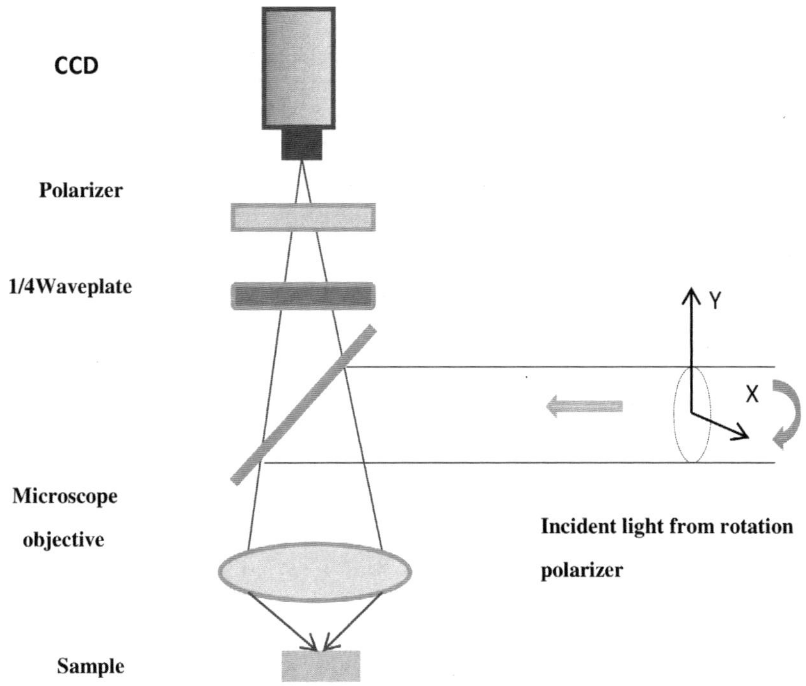

Fig.1. Diagram of modulated polarization parameter imaging system with a tilted sample.

PIMI system was built by the same as Olympus reflection microscopic system BX51M[4]. And the microscope objective is 0.9 Numerical Aperture (NA) 100X magnification under the narrow-line width(<20nm) 532nm illumination and CCD with pixel size of 3.4 micron and output range of 12 bits. The incident light with the polarizer rotating lead to the polarization of its sweeping, while the CCD images were synchronized with the position and movement of rotation polarizer automatic control. In this way, the measurement of the GaN sample with the tilted small angle has been established. As expected, (SUM) will be imaged by part of picture inside or outside focus and another part in focus. Whatever inside and outside focus is out of focus, which can lead to optical blur. In order to the standard test, the SUM on focus will be used by ensuring the existence of the imaging of focus plane, which is clearer. Furthermore, the PIMI has the sub 100 nm resolution, which is capable of wide view-field and far field measurement comparing with the Scanning Electron microscopy(SEM).

4. Results and Discussion

We have take the images of the sample through the indirect microscopy(PIMI) in different magnifications. The image results are shown in below figs.(2-4).

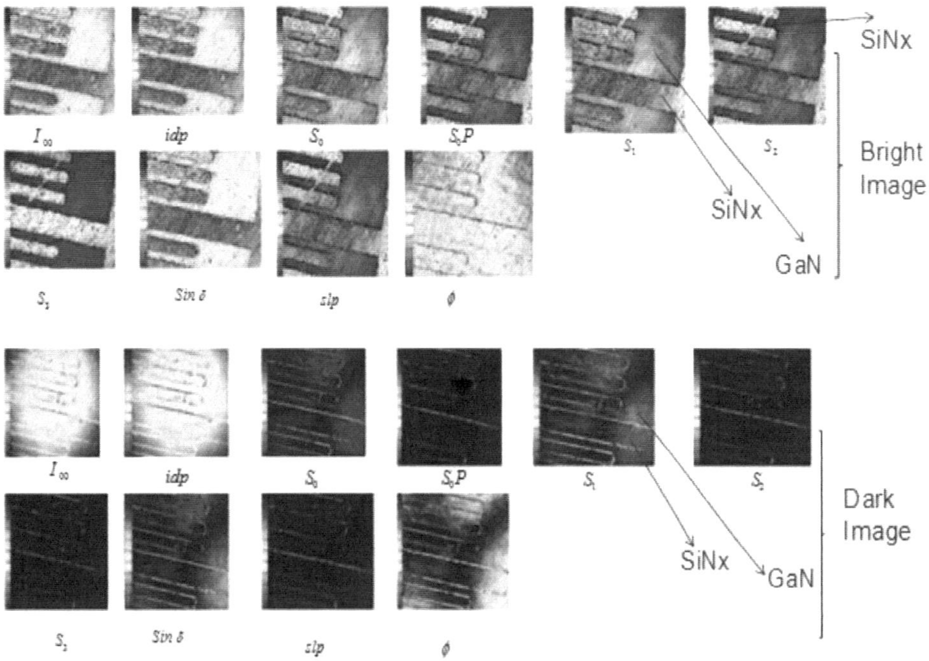

Fig.2: 10 X magnification in the same position

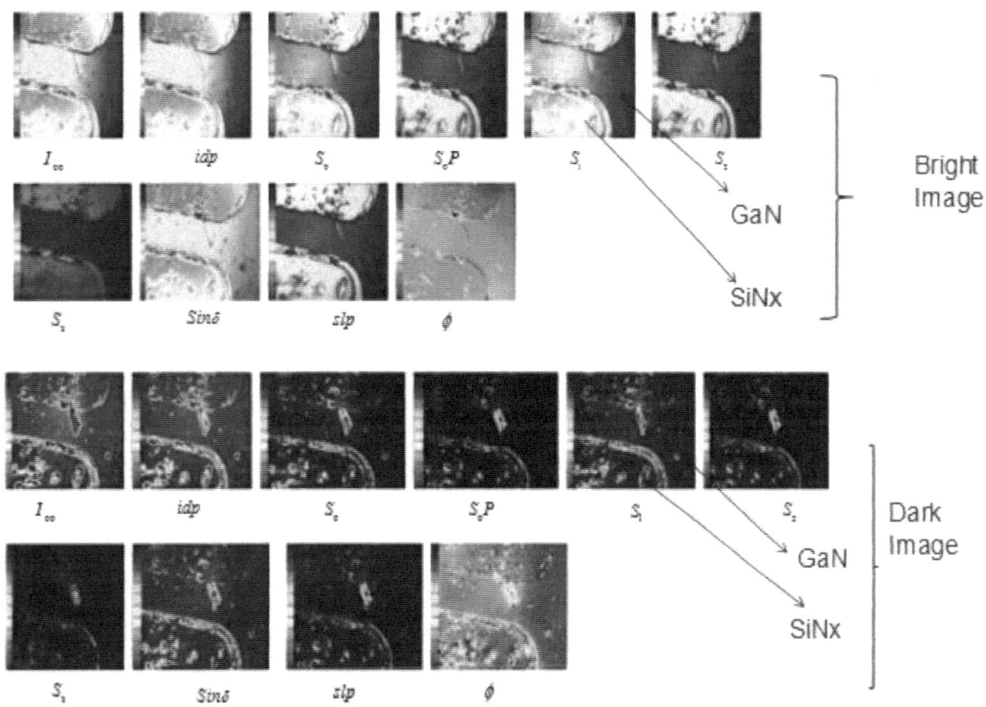

Fig.3: 50 X magnification in the same position

IOP Conf. Series: Journal of Physics: Conf. Series **844** (2017) 012005 doi:10.1088/1742-6596/844/1/012005

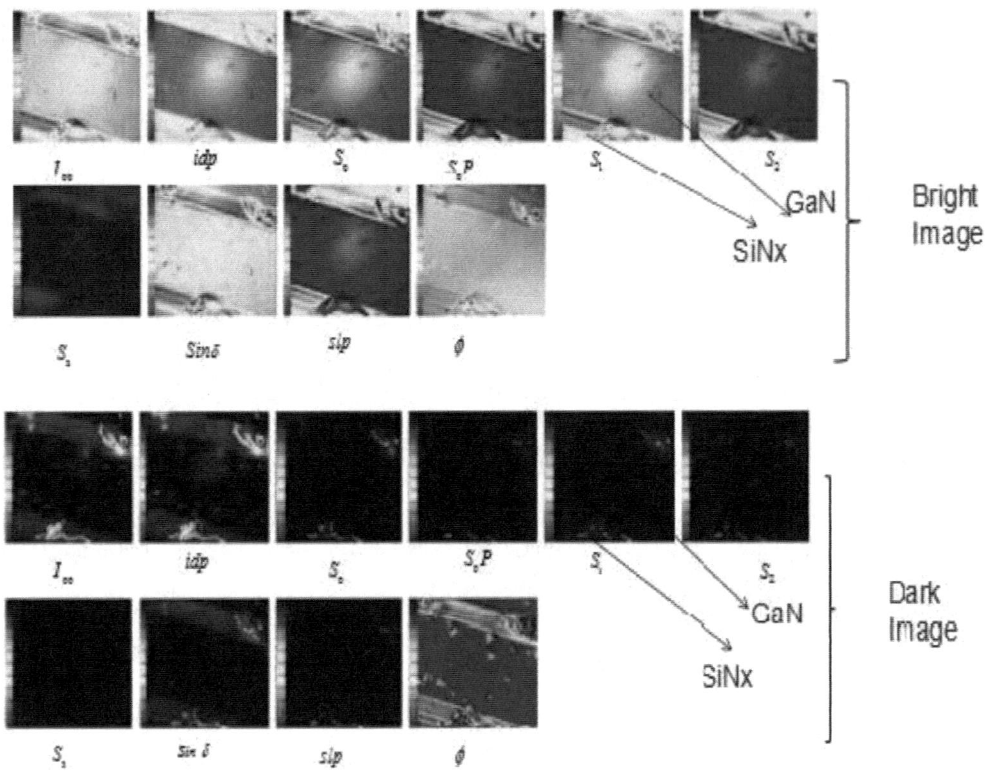

Fig.4: 100 X magnification in the same position

Table-1

Sample-1	OCL	MSE	OCL/MSE	RMSE	OCL/RMSE	Contrast
Ioo	0.2481	3.2021E-04	774.8652	1.79E-02	13	0.02
Phi	0.2492	8.1357E-04	306.2633	2.56E-02	9	0.04

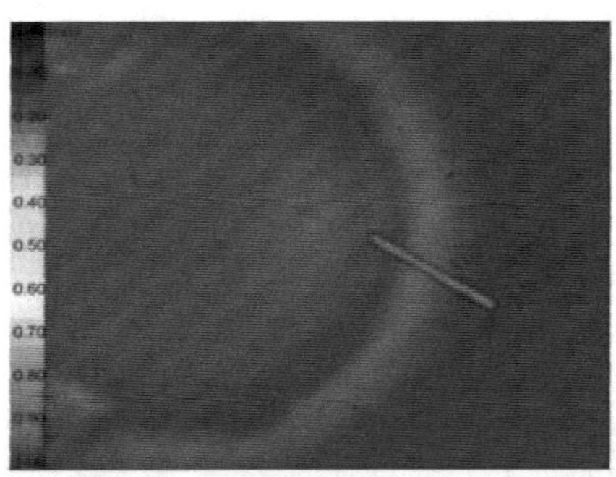

(a)(I00)

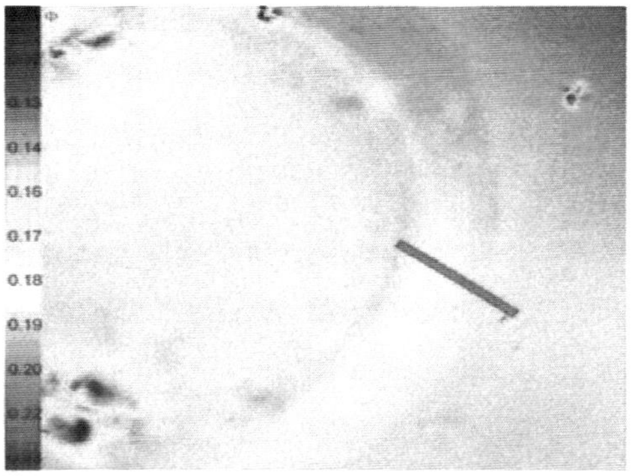

(b)Phi

Fig. 5: PIMI Image with line (a) I00 (b) Phi

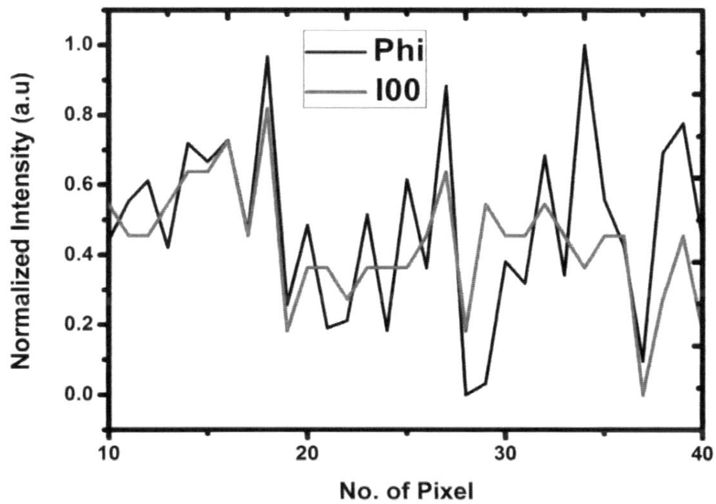

Fig.6: Image quality and contrast variation between Phi and I00 PIMI Image.

With the help of all PIMI images, first we would to see whether the etched part is fully etched without SiNx. You may see the pattern with 5 fingers (fig.2) and one can see the area between fingers to see whether there is SiNx left or not?.And check the difference on the fingers and between the finger, if there is SiNx on the finger, there should be no SiNx between finger, or its opposite. According to all issues, we have done the measurements of this sample through PIMI on 10X,50X and 100X magnification both bright and dark fields.I00 is image of conventional images and from S_0 to ϕ are the stokes parameter of the PIMI. From the above results of GaN images, it will be easy to confirmed that if SiNx on the fingers in sample then GaN is between the fingers or its opposite. Form figs. 2 to 4, you can see the PIMI images with different magnifications of the same sample. In dark images has more resolution as compare to bright images. The confirmation of this new idea, we have draw a line in PIMI image of Phi and I00 shown in fig.5 and also calculated the MSE,RMSE and contrast in table-1.Further, we have draw the curve between Phi and I00, which can seen in fig.6.. Finally we got the contrast of phi images is better than I00 images, and we have also compared the results between bright and dark field, found that the resolution of PIMI is more accurate is near about 80 nm. And we achieve the resolution of the conventional microscopy is 322.47nm for GaN sample whose ideal resolution is 296nm using the same objective parameter. In this way, we can also calculate the resolution through the line methods of any of the sample images.

5. Conclusion:

In this paper, we discussed how GaN have been used for better imaging surpassing and complementing the currently available optical microscope (PIMI). GaN have shown ability to control electromagnetic wave at the nanoscale, allowing nanometers order of imaging resolution to come true. Therefore, with the development of nanofabrication and nano manufacturing methods, and the integration of new creative ideas, overcoming the results and limitations mentioned in this paper will be the continuous efforts to make GaN-based imaging techniques to be a next generation of imaging technology replacing current optical microscopy, which thus can be called nanoscopy.

Reference:

[1]. E Abbe, Beiträge zur Theorie des Mikroskops under mikroskopischen Wahrnehmung. Arch. Mikroskop. Anat. 9(1), 413–418 (1873)

[2]. MV Ardenne, U.K. Patent 511204-A, 1938.

[3]. R Reinhold, Patent DE 906737, 1931

[4]. X. Liu, B. Qiu, Q. Chen, Z. Ni, Y. Jiang, M. Long, and L. Gui, "Characterization of graphene layers using super resolution polarization parameter indirect microscopic imaging," Optics express **22**, 20446-20456 (2014).

[5]. Y.-s. Zhou, C.-r. Li, L.-l. Tang, C.-x. Gao, L. Ren, and L.-l. Ma, "Permanent target for synthetic aperture radar image resolution assessment," in *2015 IEEE International Geoscience and Remote Sensing Symposium (IGARSS)*(IEEE2015), pp. 4284-4287.

[6]. Y. Takahashi, T. Akashi, T. Shimakura, T. Tanigaki, T. Kawasaki, H. Shinada, and N. Osakabe, "Resolution Assessment of an Aberration Corrected 1.2-MV Field Emission Transmission Electron Microscope," Microscopy and Microanalysis **21**, 1865-1866 (2015).

[7]. J. Montoya, A. Ferrero, L. Yu, S. Leng, and C. McCollough, "WE-FG-207B-09: Experimental Assessment of Noise and Spatial Resolution in Virtual Non-Contrast Dual-Energy CT Images Across Multiple Patient Sizes and CT Systems," Medical Physics **43**, 3836-3836 (2016).

[8]. G. Brown, I. Daniels, C. Richardson, P. Revell, D. Peppercorn, and M. Bourne, "Techniques and trouble-shooting in high spatial resolution thin slice MRI for rectal cancer," The British journal of radiology (2014).

[9]. F. Huth, A. Govyadinov, S. Amarie, W. Nuansing, F. Keilmann, and R. Hillenbrand, "Nano-FTIR absorption spectroscopy of molecular fingerprints at 20 nm spatial resolution," Nano letters **12**, 3973-3978 (2012).

[10]. A. Arbabi, Y. Horie, M. Bagheri, and A. Faraon, "Dielectric metasurfaces for complete control of phase and polarization with subwavelength spatial resolution and high transmission," Nature nanotechnology (2015).

[11]. Q. Zhan and J. R. Leger, "High-resolution imaging ellipsometer," Appl. Opt. 41(22), 4443–4450 (2002).

[12]. T. G. Browna, M. A. Alonso, and A. Vella, "Focused beam scatterometry for deep sub wavelength metrology,"Proc. SPIE 8949, 89490Y1 (2014).

[13]. A.Echalier, R. L. Glazer, V. Fülöp, and M. A. Geday, "Assessing crystallization droplets using birefringence,"ActaCrystallogr. D Biol. Crystallography. 60(4), 696–702 (2004).

[14]. C. Ye, S. Liu, X. Teng, Q. Fang, and G. Li, "Morphological and optical characteristics of nanocrystalline TiO_2 thin films by quantitative optical anisotropy and imaging techniques," Meas. Sci. Technol. 17(2), 436–440 (2006).

Enhancement transmission filter using a two-dimensional subwavelength periodic membrane

Jianyu Zhou[1], Tian Sang, Junlang Li, Rui Wang, La Wang and Jian Gao

Department of Photoelectric Information Science and Engineering, School of Science, Jiangnan University, Wuxi 214122, China

E-mail:sangt@jiangnan.edu.cn

Abstract. Enhancement transmission filter using a two-dimensional (2-D) subwavelength periodic membrane is proposed. It can be found that strong refractive-index modulation of the silicon periodic membrane can support the excitation of multiple guided-mode resonances (GMRs) in a reflection band, and every GMR relates a transmission peak on its edge, therefore the overlapping of the edges of these resonances can be tailored to create enhancement transmission. Thelocation of the transmission peak is shifted linearly with a slop of 1.51 as the period is varied. Enhancement transmission with multiple channels near 1310 nm can also be achieved using the interaction of the nondegenerate GMRs at oblique incidence.

1. Introduction

Guided-mode resonance (GMR) gratings have attracted considerable interests due to their simple structures with versatile spectral characteristics [1,2]. As the incident light illuminates on a periodic surface structure, the leaky waveguide modes can be resonantly excited as the phase-matching condition is satisfied. At resonance, the incident light is totally reflected with highly angular and spectral selectivity maintained. This phenomenon is also called as leaky mode resonance [3,4]. It has been proposed that optical elements such as optical filters [4,5], optical switch devices [6], photonic detectors [7-8], sensors [9], and ultrabroadband mirrors [10,11,12] can be obtained by using the resonant leaky modes.

For all we know, most research to date has mainly focused on the reflection properties of these elements, with the resonance effect providing a reflection peak or a reflection band [1,2,3,4,5,6,7,8,9,10,11,12]. Much less attention has been paid to the resonant transmission properties. In 1995, Magnusson and Wang [13] put forward the concept of resonant bandpass filtering by integrating the resonant waveguide gratings into a dielectric multilayer structure. Later, Tibuleac and Magnusson [14,15] showed that these diffractive thin-film structures yield high efficiency narrowband transmission. Last few years, it has been shown that by tailoring the overlapping of the edges of the GMRs, the enhancement transmission filter can be achieved in one-dimensional (1-D) periodic membrane structure [16]. Unfortunately, the 1-D periodic structure is sensitive to the polarization, which limits its application in unpolarized filtering.

IOP Conf. Series: Journal of Physics: Conf. Series **844** (2017) 012006 doi:10.1088/1742-6596/844/1/012006

In this paper, the enhancement transmission filter using a 2-D subwavelength periodic membrane is proposed. It can be found that strong refractive-index modulation of the silicon periodic membrane can support the excitation of multiple GMRs, and the interaction and coupling of the GMRs can be used to obtain the broadband reflection. Furthermore, the overlapping of the edges of GMRs can be tailored to create enhancement transmission by slightly changing the structural parameters, and the geometric symmetry of the 2-D construction [17] ensures the nonpolarizing property of the enhancement transmission filter at normal incidence. The location of the transmission peak is shifted linearly as the period is varied. Enhancement transmission with multiple channels near 1310 nm can also be achieved at oblique incidence for both the TE and TM polarizations.

2.Design of broadband mirror

As illustrated in figure 1, the general structure of the 2-D subwavelength periodic membrane consisting of the cubic arrays, a homogeneous layer and the substrate. The material of the periodic membrane is silicon and air, which can support the excitation of the leaky mode resonances in a reflection band [18,19]. The geometric symmetry of the 2-D construction at normal incidence is used to ensure the nopolarizing property of the structure in our design. For the design of broadband mirror, the optimal parameters are: $n_c=1$, $n_s=1.48$, $n_g=n_h=3.48$, the membrane thicknesses $d_g=0.4377$ μm, $d_h=0.0534$ μm, the periods $\Lambda=\Lambda_x=\Lambda_y=0.63$ μm, the filling factors $F=F_x=F_y=0.6$, $\theta=\delta=0°$, φ can be taken any angle for the nonpolarizing property. (θ is the incident angle, δ denotes the angle between the incident surface and the y-axis, $\varphi=0°$ and $\varphi=90°$ denote the TM polarization and the TE polarization, respectively.)

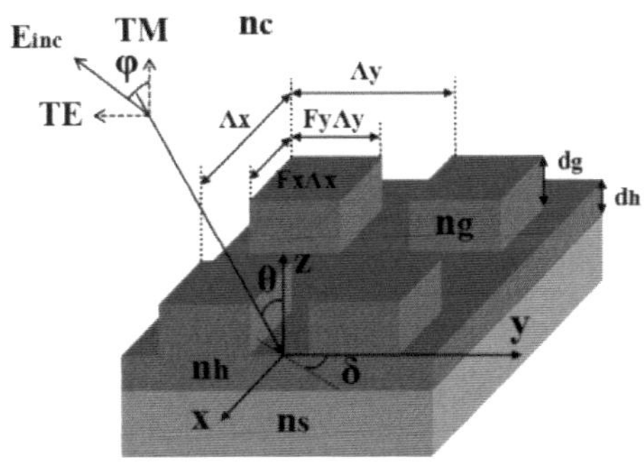

Figure 1. Schematicdiagram of the 2-D subwavelength periodic membrane.

For the subwavelengthperiodic membrane structure, the range condition for allowed GMRs can be simplified to

$$\frac{m^2+n^2}{N^2}<\frac{\Lambda^2}{\lambda^2} \qquad (1)$$

where *m* is the diffraction order in the x-axis, *n* is the diffraction order in the y-axis, λ is the free space wavelength. $N^2=n_c^2$ in the cover region, and $N^2=n_s^2$ in the substrate region. The period Λ is sufficiently small, such that only the zero-order transmitted and reflected waves propagate under our design [20].

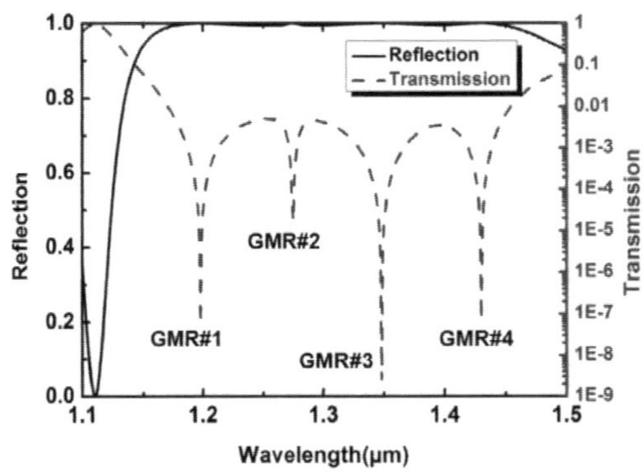

Figure 2. Spectrum of the unpolarized broadband mirror of the 2-D subwavelength periodic membrane at normal incidence. The parameters are: $n_c=1$, $n_s=1.48$, $n_g=n_h=3.48$, $d_g=0.4377$ μm, $d_h=0.0534$ μm, $\Lambda=\Lambda_x=\Lambda_y=0.63$ μm, $F=F_x=F_y=0.6$, $\theta=\delta=0°$.

Figure 2 is the spectrum of the unpolarized broadband mirror of the 2-D subwavelength periodic membrane at normal incidence. As illustrated in figure 2, the broadband reflection spectrum (R>99%) can be obtained with the bandwidth of ~285 nm at the central wavelength of 1310 nm, and this unpolarized broadband mirror can be used in many fields such as an alternative to distributed Bragg reflectors (DBRs) especially for vertical-cavity surface-emitting laser (VCSEL) applications [17].

Furthermore, the locations of the four GMRs in the broadband reflection spectrum can also be indicated by using the transmission spectrum in logarithmic scale. The GMRs are almost evenly distributed on the broadband at the locations of 1196 nm, 1274 nm, 1348 nm, and 1431 nm, respectively, and the reflection peak of each is much larger than 99%. Thus it can be drawn that the interaction and coupling of these GMRs with large bandwidths, resulting in the unpolarized broadband reflection spectrum. The spectral response, as for others presented in this paper, are based on rigorous coupled-wave analysis [21].

3. Enhancement transmission filter

Since the four GMRs can be excited in the spectral band of the 2-D subwavelength periodic membrane, the interaction of these resonances can be tailored by slightly varying the structure parameters.

Figure 3 is the transmission spectrum of the nonpolarized enhancement transmission filter at 1310 nm which can be achieved by slightly changing the parameters of the broadband mirror. The optimal parameters of the nonpolarized enhancement transmission filter are: $n_c=1$, $n_s=1.48$, $n_g=n_h=3.48$, $d_g=0.5$ μm, $d_h=0.054$ μm, $\Lambda=\Lambda_x=\Lambda_y=0.635$ μm, $F=F_x=F_y=0.6$, $\theta=\delta=0°$.

As can be seen in figure 3, an nonpolarized transmission peak with very high transmissivity of 99.1% at 1310 nm arises from the overlapping of the edges of the GMRs which can be seen by the transmission spectrum in logarithmic scale. The full width half maximun of the transmission peak is about 15 nm which can be measured easily, thus the quality factor Q of the transmission filter defined as $\lambda_{peak}/\Delta\lambda$ is about 87.3. The GMR#2, GMR#3 and GMR#4 are overlapped with each other during the process of parameters adjustments. The new resonance GMR#2,3,4 and GMR#1 distributed on both sides of the transmission peak form two broad reflection background sidebands, and the overlapping of the edges of GMR#1 and GMR#2,3,4 creates the transmission peak.

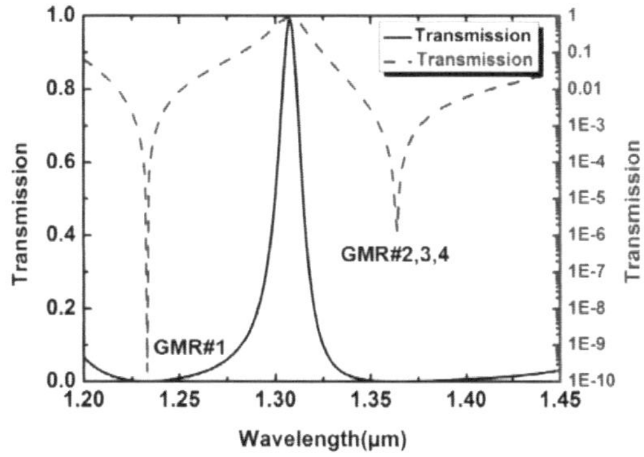

Figure 3. Spectrum of the nonpolarized enhancement transmission filter. The parameters are: n_c=1, n_s=1.48, n_g=n_h=3.48, d_g=0.5 μm, d_h=0.054 μm, $\Lambda=\Lambda_x=\Lambda_y$=0.635 μm, $F=F_x=F_y$= 0.6, θ=δ=0°.

For the practical unpolarized enhancement transmission filter, it's expected that the transmission peak can cover a wide range of wavelength band. In the process of changing all parameters, it's found that the period value changes almost linearly with the position of the transmission peak. Figure 4 is the color-coded transmission map of the unpolarized enhancement transmission filter as a function of period, the other parameters are the same as those in figure 3.

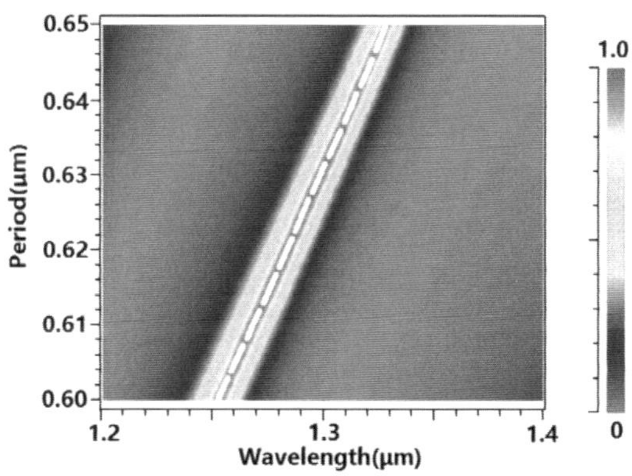

Figure 4. Color-coded transmission map of the unpolarized enhancement transmission filter as a function of period, the other parmeters are the same as those in figure 3.

As shown in figure 4, in the wavelength range of 1250 nm to 1330nm, the linear full coverage of the transmission peak can be obtained only by adjusting the period in a suitable range from 0.6 μm to 0.65 μm. The white dash line indicates the position of the transmission peak which displays a redshift with the increase of period, and the bandwidth of the transmission peak remains 15 nm, two flat low sidebands distributed on both sides of the transmission peak are also maintained. The analytic relationship between λ_{peak} and period in a suitable range can be obtained by fitting the white dash line in figure 4, that is:

$$\lambda_{peak}=1.51*\Lambda+0.3475 \quad (2)$$

The location of the transmission peak is shifted linearly with a slop of 1.51 by fitting the data as period is varied, thus the transmission peaks can be tuned to the design values by selecting the proper period.

For the 2-D subwavelength periodic membrane, the geometric symmetry of the structure is broken at oblique incidence, and the transmission spectra will show different characteristics. Figure 5(a) and 5(b) are the transmission spectra of TM and TE polarizations at the incident angle of 10°, respectively. It can be seen that under the TM polarization, two transmission peaks with relative wide bandwidths appear while four sharper transmission peaks appear under TE polarization. Similar to the 1-D grating structure [22], it is speculated that the multiple enhancement transmission peaks arise from the interaction of the nondegenerate GMRs under oblique incidence. The interaction of multiple nondegenerate GMRs squeeze out multiple transmission peaks, which is helpful for the design and application of multi-channel transmission filtering.

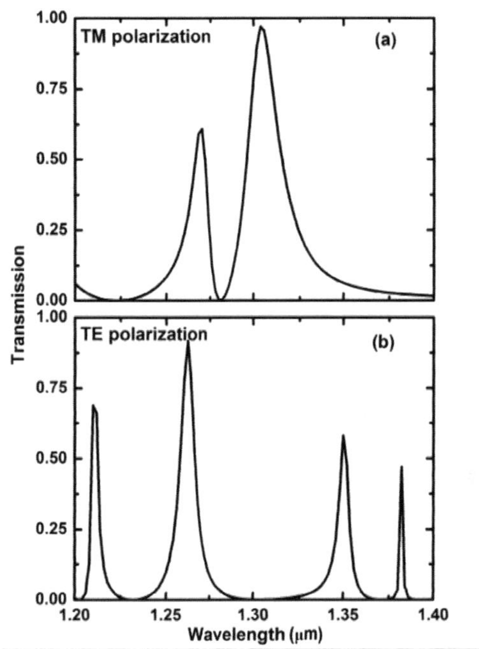

Figure 5. (a) Spectrum of TM polarization at the incident angle of 10° (b) Spectrum of TE polarization at the incident angle of 10°, the other parameters are the same as those in figure 3.

4.Conclusion

The unpolarized enhancement transmission filter using a 2-D subwavelength periodic membrane is proposed. It can be found that strong refractive-index modulation of the silicon periodic membrane can support the excitation of multiple GMRs, and the interaction and coupling of the GMRs can be used to obtain the broadband reflection. Furthermore, the overlapping of the edges of GMRs can be tailored to create enhancement transmission by changing the parameters slightly, and the geometric symmetry of the 2-D construction ensures the nonpolarizing property of the enhancement transmission filter at normal incidence. The location of the transmission peak is shifted linearly with a slop of 1.51 as the period is varied. Enhancement transmission with multiple channels near 1310 nm can also be achieved using the interaction of the nondegenerate GMRs under oblique incidence.

Acknowledgment

This work is supported by the Fundamental Research Funds for the Central Universities (Grant nos. JUSRP115A15, JUSRP51628B), Jiangsu Provincial Research Center of Light Industrial Optoelectronic Engineering and Technology (Grant no. BM2014402), and National Natural Science Foundation of China (Grant nos. 11404143, 11504139).

Reference

[1] Wang S S and Magnusson R 1993 *Appl. Opt.* **32** 2606
[2] Hane K, Kobayashi T, Hu F R and Kanamori Y 2006 *Appl. Phys. Lett.* **88** 141109
[3] Ding Y and Magnusson R 2004 *Opt. Express.* **12** 1885
[4] Sang T, Wang Y, Li J, Zhou J, Jiang W, Wang J and Chen G 2017 *Opt. Commun.* **382** 138
[5] Readerharris P, Ricciardi A, Krauss T and Falco A D 2013 *Opt. Express.* **21** 1002
[6] Ngo Q M, Hoang T T, Nguyen D L, Vu D L and PhamV H 2013 *J. Opt.* **15** 527
[7] Tu Y K, Tsai M Z, Lee I C, Hsu H Y and HuangC S 2016 *Analyst.* **141** 4189
[8] Zhu A Y, Zhu S and Lo G Q 2014 *Opt. Express.* **22** 2247
[9] Boonruang S and Mohammed W S 2015 *Appl. Phys. Express.* **8** 092004
[10] Sang T, Wang L, Ji S, Ji Y, Chen H and Wang Z 2009 *J. Opt. Soc. Am. A.* **26** 559
[11] Chen L, Huang M C Y, Mateus C F R, Changhasnain C J and Suzuki Y 2006 *Appl. Phys.Lett.* **88** 031102
[12] Magnusson R and Shokoohsaremi M 2008 *Opt. Express.* **16** 3456
[13] Magnusson R and Wang S S 1995 *Appl. Opt.* **34** 8106
[14] Tibuleac S and Magnusson R 1997 *IEEE Photon. Tech. Lett.* **9** 464
[15] Tibuleac S and Magnusson R2001 *Opt. Lett.* **26** 584
[16] Sang T, Wang Z, Zhou X and Cai S 2010 *Appl. Phy. Lett.* **97** 071107
[17] Bekele D A, Park G C, Malureanu R and Chung I S 2015 *IEEE Photon. Tech. Lett.* **27** 1733
[18] Magnusson R and ShokoohsaremiM 2008 *Opt. Express.* **16** 3456
[19] Shokoohsaremi M and Magnusson R 2014 *Opt. Lett.* **39** 6958
[20] Magnusson R 2012 *Opt. Lett.* **37** 3792
[21] Moharam M G, Grann E B, Pommet D A and Gaylord T K 1995 *J. Opt. Soc. Am. A.* **12** 1068
[22] Sang T, Wang Z, Wang L, Wu Y and Chen L 2005 *J. Opt. A.* **8** 62

Experimental demonstration of directly modulated buried heterostructure DFB semiconductor laser based on reconstruction equivalent chirp technique

Guowang Zhao[1], Xiang Tian[1], Yuechun Shi[1], Yunshan Zhang[23*], Jilin Zheng[4], Linjie Zou[1], Xiangfei Chen[1],

[1]National Laboratory of Solid State Microstructures and College of Engineering and Applied Sciences, Nanjing University, Nanjing, China 210093

[2]School of optoelectronic engineering, Nanjing University of Posts and Telecommunication, Nanjing, China 210023

[3]Nanjing University (Suzhou) High-Tech Institute, Suzhou, China 215123

[4]Institute of Communication Engineering, PLA University of Science and Technology, Nanjing, China 210007

[*]Email Address: yszhang@njupt.edu.cn

Abstract: A directly modulated buried heterostructure distributed feedback (DFB) semiconductor laser based on the reconstruction equivalent chirp (REC) technique in the 1.3 μm wavelength domain is experimentally demonstrated. The packaged laser shows a good single longitudinal mode (SLM) property, with the side mode suppression ratio (SMSR) higher than 40 dB when the injection current is higher than 20 mA, as the threshold current is 10 mA. The direct modulation performance of the laser is also delightful, with the modulation bandwidth up to 10GHz at a low injection current (40mA), and the measured spurious free dynamic range (SFDR) up to 86 dB·Hz$^{2/3}$.

Key words: directly modulated; buried heterostructure; DFB semiconductor laser; reconstruction equivalent chirp technique.

1. Introduction

Distributed feedback (DFB) semiconductor laser is one of the key components in the coherent optical communication system. DFB lasers with good single longitudinal mode (SLM) property and direct modulation performance are widely used to make the communication system much simpler [1-3]. The lasing wavelength and the SLM property of DFB lasers are mainly determined by the internal brag grating structure. However, in the traditional way, the fabrication of complex grating structures is accomplished with the electron beam lithography (EBL) method, which is known as complex and time consuming [4].

A method named reconstruction equivalent chirp (REC) technique has been used to fabricate the grating structures in DFB lasers [5-7]. According to the REC technique, complex grating structures, such like phase shifted gratings and apodized gratings, can be equivalent as a combination of uniform Bragg gratings and designed sampled gratings [8, 9]. Therefore, the fabrication of the grating structures based on REC technique can be realized by one step of holographic exposure to fabricate uniform grating structures, and another step of photolithographic to fabricate sampled gratings. However, due to the sampling structure, the effective index coupling coefficient of DFB lasers based on REC technique reduces to 1/π of which of conventional DFB lasers, causing the threshold current much higher and the direct modulation performance worse [5]. As a consequence, a buried heterostructure (BH) waveguide is put in use to lower the threshold current and enhance the direct modulation performance [10, 11]. In this article, a 1310nm DFB laser with low threshold current of 10 mA is experimentally demonstrated. The side mode suppression ratio (SMSR) is higher than 40 dB at the injection current of 20 mA, which shows a good SLM property. The direct modulation characteristic has also been measured. The modulation bandwidth is 10GHz at a low injection current of 40 mA, and the spurious free dynamic range (SFDR) is 86 dB·Hz$^{2/3}$ at 10 GHz. The experiment results mean that the DFB lasers with BH structure based on REC technique are qualified for the fiber-communication system.

2. Principle and Design

According to the REC technique, a complex grating structure can be equivalent as a combination of a designed sampled grating and a seed uniform grating. If we set the Bragg wavelength of the +1st order sub-grating as the working channel, the relationship between the lasing wavelength λ and the sampling period P can be expressed as,

$$\frac{1}{\lambda/_{2n}} = \frac{1}{\Lambda_0} + \frac{1}{P}$$

Where n is the effective index at lasing wavelength λ, and Λ_0 is the seed grating period. As a uniform grating, the Λ_0 is fixed, and then we can control the lasing wavelength by changing the sampling period. In order to achieve a good SLM property, we lead in a quarter-wave equivalent phase shift (EPS). As shown in Figure 1, the ESP θ can be achieved by bringing a shift ΔP in the sampling period at the center of the laser cavity. We can get a quarter-wave ESP by setting the ΔP equal to $P/2$, as the ESP θ is equal to $2\pi\Delta P/P$. The sampling period is usually in a precision of micrometer order, which means that the REC technique is much more conventional than the traditional ways.

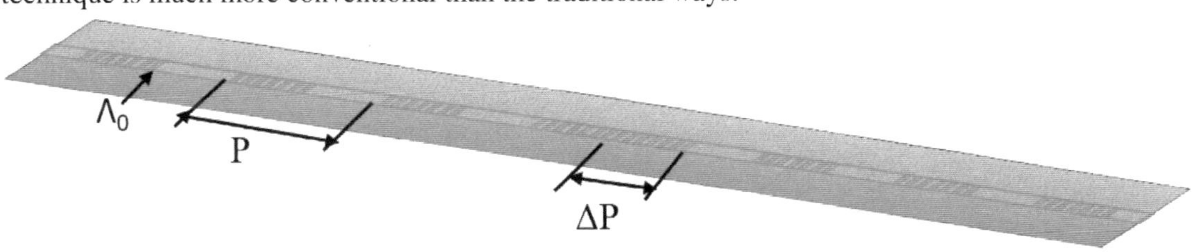

Figure 1 The schematic of the grating structure with equivalent π phase-shift

The procedure to fabricate DFB laser with BH structure based on REC technique is almost the same with the traditional method [12, 13]. What is different appears when the grating structure is fabricated. According to the REC technique, after growing a multiple quantum-well (MQW) active structure by metal-organic vapor phase epitaxy (MOVPE), the uniform gratings, whose Bragg wavelength was about 1375 nm, were fabricated by holographic exposure. And then the sampled grating structures with the period of 4.15 μm were patterned by traditional photolithographic. And then a 3 μm wide ridge waveguide vertical to the grating lines was formed by wet etching to define the active regions. After this, a p-InP layer and an n-InP layer were grown around the ridge by MOCVD to form the current block layer. And then the p-InP cladding and InGaAs contact layers were regrown by MOCVD. In this article, two isolation ditches were etched along the ridge after the contact window was opened to reduce

IOP Conf. Series: Journal of Physics: Conf. Series **844** (2017) 012007 doi:10.1088/1742-6596/844/1/012007

the capacitance. High reflection film and Antireflection film were coted on the laser facets.

3. Experimental Results and Discussion

A buried heterostructure DFB semiconductor laser has been fabricated at the wavelength around 1310 nm. The characteristics of the laser has been measured experimentally. The *P-I* curve at the temperature of 25°C is shown in Figure 2. The threshold current is about 10 mA and the slope efficiency is about 0.4 W/A. The spectra of the laser has also been measured. As shown in Figure 3(a), the lasing wavelength is 1308.18 nm and the SMSR is 48.31 dB with the injection current of 40 mA. Figure 3(b) shows the measured SMSRs at different injection currents at 25°C. According to the measurement, the SMSR is higher than 40 dB when the injection current is higher than 20 mA, which shows a good SLM property.

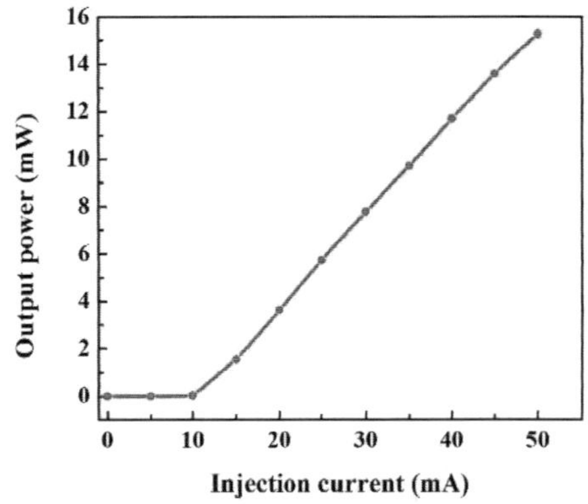

Figure 2 *P-I* curve measured at the temperature of 25℃

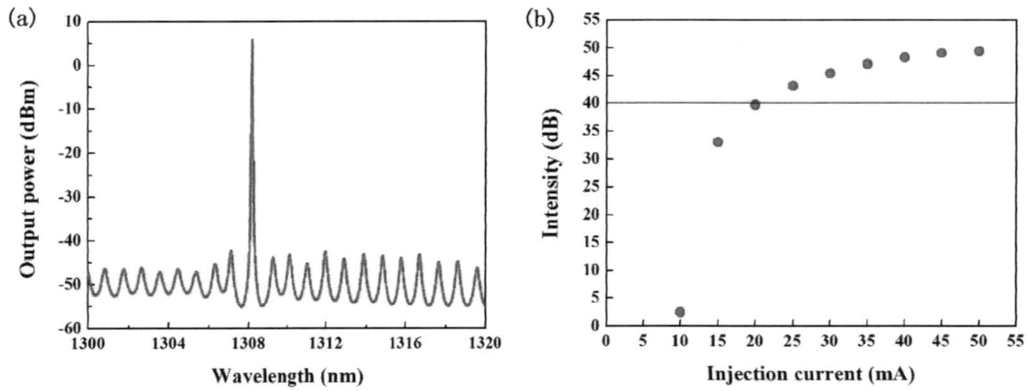

Figure 3 (a) Spectra of the laser measured with the injection current of 40mA; (b) SMSRs measured at different injection currents at 25°C

The direct modulation performance of the laser is also researched in this article. The small signal frequency responses (SSFRs) at different injection currents at 25°C are shown in Figure 4. As we can see from the diagram, the 3 dB modulation bandwidth increases obviously as the injection current increases. The 3 dB bandwidth is larger than 10 GHz when the injection current is 40 mA.

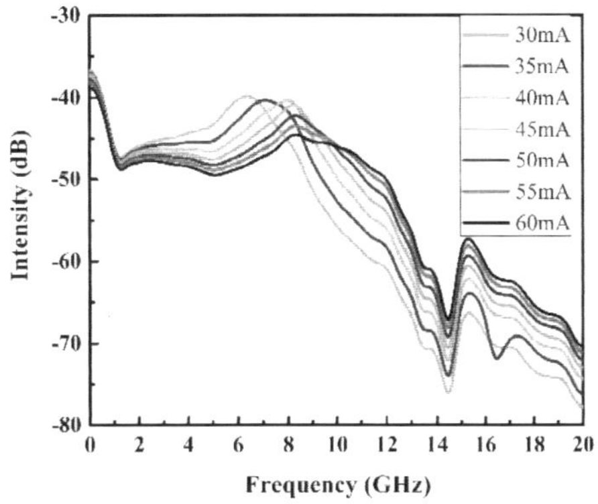

Figure 4 The SSFRs measured at different injection currents at 25°C

The third-order intermodulation distortion (IMD3) is a significant property for the laser to estimate the direct modulation linearity. Two RF signals at 10 and 10.2 GHz are used to measure the IMD3 operating at 10 GHz with an injection current of 40 mA. The input power of the two RF signals is adjusted to be the same, varied from -6.12 to -0.63 dBm. The measurement is shown in Figure 5. According to the tested results, the noise floor with a 1 Hz bandwidth is reckoned to be -140 dBm/Hz. As shown in the diagram, the third-order intermodulation increases much faster than the fundamental signal. The tested SFDR is 86 dBzHz$^{2/3}$, showing a good linearity.

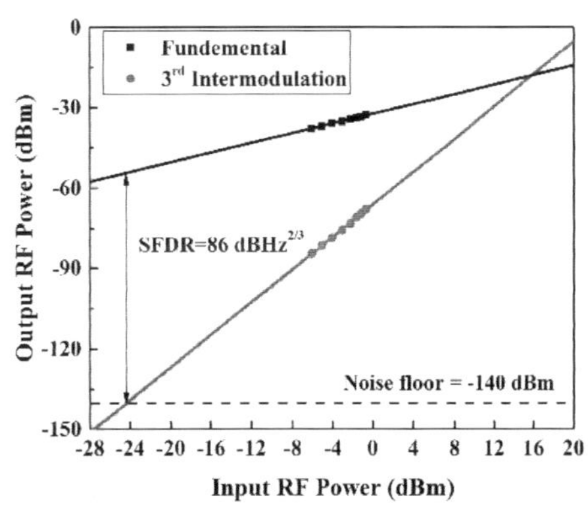

Figure 5 The SFDR of the laser modulation at 10 GHz

4. Conclusion

In this article, a DFB semiconductor laser with BH structure based on REC technique has been reported. A low threshold current and high output power because of the BH structure has been obtained. Good SLM property has been achieved with the SMSR higher than 40 dB when the injection current is higher than 20 mA. The direct modulation performance has also been researched in this article. The modulation bandwidth is higher than 10 GHz when the injection current is 40 mA, and the SFDR is 86 dB·Hz$^{2/3}$ at an operating frequency of 10 GHz. The experiment results mean that the DFB lasers with BH structure based on REC technique is qualified for the fiber-communication system.

Acknowledgment

This work was supported by the National Nature Science Foundation of China (Grant *No.61504058, 61504170, 61306068, 61435014, 11574141*), Jiangsu Province Natural Science Foundation of China (Grant *No.BK2014014, BK20141168*), and National "863" project of China (Grant *No.2015AA016902*).

References

1. Michael F, Michael Tran, Francois L, et al. Low RIN 1.55-μm directly modulated DFB lasers for analog signal transmission. IEEE Photonic Tech L, 2012, 24: 116–118.
2. Suematsu Y, Iga K. Semiconductor lasers in photonics[J]. Journal of Lightwave Technology, 2008, 26(9): 1132-1144.
3. Soda H, Kotaki Y, Sudo H, et al. Stability in single longitudinal mode operation in GaInAsP/InP phase-adjusted DFB lasers[J]. IEEE Journal of Quantum Electronics, 1987, 23(6): 804-814.
4. Vieu C, Carcenac F, Pepin A, et al. Electron beam lithography: resolution limits and applications[J]. Applied Surface Science, 2000, 164(1): 111-117.
5. Y Dai, X Chen. DFB semiconductor lasers based on reconstruction-equivalent-chirp technology[J]. Optics express, 2007, 15(5): 2348-2353.
6. J Li, H Wang, X Chen, Z Yin, Y Shi, Y Lu, Y Dai and H Zhu,, Experimental demonstration of distributed feedback semiconductor lasers based on reconstruction-equivalent-chirp technology[J]. Optics express, 2009, 17(7): 5240-5245.
7. Y Shi, X Chen, Y Zhou, et al. Experimental demonstration of eight-wavelength distributed feedback semiconductor laser array using equivalent phase shift. Opt Lett, 2012, 37: 3315–3317.
8. Y Shi, S Li, L Li, R Guo, T Zhang, L Rui, W Li, L Lu, T Song, and Y Zhou. Study of the multi-wavelength DFB semiconductor laser array based on the reconstruction-equivalent-chirp technique[J]. Journal of Lightwave Technology, 2013, 31(20): 3243-3250.
9. Dai Y, Chen X, Sun J, et al. High-performance, high-chip-count optical code division multiple access encoders-decoders based on a reconstruction equivalent-chirp technique[J]. Optics letters, 2006, 31(11): 1618-1620.
10. Takemasa K, Kubota M, Wada H. 1.3-μm AlGaInAs-InP buried-heterostructure lasers with mode profile converter[J]. IEEE Photonics Technology Letters, 2000, 12(5): 471-473.
11. Matsuo S, Fujii T, Hasebe K, et al. Directly modulated buried heterostructure DFB laser on SiO 2/Si substrate fabricated by regrowth of InP using bonded active layer[J]. Optics express, 2014, 22(10): 12139-12147.
12. Li S, Li R, Li L, et al. Dual wavelength semiconductor laser based on reconstruction-equivalent-chirp technique[J]. IEEE Photonics Technology Letters, 2013, 25(3): 299-302.
13. Shi Y, Li S, Li J, et al. An apodized DFB semiconductor laser realized by varying duty cycle of sampling Bragg grating and reconstruction-equivalent-chirp technology[J]. Optics Communications, 2010, 283(9): 1840-1844.

IOP Conf. Series: Journal of Physics: Conf. Series **844** (2017) 012008 doi:10.1088/1742-6596/844/1/012008

Fabrication of Pattern Poled Lithium Niobate Film and its Nonlinear Optical Applications

M Xu[1], M Wang[1], Z Chen[1], J Tang[1], G Shao[1], Y Ming[1], G Cui[1] and Y Lu[1]

[1]National Laboratory of Solid State Microstructures and College of Engineering and Applied Sciences, Nanjing University, Nanjing 210093, China

E-mail: cuiguoxin@nju.edu.cn

Abstract. We develop an approach to fabricate arbitrary ferroelectric domain patterns on lithium niobate film (30-50 μm thick) by applying a structured external field at room temperature. The fabricating method can be operated easily to reach 1 μm linewidth resolution. The ferroelectric domain inversion is stable and uniform. Nonlinear diffraction is generated when the fundamental wave pumps to film. Various nonlinear wavefronts are obtained such as the frequency converted optical vortex beam. A nonlinear holographic concept is proposed to explain the physical phenomena and guide the corresponding domain design. The applications in optical field manipulation and novel photonic states generation are discussed.

1. Introduction

Lithium niobate (LN), a typical optoelectronic material, has been widely used in integrated optics due to its excellent nonlinear, electro-optic and acousto-optic properties [1-4]. Recently, the ferroelectric domain engineering on LN have attracted much attention because of the demand to realize nonlinear wave mixing process, including collinear and noncollinear situations in quasi-phase-matching (QPM) manners [5-9]. While it is difficult to carry out uniform micron linewidth ferroelectric domain inversion on typical 500-μm-thick LN crystal (the coercive field is about 21 kV/mm.) at room temperature as the result of the domain growth from +c facet to -c facet. At the same time, the domain structures will deteriorate soon for submicron LN film (540 nm thick) [10].

With the development of domain engineering applications, ferroelectric domain structure with micron or sub-micron linewidth is desired [11-14]. Here we develop a method to fabricate arbitrary ferroelectric domain patterns on LN film (30-50μm thick) by applying a structured external field at room temperature [15-16]. The ferroelectric domain inversion is stable and uniform with 1 μm linewidth resolution.

Based on the pattern poled LN film, it is convenient to observe nonlinear diffraction of vortex beam following the nonlinear holographic principle [8,17]. Furthermore, we can effectively tailor nonlinear wavefronts with designed ferroelectric domain through computer-generation holograms (CGH) [18-20]. Also, suitable domain structure could be flexibly introduced into LN for spontaneous parametric down-conversion [21].

2. Fabrication of pattern poled LN film

We develop an approach to fabricate arbitrary ferroelectric domain patterns on LN film (30-50 μm thick) by applying a structured external field at room temperature. First of all, we find out the +c facet of LN film, and fix the LN film on the glass substrate for protection. Next, the +c facet of LN film is covered with a 1-μm-thick BCI-3511 photoresist layer by spin coating to obtain the desired pattern with maskless photolithography (SF-100, Intelligent Micro Pattern Co.). We then deposit a 120 nm thick chromium(Cr) film with electron beam evaporation (ZZSX-500, Beijing Beiyi Innovation Vacuum Technology Co.) and lift off to get a pattern electrode. Finally, the ferroelectric domain can be inverted on the designed Cr electrode pattern with the setup of Fig.1 [16]. The layer of NaCl solution can realize the ohmic contact between the ITO glass substrate and the LN film. The probes get in touch with the pattern electrode and the ITO glass substrate respectively. A pulse external electric field (the electric density E is slight higher than the coercive field, the pulse period is 1Hz, and period number is 10.) is applied to carry out the ferroelectric domain inversion.

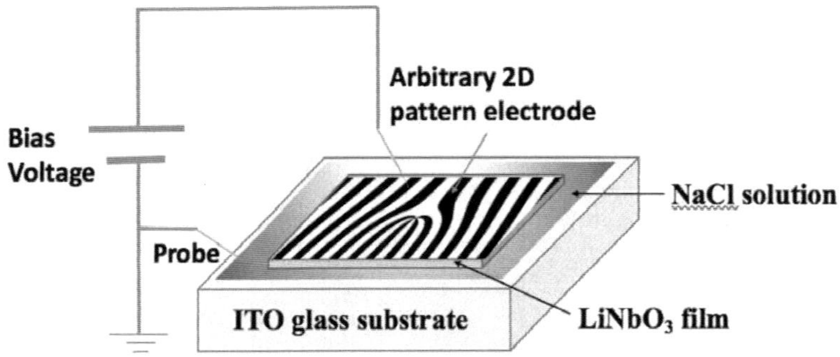

Figure 1. Setup for ferroelectric domain inversion

The most reliable technique to verify ferroelectric domain inversion is selective etching of the crystal surface. After cleaning the Cr electrode, we put the pattern poled LN film into the solution mixed by HF and H_2O_2 (1:1 in volume) at 70°C for one hour. The microscope images of LN film surfaces are recorded as shown in Fig.2 (a), (b). The minimum linewidth of each domain is about 1μm.

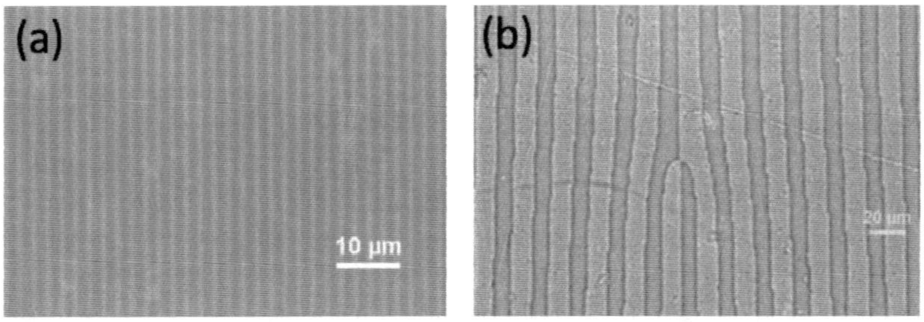

Figure 2. Microscope images of LN film surface after selective etching

There are advantages of this method to fabricate pattern poled LN film compared to typical 500-μm-thick LN crystal. This method can be operated easily and does not need extreme conditions. All the poling experiments discussed in this paper are carried out at room temperature and atmospheric pressure. At the same time, arbitrary 2D domain pattern can be generated and the ferroelectric domain inversion is stable and uniform with 1 μm linewidth resolution. The domain can consistently reach the -c facet of the film with the patterns on +c facet.

3. Nonlinear optical process of pattern poled LN film

3.1. The experimental setup

We can observe different nonlinear phenomena originated from various pattern poled LN films on a screen in the far field using the experimental setup shown in Fig.3. The fundamental pump beam from a fs-laser (MIRA-HP & OPO, Coherent Co.), delivering 200 fs, 700 mW pulses at 1064 nm with repetition rate of 76 MHz, is directed at a small angel with respect to the Z axis. The laser beam is focused with a lens (f = 50 mm) in the plane of the pattern poled LN film. A shortpass filter is used to block the fundamental wave (FW).

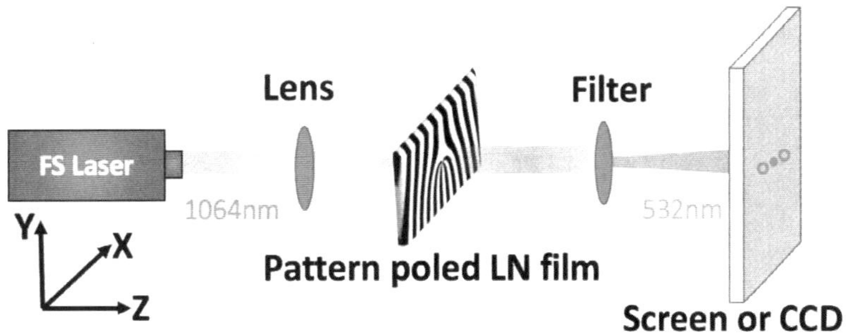

Figure 3. The experimental setup for nonlinear optical process of pattern poled LN film

3.2. Transverse QPM and nonlinear diffraction

In this paper, we will focus on second-harmonic generation (SHG), where both input beams have the same fundamental frequency (FF), $\omega_1 = \omega_2$, and the second-harmonic (SH) beam has the frequency of $\omega_3 = \omega_1 + \omega_2 = 2\,\omega_1$. We can use the coupled wave equation to describe the interaction between the fundamental and second-harmonic beams , which is defined in the form of: $E_i = A_i \exp[-i(k_i \cdot z - \omega_i t)], (i = 1\, or\, 3)$, where A_i and k_i are the amplitudes and wavevectors of the fundamental wave (FW) and SH wave respectively. The equations are derived under the slowly varying amplitude approximation, and the propagation direction is set to be along the Z-axis of the LN film [22]:

$$\frac{dA_1}{dz} = \frac{2i\omega_1^2}{k_1 c^2} d_{eff} A_3 A_1^* e^{-i\Delta kz}, \tag{1}$$

$$\frac{dA_3}{dz} = \frac{i\omega_3^2}{k_3 c^2} d_{eff} A_1^2 e^{i\Delta kz}, \tag{2}$$

where d_{eff} is the nonlinear susceptibility coefficient and $\Delta k = 2k_1 - k_3$. If $\Delta k \neq 0$, the SH waves generated at different positions in the film are not in phase. This results in the destructive interference described by the factor $sinc^2(\frac{1}{2}\Delta kz)$.

We can partly satisfy the phase matching condition by modulating the sign of nonlinear coefficient along the X-axis. By inverting the ferroelectric domain orientation, the nonlinear coefficient is:

$$d(x) = d_{ij} \cdot Sign[\cos(2\pi x/\Lambda)], \tag{3}$$

where Λ is the period. And the nonlinear coefficient can also be written as an expansion of Fourier series:

$$d(x) = d_{ij} \textstyle\sum_{m=-\infty}^{\infty} C_m \exp(iG_m x), \tag{4}$$

where $C_m = \frac{2}{m\pi}\sin(\frac{m\pi}{2})$ and $G_m = m\frac{2\pi}{\Lambda}$ are the *m*th Fourier coefficient and reciprocal vector. When the expression is substituted into Eqs. (1) and (2), we can see that the *m*th reciprocal vector G_m will compensates for the transverse phase mismatch as shown in Fig. 4 (c). The *m*th diffraction angle of nonlinear Raman-Nath diffraction is $sin(\alpha_m) = \frac{G_m}{k_2}$. And according to Snell's law, the external radiation angles are $\sin(\theta_m) = \frac{m\lambda_2}{\Lambda}$, where λ_2 is the SH wavelength. The 1st order nonlinear SH spots we measured are shown in Fig. 4 (d), which are in agreement with the theoretical calculations.

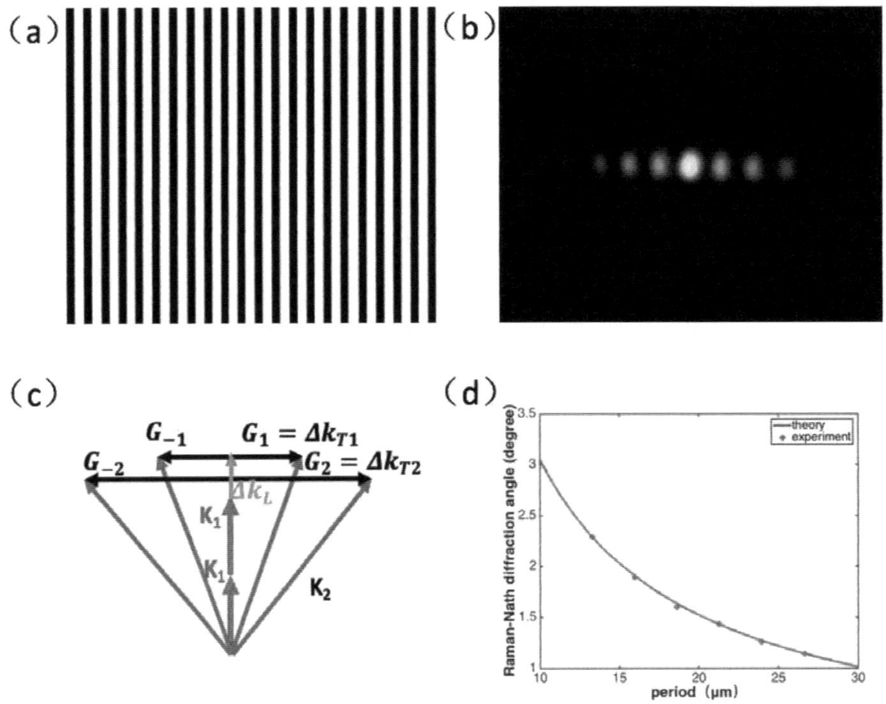

Figure 4. (a) periodic grating structure (b) nonlinear Raman-Nath diffraction (c) transverse QPM scheme (d) The 1st order nonlinear SH angels

3.3. Nonlinear holography

Holograms can be numerically computed, if the amplitude and phase of the desired wave are set as $A(x,y)$ and $\varphi(x,y)$. The amplitude transmittance of the hologram is [23]:

$$T(x,y) = 0.5\left\{1 + A(x,y)\cos\left[\frac{2\pi x}{\Lambda} - \varphi(x,y)\right]\right\}. \quad (5)$$

The desired wave is shaped in the first diffraction order as the Fourier transform of $A(x,y)\exp(i\varphi(x,y))$. We can introduce the concept of computer-generated holography (CGH) into the nonlinear optics using a binary hologram [24]:

$$T(x,y) = sign\left\{\cos\left(\frac{2\pi x}{\Lambda} + \varphi(x,y)\right) - \cos[\arcsin(A(x,y))]\right\}, \quad (6)$$

which can also be described by a Fourier series:

$$T(x,y) = \sum_{m=-\infty}^{\infty}\left\{\frac{\sin[m\cdot arcsin(A(x,y))]}{m\pi}\exp\left[im\left(\frac{2\pi x}{\Lambda} + \varphi(x,y)\right)\right]\right\}. \quad (7)$$

It is obviously to see the shaped wave in the first diffraction order. So the nonlinear coefficient can be modulated to satisfy transverse QPM, and the holographic information is recorded by the ferroelectric domain inversion. We designed some patterns in our experiments. The modulation of the second-order nonlinearity coefficient of LN film is [17]:

$$d(\boldsymbol{r},\phi) = d_{ij}\cdot sign[\cos(2\pi g_{(r,\phi)})], \quad (8)$$

where $\boldsymbol{r} = x\hat{\mathbf{x}} + y\hat{\mathbf{y}}$, $\phi = tan^{-1}(y/x)$ is the azimuthal angel, d_{ij} is an element of the quadratic susceptibility tensor, $g_{(r,\phi)}$ can be arbitrary two dimensional function such as $g_{(r,\phi)} = |\boldsymbol{r}|/\Lambda$ for a concentric circles structure, or $g_{(r,\phi)} = |\boldsymbol{r}|\cos(\phi)/\Lambda + l_c\phi/2\pi$ for a fork grating structure with the topological charge (l_c), where Λ is the poling period, as shown in Fig.5 (a), (b). According to Eq. (7), we can effectively tailor nonlinear wavefronts as shown in Fig.5 (d), (e) with designed ferroelectric domains based on the nonlinear holographic method. Here $g_{(r,\phi)}$ can also be discrete lattice structure as Fig.5 (c), and the diffraction pattern is formed by multiple QPM situations, which is recorded by CCD in Fig.5 (f).

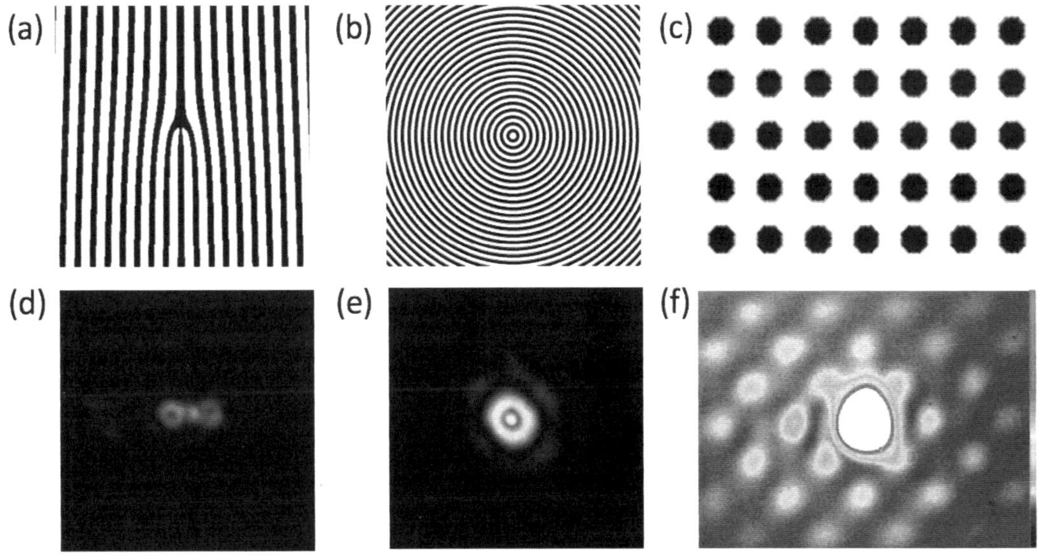

Figure 5. (a), (b) and (c) Different designs of ferroelectric domain pattern. (d), (e) and (f) corresponding nonlinear holographic patterns recorded by camera and CCD

4. Conclusion

In summary, we present a convenient method to fabricate stable arbitrary ferroelectric domain patterns on LN film at room temperature and atmospheric pressure. Nonlinear SH wave can be shaped following the nonlinear holographic principle based on pattern poled LN film, when the fundamental wave pumps to film. Various nonlinear wavefronts are obtained such as the frequency converted optical vortex beam. We can also design arbitrary domain structure in the continuous or discrete cases. Furthermore, suitable domain structures could be flexibly introduced into LN for spontaneous parametric down-conversion, which can be used in optical field manipulation and novel photonic states generation.

Acknowledgment

This work was supported by the National Natural Science Foundation of China (NSFC) (No.61490714, No. 61435008 and No. 11604144).

References

[1] Song X, Xu F and Lu Y 2011 *Opt. Lett.* **36** 4434.
[2] Shao G, Song X, Xu F and Lu Y 2012 *Opt. Express* **20** 19343.
[3] Song X,Wang Q, Yu Z, Xu F and Lu Y 2011 *J. Nonlinear Opt. Phys. Mater.* **20** 129.
[4] Hu X, Ming Y, Zhang X, Lu Y and Zhu Y 2012 *Appl. Phys. Lett.* **101** 151109.
[5] Arie A and Voloch N 2010 *Laser Photon. Rev.* **4** 355.
[6] Shapira A, Naor L and Arie A 2015 *Sci. Bull.* **60** 1403.
[7] Liu H, Li J, Zhao X, Zheng Y and Chen X 2016 *Opt. Express* **24** 15666.
[8] Fang X *et al.* 2015 *Appl. Phys. Lett.* **39** 8185.
[9] Fang X *et al.* 2016 *Opt. Lett.* **41** 1169.
[10] Shao G, Bai Y, Cui G, Li C, Qiu X, Geng D, Wu D and Lu Y 2016 *Aip Advances* **6** 075011.
[11] Sheng Y, Ma D, Ren M, Chai W, Li Z, Koynov K and Krolikowski W 2011 *Appl. Phys. Lett.* **99** 031108.
[12] Li L, Li T, Wang S, Zhang C and Zhu S 2011 *Phys. Rev. Lett.* **107** 126804.

[13] Jin H, Xu P, Luo X, Leng H, Gong Y, Yu W, Zhong M, Zhao G and Zhu S 2013 *Phys. Rev. Lett.* **111** 023603.
[14] Shapira A and Aire A 2011 *Opt. Lett.* **36** 1933.
[15] Yamada M, Nada N, Saitoh M and Watanabe K 1992 *Appl. Phys. Lett.* **62** 435.
[16] Mohageg M, Strekalov D, Savchenkov A, Matsko A, Ilchenko V and Maleki L 2005 *Opt. Express* **13** 3408.
[17] Bloch N, Shemer K, Shapira A, Shiloh R, Juwiler I and Arie A 2012 *Phys. Rev. Lett.* **108** 233902.
[18] Shapira A, Juwiler I and Arie A 2011 *Opt. Lett.* **36** 3015.
[19] Shapira A, Shiloh R, Juwiler I and Arie A 2012 *Opt. Lett.* **37** 2136.
[20] Shapira A, Juwiler I and Arie A 2013 *Laser Photon. Rev.* **7** 25.
[21] Ming Y, Tang J, Chen Z, Xu F, Zhang L and Lu Y 2015 *IEEE J. Sel. To. Quant.* **21** 6601206.
[22] Yariv A and Yeh P 1983 *Optical Waves in Crystals* (New York: John Wiley & Sons) Chap.12.
[23] Burch J 1967 *IEEE Proc.* **55** 599.
[24] Lee W 1979 *Appl. Opt.* **18** 3661.

IOP Conf. Series: Journal of Physics: Conf. Series **844** (2017) 012009 doi:10.1088/1742-6596/844/1/012009

Inverse Design of a SOI T-junction Polarization Beamsplitter

Zi Ye, Jifang Qiu, Chong Meng, Li Zheng, Zhenli Dong, and Jian Wu*

State Key Laboratory of Information Photonics and Optical Communications,

Beijing University of Posts and Telecommunications, Beijing 100876, China

JennyYe@bupt.edu.cn, jifangqiu@bupt.edu.cn, cmeng@bupt.edu.cn, al6y@bupt.edu.cn, imawmtb-99@foxmail.com, *jianwu@bupt.edu.cn

Abstract. A SOI T-junction polarization beamsplitter with an ultra-compact footprint of 2.8×2.8μm^2 is designed based on the method of inverse design. Simulated results show that the conversion efficiencies for TE and TM lights are 73.34% (simulated insertion loss of 2dB) and 80.4% (simulated insertion loss of 1.7dB) at 1550nm, respectively; the simulated extinction ratios for TE and TM lights are 19.3dB and 13.99dB at 1558nm, respectively.

1. Introduction

Recent development in optimization algorithm and computing power provides a broader way to improve the performance of on-chip integrated devices. Unfortunately, most optical devices are still designed by hand now. There is a creative device design method called inverse design [1], which can take advantage of the enormous available design space to design arbitrary high performances linear optical devices with ultra-compact footprints.

The polarization beamsplitter is an essential component in integrated on-chip photonics systems. It is widely used to separate different polarization states. However, most traditional PBS devices are fairly large [2], and there are strict requirements for wavelength sensitive due to the underlying phase-matching principle [3-4]. In contrast, inverse design method can design free-form PBS device with small footprints and wavelength insensitive performance. T-junction waveguide has been used to get ultra-compact footprints and high integration of on-chip silicon photonic devices. The original T-junction waveguide can be freely optimized enables ultimate devices that can be highly functional.

In this paper, we designed a compact T-junction PBS on a silicon-on-insulator (SOI) platform, with a compact footprint of 2.8×2.8μm^2, by using inverse-design method. The program exhibits TE and TM conversion efficiencies of 73.34% and 80.4% at 1550nm, and the FDTD solver shows low insertion losses (about 2dB for TE and 1.7dB for TM) and high extinction ratios, which is over 12dB in a wavelength range from 1545nm to 1567nm for TM and over 17dB for TE in entire C-band.

2. Device design

Fig.1 shows the initial scheme of the T-junction polarization beamsplitter, which consists of one input waveguide, two output waveguides and a square design region. The size of the design region is 2.8×2.8μm^2 with 70×70 pixels, the shape of each pixel is square with a size of 40×40nm^2. The input and output waveguides, with width of 500nm, are located at the center of their respective edge of the design region. The thickness of Si layer of the SOI wafer is 220nm. Refractive indices of $n_{Si} = 3.5$, $n_{SiO_2} = 1.5$ are used. The fundamental TE-polarized light and the fundamental TM-polarized light output through the bottom and top output waveguides.

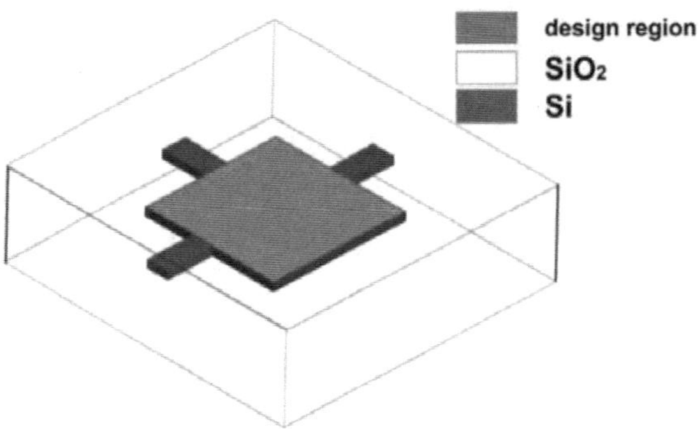

Figure 1. Initial scheme of the T-junction PBS.

Initially, we defined the waveguide mode and the mode conversion efficiency between input and output mode at a certain wavelength in order to specify the desired performance of our device. Next, inverse design optimization process was used to decide the permittivity distribution of design area [5-6]. The optimization process contains two stages, including: 1) an initial distribution of the permittivity in the design region is obtained by using alternating direction method of multipliers (ADMM) algorithm, which however is continuously distributed, impossible for fabrication; 2) the continuous permittivity is transformed to binary index by level-set method to ensure that device can be fabricated. With the help of GPU-based high performance computing, the T-junction PBS was designed in nearly 20h.

3. Results

The ultimate designed T-junction PBS structure is shown in Figure 2(a). The input waveguide named port1 is either fundamental TE or TM polarized light. Then the TE and TM polarized lights couple into the bottom (named port2) and top (named port3) output waveguides, respectively. Figure 2(b)(c) illustrate the electric field intensity distributions for TE (b) and TM (c) polarized light at the design wavelength of 1550 nm. The inverse design programming figures out the conversion efficiencies into the port2 and port3 are 73.34% and 80.4% at 1550nm respectively.

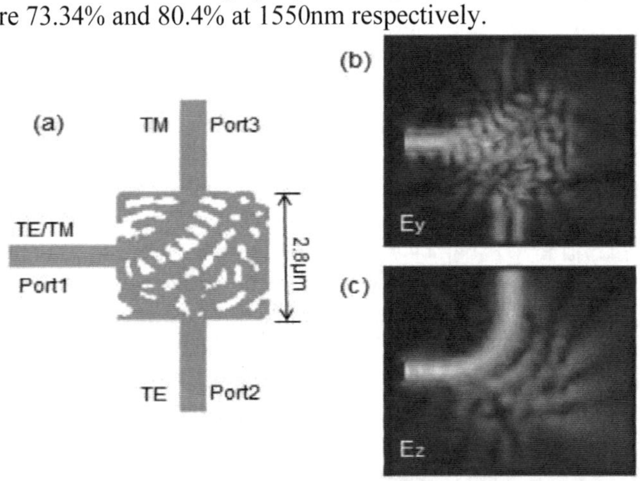

Figure 2. Simulation results of the T-junction PBS. (a) Top view of the device. (b) (c) Electric field intensity distributions for TE (E_y-dominant) and TM(E_z-dominant) inside the device.

To further verify this inverse-designed PBS, we used Lumerical Solutions Software to analyze the performance of the PBS by importing ultimate designed structure into the FDTD solver in Lumerical. The wavelengths of input lights with TE/TM-polarized varied from 1520nm to 1580nm. Figure 3 shows the broadband fluctuation of insertion losses and extinction ratio over 60nm. Simulation results show that the insertion losses of TE and TM lights are 1.93dB and 1.84dB and the extinction ratios are 19.3dB and 13.99dB at 1558nm. Figure 3(b) exhibits extinction ratios are higher than 12dB in a wavelength range of 22nm for TM, and higher than 17dB in total C-band for TE. The T-junction PBS verifies that inverse design can generate new generation of ultra-compact optical devices with novel functionality.

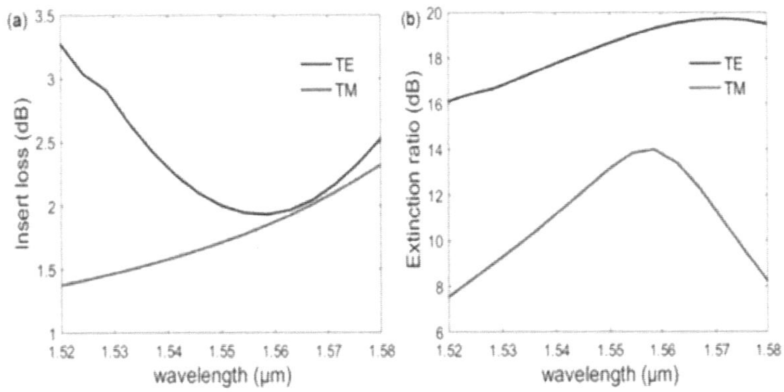

Figure 3 T-junction PBS insertion losses (a) and extinction ratios (b) as a function of wavelength.

Acknowledgment

This work is supported by NSFC program (61331008, 61335009, 61475022, 61307055, 61571067, 61505011, 61471054), 863 program (2013AA014202, 2015AA015503).

References

[1] Jesse Lu and Jelena Vučković, "Nanophotonic computational design," Opt. Express 21(11),13351-13367 (2013).
[2] Junbo Feng and Zhiping Zhou, "Polarization beam splitter using a binary blazed grating coupler," Opt. Lett. 32(12), 1662-1664 (2007).
[3] Daoxin Dai, Zhi Wang, and John E. Bowers, "Ultrashort broadband polarization beam splitter based on an asymmetrical directional coupler," Opt. Lett. 36(13), 2590-2592 (2011).
[4] Fukuda, H. et al. "Ultrasmall polarization splitter based on silicon wire waveguides," Opt. Express 14(25), 12401-12408 (2006).
[5] Alexander Y. Piggott, Jesse Lu, Konstantinos G. Lagoudakis, Jan Petykiewicz, Thomas M. Babinec and Jelena Vučkovi, "Inverse design and demonstration of a compact and broadband on-chip wavelength demultiplexer," Nature Photon, 9, 374-377(2015).
[6] Jesse Lu and Jelena Vučković, "Objective-first design of high-efficiency, small-footprint couplers between arbitrary nanophotonic waveguide modes," Opt. Express 20(7), 7221-7236 (2012).

IOP Conf. Series: Journal of Physics: Conf. Series **844** (2017) 012010 doi:10.1088/1742-6596/844/1/012010

High-speed Balanced Homodyne Detector for Quantum Information Applications

Yunsong Wang, Xinyu Chen, and Lijian Zhang

College of Engineering and Applied Sciences, Nanjing University, Nanjing, China

E-mail: yunsong_wang@163.com

Abstract. We have developed a balanced homodyne detector capable of working with femtosecond pulsed light at a 76MHz repetition rate and 800nm wavelength. It exhibits a common mode rejection ratio of 55dB, and can achieve the shot-noise limit. Provided with good performance, our detector can be used in high speed quantum information applications.

1. Introduction

With the powerful ability of demonstrating complete characterization of a quantum light state, high-speed balanced homodyne detection is essential for foundational analysis of quantum optics on many areas, such as quantum process tomography, quantum key distribution and quantum precision measurement. By measuring the quadrature operators of the input signal state, the balanced homodyne detector can then obtain the quasi probability distribution or in other words, the Wigner function after kinds of quantum state tomography [1]. The first experiment determining the quantum state of light field was proposed by Smithy, Beck, Cooper, Raymer, et al in 1993 [1,2]. Nowadays, the quantum state tomography with balanced homodyne detector has become a standard method in quantum information field.

Over the past decades, remarkable effort has been made to make a balanced homodyne detector with satisfactory performance applying to various applications [3]. Conventionally, researchers use a circuit with a dual-stage amplifier to gain higher magnification and stability, but this design will bring extra noise as well. In our experiments, a single-stage simple-construct amplifier homodyne detector was also found able to satisfy the performance requirements of homodyne detectors: (a) low enough electrical noise competent to beat the shot-noise limit which is the key requirement to make a quantum measurement; (b) high enough bandwidth to distinguish 76 MHz repetition rate femtosecond optical pulse; (c) high common mode rejection ratio (CMRR) to make a balance between the two photodiodes [1].

2. Theoretical analysis

Depicted in the Fig. 1 is a simple layout of balanced homodyne detector. The signal $\hat{\rho}_s$ and the much more powerful local oscillator $|\alpha\rangle_{LO}$ are combined on the 50:50 beam splitter, resulting in two output lights $\hat{b}_1$and $\hat{b}_2$ which are collected by two photodiodes respectively. The photocurrent from D1 and D2 are subtracted finally to get a difference current $\hat{I}$ which will collected by our circuit and oscilloscope.

IOP Conf. Series: Journal of Physics: Conf. Series **844** (2017) 012010 doi:10.1088/1742-6596/844/1/012010

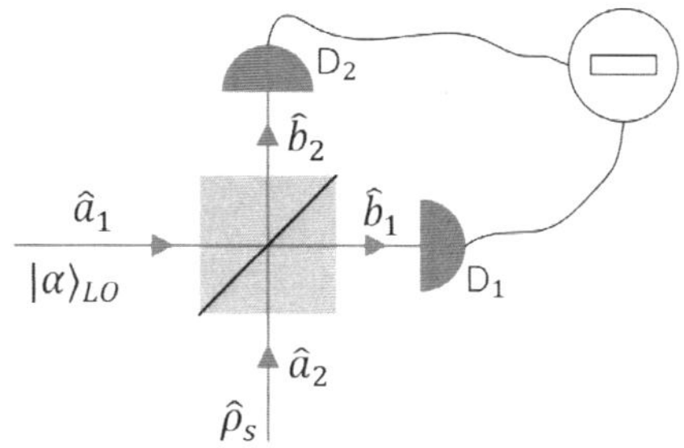

Figure 1. Principle diagram of the balanced homodyne detector.

We use the annihilation operators of the input and output field here for convenience. The transformation over a 50:50 beam splitter follow the rule:

$$\hat{b}_1 = \frac{1}{\sqrt{2}}(\hat{a}_1 + \hat{a}_2), \tag{2.1}$$

$$\hat{b}_2 = \frac{1}{\sqrt{2}}(\hat{a}_1 - \hat{a}_2); \tag{2.2}$$

According to the rules in equation (2.1) and (2.2), we can derive the difference photocurrent operator $\hat{I}$:

$$\hat{I} = \hat{b}_1^\dagger \hat{b}_1 - \hat{b}_2^\dagger \hat{b}_2 \tag{2.3}$$

$$\hat{I} = \hat{a}_1^\dagger \hat{a}_2 + \hat{a}_2^\dagger \hat{a}_1 \tag{2.4}$$

Considering the local oscillator $\hat{a}_1$ is a strong coherent state $|\alpha\rangle_{LO}$ which has the eigenvalue relations with the annihilation representation:

$$\hat{a}|\alpha\rangle = \alpha|\alpha\rangle, \ \langle\alpha|\hat{a}^\dagger = \langle\alpha|a^*; \tag{2.5}$$

where $\alpha = |\alpha|e^{i\theta}$,and θ is the local oscillator phase relative to the signal, we can infer the expected photocurrent difference $\langle\hat{I}\rangle$:

$$\langle\hat{I}\rangle = |\alpha|\langle(\hat{a}_2^\dagger + \hat{a}_2)cos\theta + (\hat{a}_2^\dagger - \hat{a}_2)sin\theta\rangle \tag{2.6}$$

Replacing the annihilation operators with the quadrature operators defined by equation (2.7) and equation (2.8), we have another representation of $\langle\hat{I}\rangle$ in equation (2.9),

$$\hat{X} = \frac{1}{2}(\hat{a}_2^\dagger + \hat{a}_2) \tag{2.7}$$

$$\hat{Y} = \frac{1}{2}(\hat{a}_2^\dagger - \hat{a}_2) \tag{2.8}$$

$$\langle\hat{I}\rangle = 2|\alpha|\langle\hat{X}cos\theta + \hat{Y}sin\theta\rangle \tag{2.9}$$

From equation (2.9), we can get the phase space information of signal mode $\hat{\rho}_s$ in a phase θ, and by scanning the local oscillator through 2π, the complete information about the signal quadrature distribution is confirmed from which the quasi probability distribution can be inferred[1].

3. Experiment Design

IOP Conf. Series: Journal of Physics: Conf. Series **844** (2017) 012010 doi:10.1088/1742-6596/844/1/012010

The schematic of detector circuit is shown in figure 1(a), two p-i-n photodiodes (Hamamatsu S3883) are wired in series to gain the difference current which is then sent to the high-bandwidth, low-noise operational amplifier (Texas Instruments OPA847) to gain a desirable amplification. Then, the voltage output is collected by oscilloscope (Agilent MSO-X 3034A) through a 50 Ohm coax. The choice of the two photodiodes are extremely strict to reach perfect subtraction, the photodiodes should exhibit similar temporal response characteristics and quantum efficiency.

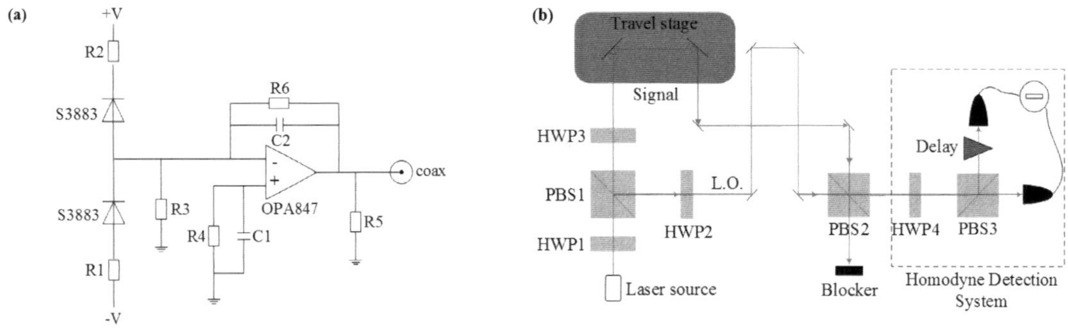

Figure 2. (a) Schematic of detector circuit which shows the main component only. (b) Optical setup of balanced homodyne detector. We use a 78MHz repetition rate, 830nm wavelength, mode-lock Ti:Sapphire laser from Coherent.

The optical setup is shown in figure 1(b), half-wave plate 1 (HWP1), HWP2 and polarization beam splitter 1 (PBS1) control the local oscillator (LO) power, while HWP1, HWP3 and PBS1 control signal power. The LO and signal are sent into PBS2 separately, care was taken to ensure both temporal and spatial mode of LO and signal should be overlapped. So, we use a travel stage as illuminated in figure to control temporal overlap and use a fiber coupler (not shown here) to control spatial overlap. The combination of HWP4 and PBS3 is used to make up a 50:50 beam splitter, HWP4 is nominally set at 22.5° to the horizontal. Finally, the two beams splat by PBS3 go into the two photodiodes respectively. To achieve a perfect balance, the route paths after PBS3 should be equal and the spot size focused on the photodiodes should be equal too.

Among the ten photodiodes (S3883) at our disposal, we have to pick the two matching best, which indicates how well the common mode component is being cancelled. There are two most important factors we took into consider: the quantum efficiency and the pulse delay. We measured the output of the detector with each photodiode on one socket while the other socket in idle and pick the two photodiodes whose output are the most similar.

4. Performance and Characterization

The most critical requirement of the homodyne detector is shot-noise limit performance which shows the ability of detecting quantum field of light instead of the classical field [3]. From Eq. (2.5), we can calculate the variance of the output current as shown in equation (4.1), where $\boldsymbol{\hat{X}_\theta = \hat{X}cos\theta + \hat{Y}sin\theta}$:

$$var[\hat{I}] = 4P_{LO}var[\hat{X}_\theta] \tag{4.1}$$

In the non-ideal condition, there is also some constant noise from the circuit itself $var[e]$, so the final variance is:

$$var[\hat{I}] = 4P_{LO}var[\hat{X}_\theta] + var[e] \tag{4.2}$$

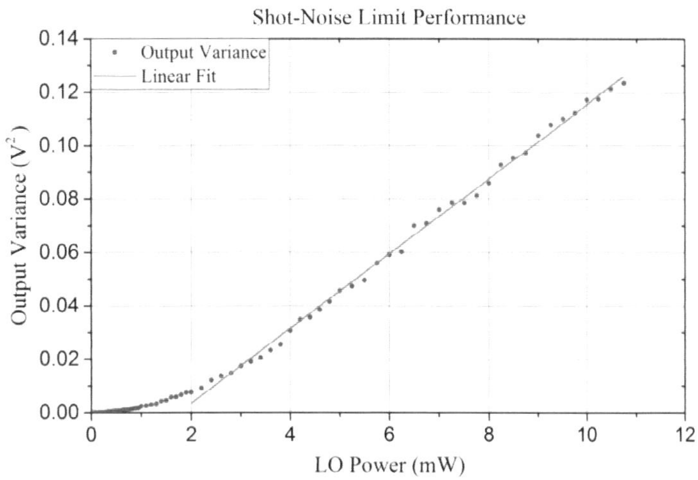

Figure 3. Plot of detector output variance as a function of the LO power. A balanced homodyne detector which is set in the linear range will work well in quantum field detection [3]

It's clear that in a well worked balanced homodyne detector the output variance should be proportional to the local oscillator power. With the signal route blocked, we get the output variance on different LO powers, one thing notable here is that the output variance is variance of the integral of detector output over the laser pulse duration. Shown in Fig.3 is the shot-noise limit performance of our detector, apparently, it has a long enough linear range to detect quantum field light.

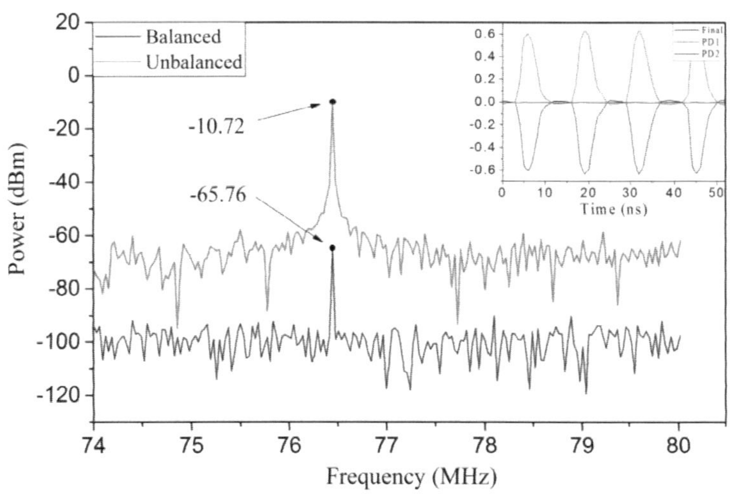

Figure 4. Inset figure is the time domain of the output of each photodiode and the final output after balancing. The main figure is the frequency spectrum of balanced output (red line) and unbalanced output (blue line).

The CMRR is another parameter we must pay attention to, it represents the ability of efficiently performing the difference between the amplitude fluctuations of the beams impinging on the two photodiodes. The time-domain outputs of balanced and unbalanced (one of the photodiodes is blocked) condition is plotted in small figure of the Fig. 4, and we plot the frequency domain in the larger figure, the value at 76MHz of unbalanced output is -10.72dBm and the balanced output is -65.76dBm. According to its definition in Eq. (4.3), the CMRR is the difference of output power between the balanced and the unbalanced frequency spectrums value [5], thus is 55.04dB.

$$CMRR_{dB} = 20log_{10}\frac{V_{unbalanced}}{V_{balanced}} \tag{4.3}$$

We compare the performance of our detector with the others, it can be seen that our detector achieves both the high speed and high CMRR. Shown in table.1 is the performance comparison between some other high-speed homodyne detectors.

Table 1. Comparison between other detectors.

	Ours	[4]	[5]	[6]
Wavelength (nm)	**800**	1550	780	786
Repetition rate (MHz)	**76**	32	76	82
CMRR (dB)	**55**	46.0	52.4	42

5. Summary

In conclusion, we demonstrate a simpler single-stage balanced homodyne detector, all of whose components are usual and easy to get. We present most of the key details essential for the detector building and some parameters we attach importance to. With a CMRR of 55dB and long enough linear area in its shot-noise limit figure, our detector is capable of quantum field detection.

6. Acknowledgment

This work is supported by the National Natural Sci- ence Foundation of China (Grant No. 11404040, No. 91536113 and No. 11474159).

References

[1] Lvovsky, Alexander I., and Michael G. Raymer 2009 *Reviews of Modern Physics* **81** 299.
[2] Smithey, D. T., et al. 1993 *Physica Scripta* T**48** 35.
[3] Cooper, Merlin, Christoph Söller, and Brian J. Smith 2013 *Journal of Modern Optics* **60** 611-6.
[4] Chi, Yue-Meng, et al 2011 *New Journal of Physics* **13** 013003.
[5] Kumar, Ranjeet, et al 2012 *Optics Communications* **285** 5259-67.
[6] Li-Jian, Zhang, and Xiao Min 2013 *Chinese Physics B* **22** 110310.
[7] Zavatta, A., S. Viciani, and M. Bellini 2005 *Laser Physics Letters* **3** 3.
[8] Esposito, Martina, et al 2014 *New Journal of Physics* **16** 043004.
[9] Haderka, Ondřej, et al 2009 *Applied optics* **48** 2884-9.
[10] Duan, Huang, et al 2013 *Chinese Physics Letters* **30** 114209.
[11] Roumpos, Georgios, and Steven T. Cundiff 2013 *JOSA B* **30** 1303-16.
[12] Nielsen, Michael A., and Isaac Chuang 2002 "Quantum computation and quantum information."
[13] Hansen, H., Aichele, T., Hettich, C., Lodahl, P., Lvovsky, A. I., Mlynek, J. and Schiller, S. 2001 *Optics Letters*, **26**, 1714-16.

IOP Conf. Series: Journal of Physics: Conf. Series **844** (2017) 012011 doi:10.1088/1742-6596/844/1/012011

Design of Three-mode Multi/Demultiplexer Based on 2-D Photonic Crystals for Mode-Division Multiplexing Transmission

PeiDong Fu, Heming Chen *

Department of Opto-Electronics, Nanjing University of Posts and Telecommunications, Nanjing 210023, China

Bell Honors School, Nanjing University of Posts and Telecommunications, Nanjing 210023, China

*chhm@njupt.edu.cn

Abstract :A three mode division multiplexer and demultiplexer (MMUX/DEMMUX) base on 2-D photonic crystal at 1550 nm was designed. Two asymmetrical directional couplers are included in this MMUX/DEMMUX, which can achieve modes conversion function of TE_0, TE_1 and TE_2 modes. In order to avoid phase mismatching in bus waveguide, taper directional is applied at waveguide junction which can reduce the insertion loss effectively. Plane waves method (PWM) and finite difference time domain (FDTD) methods were used to simulating the performance of MMUX/DEMMUX. Numerical simulations show that the designed device has the potential for high-capacity MDM optical communication systems with a low insertion loss (<0.27dB) and a low mode crosstalk (< −25.4 dB).

1. Introduction

In order to increases the capacity of optical networks effectively, WDM technology [1,2] has become an important form of optical network technologies. However, it is well known that the capacity of WDM is approaching the Shannon limit of sing-mode optical fiber system. One of the promising approach for increasing the capacity is mode-division multiplexing (MDM) transmission in which each optical mode is exploited as an independent channel for transferring optical data.

In the mode multiplexed optical transmission system, MMUX/DEMMUX has become a hot research area as key components, which can be realized by using PLC-based asymmetrical parallel waveguides (APW) [3-7], asymmetrical Y-junction [8–10], as well as few-mode fibers [11–13]. MMUX/DEMMUX based on chip Y-splitter were designed by Jeffrey B. Driscoll in Columbia University[8], which can realize (de)multiplex two modes of the same polarization with a low insert loss ranging from 0.1to 0.7dB. However, these devices are very sensitive to process errors about the angle between Y-junction arms. Yue-De in University of Hong Kong have successfully multi/demultiplexed three modes by quasi phase-matching (QPM) the modes of asymmetrical parallel waveguides[3], but its worst channel crosstalk is only -10dB.

In this paper, a three mode multiplexer and demultiplexer using asymmetrical parallel waveguides (APW) base on the 2-D Si square lattice photonic crystal is proposed. In this MMUX/DEMMUX, the single-mode waveguide formed by removing a row of Si dielectric cylinders. Adjusting the size of two rows of Si dielectric cylinders forms TE_0 and TE_1 modes waveguide which can satisfy the phase matching condition. The multi-mode waveguide of TE_0, TE_1 and TE_2 modes formed by adjusting the size of three rows of Si dielectric cylinders.

In order to inhibit mismatch problem between the different multi-mode waveguide, taper waveguide is adopted between each bus waveguide. Therefore, TE_0, TE_1 and TE_2 modes MMUX/DEMMUX can be achieved which has a great value of the mode multiplexed optical communication system with advantages of small size, low insert loss, low mode crosstalk and easily integrated.

2. Structural model and MDM mechanism

2.1 Structural model

The structure of 2-D MMUX/DEMMUX based on photonic crystal, which has 95 × 21 Si dielectric cylinder compositions is shown in Figure 1. There are two asymmetrical parallel waveguides (APW) as mode conversion function in the structure. One of APW(P_{01}) is formed by putting a rectangular single-mode waveguide (G1) close to the multi-mode waveguide (G2), another APW(P_{02}) is formed by putting a rectangular single-mode waveguide (G3) close to the multi-mode waveguide (G4). G1&G3 is formed by removing a row of Si dielectric cylinders. G2 is formed by adjusting the size of two rows of Si dielectric cylinders which supports two guided modes (TE_0 and TE_1). G3 is formed by adjusting the size of three rows of Si dielectric cylinders which supports two guided modes (TE_0, TE_1 and TE_2). In order to ensure that all the various modes of bus waveguide couples effectively, we proposed a taper waveguide structure, which includes waveguide width of the gradient and Si dielectric cylinder radius of gradient.

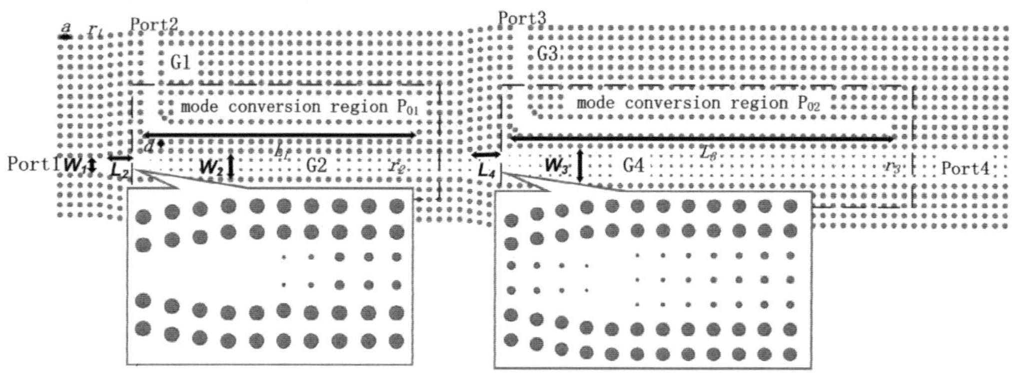

Fig.1. The structure of three-mode photonic crystal MMUX/DEMMUX

Structural parameters of the device are set as below: lattice constant is a = 0.55μm, Si dielectric cylinders radius is r_1 = 0.1μm, Si refractive index is n_1 = 3.4, single-mode waveguide width is $w_1 = 2a$ = 1.1μm. The multi-mode waveguide width and silicon dielectric cylinder radius are set as $w_2 = 3a$ = 1.65μm, r_2 = 0.0463μm The AWP interaction length is set to be $L_1 = 26a$ = 14.3μm, gap of the AWP is $d = a$ = 0.55μm. Taper waveguide length is L_2 = 1.65μm, the width of which increases linearly with a slope equal to a / 3. Structural parameters TE_0 and TE_2 mode conversion area are set as below: multi-mode waveguide Si dielectric cylinder radius is r_3 = 0.0535μm, width is $w_3 = 4a$ = 2.2μm. The AWP interaction length and gap are set to be $L_3 = 36a$ = 19.8μm, $d = a$ = 0.55μm. the width of which increases linearly with a slope equal to a / 4.

2.2 MMUX/DEMMUX mechanism

Mode conversion is realized by quasi-phase matching principle [14-17] in this paper. In the mode conversion region P_{01}, the propagation constant of TE_0 in rectangular single-mode waveguide should be equal to that of TE_1 in the multi-mode waveguide. And in mode conversion area P_{02}, propagation constant of TE_0 in rectangular single-mode waveguide equals to that of TE_2 in the multi-mode waveguide. Normalized propagation constant of TE_0 at 1.55μm wavelength is 0.206 in this structure. The curve of propagation constant and radius of Si dielectric cylinder in multi-mode waveguide (G2) by simulation is shown in Fig.2(a), The curve of propagation constant and radius of Si dielectric cylinders in multi-mode waveguide (G4) by simulation is shown in Fig.2(b). As shown in the Fig.2(a)-(b), in order to satisfy the phase matching condition, we set the radius of Si dielectric cylinders to be 0.0463μm and 0.0535μm in G2 and G4, respectively. The light wave transmitted in the mode conversion region P_{01} and P_{02} can be seen in Fig.2(c)-(d). The coupling length in P_{01} and P_{02} are 26*a* and 36*a*, respectively.

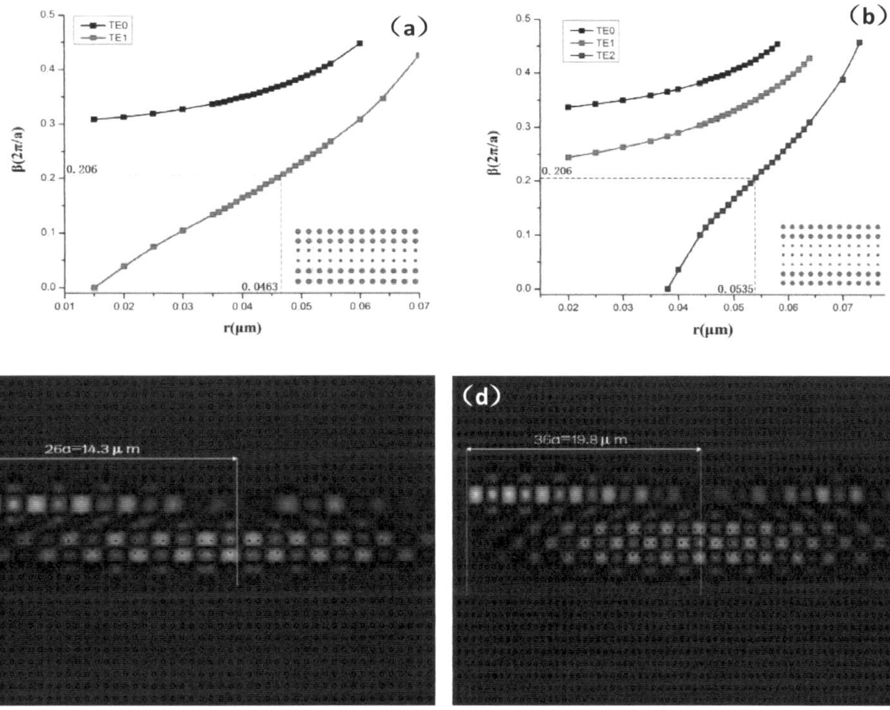

Fig.2. (a) propagation constant curve of G2 (b) propagation constant curve of G4
(c) the steady field intensity distribution for mode conversion TE_0 andTE_1
(d) the steady field intensity distribution for mode conversion TE_0 andTE_2

When TE_0 mode at 1550nm is launched from port1, port2 and port3, respectively, TE_0 will convert to TE_1 and enter into bus waveguide in the mode conversion region P_{01}, and TE_0 will convert to TE_2 in P_{02}. Finally, they output from port4 together. The structure can be also used as a DEMMUX because the parallel waveguide has a symmetric property. During demultiplexing process, the TE_0, TE_1 and TE_2 modes are excited from port4, and all converted to TE_0 modes and output from port1, port2 and prot3, respectively. Based on the above analysis, stable mode fields of MMUX/DEMMUX are as shown in Fig.3(a)-(c).

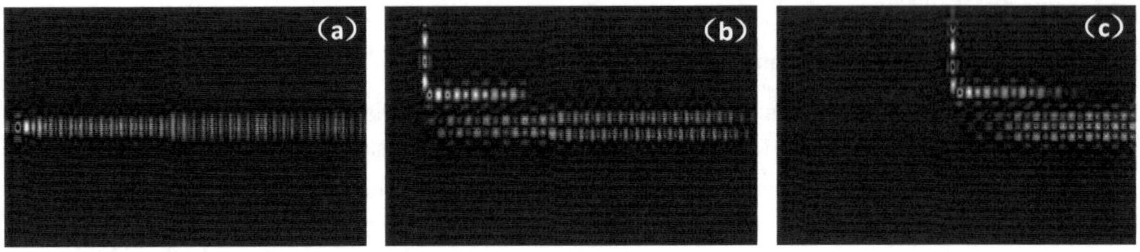

Fig.3. the steady field intensity distribution for three-mode MMUX/DEMMUD (a)TE_0 (b)TE_0 and TE_1(c)TE_0 and TE_2

3. Properties and analysis

3.1 The simulation of MMUX

In this paper, the capability of the device is analyzed using 2-D FDTD method. For MMUX function, we input CW light of TE_0 at 1550nm into port1, and light intensity are observed at port2, port3 and port4 respectively, the intensity of port4($T_{1\text{-}4}$), port2($S_{1\text{-}2}$) and port3($S_{1\text{-}3}$) are shown in Fig.4(a)-(c). Similarly, the intensity of port4($T_{2\text{-}4}$), port1($S_{2\text{-}1}$) and port3($S_{2\text{-}3}$) are shown in Fig.4(b)-(f), when CW light of TE_0 at 1550nm is put into port2. The intensity of port4($T_{3\text{-}4}$), port1($S_{3\text{-}2}$) and port2($S_{3\text{-}1}$) are shown in Fig.4(g)-(i), when CW light of TE_0 at 1550nm is put into port3. It could be seen that transmittance of the device is higher than 96% and the worst crosstalk is less than -25.4dB.

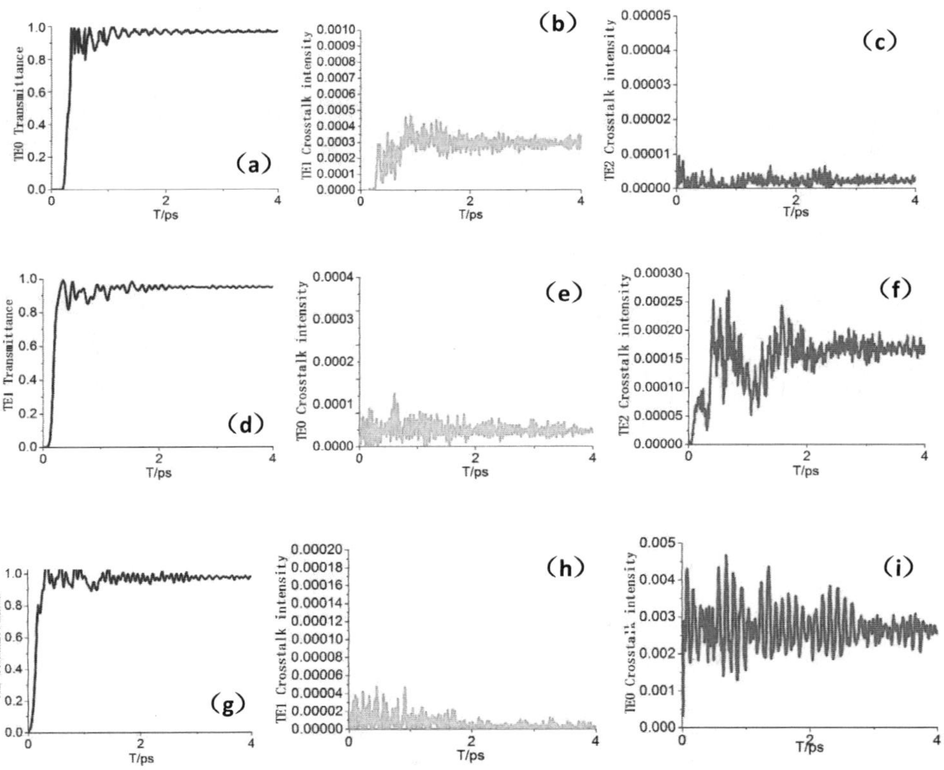

Fig.4. the transmission rate of three-mode MMUX (a)$T_{1\text{-}4}$(b)$S_{1\text{-}2}$(c)$S_{1\text{-}3}$(d)$T_{2\text{-}4}$(e)$S_{2\text{-}1}$(f)$S_{2\text{-}3}$(g)$T_{3\text{-}4}$(h)$S_{3\text{-}2}$(i)$S_{3\text{-}1}$

3.2 The simulation of DEMMUX

For DEMMUX function, we put CW light of TE_0 mode at 1550nm into port4, and light intensity are observed at port1, port2 and port3, respectively. the intensity of port1($T_{4\text{-}1}$),

IOP Conf. Series: Journal of Physics: Conf. Series **844** (2017) 012011 doi:10.1088/1742-6596/844/1/012011

port2(S_{4-2}) and port3(S_{4-3}) as shown in Fig.5(a)-(c). Similarly, the intensity of port2(T_{4-2}), port1(S_{4-1}) and port3(S_{4-3}) when CW light of TE_1 at 1550nm is put into port4, as shown in Fig.5(b)-(f). It is observed that transmittance is higher than 94%. The intensity of port3(T_{4-3}), port1(S_{4-2}) and port2(S_{4-1}) when CW light of TE_2 at 1550nm is put into port4, as shown in Fig.5(g)-(i). we can obtain that the DEMMUX performs with a lower crosstalk of 10^{-4} order of magnitude and the worst transmittance of 94%.

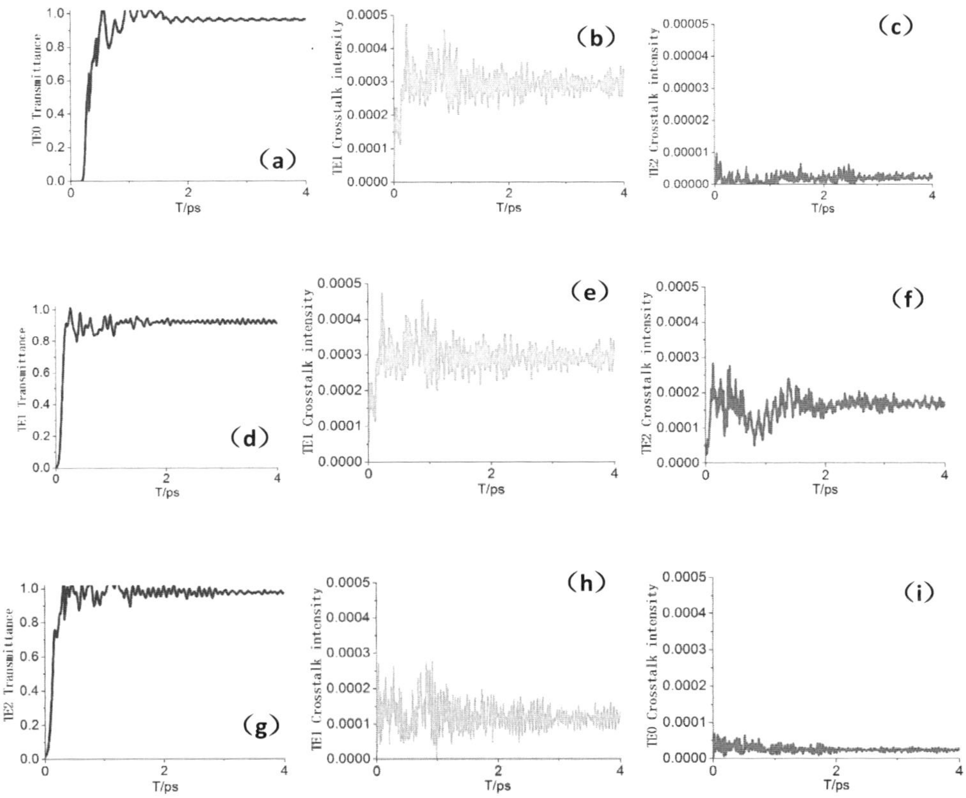

Fig.5. the transmission rate of three-mode DEMMUX(a)T_{4-1}(b)S_{4-2}(c)S_{4-3}(d)T_{4-2}(e)S_{4-1}(f)S_{4-3}(g)T_{4-3}(h)S_{4-2}(i)S_{4-1}

3.3 Performance index Analysis of MMUX/DEMMUX

Insertion loss and crosstalk between channels are important factors of MMUX/DEMMUX. Channel crosstalk is defined as:

$$C=10\lg\frac{S}{T} \quad (1)$$

where S represents the radiation intensity from transmission channel into the adjacent channel, T represents transmission intensity of the transmission channel. Insertion loss is defined as:

$$CT=-10\lg\frac{P_o}{P_i} \quad (2)$$

The Po indicates the power of the inputting light, Pi represents the transmission power, in formula (2). The performance of crosstalk in MMUX function and DEMMUX function is shown as table1(a)-(b).

Table 1. the crosstalk of the device (a)MMUX (b) DEMMUX

(a)

Crosstalk	TE_0	TE_1	TE_2
TE_0		-35dB	-42.8dB
TE_1	-43dB		-37.3dB
TE_2	-25.4dB	-39.9dB	

(b)

Crosstalk	TE_0	TE_1	TE_2
TE_0		-31.3dB	-43.2dB
TE_1	-42.7dB		-37.2dB
TE_2	-43.8dB	-39.1dB	

In MMUX function, the insertion loss of three TE_0 modes are 0.17dB, 0.13dB and 0.09dB, respectively. In DEMMUX function, the insertion loss of TE_0, TE_1 and TE_2 modes are 0.17dB, 0.27dB and 0.13dB, respectively. The simulation results show that insertion loss of this device is less than 0.27dB and channel crosstalk is less than -25.4dB.

4. Conclusion

In this paper, a new three mode MMUX/DEMMUX base on 2-D photonic crystals at 1550 nm is designed. The MMUX/DEMMUX can multi/demultiplex TE_0, TE_1 and TE_2 modes at 1550nm with introducing two pairs of parallel asymmetrical waveguide. The parameters of this device are analyzed using FDTD method. Numerical simulations show that the insertion loss is less than 0.27dB, channel crosstalk is less than −25.4dB. At the same time, dimensions of the device are only 53μm×12μm. It has a great value to mode division multiplexing system for its ultra-compact size and excellent performance.

Reference

1. T. Niemi , L.H. Frandsen, K.K Hede, "Wavelength-division demultiplexing using photonic crystal waveguides," IEEE Photonics Technology Letters **18**, 226-228(2006).
2. K. Nozaki, E. Kuramochi, A. Shinya, and M. Notomi, "25-channel all-optical gate switches realized by integrating silicon photonic crystal nanocavities," Opt. Express **22**, 14263–14274 (2014).
3. Y.D Yang, Y Li, Y.Z Huang, "Silicon nitride three-mode division multiplexing and wavelength-division multiplexing using asymmetrical directional couplers and microring resonators," Opt. Express **22**, 22172-22183(2014).
4. Y. Kawaguchi, K. Tsutsumi, "Mode multiplexing and demultiplexing devices using multimode interference couplers," Electronics Letters **38**, 1701-1702(2003).
5. T. Uematsu, Y. Ishizaka, Y. Kawaguchi, "Design of a Compact Two-Mode Multi/Demultiplexer Consisting of Multimode Interference Waveguides and a Wavelength-Insensitive Phase Shifter for Mode-Division Multiplexing Transmission," Journal of Lightwave Technology **30**, 2421-2426(2012).
6. N. Hanzawa, K. Saitoh, T. Sakamoto, T. Matsui, K. Tsujikawa, T. Uematsu, and F. Yamamoto, "PLC-based four-mode multi/demultiplexer with LP11 mode rotator on one chip,"J. Lightwave Technol. **33**, 1161–1165 (2015).
7. Y. Ding, J. Xu, F. Ros, "On-chip two-mode division multiplexing using tapered directional coupler-based mode multiplexer and demultiplexer," Opt. Express **21**,

10376-82(2013).

8. J.B Driscoll, R.R Grote, B. Souhan, "Asymmetric Y junctions in silicon waveguides for on-chip mode-division multiplexing," Optics Letters **38**, 1854-6(2013).
9. W. Chen, P. Wang, J. Yang, "Mode multi/demultiplexer based on cascaded asymmetric Y-junctions," Opt. Express **21**, 25113-9(2013).
10. N. Riesen, J.D Love, "Design of mode-sorting asymmetric Y-junctions," Applied Optics **51**, 2778-83(2012).
11. C.P. Tsekrekos, D. Syvridis, "All-Fiber Broadband Mode Converter for Future Wavelength and Mode Division Multiplexing Systems," IEEE Photonics Technology Letters **24**, 1638-1641(2012).
12. C.P. Tsekrekos, D.Syvridis, "Symmetric Few-Mode Fiber Couplers as the Key Component for Broadband Mode Multiplexing," Journal of Lightwave Technology **32**, 2461-2467(2014).
13. Y. Shuchang, S. Fu, M. Zhang, "Demodulation and multi-input multi-output equalization for mode division multiplexing system using a novel few-mode fiber," Acta Physica Sinica **62**, 1691-1702(2013).
14. Z. Wen, Z.Y. Yang Z, "Multi/demulti-plexer based on transverse mode conversion in photonic crystal waveguides," Opt. Express **23**, 24770-84(2015).
15. V. Berger, "Nonlinear photonic crystals," Physical Review Letters **81**(19):4136-4139(1998).
16. B.A. Dorin, W.N. Ye. "Two-mode division multiplexing in a silicon-on-insulator ring resonator," Opt. Express **22**, 4547-58(2014).
17. S. Bagheri and W. M. J. Green, "Silicon-on-insulator mode-selective add-drop unit for on-chip mode-division multiplexing," in *Proceedings of IEEE International Conference on Group IV Photonic* (San Francisco, USA, Sep.2009), pp. 166-168.

Acknowledgments
This work supported by the National Natural Science Foundation of China (Grant Nos. 61077084, 61571237) and the Natural Science Foundation of Jiangsu Province (Grant No.BK 20151509).

IOP Conf. Series: Journal of Physics: Conf. Series **844** (2017) 012012 doi:10.1088/1742-6596/844/1/012012

Nonlinear optical absorption and ultrafast optical response of un-doped and P-doped nanocrystalline Si/SiO_2 multilayers

P Zhang[1], D K Li[2], L Y Jiang[1], and J Xu[2*]

[1]Henan Key Lab of Information-based Electrical Appliances, College of Electrical and Information Engineering, Zhengzhou University of Light Industry, Zhengzhou, 450002, China
[2]National Laboratory of Solid State Microstructures, School of Electronic Science and Engineering and Collaborative Innovation Centre of Advanced Microstructures, Nanjing University, Nanjing, 210093, China

*junxu@nju.edu.cn.

Abstract. The nonlinear optical properties and ultrafast optical response of Si/SiO_2 multilayers with and without phosphorus doping are comparatively studied in the present work. The enhanced nonlinear absorption and fast recombination process were observed after phosphorus doping, the defect states generated by high phosphorus doping was the possible origination. It was suggested that the nonlinear optical repose as well as the ultrafast dynamic process could be changed greatly via phosphorous doping, which provided a new approach to improve the performance of Si-based photonic devices.

1. Introduction

Nanocrystalline-Si (nc-Si) materials have been recently attracted much attention since they exhibit the intense photoluminescence as well as the large optical nonlinearities, which can be potentially applied in the future Si-based photonic devices [1,2]. However, the large nonlinear absorption and slow recombination process degrade extremely the photonic devices' performance, such as the modulation rate, gain bandwidth, wavelength conversation efficiency and so on [2]. In our previous work, the enhanced photoluminescence intensity and optical nonlinearities were observed, which was ascribed to the interface/surface states in nc-Si dots. It was found that the linear and nonlinear optical properties can be changed by altering the nc-Si dot size or post-annealing temperature [3,4]. Moreover, we found doping in nc-Si is another effective method to modify the chemical environment of nc-Si, and the density of interface/surface states could be continuously controlled, which made the nonlinear absorption coefficient reduced by 4 folds through phosphorus (P) doping [5]. Nevertheless, it was reported the nonlinear absorption increased after P doping, the impurity-related states below the conduction band induced by P doping was the major reason [6]. Meanwhile, the doping effect on the transient dynamics of nc-Si is still unclear.

Therefore, it is interesting to comprehensive understand the influences of P doping on nonlinear optical properties and transient process of nc-Si materials. In this work, we study the nonlinear optical properties and ultrafast optical response of Si/SiO_2 multilayers with and without P doping. It was found that P doping may tune the interfacial environment and induce new energy level, which could modify the nonlinear optical behavior and the initial carries dynamics of nc-Si/SiO_2 multilayers.

2. Experimental

The phosphorous-doped nc-Si/SiO_2 multilayers structures were fabricated by annealing the phosphorous-doped amorphous stacked structures deposited in conventional plasma-enhanced chemical-vapor system. The details of preparation conditions can be found elsewhere [3-5]. The thickness of the amorphous sublayers was about 4 nm, and the P doping concentration in nc-Si could be controlled by changing the gas flow ratio R (R = [PH_3] / [SiH_4]). Here after, we denoted the samples prepared by R = 0 and 5 as sample 1 and sample 2, respectively. The Z-scan technique was applied to measure the nonlinear optical absorption coefficient, a 100-fs mode-locked Ti-sapphire laser (Coherent Libra-HE) with a wavelength of 800 nm at 1 kHz was used as the excitation source. The pump and probe setup was utilized to monitor the temporal evolution of the photo-generated carriers. The mode-locked Ti: sapphire laser with fundamental frequency at 800 nm with 100-fs pulse duration were separated into two parts, one is the pump laser at 400 nm (3.1 eV) generated by frequency-doubled the fundamental frequency laser through a BBO crystal, the other one acts as the probe beam at 800 nm.

3. Discussion and results

The formation of nc-Si after 900 °C annealing is confirmed by cross-sectional transmission electron microscopy (TEM) and the Raman spectra. Figure 1(a) shows the TEM image of nc-Si/SiO_2 multilayers, the dot size is about 4 nm which is uniformly distributed in the sample separated by the SiO_2 layers. It is found that the crystallinity is increased with increasing the doping concentration and annealing temperature [5], as shown in Figure 1(b).

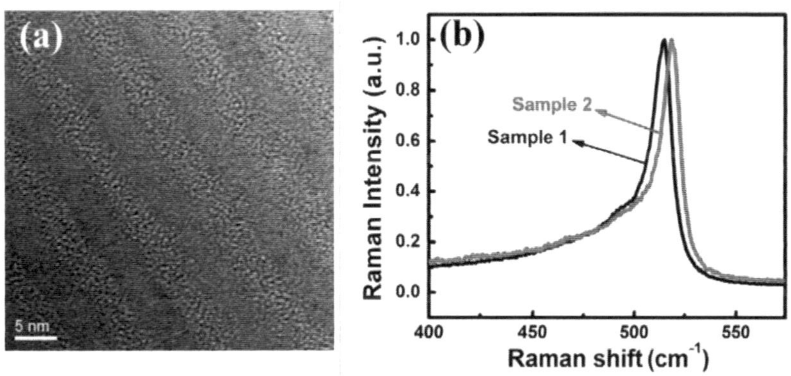

Figure 1. (a) Cross-sectional TEM image and (b) Raman spectra of nc-Si/SiO_2 multilayers annealed at 900 °C.

Figure 2 shows the normalized Z-scan transmittance traces of samples 1 and 2 in the open aperture configuration under laser intensity $I_0 = 4.78\times10^{10}$ W/cm^2. It is clear that both the Z-scan traces show a dip at the focal point, indicating nonlinear absorption (NLA) is the reverse absorption (RSA) process. In our previous work, it is found that the nonlinear optical absorption is changed with the annealing temperature as well as the nc-Si dot size. When the sample changes from amorphous phase to nanocrystalline state or increase the size of nc-Si dot, the nonlinear absorption turns from the reverse saturation absorption (RSA) to the saturation absorption (SA) process. It is proposed that the localized states at the nc-Si/SiO_2 interfacial region play the key role in the observed switching behaviours [3,4]. When the excitation intensity I_0 is about 3.54×10^{11} W/cm^2, the sample 1 shows the saturation absorption (SA) properties which is ascribed to the phonon-assisted one-photon transition process between the valence band and interface states in the gap [4]. We considered that the RSA property at low intensity ($I_0 = 4.78\times10^{10}$ W/cm^2) is originated from the two step absorption process via the interface states [3].

IOP Conf. Series: Journal of Physics: Conf. Series **844** (2017) 012012 doi:10.1088/1742-6596/844/1/012012

The calculated NLA coefficient for samples 1 and 2 is 1.78×10^{-7} cm/W and 2.27×10^{-7} cm/W, respectively, indicating that the density of states associated with two step process increased after high P doping, which is different from the P-doping in the nc-Si with the average size of 2 nm, where the NLA coefficient monotonously decreases with increasing P doping concentrations [5]. It is found that P atoms prefer to locate at the surface of nc-Si and passivate the dangling bond when the size of the nc-Si is small, while the substitutional doping may occur when the size of nc-Si is large. However, it is also worth noting that there is no obvious infrared absorption associated with the intra-valence-band transition can be observed at the high doping concentration. Therefore, the impurity-related states are not responsible for the observed nonlinear absorption phenomena. According to the ESR measurements, a kind of new defect states will be generated after high P doping concentration due to the damage of the Si crystalline lattice [7], which may play a great role in the increase of the nonlinear absorption coefficient.

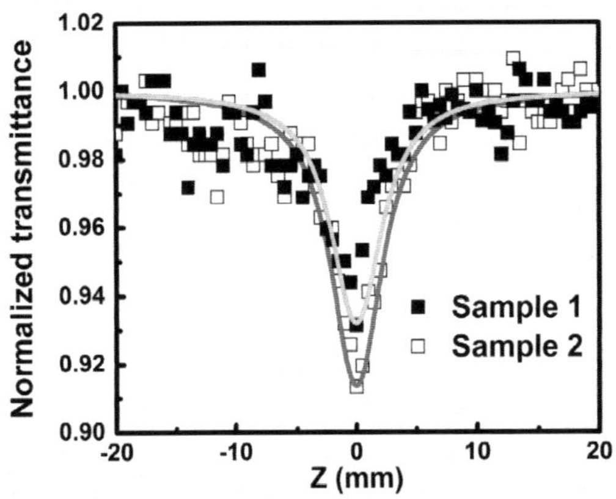

Figure 2. Open aperture Z-scan traces of samples 1 and 2 annealed at 900 °C. The excitation wavelength and intensity are 800 nm and $I_0 = 4.78\times10^{10}$ W/cm^2, respectively. The solid lines are the fitting curves of the experimental data.

To further investigate the P doping effect on the transient dynamic process of Si/SiO$_2$ multilayers, the pump and probe measurements were carried out. Figure 3 (a) and (b) show the dynamics of transient transmission of samples1 and 2 under 1 mW excitation. It is evident that all signals of transient transmission (ΔT) is positive, demonstrating the occurrence of transient bleach, which may originate from the occupation of the interface-related states by the photo-generated carriers. The ultrafast built-up time is less than 1 ps, illustrating that the excited-carries generated by pump beam ware trapped rapidly by the interface states of nc-Si/SiO$_2$ multilayers.

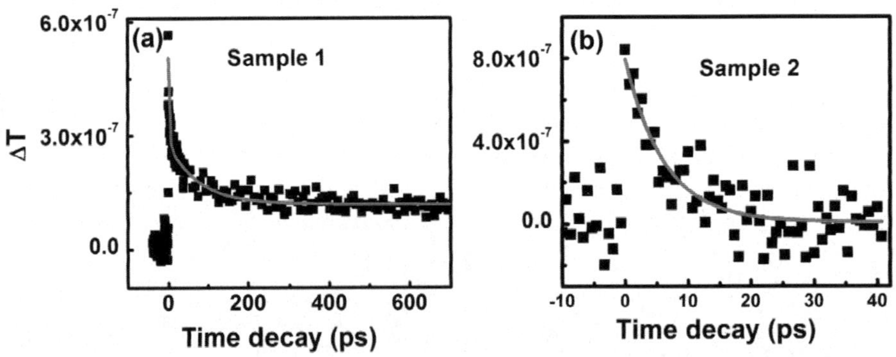

Figure 3. Dynamics of transient transmission of (a) sample 1 and (b) sample 2 annealed at 900 °C under pump power at 1 mW.

A two exponential decay function($y(t) = A + B_1 e^{-t/\tau_1} + B_2 e^{-t/\tau_2}$) was used to fit the decay time of sample 1, the obtained lifetime τ_1 and τ_2 are roughly about 2.9 and 78.2 ps, respectively. The fast component is attributed to the Auger recombination process, while the slow one may arise from the recombination process of electron-hole localized at interfaces states. Due to the limitation of pump–probe decay line, the long-lived lifetime can't be completely measured. However, the transient carrier dynamic of sample 2 can be well described by the single-exponential function ($y(t) = A + Be^{-t/\tau}$) with decay time of $\tau = 6.2$ ps, which may be ascribed to the process of strong Auger process and /or photo-excited carries trapped by defect states generated via high P doping. In this case, the long-lived lifetime is not obvious because of the generated carriers trapped rapidly by the nonradiative defect states.

4. Conclusion

The un-doped and P-doped Si/SiO_2 multilayers with dot size of 4 nm were fabricated in PECVD system, the optical nonlinearities were investigated by the open aperture Z-scan technique, the NLA coefficients β increased through P-doping, which is ascribed to the new crystal defects states induced by high P doping. Meanwhile, the ultrafast optical response of Si/SiO_2 multilayers was explored by using pump and probe technique. It is shown that the carrier trapping process due to the nonradiative defect states influences the ultrafast optical response obviously.

Acknowledgment

This work is supported by "973 Project (2013CB632101), "333 Project" of Jiangsu Provence (BRA2015284) and the Doctoral Foundation of Zhengzhou University of Light Industry (2014BSJJ041).

References

[1] Priolo F, Gregorkiewicz T, Galli M and Krauss T F 2014 *Nat. Nanotechnology* **9** 19–32.
[2] Leuthold J, Koos C and Freude W 2010 *Nat. Photonics* **4** 535–44.
[3] Zhang P, Zhang X, Lu P, Xu J, Xu X, Li W and Chen K 2014 *Appl. Surf. Sci.* **292** 262–6.
[4] Zhang P, Zhang X, Xu J, Mu W, Xu J, Li W and Chen K 2014 *Nanoscale Res. Lett.* **9** 28.
[5] Zhang P, Zhang X , Xu S, Lu P, Tan D, Xu J, Wang F, Jiang L, and Chen K 2017 *Opt. Mater. Express* **7** 304–12.
[6] ImakitaK, Ito M, Fujii M and Hayashi S 2009 *Opt. Express* **17** 7368–76.
[7] Lu P, Mu W, Xu J, Zhang X, Zhang W, Li W, Xu L and K. Chen 2016 *Sci. Rep.* **6** 22888.

40nm Quasi-Continuous Wavelength Tuning of V-cavity Semiconductor Laser

Tize Li, Jianjun Meng and Jian-Jun He

Centre for Integrated Optoelectronics, State Key Laboratory of Modern Optical Instrumentation, Zhejiang University, Hangzhou, China 310027.

jjhe@zju.edu.cn

Abstract. Full band quasi-continuous wavelength tuning with ~36dB side mode suppression ratio of V-cavity semiconductor laser is reported. A simple and general tuning algorithm employing only two electrodes to control the frequency combs shifting synchronously is presented and demonstrated, respectively.

1. Introduction

Tunable lasers with large quasi-continuous tuning is important for biomedical and spectroscopy applications. So as to achieve good performance, many structures have been presented, such as sampled grating distributed Bragg reflector (SGDBR) laser, modulated grating Y-branch (MG-Y) laser, digital super-mode (DS) DBR laser and distributed feedback (DFB) laser array. However, all these grating-based structure require multiple epitaxial growth and complex tuning algorithm. Recently, a compact V-cavity tunable semiconductor laser was proposed [1]. Single-electrode controlled wavelength tuning of 16 and 26 consecutive channels with 100GHz spacing was demonstrated experimentally with an excellent side-mode-suppression-ratio (SMSR) of about 40 dB and 37dB, respectively [2]. Temperature induced gain spectrum shift is employed in combination with the Vernier tuning mechanism to extend the digitally wavelength tuning range to cover the full C-band [3]. The V-cavity tunable laser is an all-active device with no grating or ring resonators, and therefore it does not require any epitaxial regrowth. The fabrication process is similar to simple Fabry-Perot lasers and the device length is less than 0.5 mm.

In this paper, we proposed an easy current control algorithm which only need two electrodes to achieve quasi-continuous tuning in V-cavity tunable laser. At fixed temperature, 16nm quasi-continuous tuning range is obtained. By 49°C temperature variation, the quasi-continuous tuning range can be extend to 40nm with ~36 dB SMSR, covering the full C-band.

2. Tuning algorithm

The V-cavity laser was designed and fabricated in InGaAsP/InP multiple quantum well structure. It comprises two F-P cavities with different optical path lengths, which form V-shaped branches with a reflective 2 ×2 half-wave coupler, as shown in Figure.1. The length of the short cavity (fixed gain cavity) is designed to be 466 μm to match its resonant wavelengths to the ITU grid of 100GHz spacing. The long cavity (channel selector cavity) is 5% longer so that the Vernier effect can be used to extend the tuning range. Three deep etched facets are used as the cavity mirrors [2]. The half-wave coupler consists of a non-imaging multimode interference (MMI) structure with an optimal coupling

IOP Conf. Series: Journal of Physics: Conf. Series **844** (2017) 012013 doi:10.1088/1742-6596/844/1/012013

coefficient to ensure high side-mode-suppression-ratio (SMSR). Three electrodes separated by isolation gaps are fabricated on the top side and common electrode is deposited on back side. The digital wavelength tuning is accomplished through thermos-optic effect, thereby tuning the lasing wavelength to one of the resonant wavelengths of the short cavity by aligning it to one of the resonant wavelengths of the long cavity [3].

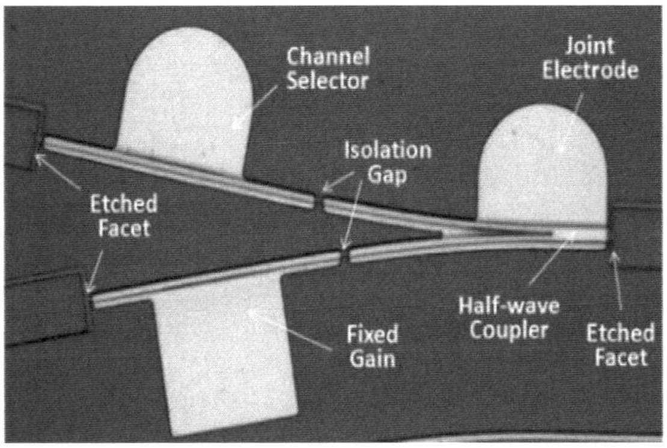

Figure 1. Optical microscope image of V-cavity laser.

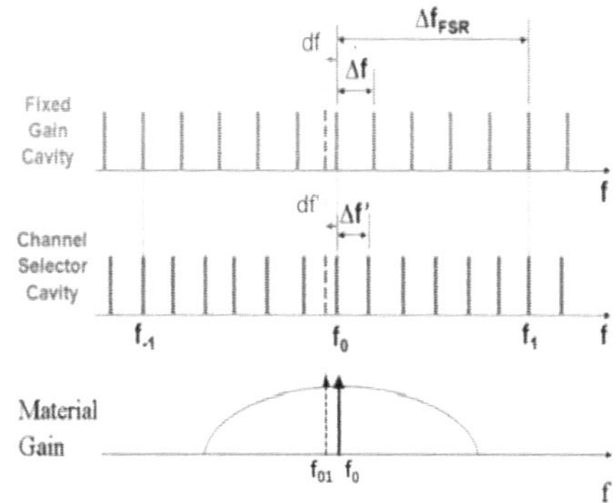

Figure 2. Schematic diagram of the synchronous frequency combs shift.

As shown in Figure.2, in order to achieve quasi-continuous tuning in V-cavity laser, the resonant frequency combs of the two cavities should be tuned synchronously with the same tuning speed (df=df') to avoid mode hop [4]. We only need to tune the wavelength combs of two cavities synchronously by one channel spacing (i.e.100 GHz) before jumping to the setting currents for the next channel where the next resonant peak of long cavity coincides with the next resonant peak of short cavity. Splicing multiple fine tuning sections of 100GHz spacing, we are able to obtain a large quasi-continuous tuning range.

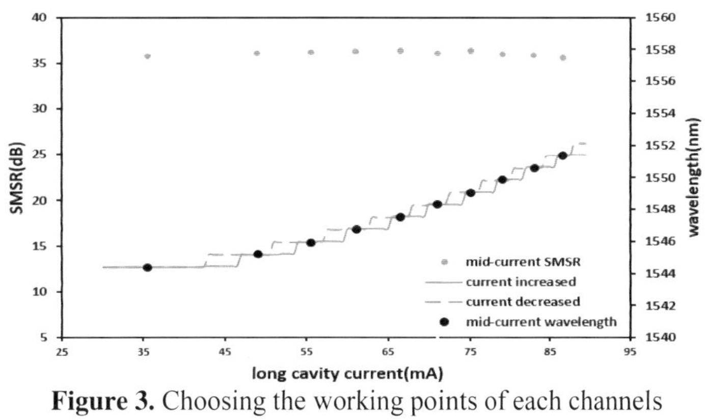

Figure 3. Choosing the working points of each channels

Figure.3 show the wavelength as a function of increased current and decreased current in long cavity respectively, with the short cavity current fixed. Digital tuning with 100GHz spacing is visible. The effects of hysteresis are also obviously visible. To avoid mode hop and ensure high SMSR during the quasi-continuous tuning, we choose the mid-currents of the overlapping regions of current increased and current decreased single electrode tuning channels as the setting currents. These currents are supposed to obtain the highest SMSR in each of channels. In this example, all mid-currents implemented ~35 dB SMSR. Changing the biased current of short cavity, we can obtain more setting currents with great SMSR by above-mentioned method.

By using the setting currents, we can construct the synchronous saw-tooth current functions for two tuning electrodes. Quasi-continuous tuning with good SMSR is able to implement by the current function. Current linear interpolation can be used to obtain smaller wavelength interval in quasi-continuous tuning.

The two electrodes tuning algorithm is also suitable for other V-cavity lasers theoretically, such as V-cavity laser with quantum well intermixed [5] and V-cavity laser with integrated thin-film heaters [6] .These V-cavity lasers are supposed to gain great promotion in power flatness and tuning speed.

3. Experimental results and discussions

We use the V-cavity laser in a XMD package with a thermal-electric cooler (TEC) for temperature control. The temperature is fixed at 35°C, the coupler electrode is biased at 15 mA. Two synchronous saw-tooth waves are injected to short cavity and long cavity respectively. As shown in Figure.4, short cavity current varies from 35 mA to about 82 mA for each fine tuning sections, long cavity current increases about 50 mA for each fine tuning sections and the lowest current of fine tuning section slightly increases about 5 mA. In a word, tuning two currents synchronously for 40~50 mA will induce a continuous tuning with 100GHz.

Since the temperature coefficient of the gain spectrum shift to longer wavelength is about 0.5nm/K and the thermal-optic coefficient of the laser is about 0.1nm/K [3], the gain spectrum shift is 5 times faster than frequency combs shift when current increase, so that the mode at the short wavelength end of gain spectrum would hop to the next free spectral range (FSR) during the current fine increasing process. In order to achieve the continuous tuning for a free spectral range (~16nm), we set the short cavity current to vary from about 80 mA to about 120mA. The gain window would be aligned to next FSR. Long cavity injected current still follows the above-mentioned rule. Tuning range would be extended to cover 16nm with ~36 dB SMSR. The overlapped emission spectra of 16nm tuning with 0.06nm spacing is shown in Figure.5. The wavelength tuning range is limited by current range. Because of performance degradation at high current injection, we set the current less than 150 mA for each electrodes. According to our experiment, 16 nm is almost the largest range at a fixed temperature.

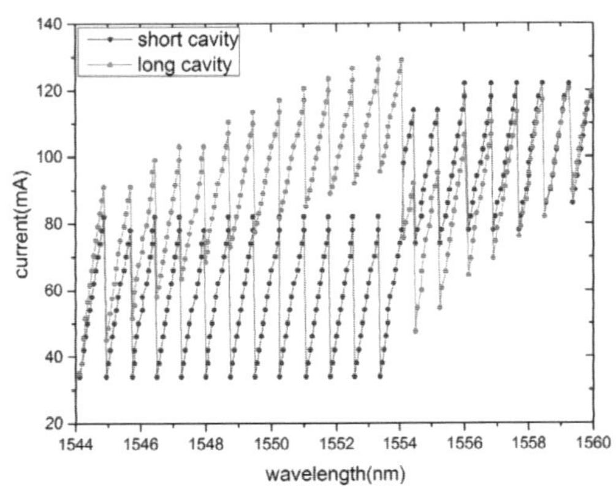

Figure 4. Measured quasi-continuous tuning curve.

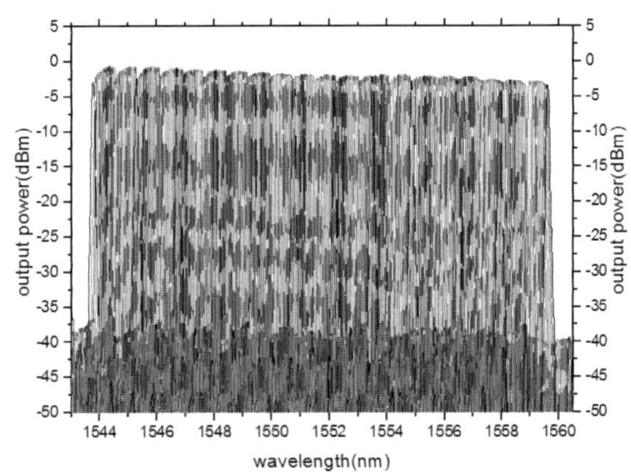

Figure 5. Overlapped laser spectra.

Because of 0.5 nm/K gain spectrum shift, the temperature varies 30°C can extend tuning range for about 15nm. We set three temperature steps at 16°C, 35°C and 65°C, and obtained quasi-continuous tuning range of 10.1nm, 15.8nm, 14.4nm, respectively, for a total of 40.3nm. The currents of two cavities both vary from 20 mA to 150 mA. Three wavelength tuning ranges are shown in Table.1. The laser which working at higher temperature outside the range from 16°C to 65°C cannot realize good performance, nor does it work at lower temperature. The wavelength and SMSR performance of full

C-band wavelength quasi-continuous tuning are shown in Figure.6 and Figure.7, respectively. The SMSR over full band mostly above 33 dB and can be as high as 39 dB.

Table 1. Wavelength quasi-continuous tuning range and SMSR of three temperature steps

Temperature(°C)	wavelength(nm)	SMSR(dB)
16	1533.997-1544.126	35.3-39.0
35	1544.126-1559.946	33.2-37.7
65	1559.946-1574.332	33.6-37.3

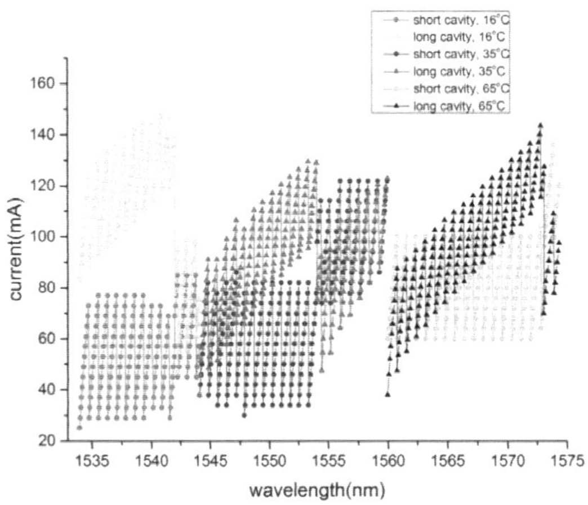

Figure 6. Quasi-continuous tuning curve for full band.

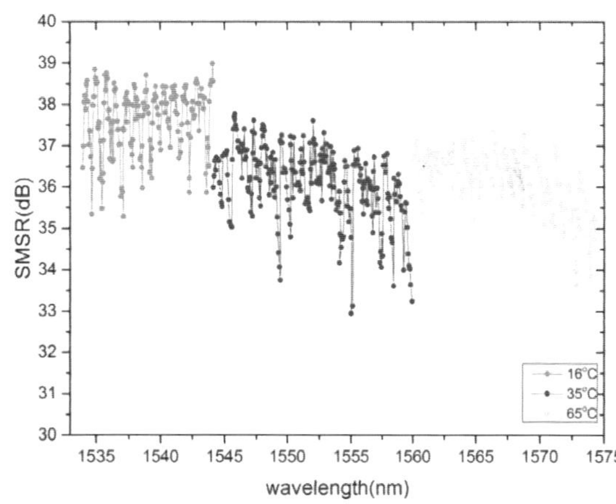

Figure 7. SMSR of full band quasi-continuous tuning.

4. Conclusion

In conclusion, we have realized the quasi-continuous tuning of the all-active V-cavity tunable laser. 16nm tuning range is achieved at a fixed temperature, 40nm tuning range is achieved at three temperature steps, with good single mode performance respectively. Simple tuning algorithm with two electrodes control allows the V-cavity semiconductor laser to be potentially used for dense wavelength division multiplexing (DWDM) networks, biomedical sensing and spectroscopy applications.

5. Acknowledgment

This work is supported by the National High-Tech R&D Program of China (Grant no. 2013AA014401), and the National Natural Science Foundation of China (grant No. 61535010 and 61377038).

References

[1] He J and Liu D 2008 *Opt. Express* **16** 3896-911.
[2] Jin J, Wang L, Yu T, Wang Y and He J 2011 *Opt. Letters* **36** 4230-32.
[3] Zhang S, Meng J, Guo S, Wang L and He J 2013 *Opt. Express* **21** 13564-71.
[4] Wu L, He J, and Gallagher D 2015 *J. Opt. Soc. Am. B* **32** 309-17.
[5] Zhang X, He J, Liu N and J.Dubowski J 2015 *Opt. Express* **23** 26336-41.
[6] Deng H, Meng J, Xiong X and He J 2016 *IEEE Photonics J* **8** 152908.

IOP Conf. Series: Journal of Physics: Conf. Series **844** (2017) 012014 doi:10.1088/1742-6596/844/1/012014

Research on influence of parasitic resistance of InGaAs solar cells under continuous wave laser irradiation

Guangji Li[1], Hongchao Zhang[1], Guanglong Zhou[1], Jian Lu[1*] and Dayong Zhou[2]

[1] School of Science, Nanjing University of Science and Technology, Nanjing Jiangsu 210094, China

[2] Shanghai Institute of Space Power-sources, Shanghai 200245, China

Corresponding author's e-mail address: lujian@njust.edu.cn (Jian Lu)

Abstract. InGaAs solar cells were irradiated by 1060-1080nm continuous wave (CW) laser, and studied the laser-electrical conversion and damage experiment with the power density as 97mW/cm^2 and 507W/cm^2 respectively. The result indicated that there is no obvious damage phenomenon but air layer appeared in the damaged region, and there is no direct relationship between the area and the extent of damage. Moreover, the p-n junction in the damage zone was destroyed, lost the ability of photoelectric conversion. The region acts as a resistance between the two electrodes, resulting in an increase in the leakage current of the solar cells and a decrease in the parallel resistance, which is the main reason leading to the decline of open circuit voltage, short circuit current and conversion efficiency. This paper would provide a reference for wireless energy transmission and the subsequent laser damage of solar cells.

1. Introduction

Laser wireless energy transmission has been a hot topic, as early as in 1991, Landis et al.[1] proposed a hypothesis which based on ground laser emission system and supply energy for space vehicles. In order to reduce the influence of atmospheric turbulence and enhance the beam quality, the laser emission system was built in the cloud-free area, such as the top of the mountain, explained the feasibility theoretically. In 1997, Yugami et al.[2] carried out laser energy transmission experiments with different laser systems at ground level 500m, and tested the long time energy transfer efficiency in different weather conditions. At present, real-time charge for UAV from laser in an effective range has been realized[3].

For the receiving end of the energy, the research and design of the solar cells have also attracted much attention of scholars, especially on the photoelectric conversion efficiency of specific wavelength. Up to now, the reported laser-electrical conversion efficiency of GaAs solar cell with wavelength of 808nm has reached 53.23%[4]. More research in the field domestic have been on laser damage to solar cells. Qiu et al.[5-6] studied the damage of silicon solar cells using continuous and pulse laser respectively. They analyzed the reason of damage morphology and compared the difference between the two damage morphology. Xue et al.[7] studied the relationship between the laser energy and the damage depth of GaAs/Ge solar cells by femtosecond laser. Currently, the research on laser induced damage is limited to the damage morphology and photoelectric conversion efficiency, and few on other parameters of solar cells.

In this paper, the damage experiment of single junction InGaAs solar cell is carried out by using CW laser with wavelength of 1060-1080nm. The results suggest that in addition to the decrease of open circuit voltage, short circuit current, photoelectric conversion efficiency of the solar cells, the parallel resistance is also significantly reduced. Highlights in this research: 1. Damage solar cells in small area by CW, and qualitatively explain the relationship between damage degree and damage area; 2. Measure the I-V characteristic curves of solar cells under laser irradiation, and study the effects of damage on the conversion efficiency of solar cells; 3. The laser with a wavelength of 1060-1080nm is in the atmospheric window, which provides a reference for the study of the wireless energy transmission in the atmosphere and the further study on damage to the solar cells; 4. Extract the electrical performance parameters of the solar cells according to I-V curve, analyze the reason causing the decrease of electrical properties of solar cells in the perspective of the parasitic resistance by comparing the open circuit voltage, short circuit current and photoelectric conversion efficiency.

2. Experiment

The solar cell sample is provided by the Shanghai Institute of Space Power-sources, which is 2×2cm^2 indium gallium arsenide (InGaAs) solar cells. The physical structure is as the following figure 1. The grid electrode and the back electrode are Ag and Al respectively, and the width of the band gap in solar cell is E_g=1.0eV.

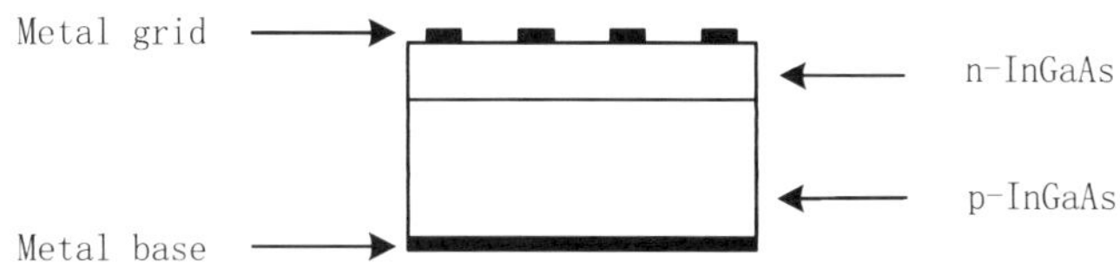

Figure 1. Physical structure of solar cell.

Laser-electrical conversion experiments are as follows (figure 2). The experiment system include continuous wave fiber laser, beam expander, source meter and temperature controller. The wavelength of the laser is 1060-1080nm, and its power is continuous controllable under 200W. The beam expander device is composed of convex lens L1 and L2, and the diameter of lens L1 is 25mm, the focal length is 40mm, the diameter of lens L2 is 50mm, the focal length is 150mm. It can be achieved that the continuous adjustment of the spot size on the surface of the solar cell through adjusting the distance between the lens L1 and L2. The positive and negative electrodes of solar cell are connected to the source meter (Keithley 2450), which is used to measure the I-V characteristic curve of the solar cell. The thermostat, cling to the back surface of the sample, controls the temperature at 25℃. Measure the laser power received by the solar cell with the German STPS-LEM2420 power meter.

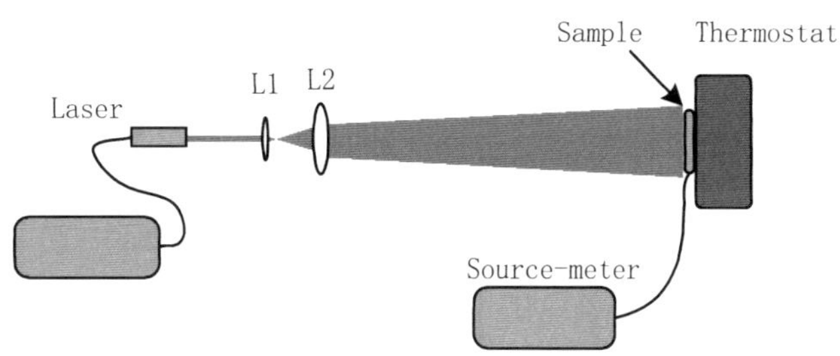

Figure 2. Schematic of measuring the I-V curve of InGaAs solar cells with 97mW/cm^2 laser irradiation.

During the experiment, the laser beam covers the whole solar cell. Adjust the laser parameters, making the power density of the sample 97mW/cm^2, and record the I-V curve of solar cell.

Damage experiment is shown in figure 3, and diameter of lens L3 is 50mm, focal length is 150mm. Adjust the position of lens L3 in order that the diameter of the spot on the sample surface is 2mm, and the power density is 507W/cm², irradiated lasting 17s. Survey the I-V curve of sample again using figure 2 after irradiation.

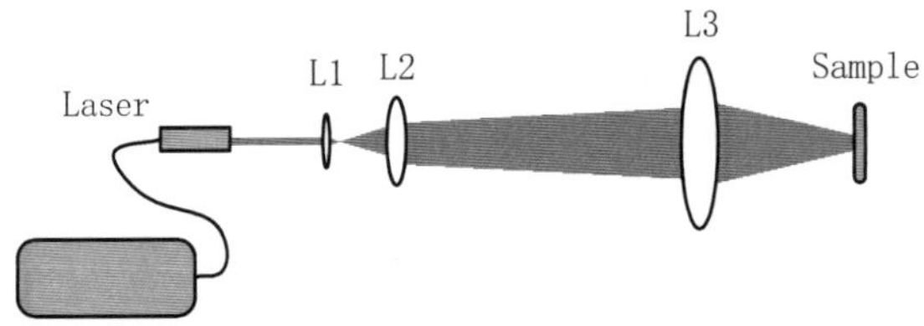

Figure 3. Schematic of InGaAs solar cells irradiated by laser with the power density of 507W/cm².

3. Result

The sample damaged by laser is in the following figure 4. The damage area is consistent with the light spot size, which is circular and slightly lighter than around area. Besides, there is air layer appeared inside the damage area. In this region, some of the metal wire is thickened due to melting then solidification, thus there is no distinct damage phenomenon in this area. The area laser irradiated occupies 0.79% of the total area. The I-V curve of the sample before and after laser irradiation is as follows (figure 5).

Figure 4. Morphology of solar cell after laser with 507W/cm² power density irradiation and lasting 17s.

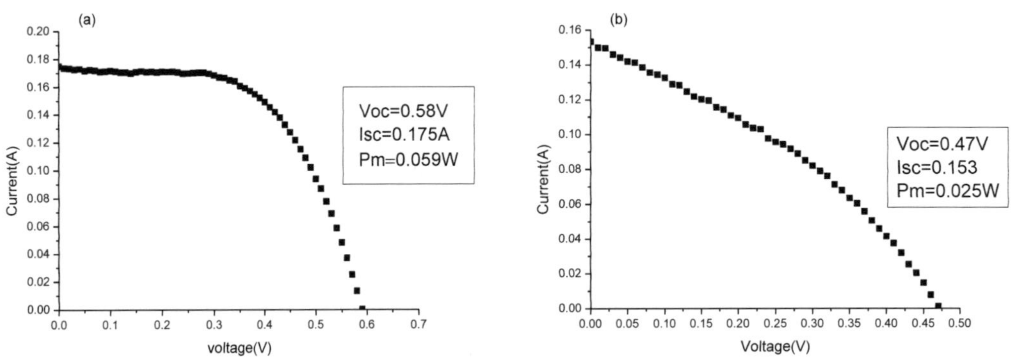

Figure 5. I-V curve of solar cells irradiated by laser with the density of 97W/cm². (a) before; (b) after.

Figure 5 shows the I-V curves of solar cell recorded by the source meter. Figure (a) shows I-V curve before damage, from which the current flowing through the solar cell remain unchanged and then decreases rapidly along with the increase of the load voltage, and the open voltage is V_{oc}=0.58V, short circuit current is I_{sc}=0.175A, maximum output power is P_m=0.059W, thus it can be derivated that laser-electrical conversion efficiency of solar cells is η=15.32%. The figure (b) illustrates the I-V curve of the solar cell after laser irradiation. The current through the solar cell declines almost linear with the increase of the external voltage. It can be extracted that the open circuit voltage V_{oc} =0.47V, short circuit current I_{sc} =0.153A, and maximum output power P_m =0.025W from the curve, thus, the laser-electrical conversion efficiency of solar cells after laser irradiation is η=6.42%. Comparing figure (a) and (b), we know that the three parameters above exhibit decrease to different extent after damage, reduced to 81.0%, 87.4% and 41.9% respectively. Obviously, the output power is affected mostly by the damage from CW laser. From I-V curves in figure 5 we compute the other parameters[8] of the solar cells before and after the damaged shown in table 1.

Table 1. Comparison of parameters before and after high power laser irradiation (the voltage V_m, current I_m, series R_{ms} and parallel R_{msh} resistance at the maximum power point, laser-electrical conversion efficiency η).

parameter	V_m(V)	I_m(A)	R_{ms}(Ω)	R_{msh}(Ω)	η(%)
Before	0.41	0.145	0.53	64.41	15.32
After	0.28	0.089	0.51	4.30	6.42

Following in table 1, the parameters of solar cell show decrease to different extent. Where, the voltage and current at maximum power point decreased to 68.3% and 61.4%; series resistance and parallel resistance decreased to 96.2% and 6.7%, respectively; and the laser-electrical conversion efficiency decreased to 41.9%. It is clear that the decrease of parallel resistance is the most, while the series resistance is the least.

The single-diode model of solar cells is expressed by

$$I = I_{ph} - I_0 \left\{ \exp\left[\frac{q(V+IR_s)}{nKT} \right] - 1 \right\} - \frac{V+IR_s}{R_{sh}} \tag{1}$$

Where, I and V are the current and voltage, R_s and R_{sh} are the series and parallel resistance, I_{ph} is the photocurrent, I_0 is the reverse saturated current, q is the elementary charge, n, K and T are the ideality factor, Boltzmann constant and Kelvin temperature, respectively. From equation (1), we can see that the electrical properties of a solar cell is determined by series and parallel resistance under the condition of constant temperature and irradiation.

It can be seen from the above analysis, the decrease of parallel resistance is the main reason for the decline of electrical performance of the solar cell damaged by laser.

4. Conclusion

There can be air layer inside the cell, but no obvious damage phenomenon after CW laser with high power irradiation on the surface of solar cells. The damage area is 0.79% of the total cell surface area while the laser-electrical conversion efficiency decreased by 58.6%. From which we can know that the p-n junction is destroyed in this region and there is no direct relationship between the decrease of solar cells' performance and the area of damage region.

The p-n junction of the damaged area is destroyed, then the cylindrical region between the two electrodes acts as resistance, which is smaller, thus leads to the increase of the leakage current and decrease of the parallel resistance of the cell. Therefore, laser induced damage of solar cells directly leads to the decrease of parallel resistance. While the change of parallel resistance has an immediate influence on the output of electrical properties of solar cell.

The series resistance decreased by 3.8%, so CW laser with high power has little influence on series resistance of solar cell.

This research can provide reference for the research of the interaction between InGaAs solar cell and laser, and the study of laser wireless energy transmission.

Acknowledgment
This work is supported by the "Jiangsu Natural Science Foundation (No.BK20130751)".

References

[1] G A Landis 1991 Space Power by Ground-Based Laser Illumination *IEEE Aerospace and Electronics Systems Magazine*. **6** 3-7.

[2] H Yugami, Y Kanamori, H Arashi, M Niino, A Moro, K Eguchi, Y Okada, A Endo and Engineers Amer Inst Chem 1997 Field Experiment of Laser Energy Transmission and Laser to Electric Conversion *32nd Intersociety Energy Conversion Engineering Conference (IECEC-97)* (New York, American: IEEE 1997) 625-30.

[3] Shane McGlaun (Blog) 2012 Lockheed Martin Stalker Uav Powered by Laser Light for 48 Hours *DAILYTECH* (2012) <http://www.dailytech.com/article.aspx?newsid–25156> Accessed July 12.

[4] T Q Shan and X L Qi 2015 Design and Optimization of Gaas Photovoltaic Converter for Laser Power Beaming *Infrared Physics & Technology*. **71** 144-50.

[5] Dong-dong Qiu, Rui Wang, Xiang-ai Cheng and Yong-jiang Sun 2011 Wave Band Effect of Solar Cells under Irradiation of Cw Laser *Laser Technology*. **35** 632-635, 683.

[6] Dongdong Qiu, Rui Wang, Xiang'ai Cheng, Zhen Zhang and Tian Jiang 2011 Mechanisms Research on Continous Wave Laser Induced Damage to Solar Cells *Chinese Journal of Lasers*. **38** 03020061-4.

[7] Qing Xue, Wenhui Wu, Yunxia Ye, Haixia Liu, Ruifang Chen and Yinqun Hua 2015 Property Degradation of Gaas/Ge Solar Cells after Femtosecond Laser Irradiation *Laser & Optoelectronics Progress*. **52** 0414051-7.

[8] Priyanka Mohan Lal and S N Singh 2007 A New Method of Determination of Series and Shunt Resistances of Silicon Solar Cells *Solar Energy Materials and Solar Cells*. **91** 137-42.

IOP Conf. Series: Journal of Physics: Conf. Series **844** (2017) 012015 doi:10.1088/1742-6596/844/1/012015

The impact of purging on the quality of AlGaN/GaN multiple quantum wells grown on AlN/sapphire template

Shuchang Wang[1,2], Xiong Zhang[2], Zili Wu[2], Jianguo Zhao[2], Qian Dai[2] and Yiping Cui[2]

[1]*College of Physics and Electronic Engineering, Changshu Institute of Technology, Changshu 215500, China*
[2]*Advanced Photonics Center, School of Electronic Science and Engineering, Southeast University, Nanjing 210096, China*

E-mail: scw@cslg.edu.cn

Abstract. The ultraviolet (UV) emitting AlGaN/GaN multiple quantum wells (MQWs) were grown on low dislocation density AlN/sapphire templates by metal-organic chemical vapor deposition (MOCVD). The impact of the purg time on the interface quality of the AlGaN/GaN quantum well was studied. The high resolution x-ray diffraction (XRD) measurement results demonstrate that the density of dislocations was reduced significantly with the purge time after growth of AlGaN barrier layer and GaN well layer was determined to be 4 min and 2 min, respectively. The mechanism of defect formation in quantum wells was investigated by scanning electron microscope (SEM) measurement.

1. Introduction
In recent years, researches toward ultraviolet (UV) emission range (240-360 nm) using III-nitride materials are gaining momentum for fabricating compact, robust and high-efficiency UV light-emitting diodes (LEDs). The UV-LEDs are targeted for a wide range of potential applications, such as water purification, sterilization, biological detection, medical treatment, and high density optical recording [1]. As is well known, InGaN-based MQW structures show strong blue or near-UV light emissions despite the existence of a large number of threading dislocations (TDs) over 10^9 cm^{-2}. AlGaN-based emitters do not have such a useful effect, different from the InGaN-based LEDs [2]. Recently, UV-LEDs using an AlN substrate have been reported [3], but the diameter and cost of AlN substrates make it still unsuitable for LED application. However, recently epitaxially laterally overgrown (ELO) AlN/sapphire has proven to have a low enough defect density to be suitable for the growth of efficient UV-LEDs [4]. In this paper, we have investigated the growth of AlGaN/GaN MQWs on low dislocation density AlN/sapphire templates by metal-organic chemical vapor deposition (MOCVD) and the impact of the purge time on the interface quality of the AlGaN/GaN MQWs was studied.

2. Experiments
The AlGaN/GaN MQWs were grown on AlN/sapphire template in a low pressure MOCVD system. Trimethyl-aluminum (TMA), trimethyl-gallium (TMG), and ammonia (NH_3) were used as the precursors for Al, Ga, and N, respectively. Silane (SiH_4) was used as the n-type doping source. Prior to

the growth, the sapphire substrates were heated to 1100 °C in H_2 ambience to remove surface contamination. A 20 nm-thick low-temperature (LT)-grown AlN nucleation layer was firstly deposited on the sapphire substrate at 600 °C. The temperature was then raised to 1040 °C to grow the high-temperature (HT) AlN interlayer. This process was followed by the growth of a 400 nm-thick un-doped $Al_xGa_{1-x}N$ epitaxial layer at a growth rate of 40 nm/min. Then a 1.5 μm-thick Si-doped $Al_{0.28}Ga_{0.72}N$ epitaxial film was grown on the un-doped AlGaN epitaxial layer at a growth temperature of 1140 °C and a growth chamber pressure of 40 Torr, following which a five-period AlGaN/GaN MQWs were grown. The AlGaN quantum barriers (QBs) had a fixed nominal AlN molar fraction of 0.28. For the growth of MQW structures, the growth temperature was maintained at 1150 °C, and the other MOCVD parameters were also kept at constant conditions, except for the H_2 purging time, which in the three samples (A, B, and C) varied from 1 min to 4 min, the details as shown in Table 1.

Table 1. The purge process parameters and the XRD results of the MQW samples.

	H_2 purge time / min		FWHM / arcsec	TDs / $\times 10^8$ cm^{-2}
	AlGaN barrier	GaN well		
Sample A	1	1	1198	21.5
Sample B	2	2	720	16.7
Sample C	4	2	452	7.23

Optical interference microscope was used to probe the surface morphology, and high resolution x-ray diffraction (XRD) was used to characterize the crystalline quality of the AlGaN/GaN MQWs. The microscopic structural properties of the MQWs were investigated by cross sectional scanning electron microscopy (SEM). Photoluminescence (PL) measurements were carried out at a room temperature (RT) by using a 224 nm wavelength laser as an excitation light source.

3. Results and discussion

The optical interference microscopy pictures taken for sample A, B, and C are shown in Fig. 1. The surface of sample C, with H_2 purge time after growth of AlGaN barrier layer and GaN well layer 4 min and 2 min, is more smooth than sample A and B. On the other hand, there seems to be more point defects but no remarkable cracks on the surface of sample A as compared to sample B and C. From these optical interference microscopy images, it was concluded that the crystalline quality of the AlGaN/GaN MQWs are strongly dependent on the purge time.

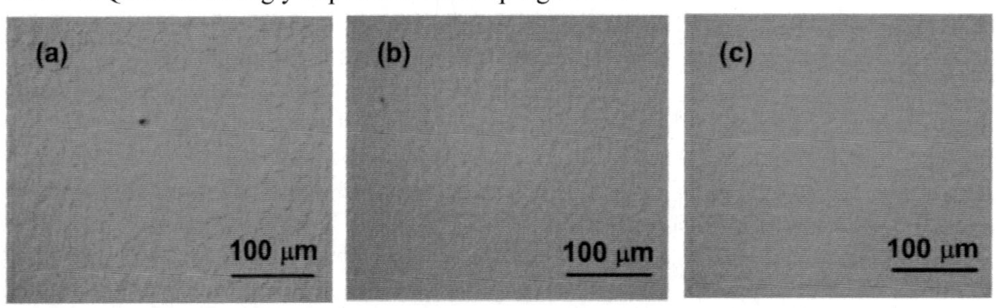

Figure 1. The optical interference microscopy pictures taken for AlGaN/GaN MQW: (a)sample A, (b) sample B, and (c) sample C.

It is evident that the size of the point defects on the surface of the samples decreased with increasing the purge time from 1 min for sample A to 4 min for sample C. The minimum size for the hexagonal pyramids was achieved with sample C which demonstrated the smoothest surface morphology.

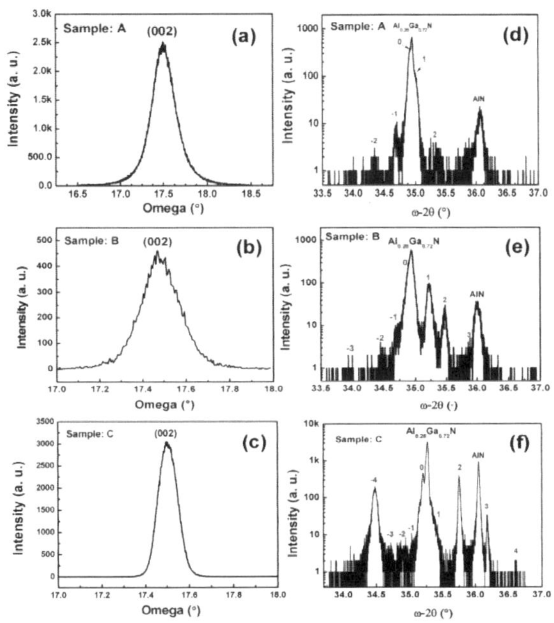

Figure 2. XRD scan omega rocking curves and ω-2θ curves of the AlGaN/GaN MQWs grown on AlN/sapphire template.

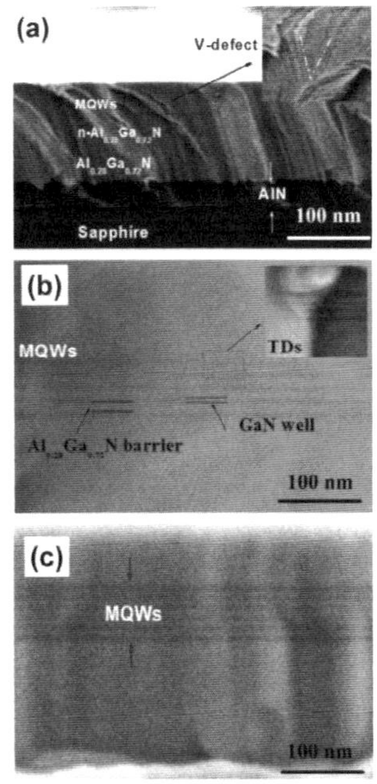

Figure 3. Cross-section SEM images for AlGaN/GaN MQW: (a)sample A, (b) sample B, and (c) sample C, taken along the [0001] c-axis. The enlarged SEM image of the MQWs marked by the dotted box in (a) and (b).

Figure 2 shows the measured high resolution XRD curves for the (002) crystal plane of the AlGaN/GaN MQW samples. The XRD scans show multiple satellite peaks, which indicate the formation of superlattice structures, especially in sample C. From Fig. 2(f) the XRD angles of the 0th, +1 and +2 satellite peaks, the thickness of barrier and well were estimated, (with AlGaN barrier thickness of 15 nm and GaN well thickness of 10 nm). The XRD full width at half maximum (FWHM) was measured to be 1198, 720, and 452 arcsec for samples A, B, and C, respectively. This agrees well with the optical microscope measurement result. Table 1 shows FWHM of the XRD of AlGaN/GaN MQWs collected from each sample. The peaks around 35.07° are attributed to the Si-doped n-AlGaN epitaxial films, while the peaks located at 36.26° are originated from the AlN interlayers. Using the same analysis process described in Ref. [5], the density of threading dislocations (TDs) in the samples are summarized in Table 1.

Figure 3 shows the cross-section SEM images of sample A, B, and C, respectively. The TDs, which originate from the island growth and coalescing at the early stage of film growth, can be clearly seen as thin lines aligning in parallel with [0001] c-axis orientation and propagate from below the AlGaN template layer towards the AlGaN/GaN MQWs, as indicated by the arrow in Fig. 3(a). It was noted that other types of defects such as stacking faults were formed in the initial growth stage of GaN well layers. The interfaces of the AlGaN/GaN MQWs are clear, but the width of GaN well has little fluctuation, as shown in Fig. 3(b). The results obtained from cross-section SEM is consistent with that obtained from optical interference microscopy.

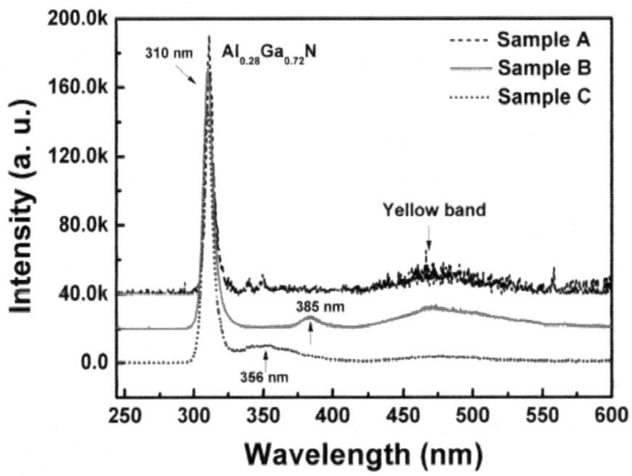

Figure 4. Room-temperature PL spectra of the AlGaN/GaN MQW samples.

Figure 4 shows the normalized RT PL spectra for sample A, B, and C. The main emission peak around 3.96 eV in the PL spectra, corresponding to a wavelength of 310 nm, is attributed to the recombination of localized excitons, as observed in the GaN-based semiconductor alloys. It was found that the PL peak position slightly blue-shifted from 470 to 356 nm from sample A to sample C. The reason is that as the purge time is increased, the defect density decreased. As a result that the rate of the nonradiative recombination is reduced obviously, which led to the blue shift of the band edge emission.

4. Conclusions

In this study, AlGaN/GaN MQWs have been successfully grown on low dislocation density AlN/sapphire templates by MOCVD. The impact of the purge time on the interface quality of the AlGaN/GaN MQW was studied. The optimized value of purge time after growth of AlGaN barrier layer and GaN well layer was determined to be 4 min and 2 min, respectively. The mechanism of defect formation in quantum wells was investigated. The results achieved in this study are helpful for fabricating UV-LEDs with enhanced device performance.

Acknowledgments

The work was supported by Foundation of Changshu Institute of Technology (Grant No. XZ1514) and the National Natural Science Foundation of China (Grant No. 61367004). This work at Southeast University was supported by the Key Research and Development Project of Science and Technology Department of Jiangsu Province, People's Republic of China (Grant No. BE2015159).

References

[1] S. Wang, X. Zhang, Z.C. Feng, Y. Cui, Surface chemical and local electronic properties of $Al_xGa_{1-x}N$ epi-layers grown by MOCVD, 2014 *Opt. Express* **22** 17440-17447

[2] J. Kwak, J. Lim, M. Park, S. Lee, K. Char, C. Lee, High-power genuine ultraviolet light-emitting diodes based on colloidal nanocrystal quantum dots, 2015 *Nano Lett.* **15** 3793–3799

[3] S. Hagedorn, A. Knauer, A. Mogilatenko, E. Richter, M. Weyers, AlN growth on nano-patterned sapphire: A route for cost efficient pseudo substrates for deep UV LEDs, 2016 *Phys. Status Solidi A* **213** 3178-3185

[4] U. Zeimer , V. Kueller , A. Knauer , A. Mogilatenko , M. Weyers , M. Kneissl, High quality AlGaN grown on ELO AlN/sapphire templates, 2013 *J. Cryst. Growth* **377** 32-36

[5] B.H. Kong, Q. Sun, J. Han, I.H. Lee, H.K. Cho, Classification of stacking faults and dislocations observed in nonpolar a-plane GaN epilayers using transmission electron microscopy, 2012 *Appl. Surf. Sci.* **258** 2522-2528

IOP Conf. Series: Journal of Physics: Conf. Series **844** (2017) 012016 doi:10.1088/1742-6596/844/1/012016

Valley polarization in silicene induced by circularly-polarized resonance light

Jie Cao*
College of Science, Hohai University, Nanjing 210098, China

Fenghua Qi
School of Electronic Engineering, Nanjing Xiaozhuang University, Nanjing 211171, China
(Dated: May 18, 2017)

In the presence of circularly-polarized resonance light, silicene develops dynamical band gaps in its quasi-energy band structure. Using numerical calculations, our results show that the gap appearing at $\hbar\omega/2$, where $\hbar\omega$ is the photon energy. More importantly, we find that these gaps are non-symmetric for two inequivalent valleys. Therefore we can introduce light-controlled valley polarization in these dynamical band gaps. Different valleytronic devices can be realized using this technique.

PACS numbers: 73.23.-b, 72.80.Vp

I. INTRODUCTION

Silicene is a promising material which consists of a honeycomb lattice of silicon atoms with buckled sublattices made of A and B sites as shown in Fig. 1(a). It has been synthesized recently and has attracted much attention. The states near the Fermi energy are π orbitals residing near the K and K' valleys at opposite corners of the hexagonal Brillouin zone which is shown in Fig.1(b). Because of the large separation in the momentum space of two valleys, inter-vally scattering is strongly suppressed when the sharp potential changes in distances of atomic order are absent.[1,2] This suggests a potential usage of the degree of freedom of valley index as a quantum identification of electron states, similar to the role of the spin degree of freedom in spintronics. For intrinsic silicene, these two valleys are degenerate, and the electronic quantities of silicene, such as the conductance, are usually twice the contributions from one valley due to the valley degeneracy. Valley polarization can only be entailed by applying external manipulations.[3-10]

A powerful method to modulate a band structure is to apply a coherent laser beam into the system. It has been proven to be very critical for topological phase transitions in the context of 2D[11,12] and 3D[13] systems. However, most of these proposals are focused on the change of the band gaps at the K and K' points. It requires a very strong off-resonant light which might actually damage the material. In this paper, we have focused on the on-resonant region where the direct absorbtion of photons are allowed as shown in Fig. 1(c) and (d). In this region, when we increase the strength of the light, the biggest modulation of the band structure first appears at the resonance circle instead of the K/K' points and the strength needed in on-resonance case is much less than in the off-resonance case.

In this paper, we use the Floquet theorem to calculate the band structure of the silicene manipulated by the on-resonance light. We find that on the resonance circle, dynamical gaps are opened for both K and K' valleys. However, due to the intrinsic spin-orbital coupling, the dynamical gaps are non-symmetric for two inequivalent valleys. Within the dynamical gaps, we can define valley polarization if the fermi level only cross one valley whether in the valence band or the conduction band.

II. EFFECTIVE MODEL UNDER THE APPLICATION OF RESONANT PERTURBATION

The irradiating field is represented as a time-dependent vector potential which will modify the momentum via Peierl's substitution $\hat{\mathbf{P}} \to \hat{\mathbf{P}} + e\mathbf{A}$, where $e > 0$ is the magnitude of the electron charge and the vector potential can be written as $\mathbf{A}(t) = [A_x(t), A_y(t)]$. The Hamiltonian of irradiated silicene is then given by

$$H(t) = V_F\left(\tau(p_x - eA_x)\sigma_x + (p_y - eA_y)\sigma_y\right) + \lambda\tau\sigma_z \quad (1)$$

instead of the static Hamiltonian given in reference [3,4,16], where $\lambda = 3.9meV$ is the spin-orbit splitting term, $V_F \approx 5\times 10^5 m/s$ is the Fermi velocity and p_x, p_y are the components of the momentum relative to the K/K' point. $\sigma_i, (i = x, y, z)$ are Pauli matrixes for the two sublattices. And $\tau = 1(-1)$ is the valley index represents valley K (K'). The Rashba SOC is omitted since it is a factor of 10 smaller than the intrinsic SOC[3].

* Author to whom correspondence should be addressed. Electronic mail: caojielunwen00@163.com

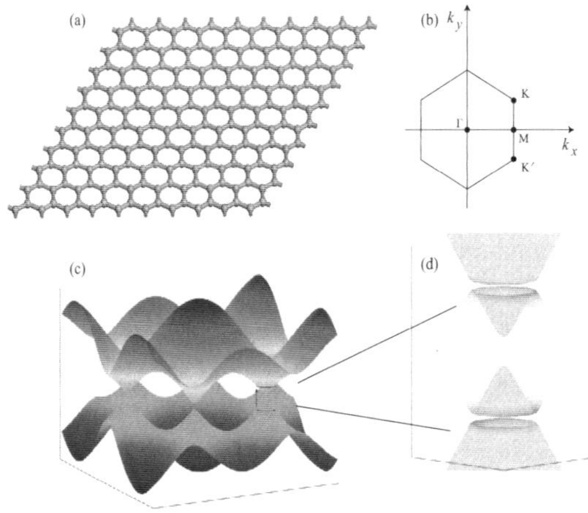

FIG. 1: (a) and (b) show the basic structure of silicene in the real space and the momentum space. (c) shows the band structure of intrinsic silicene. If we apply a on-resonance light, direct photon absorption is allowed. (d) shows that the modulation of the band structure takes place on the resonance circle. We can see that dynamical gaps are opened on the resonance circle.

The information about the effect of the radiation is encoded in the time-dependent vector potential, such as the polarization of the light, the field strength and the radiation frequency. In the case of linearly polarized light, the vector potential can be written as $\mathbf{A} = A\cos(\omega t)[\cos\theta, \sin\theta]$, where θ is the angle of polarization in the plane of silicene and ω is the frequency. The strength of the vector is given by the parameter A. Furthermore, for the circularly polarized light, a phase difference between A_x and A_y is introduced $\mathbf{A} = A[\cos\omega t, \pm\sin\omega t]$ where the plus (minus) sign refers to right(left)-handed polarization.

Here we are only interested in the circularly polarized light and focus on the resonance region, i.e, $\hbar\omega > 2\lambda$. In this region, photon absorption take place on the resonance circles $E_\pm(k_0) = \pm\hbar\omega/2$ where $E_\pm(k)$ is the eigenvalue of the intrinsic silicene without radiation. When the strength of radiation is increased, dynamic gaps first appear on this resonance circles. It is well known that the coupling between the circularly polarized light and silicene is strongly valley dependent, spin-valley physics [4,7] are developing rapidly. In our case, we find that the values of dynamical gaps are also valley dependent. In the following, we will calculate the values of these gaps numerically using Floquet theorem, and we find that the gaps for different valley are quite asymmetric.

III. NUMERICAL RESULTS USING FLOQUET THEOREM

We employ the Floquet theorem[17] to solve the time-dependent problem numerically. The full time-dependent wave function $\Psi(k, t)$ as the product of a periodic function $\Phi(k, t)$ (with period $T=2\pi/\omega$) and a time evolution factor

$$\Psi_k(x, t) = e^{-i\varepsilon_k t/\hbar}\Phi_k(x, t) \tag{2}$$

where $\Phi(t + T)=\Phi(t)$. By defining the Floquet operator $F_k(t)=H_k(t) - i\hbar\frac{\partial}{\partial t}$, the Schrödinger equation can be written as $F_k(t)\Psi_k(x, t)=0$. Substituting equation (2) into the Schrödinger equation yields $H_k(t)\Phi_k(x, t) - i\hbar\frac{\partial}{\partial t}\Phi_k(x, t)=\varepsilon_k\Phi_k(x, t)$. The periodic wavefunctions Φ are called 'Floquet states', and ε_k is called the quasienergy.

In order to solve the Schrödinger equation for ε_k and $\Phi_k(x, t)$, the expanded Hilbert space $\mathcal{R}\otimes\mathcal{T}$ of squareintegrable function of space and function of time with period T is needed. In this space, the inner product is defined as the usual spatial inner product with the average over one period:

$$\langle\langle n|m\rangle\rangle = \frac{1}{T}\int_0^T \langle n|m\rangle dt \tag{3}$$

where n and m represent states in $\mathcal{R}\otimes\mathcal{T}$ [17]. Hence the wavefunction Φ satisfy that

$$\langle\langle\Phi_{\alpha k}(t)|\Phi_{\beta k}(t)\rangle\rangle = \delta_{\alpha\beta} \tag{4}$$

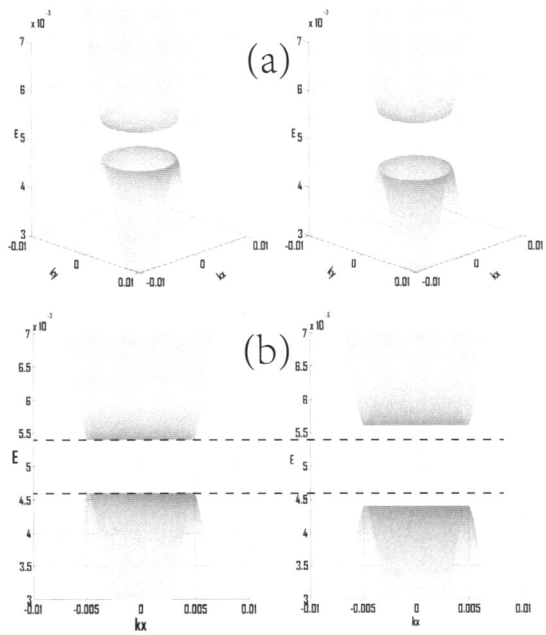

FIG. 2: We use the left circularly polarized light and set $\hbar\omega = 10meV$ and $V_F eA = 1meV$. In Fig.(a) we can see that isotropic dynamical gaps are opened at $E = \hbar\omega/2$ and the magnitude of the gaps are asymmetric for two different valleys. In Fig.(b) we compare the difference of the two gaps and find that in our case the gap opened in K valley is a little smaller than the gap opened in K' valley.

Since the functions $\Phi(k,t)$ are periodic, one can expand $\Phi(k,t)$ with the complete set of eigenfunctions (φ) of the non-radiative Hamiltonian,

$$\Phi_{\alpha k}(x,t) = \sum_{\gamma} \sum_{n=-\infty}^{+\infty} B_{k,n}^{\alpha,\gamma} \varphi_k^{\gamma}(x) e^{in\Omega t} \tag{5}$$

where α labels the Floquet state and γ labels the eigenstates of Hamiltonian $H(\mathbf{k})$. B is the expanding coefficients. The perturbation term does not couple states with different momenta, therefore the momentum is still a good quantum number in the Floquet states. The details of the derivation is discussed in appendix A.

In the numerical calculation, we use a left-circularly-polarized light and we set $\hbar\omega = 10meV$ and $V_F eA = 1meV$. In Fig. 2(a), we have drawn the 3-dimension dispersion relationship in the vicinity of both K and K' valley. We can intuitively see that the left circularly-polarized light has opened isotropic gaps at $E = \hbar\omega/2$ in both valleys, however, the magnitude of the gaps are different for K and K' valleys. In our case, the gap opened in K valley is larger than the gap in K' valley. If we use a right circularly-polarized light, the conclusion is reversed. In Fig. 2(b), we compare the gap difference more specifically. The gap opened in K valley is about $0.8meV$ while the gap in K' valley is about $1.2meV$. If the fermi energy is higher than $5.4meV$ and lower than $5.6meV$ (or higher than $4.4meV$ and lower than $4.6meV$), the conducting electrons are only contributed by K valley. In such case we can say that the current is valley polarized.

In Fig. 3, we increase the strength of the polarized light to $V_F eA = 4meV$ and we can see that the magnitude of the gaps opened on the resonance circle are also increased. Moreover, the asymmetry of different valleys become much clearer. In Fig. 3(b), we show that the gap opened in K valley is about $2.4meV$ while the gap in K' valley is about $3.8meV$. We can define valley polarization if the fermi energy is higher than $6.1meV$ and lower than $6.8meV$ (or higher than $3.2meV$ and lower than $3.9meV$). The conducting electrons are only contributed by K valley.

IV. SUMMARIES

In this paper, we have numerically calculated the Floquet band structure of silicene manipulated by the on-resonance light. We find that on the resonance circle, dynamical gaps are asymmetric for K and K' valleys. Hence, we can define valley polarization

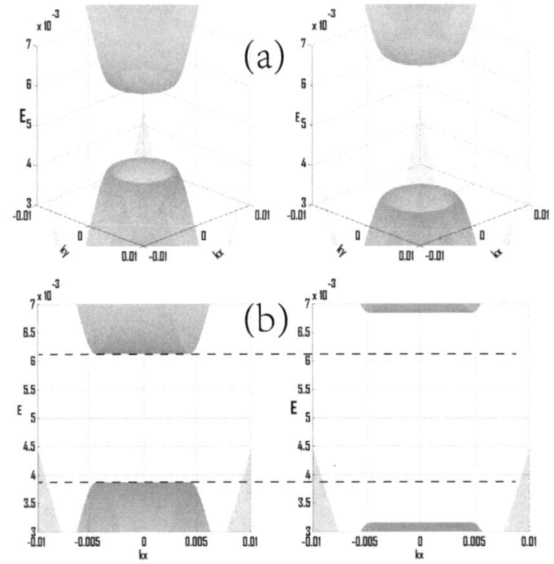

FIG. 3: We use the left circularly polarized light and set $\hbar\omega = 10meV$ and $V_F eA = 4meV$. In Fig.(a) we can see that the isotropic gaps are clearly bigger than the $V_F ea = 1meV$ case. In Fig.(b) we show that the gap difference between two valleys are also increase.

if Fermi level only penetrate one valley. If the light is left circularly-polarized, we can define K valley polarized current; if the light is right circularly-polarized, we can define K' valley polarized current. We can tune the valley polarization by changing the chirality of the light. Moreover, the advantage of this method is that the frequency of the light is much lower than the off-resonance light. One can safely increase the intensity of the light without burning the material.

V. ACKNOWLEDGEMENTS

This work was supported by the National Natural Science Foundation of China (Grant No. 11504084 and 11647164).

APPENDIX A:

We shall begin with the Schrödinger equation

$$H_k(t)\Phi_{\alpha k}(x,t) - i\hbar\frac{\partial}{\partial t}\Phi_{\alpha k}(x,t) = \varepsilon_k^{\alpha}\Phi_{\alpha k}(x,t) \tag{A1}$$

Substituting the expansion (5) into Schrödinger equation, we get

$$\sum_{\gamma}\sum_{n}(H_k(t) - i\hbar\frac{\partial}{\partial t})B_{k,n}^{\alpha,\gamma}\varphi_k^{\gamma}(x)e^{in\omega t} = \sum_{\gamma'}\sum_{m}\varepsilon_k^{\alpha}B_{k,m}^{\alpha,\gamma'}\varphi_k^{\gamma'}(x)e^{im\omega t} \tag{A2}$$

Multiply equation (A2) with $\varphi_k^{\gamma'}(x)e^{-im\omega t}$ from the left, and after a time-average over one period of driving T. Since in terms of the kets $|\varphi_k\rangle$, $\langle x|\varphi_k\rangle = \varphi_k(x)$, the Schrödinger equation becomes

$$\sum_{\gamma}\sum_{n}\langle\varphi_k^{\gamma'}|H_k^{m-n} + n\hbar\omega\delta_{mn}|\varphi_k^{\gamma}\rangle B_{k,n}^{\alpha,\gamma} = \varepsilon_k^{\alpha}B_{k,m}^{\alpha,\gamma'} \tag{A3}$$

with definition

$$\begin{aligned} H_k^{m-n} &= \tfrac{1}{T}\int_0^T dte^{-i(m-n)\omega t}H_k(t) \\ H_F^{m,n}(k) &= H_k^{m-n} + n\hbar\omega\delta_{mn} \end{aligned} \tag{A4}$$

IOP Conf. Series: Journal of Physics: Conf. Series **844** (2017) 012016 doi:10.1088/1742-6596/844/1/012016

Hence, the matrix form of the Schrödinger equation (A3) has a block-triagonal structure with only the number of angular frequencies ω in the diagonal varying from block to block. And the quasienergies ε_α can be obtained as the eigenvalues of the secular equation

$$det|H_F(k) - \varepsilon\mathbf{1}| = 0 \tag{A5}$$

Although the summation over n in Eqa. (A3) is from $-\infty$ to ∞, in real calculation, we only need to truncate the matrix $H_F(k)$ into finite dimension, the bigger the dimension of the matrix, the more accurate results one would gets.

REFERENCE

[1] S.V. Morozov et al., Phys. Rev. Lett. **97**, 016801 (2006)
[2] R.V. Gorbachev, F.V. Tikhonenko, A.S. Mayorov, D.W. Horsell, and A.K. Savchenko, Phys. Rev. Lett. **98**, 176805 (2007)
[3] Motohiko Ezawa, Phys. Rev. Lett. **109**, 055502 (2012).
[4] C. J. Tabert and E. J. Nicol, Phys. Rev. Lett. **110**, 197402 (2013).
[5] Hui Pan, Zhenshan Li, Cheng-Cheng Liu, Guobao Zhu, Zhenhua Qiao, and Yugui Yao, Phys. Rev. Lett. **112**, 106802 (2014).
[6] Yu Wang, Appl. Phys. Lett. **104**, 032105 (2014).
[7] Takehito Yokoyama, Phys. Rev. B **87**, 241409(R) (2013).
[8] Bumned Soodchomshom, Journal of Applied Physics **115**, 023706 (2014).
[9] Wen Lia, Wei-Tao Lua, Yun-Fang Lic and Hai-Hua, Physica E **88**, 284 (2017).
[10] Tayyaba Aftab, Physics Letters A **381**, 935 (2017).
[11] T. Oka and H. Aoki, Phys. Rev. B **79**, 081406(R) (2009)
[12] N. Lindner, G. Refael, and V. Gaslitski, Nat. Phys. **7**, 490 (2011)
[13] Selma Koghee, Lih-King Lim, M. O. Goerbig, and C. Morais Smith, Phys. Rev. A **85**, 023637 (2012)
[14] Awadhesh Narayan Phys. Rev. B **91**, 205445 (2015)
[15] C.-C. Liu, H. Jiang, and Y. Yao, Phys. Rev. B **84**, 195430 (2011).
[16] N. D. Drummond, V. Zólyomi, and V. I. Fal'ko, Phys. Rev. B **85**, 075432 (2012).
[17] Dittrich T, Hänggi P, Ingold G-L, Kramer B, Schön G and Zwerger W, *Quantum Transport and Dissipation* (Weinheim: Wiley) 1998 chapter 5

Impact of stoichiometry on the linear and nonlinear optical response of SnO_x thin films

Zhong-guo Li[1], Ling-yan Liang[2], Hong-tao Cao[2] and Ying-lin Song[3]

[1] College of Physics & Electronic Engineering, Changshu Institute of Technology, Changshu, China 215500, claes.lee@gmail.com

[2] Ningbo Institute of Material Technology and Engineering, Chinese Academy of Sciences, Ningbo, China 315201, lly@nimte.ac.cn

[3] Department of Physics, Harbin Institute of Technology, Harbin, China 150001, ylsong@hit.edu.cn

Abstract SnO is a promising p-type oxide semiconductor materials for applications such as transparent electronics and solar cells. However, further improvement of its performance is hindered by its diverse stoichiometry. We investigated the nonlinear and saturable absorption characteristics of pristine SnO and O-rich SnO_x films by femtosecond degenerate pump-probe measurements at 515 nm. UV-Vis absorption data indicate bandgap blueshift with increasing oxygen concentration. Pristine SnO film exhibit saturable absorption while nonlinear absorption is observed in O-rich SnO_x films. Our results shed light on the utilization of SnO in future device applications.

1. Introduction

Recently, transparent metal oxides (MOs) have garnered tremendous research attention due to their potential applications in transparent electronics, sustainable energy, nano-photonics, etc [1]. Among various oxide semiconductors, tin monoxide (SnO) has attracted numerous interests owing to its unique p-type conductivity, high carrier mobility and wide optical band gap [2-7]. However, a fundamental issue with SnO is that the stability of SnO is rather low, which can be easily either over-oxidized into SnO_2 or reduced into metallic Sn. As a result, considerable efforts have been focused on the preparation and characterization of tin monoxide with distinct stoichiometry. Previous studies revealed that stoichiometry play a critical role in the electronic properties of SnO_x materials [4,5,7]. However, few experimental studies have been conducted on the effect of stoichiometry on the linear or nonlinear optical response of SnO materials. Therefore, it is essential to study the optical response of SnO_x materials for further improving the performance of optoelectronic and solar energy devices based on SnO. In this paper, the static and dynamic optical response of pristine SnO and O-rich SnO_x ($1<x<1.5$) thin films were investigated using UV-Vis and femtosecond degenerate pump-probe measurements, respectively. Our results demonstrate that stoichiometry plays a significantly role in the optical response of tin monoxide materials.

2. Experiment

The films used in this study were prepared on quartz glasses using reactive rf magnetron sputtering under different oxygen partial pressure (P_O), with a thickness of ~200 nm. The crystallinity of the obtained films was characterized by x-ray diffraction (XRD, Bruker D8 Advance X-ray diffractometer) with Cu Kα radiation. Transmittance spectra were recorded via a UV-vis-IR spectrophotometer (Perkin-Elmer, Lambda 950). Details about the film fabrication and property characterizations can be found elsewhere [5]. The femtosecond degenerate pump-probe measurements were performed using a standard femtosecond pump-probe spectroscopy [8]. The light source was a Yb:KGW femtosecond laser (1.20 eV, 190 fs, 6 kHz). The 515 nm laser pulse used in the present work were generated by second harmonic generation with KDP crystal. The temple resolution of the measurement system was ~280 fs. To suppress many-body effects, the pump fluence was kept below 30 μJ/cm^2. All measurements were performed at room temperature.

3. Results

Figure 1(a) shows the XRD patterns of the pristine SnO and O-rich SnO_x films. Clear XRD peaks were observed in the pristine SnO film, which can be unambiguously assigned to those of the α-SnO crystal structure (P4/nmm, JCPDS card No.06-0395). However, the XRD pattern became broad and featureless for the SnO_x film prepared with P_O above 10.7%, indicating the amorphous nature of the films. Figure 1(b) presents the linear transmittance spectra of the SnO films. It is found that the transmittance spectra of the two films are similar. However, the optical bandgap of the O-rich SnO_x film slightly blueshift, which is a good agreement with the previous report [5]. Furthermore, the optical transmittance spectra clearly demonstrate that there is absorption below the band edge, which is related to the vacancies induced defect state within the band gap.

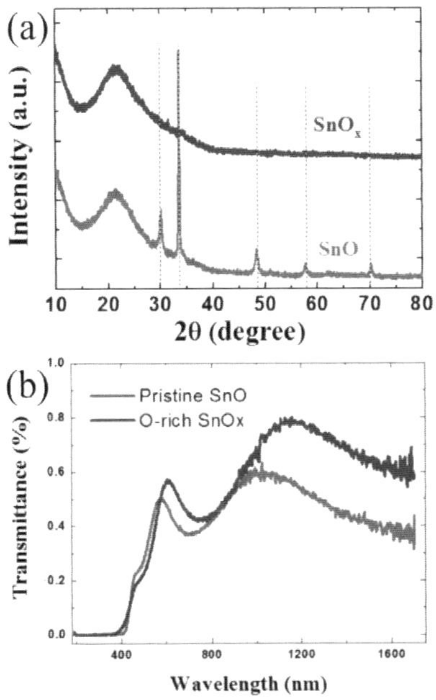

Fig. 1. (a) XRD patterns of the pristine SnO and O-rich SnO_x films. (b) The transmittance spectra of the SnO_x films.

Figure 2 illustrates the pump-probe curves of pristine SnO and O-rich SnO_x films at various time delays under 515 nm pulses excitation. Both films demonstrate a Gaussian-like dip/peak at zero time delay, which can be attributed to the coherent overlap of pump and probe pulses. Note that the transmission spectra of the two films reveal that bandtail states indeed exist in the SnO and SnO_x films, which could be excited by the pump photon (515 nm, 2.4 eV). However, the Gaussian-like function at zero time delay is an instantaneous process, which could be attributed to the convolution of the cross-correlation between the pump and probe pulses. In Fig. 2, the fitting curve of the convolution was shown in red line. The transient curve of the pristine SnO increase after pump excitation, which is related to the band-filling effects. The transient curve recover to negative value after 5 ps, qualitatively indicating the lifetime of carrier relaxation to the bandtail states. In contrast to the pristine SnO, the transient curve of the O-rich SnO_x film show a negative value follow by a decay feature, which are related to the free carrier-induced absorption mechanism (FCA). We utilized two exponential decay functions to fit the trace. And the corresponding characteristic time constants of the pristine SnO are 1.5 picosecond for the fast exponential decay function and ~13 ps for the other exponential function, respectively. These values agrees well with the pervious report [8]. However, the time constants of the O-rich SnO_x film are fitted to be around 10 ps and >50 ps, respectively. These results could be attributed to the defect state within bandgap induced by mixed valence state tin ions, which trap the photo-generated carriers. Our results clearly demonstrate that the dynamic optical response of tin monoxide rely heavily on the oxygen stoichiometry of the film.

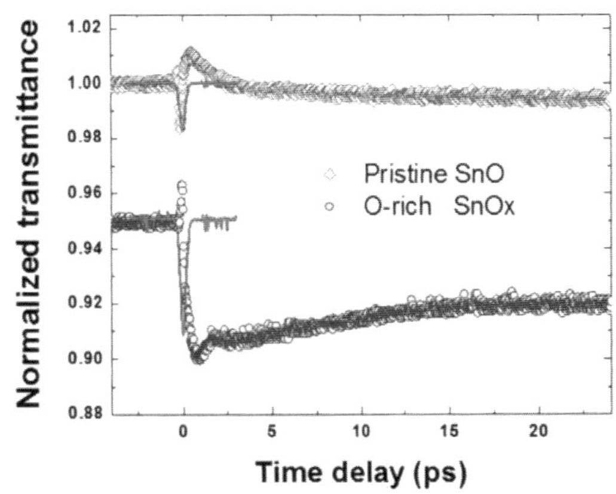

Fig. 2. Normalized pump-probe decay traces for pristine SnO and O-rich SnO_x films. The red lines represent the cross-correlation between the pump and probe pulses.

4. Conclusion

In summary, we studied the static and dynamic optical response of two SnO films with different oxygen stoichiometry. The thin film stoichiometry was found to play a critical role in determining the optical properties and dynamics of photo-generated carriers. The pristine SnO and O-rich SnO_x films show photo-bleaching and free carrier absorption after 515 nm excitation, respectively. And the carrier decay in the O-rich SnO_x film is slower than that of the pristine SnO film. Our reports may lead to some hints for improvement of the devices based on SnO materials.

Acknowledgment

This work is supported by the National Science Foundation (NSF) (61474126, 61674155); The National Key Research and Development Program of China (2016YFF0203600).

IOP Conf. Series: Journal of Physics: Conf. Series **844** (2017) 012017 doi:10.1088/1742-6596/844/1/012017

Reference

[1] X. Yu, T. J. Marks, and A. Facchetti, "Metal oxides for optoelectronic applications," Nat. Mater. 15, 383-396 (2016).

[2] Z. Wang, P. K. Nayak, J. A. Caraveo-Frescas, and H. N. Alshareef, "Recent Developments in p-Type Oxide Semiconductor Materials and Devices," Adv. Mater. 28, 3831-3892 (2016).

[3] Y. Ogo, H. Hiramatsu, K. Nomura, H. Yanagi, T. Kamiya, M. Hirano, and H. Hosono, "p-channel thin-film transistor using p-type oxide semiconductor, SnO," Appl. Phys. Lett. 93, 032113 (2008).

[4] J. A. Caraveo-Frescas, P. K. Nayak, H. A. Al-Jawhari, D. B. Granato, U. Schwingenschlogl, and H. N. Alshareef, "Record Mobility in Transparent p-Type Tin Monoxide Films and Devices by Phase Engineering," ACS Nano 7, 5160-5167 (2013).

[5] H. Luo, L. Y. Liang, H. T. Cao, Z. M. Liu, and F. Zhuge, "Structural, Chemical, Optical, and Electrical Evolution of SnO_x Films Deposited by Reactive rf Magnetron Sputtering," ACS Appl. Mater. Interfaces 4, 5673-5677 (2012).

[6] J. Zhang, X. Kong, J. Yang, Y. Li, J. Wilson, J. Liu, Q. Xin, Q. Wang, and A. Song, "Analysis of carrier transport and band tail states in p-type tin monoxide thin-film transistors by temperature dependent characteristics," Appl. Phys. Lett. 108, 263503 (2016).

[7] J. Wang, N. Umezawa, and H. Hosono, "Mixed Valence Tin Oxides as Novel van der Waals Materials: Theoretical Predictions and Potential Applications," Adv. Energy Mater. 6, 1501190 (2016).

[8] Z. G. Li, L. Y. Liang, H. T. Cao, Z. G. Xiao, X. Z. Wu, Y. Fang, J. Y. Yang, T. H. Wei, and Y. L. Song, "Ultrafast carrier dynamics in SnO_x thin films," Appl. Phys. Lett. 106, 102103 (2015).

IOP Conf. Series: Journal of Physics: Conf. Series **844** (2017) 012018 doi:10.1088/1742-6596/844/1/012018

Edge-dependent optical absorption of silicene nanoribbons in an in-plane magnetic field

Jianfei Zou[1] and Yang Ji

College of Science, Hohai University, Nanjing 210098, China

[1]E-mail: zoujianfei@hhu.edu.cn

Abstract. We theoretically study the low energy electronic structures and optical conductivities of silicene nanoribbons (SiNRs) with zigzag and armchair edges in the presence of an in-plane magnetic field. It is found that the absorption spectrum exhibits a rich structure and remarkably depends on the edge types and the width of the SiNRs. In particular, the helical edge states for the zigzag SiNRs is gapped by the magnetic field, whereas those for the wide armchair SiNRs remain gapless, which could be demonstrated by the optical absorption measurement in experiments.

1. Introduction

Silicene, a single layer of silicon atoms closely packed in a two-dimensional (2D) honeycomb lattice, has recently attracted much interest since its successful fabrication in experiments [1-3]. In contrast to its counterpart of carbon, namely graphene, silicene has a buckled structure and a relatively large spin-orbit interaction (SOI) [4]. Consequently, silicene shows unusual electronic structures, which are characterized by the spin-orbital gap at the K points in the Brillouin zone. It has been predicted that the SOI could bring about exotic phenomena, such as the quantum spin Hall effect [4,5] and spin-valley coupled optical response [6-9]. What is more, the combination of the SOI and a perpendicular electric field or exchange field in silicene sheet could give rise to a rich variety of phases: a quantum spin Hall insulator, a quantum anomalous Hall insulator, a valley-polarized metal and a band insulator [10].

Silicene nanoribbons (SiNRs) are quasi-one dimensional (1D) strips of monolayer silicene with nanometer width, which have also been synthesized in experiments [11,12]. Distinct from 2D silicene sheets, 1D SiNRs have richer physics because of the quantum confinement and edge effects. It has been found that the quantum spin Hall edge states at zigzag and armchair terminations of honeycomb lattice are fundamentally different [13,14]. Akin to graphene nanoribbons [15,16], the energy gap of silicene and its nanoribbons can be modulated by applying electric field [17], adsorption [18], and doping [19], that makes them promising for the electronic and optical applications. It is well known that the optical absorption measurement has become an important tool to reveal the electronic structures and the topological states of the materials. There have been numerous works on the optical properties of graphene nanoribbons [20-26], but a few researches on that of SiNRs. Bao and his collaborators have predicted selection rule for the optical spectrum of armchair SiNRs (ASiNRs) [27]. Shyu has studied the electric field-modulated optical properties of armchair SiNRs (ASiNRs) [28] and the magneto-optical properties of zigzag SiNRs (ZSiNRs) [29].

In this work, we study the low energy electronic and optical properties of the SiNRs with zigzag and armchair edges in the presence of an in-plane magnetic field. It is found that the effects of the in-plane magnetic field on the properties of the zigzag and armchair SiNRs, especially on the edge states, completely differ. The rest of the paper is organized as follows. In section 2, we introduce the tight-binding model for the SiNRs and the formulism for the optical absorption. In section 3, we show the numerical results of the electronic structure and the optical conductivity, as well as some theoretical analysis. Finally, a brief summary is provided in section 4.

2. Model and formulism

Now we consider the single layer SiNRs with two edges run along the x direction, as shown in figure 1. The width of a SiNR in the y direction is defined as the number of the longitudinal diatomic chains. In the tight binding approximation, the π electrons in a pristine SiNR can be described by the following Hamiltonian [30,31]

$$H_0 = -t_0 \sum_{<i,j>\alpha} c^{\dagger}_{i,\alpha} c_{j,\alpha} + \frac{i\lambda_{SO}}{3\sqrt{3}} \sum_{<<i,j>>\alpha\beta} v_{ij} c^{\dagger}_{i,\alpha} s_{z,\alpha\beta} c_{j,\beta}, \quad (1)$$

where $c^{\dagger}_{i,\alpha}$ ($c_{i,\alpha}$) is the creation (annihilation) operator of an electron with spin index α at site i in the honeycomb lattice, and $\langle i,j \rangle$ ($\langle\langle i,j \rangle\rangle$) runs over the pairs of nearest (next-nearest) neighbor sties. (s_x, s_y, s_z) is a vector of Pauli matrix for electron spin. The first term in equation (1) represents the standard nearest-neighbor hopping with the transfer energy $t = 1.6$ eV, while the second term accounts for the intrinsic SOI with its strength $\lambda_{SO} = 3.9$ meV and the factor $\upsilon_{ij} = 1\ (-1)$ if the next-nearest neighbor hopping is anticlockwise (clockwise) with respect to the positive z axis. Here we ignore the intrinsic Rashba SOI since it is very tiny in silicene (about 0.7 meV) [10]. For simplicity the change of the transfer energy at edges is also not considered.

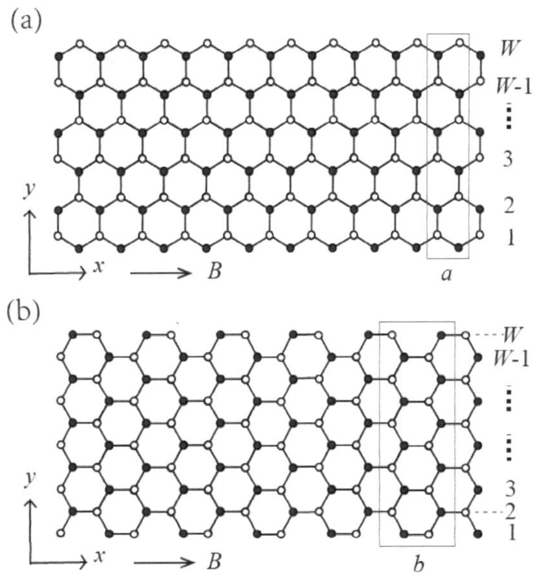

Figure 1. (Color online) Schematics of (a) zigzag and (b) armchair SiNRs. Here we set the width $W = 6$ for the zigzag and $W = 10$ for the armchair nanoribbons, respectively. A unit cell consisting of $2W$ atoms is denoted by a rectangle. The lattice constant $a = 3.86$Å for the zigzag and $b = \sqrt{3}a$ for the armchair SiNRs. An in-plane magnetic field is applied along the x direction.

As the SiNR is subjected to an externally static in-plane magnetic field along the x direction, an additional term due to Zeeman effect should be added to the Hamiltonian above, which has the form

$$H_1 = \lambda_m \sum_{i,\alpha\beta} c^{\dagger}_{i,\alpha} s_{x,\alpha\beta} c_{i,\beta}. \quad (2)$$

Here the parameter λ_m represents the Zeeman splitting induced by the in-plane magnetic field. Therefore the total Hamiltonian for the system is written as $H = H_0 + H_1$. It should be noted that the in-plane magnetic field could not cause orbital effect in the SiNR but plays a role like an exchange field in a ferromagnet. Besides, the direction of the in-plane magnetic field is irrelevant to the results of this paper. The Zeeman (exchange) field in the SiNR maybe also arise via proximity coupling to a ferromagnetic insulating substrate [10].

Since the nanoribbon is translationally invariant along the x direction, we can perform the partial Fourier transformation on the tight binding Hamiltonian H to rewrite it as $H(k_x)$, in which k_x is the Bloch wavevetor in the longitudinal direction. And then numerically diagonalizing the Hamiltonian, one can obtain the energy bands E_{n,k_x} and the corresponding eigenstates $|n,k_x\rangle$. Here the letter n denotes the band index according to the minimum energy spacing between the subband and Fermi level. We will show the bandstructures of SiNRs with different width and edge types in the next section.

Now we irradiate the SiNRs by using a beam of monochromatic light with a linear polarization along the longitudinal direction of the nanoribbon. In order to investigate the optical response of SiNRs, one should calculate the real part of the dynamical conductivity $\sigma(\omega)$, which describes the optical absorption. Within the linear response theory, it can be expressed as

$$\sigma(\omega)=\frac{\pi e^2}{A\omega}\sum_{n,n',k_x}\left|\langle n',k_x|\hat{v}_x|n,k_x\rangle\right|^2\left[f(E_{n,k_x})-f(E_{n',k_x})\right] \times\delta(E_{n',k_x}-E_{n,k_x}-\hbar\omega) \quad (3)$$

where e is the electron charge, A is the area of the sample, and ω is the irradiation frequency. $\langle n',k_x|\hat{v}_x|n,k_x\rangle$ is the matrix element for the x component of the velocity operator $\hat{v}_x=\frac{\partial H}{\hbar\,\partial k_x}$. $f(E_{n,k})$ is the Fermi-Dirac distribution function. At low temperature, one can assume $f(E_{n,k})=1$ for $E_{n,k}<E_F$ and $f(E_{n,k})=0$ for $E_{n,k}>E_F$, where E_F is the Fermi level. Here we study an undoped SiNR at low temperature such that $E_F=0$.

3. Results and discussion

In the following, we will show some numerical results of the bandstructure and optical conductivity for the zigzag and armchair SiNRs. For clear illustration of the effects of the SOI and the in-plane magnetic field, we use the reduced parameters and enlarge the strength of the SOI. Besides, the Dirac delta function in equation (3) is approximated by a Lorenz function with its broadening parameter $\eta=0.003t_0$. Hereafter we refer to a zigzag SiNR with W zigzag chains as a W-ZSiNR and an armchair SiNR with W dimer lines as a W-ASiNR.

3.1. *Zigzag nanoribbons.*

In figure 2(a) and (b), we present the low energy bandstructures of an 11-ZSiNR in the absence and presence of the in-plane magnetic field, respectively. The physical width for the 11-ZSiNR ($W'=\sqrt{3}Wa/2$) is about 3.68 nm. In figure 2(a), it is seen that the band structure with spin degeneracy is symmetric about the wavevector $k_x=\pi/a$. Moreover, the highest occupied subband ($n_v=1$) and the lowest unoccupied subband ($n_c=1$) touch at $k_x=\pi/a$ and their corresponding dispersions near the zero energy are linear. It can be demonstrated that the gapless states are quantum spin Hall edge states [4]. As the in-plane magnetic field is applied in the nanoribbon, each subband with spin degenerate in figure 2 (a) splits into two subbands in figure 2(b). More importantly, the Zeeman field leads to a gap of the edge states, which is about $0.1t_0$ as illustrated in figure 2(b). That could be reflected in the low frequency optical absorption. By the way, one should note that the out-of-plane Zeeman field only results in a shift of the subbands with spin up (down) by an half positive (negative) Zeeman energy, so that it could not open a gap in our system.

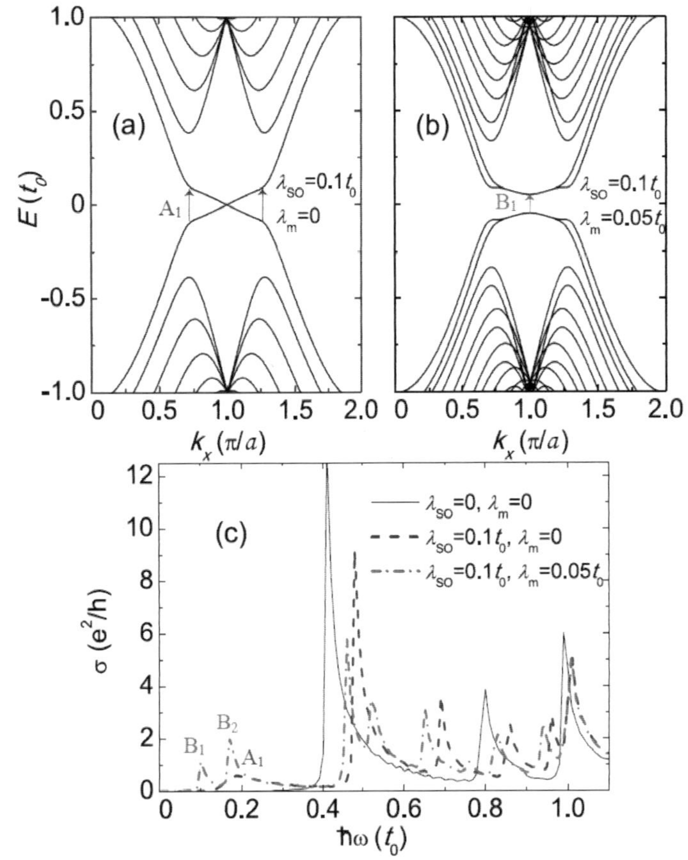

Figure 2. (Color online) Band structures of 11-ZSiNR (a) with and (b) without the in-plane magnetic field. The up arrows denote the optical transitions between the edge states. (c) Spectra of optical conductivity in different cases.

Figure 2(c) displays the corresponding optical conductivities in different cases which are denoted by the different lines. The solid line stands for the spectrum of a pristine 11-ZSiRN without SOI, which is similar to that of a zigzag graphene nanoribbon [23]. The three absorption peaks come from the optical transitions satisfying the selection rule: $\Delta n = n_c - n_v = odd$. As the SOI is considered, all the original peaks exhibit the blue shift, meanwhile three additional peaks generating from a new selection rule $\Delta n = even$ emerge [29]. It should be noted that the first new peak A_1 at $\omega = 0.2t_0$ is predominated by the optical transitions between the edge states at the vicinities of the K points ($k_x = \frac{2\pi}{3a}, \frac{4\pi}{3a}$). When the in-plane magnetic field is applied in the sample, the Zeeman field leads to a richer optical spectrum due to the band splitting. It is clearly seen that each of the first three peaks, which is in the dashed line for the free 11-ZSiRN without the Zeeman field, splits into two peaks in the dash-dotted line. In particular, the two subpeaks B_1 and B_2 emerge at $\hbar\omega = 0.10t_0, 0.17t_0$, which are contributed by the transitions at the vicinities of the center of Brillouin zone ($k_x = \pi/a$) and K points, respectively. The threshold peak B_1 at $\hbar\omega = 0.10t_0$ is the signature for the energy gap of the edge states induced by the in-plane Zeeman field.

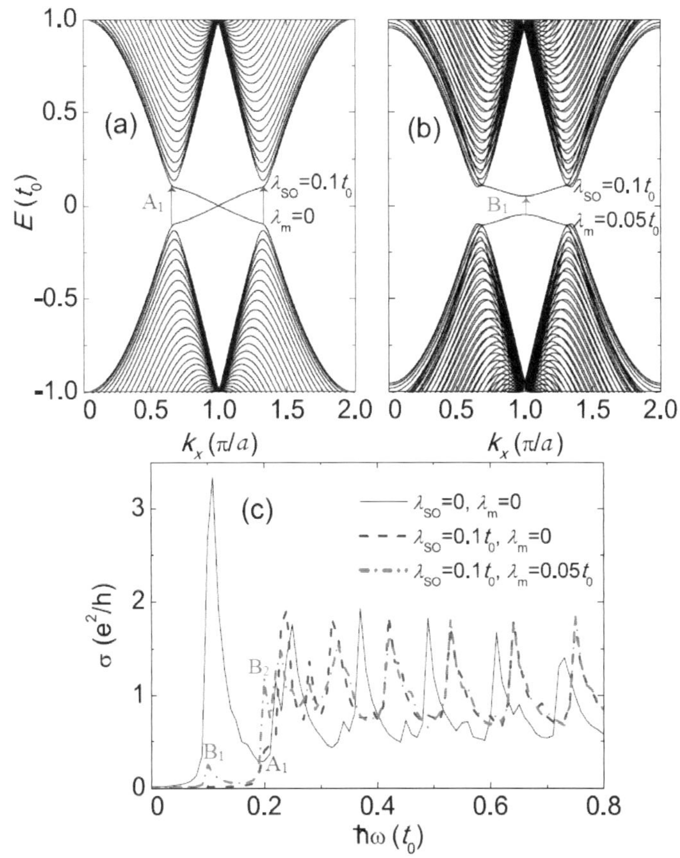

Figure 3. (Color online) Band structures of 50-ZSiNR (a) with and (b) without the in-plane magnetic field. (c) Spectra of optical conductivity in different cases. The arrows in (a) and (b) denote the first absorption shoulder and peak in (c), respectively.

We also calculate the bandstructures and optical conductivities for a 50-ZSiRN ($W' = 16.7$ nm), as illustrated in figure 3. In the wide ZSiRN without a magnetic field, the matching frequency for the optical transition between the edge states at the K points is near that for the transitions involving the bulk states, as seen in figure 3(a), such that the corresponding absorption peaks almost overlap as shown on the dashed line in figure 3(c). In fact one can see a shoulder A_1at $\hbar\omega = 0.20t_0$, which is attributed to the transition between the edge states near the K points. As the in-plane magnetic field is considered, there exists a tiny peak B_1 at $\hbar\omega = 0.10t_0$ same as that for 11-ZSiRN, which indicates the width-independent energy gap of the edge states opened by the Zeeman field (shown in figure 3(b)), while the second peak B_2 remains at the frequency $\hbar\omega = 0.20t_0$. Actually, if consider a semi-infinite silicene sheet with a zigzag edge along the x direction, we can obtain an effective Hamiltonian for the helical edge states: $H_e = \hbar v_F k_x s_z + \lambda_m s_x$, where v_F is the Fermi velocity of the edge states dependent on the SOI [14]. Consequently, one gets the dispersion for the edge states: $\epsilon = \pm\sqrt{(\hbar v_F k_x)^2 + \lambda_m^2}$ and the gap: $\Delta E_e = 2\lambda_m$, which could be identified by the threshold peak of the optical absorption.

3.2. Armchair nanoribbons.

Now we turn to the armchair nanoribbons. It has been confirmed that the armchair graphene nanoribbons can be divided into two families: one family with $W = 3p, 3p + 1$ (p is an integer) are semiconductors, while the other with $W = 3p + 2$ are semimetals. This classification is also suitable for the ASiNRs with a weak SOI. Figure 4(a) and (b) exhibit the low energy bandstructures for a semiconducting 15-ASiRN with and without the in-plane magnetic field, respectively. Its physical width ($W' = Wa/2$) is about 2.90 nm. One can find that the free 15-ASiRN is a semiconductor with the gap about $0.18t_0$, while the 15-ASiRN under the in-plane Zeeman field gets a narrower gap about $0.11t_0$ due to the breaking of the spin degeneracy.

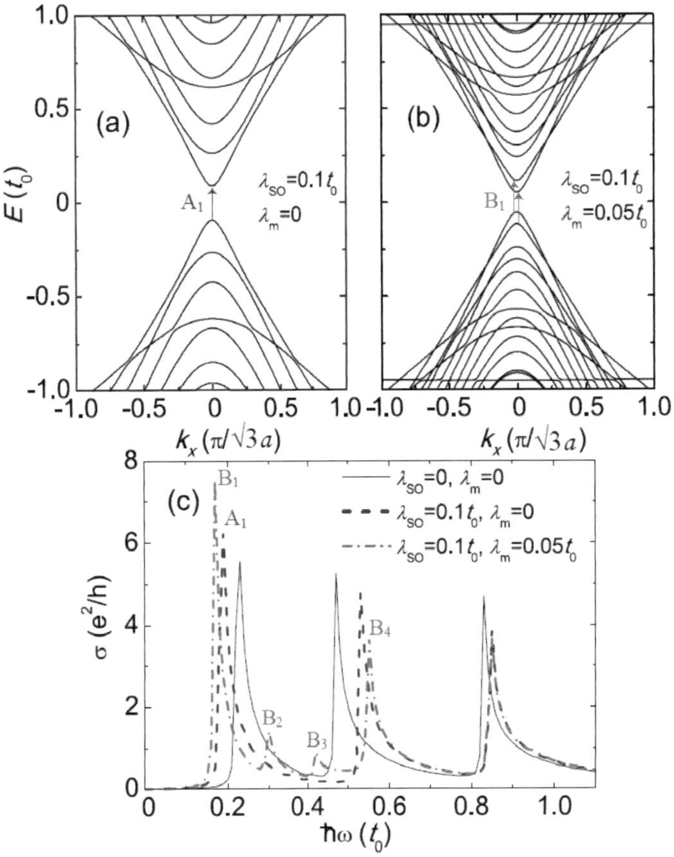

Figure 4. (Color online) Band structures of 15-ASiNR (a) with and (b) without the in-plane magnetic field. (c) Optical conductivity versus light frequency in different cases. The arrows in (a) and (b) denote the threshold absorption peaks in (c).

In figure 4(c), we plot the corresponding optical conductivities for the 15-ASiRN. In this figure, the solid line represents the spectrum for a pristine armchair nanoribbon without the SOI, which is consistent with the result for the armchair graphene nanoribbon [23]. On this line, one can see three absorption peaks resulting from the direct interband transitions: $\Delta n = 0$. As the SOI is introduced in the armchair nanoribbon, the first original peak shifts to the lower frequency, while the others shift to the higher. However there is no additional peak, which indicates that the selection rule $\Delta n = 0$ still holds in the presence of the SOI [28]. This is quite distinct from the results for the ZSiNRs. After the application of the in-plane magnetic field in the sample, more absorption peaks appear since the selection rule is altered. It is obvious that there exist two subpeaks, which are labelled as B_2 at $\hbar\omega = 0.30t_0$ and B_3 at $\hbar\omega = 0.42t_0$ between the first two adjacent principal peaks $B_1$and B_4. This two subpeaks are from the optical transitions $\Delta n = \pm 2$ ($n_v \rightarrow n_c$: $1(3) \rightarrow 3(1)$) and $\Delta n = \pm 2, \pm 4$ ($n_v \rightarrow n_c$: $2(4) \rightarrow 4(2), 1(5) \rightarrow 5(1)$) respectively. Furthermore, It is noted that the frequency $\hbar\omega = 0.17t_0$ for the first main peak B_1, which responses to the transition $\Delta n = \pm 1$ ($n_v \rightarrow n_c$: $1(2) \rightarrow 2(1)$), is larger than the energy gap $\Delta E_g = 0.11t_0$. The transition between the highest occupied subband (n_v =1) and the lowest unoccupied subband (n_c =1) is forbidden.

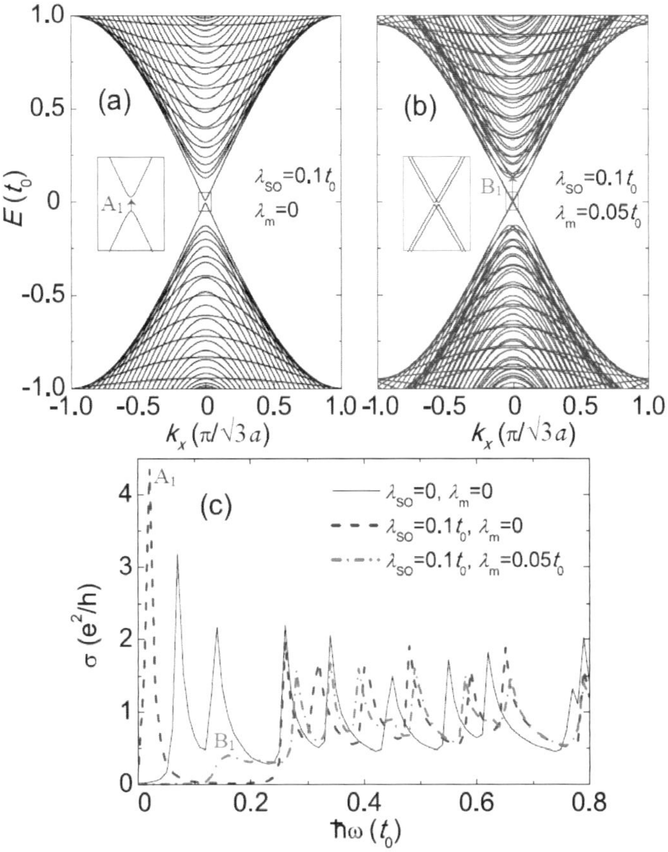

Figure 5. (Color online) Band structures of 54-ASiNR (a) with and (b) without the in-plane magnetic field. The insets zoom in on the region near the Dirac points. (c) Optical conductivity versus light frequency in different cases. The arrows in (a) and (b) denote the first absorption peak and shoulder in (c), respectively.

Figure 5 exhibits the bandstructures and the optical conductivities for a semiconducting 54-ASiRN with physical width $W' = 10.4$ nm. In figure 5(a), it is seen that a tiny gap about $0.015t_0$ for the free 54-ASiRN is opened by the weak coupling between edge states localized at opposite edges. One can infer that the gap of the edge states approaches to zero as the ASiRN gets wider. The absorption peak A_1 near the long wave limit, as illustrated in figure 5(c), is the signature of the existence of the edge states with the tiny gap. As the in-plane magnetic field is applied in the 54-ASiNRs, the gap of the edge states is close to zero. Actually it is about $0.002t_0$, as shown in the inset of figure 5(b). Consequently, one can see a wide shoulder B_2 extended from $0.15t_0$ to $0.25t_0$ in figure 5(c), which is attributed to the transitions between the edge and bulk states.

We also calculate the bandstructures and optical conductivities for the metallic 14-ASiRN and 53-ASiRN (not shown here). It is found that the absorption spectrum for the 14-ASiRN is almost unaffected by the in-plane magnetic field, although the energy bands are split. However, the absorption spectrum for the 53-ASiRN in the Zeeman field is similar to that for the 54-ASiRN, since the effect of the hybridization between the edge states at opposite boundaries in the wide ASiRNs is faint. We could infer that the edge states for the wide enough ASiRNs remain gapless if even the in-plane magnetic field is applied. This is quite distinct from the result for the ZSiRNs. Indeed, we can analytically obtain the dispersion of the gapless edge states for a semi-infinite silicene sheet with an armchair edge along the x direction. Considering the effective Hamiltonian near the K points for a silicene sheet under an in-plane magnetic field, and using the approach proposed in the previous works [14,32], one can derive an approximate dispersion for the gapless edge states: $\epsilon = \pm\hbar v_{F0}\lambda_{SO}k_x/\sqrt{\lambda_m^2 + \lambda_{SO}^2}$, where $v_{F0} = \sqrt{3}t_0 a/2$ is the bare Fermi velocity in the freestanding silicene sheet. It is concluded that the effect of the in-plane Zeeman field on the edge states at the armchair edges is to decrease their velocities but not open a gap.

4. Conclusion

We have investigated the bandstructures and optical conductivities for SiNRs including both zigzag and armchair types in the presence of an in-plane magnetic field. It is found that the Zeeman field induces a gap of the edge states for the ZSiNRs, but decreases the gap for the ASiNRs. We infer that the gapless edge states in the wide ASiNRs are robust to the in-plane magnetic field, but the ones in the ZSiNRs are not. Furthermore, the absorption spectrum becomes richer and remarkably depends on the edge types and the width of the SiRNs. Some particular absorption peaks are attributed to the transitions involving the edge states. The fundamental difference of the effects of the in-plane magnetic field on the topological edge states in the zigzag and armchair SiNRs could be identified by the optical absorption measurement in experiments. We expect that our research is useful for designing the Si-based optoelectronic nanodevices in the future.

Acknowledgment

This work was supported by the Fundamental Research Funds for the Central Universities (Grant No. 2013B00514) and the National Natural Science Foundation of China (Grant Nos. 11347127, 61404044).

References

[1] Vogt P, Padova P D, Quaresima C, Avila J, Frantzeskakis E, Asensio M C, Resta A, Ealet B and Lay G L 2012 *Phys. Rev. Lett.* **108** 155501
[2] Lin C-L, Arafune R, Kawahara K, Tsukahara N, Minamitani E, Kim Y, Takagi N and Kawai M 2012 *Appl. Phys. Express* **5** 045802
[3] Fleurence A, Friedlein R, Ozaki T, Kawai H, Wang Y and Yamada-Takamura Y 2012 *Phys. Rev. Lett.* **108** 245501
[4] Liu C-C, Feng W, Yao Y 2011 *Phys. Rev. Lett.* **107** 076802
[5] Tabert C J and Nicol E J 2013 *Phys. Rev. B* **87** 235426
[6] Ezawa M 2012 *Phys. Rev. B* **86** 161407
[7] Stille L, Tabert C J and Nicol E J 2012 *Phys. Rev. B* **86** 195405
[8] Tabert C J and Nicol E J 2013 *Phys. Rev. Lett.* **110** 197402
[9] Zou J, Tang C and Zhang A 2017 *Phys. Lett. A* **381** 1197
[10] Ezawa M 2012 *Phys. Rev. Lett.* **109** 055502
[11] Aufray B, Kara A, Vizzini S, Oughaddou H, Léandri C, Ealet B and Lay G L 2010 *Appl. Phys. Lett.* **96** 183102
[12] Padova P D *et al* 2010 *Appl. Phys. Lett.* **96** 261905
[13] Ezawa M and Nagaosa N 2013 *Phys. Rev. B* **88** 121401
[14] Cano-Cortés L, Ortix C and van den Brink J 2013 *Phys. Rev. Lett.* **111** 146801
[15] Han M Y, Özyilmaz B, Zhang Y and Kim P 2007 *Phys. Rev. Lett.* **98** 206805
[16] Zou J and Jin G 2011 *Appl. Phys. Lett.* **98** 122106
[17] Z. Ni, Liu Q, Tang K, Zheng J, Zhou J, Qin R, Gao Z, Yu D and Lu J 2012 *Nano Lett.* 12 113
[18] Quhe R *et al* 2012 *Sci. Rep.* **2** 853
[19] Ma L, Zhang J-M, Xu K-W, Ji V 2013 *Physica B* **425** 66
[20] Hsu H and Reichl L E 2007 *Phys. Rev. B* **76** 045418
[21] Liu J, Wright A R, Zhang C and Ma Z 2008 *Appl. Phys. Lett.* **93** 041106
[22] Liao W, Zhou G and Xi F 2008 *J. Appl. Phys.* **104** 126105
[23] Sasaki K-i, Kato K, Tokura Y, Oguri K and Sogawa T 2011 *Phys. Rev. B* **84** 085458
[24] Chung H C, Lee M H, Chang C P and Lin M F 2011 *Opt. Express* **19** 23352
[25] Zhu W-H, Ding G-H and Dong B 2012 *Appl. Phys. Lett.* **100** 103101
[26] Nematian H, Moradinasab M, Pourfath M, Fathipour M and Kosina H 2012 *J. Appl. Phys.* **111** 093512

[27] Bao H, Liao W, Guo J, Yang X, Zhao H and Zhou G 2015 *J. Phys. D: Appl. Phys.* 48 455306
[28] Shyu F-L 2017 *Physica E* **85** 117
[29] Shyu F-L 2017 *Physica E* **87** 178
[30] Kane C L and Mele E J 2005 *Phys. Rev. Lett.* **95** 146802
[31] Kane C L and Mele E J 2005 *Phys. Rev. Lett.* **95** 226801
[32] Brey L and Fertig H A 2006 *Phys. Rev. B* **73** 235411

IOP Conf. Series: Journal of Physics: Conf. Series **844** (2017) 012019 doi:10.1088/1742-6596/844/1/012019

Measurements of the magneto-optical properties of PbS-doped silica optical fiber

Weilong Dong, Yi Huang, Huangchao Chen, Yanhua Dong, Jianxiang Wen and Tingyun Wang

Key laboratory of Specialty Fiber Optics and Optical Access Networks, School of Communication and Information Engineering, Shanghai University, Shanghai, 200072, China

E-mail: tywang@shu.edu.cn

Abstract. The Verdet constants of PbS-doped silica optical fiber and single mode fiber (SMF-28e) have been investigated based on a magneto-optical effect measurement system at wavelengths between 660 and 1550 nm. The Verdet constant of PbS-doped fiber is 3.17 rad/Tm, 31.5% larger than that of SMF at 660 nm. The PbS-doped silica optical fiber can become a promising material for Faraday rotator.

1. Introduction

Faraday rotators are important optical components widely utilized in optical isolators, optical modulators and sensors of current and magnetic fields [1-3]. Faraday rotator can be a crystal, bulk glass or optical fiber, placed inside a magnetic field. However, crystals and bulk glasses as Faraday rotator present technological disadvantages such as difficult and high-cost growth and temperature dependence [4, 5]. The optical fiber as a Faraday rotator has certain advantages, such as small size, high sensitivity, immunity to electromagnetic interference. The PbS-doped fiber is fabricated using atomic layer deposition (ALD) and modified chemical vapor deposition (MCVD) technology, whose refractive index difference (RID) is about 0.75% and the concentration of Pb element is 0.11 at%, respectively, and its photoluminescence (PL) spectra nearly cover a broad band from 800 to 1600 nm, which matches well the whole optical communication window [6, 7]. However, magneto-optical properties of PbS-doped silica optical fibers have rarely been investigated. In this paper, the wavelength dependence of Verdet constants of PbS-doped silica optical fiber has been experimentally measured by the Stokes polarimetry. The Verdet constants of PbS-doped silica optical fiber are larger than that of SMF at different wavelengths.

2. Theory and Experiment

Faraday effect as one of the most useful magneto-optical phenomena refers to the rotation of the plane of polarization of linearly polarized light due to magnetic field induced circular birefringence of a material. The rotation is given by equation (1).

$$\theta = H \cdot L \cdot V \tag{1}$$

where θ is the angle of rotation of the plane of polarization, H is the applied magnetic field, L is the distance travelled by light through the material parallel to the magnetic field and V is Verdet constant.

The magnetic-field-induced Larmor precession of electron orbits is the simplest mechanism for Faraday effect [8]. The well-known Becquerel formula, which describes the wavelength dependence of the Verdet constant, was originally proposed by Becquerel [9], described by equation (2)

$$V = \frac{e}{2mc}\lambda\frac{dn}{d\lambda} \quad (2)$$

where e is the electron charge, m is the electron mass, c is the speed of light in vacuum and $\frac{dn}{d\lambda}$ is the dispersion of the material. The equation (2) shows that the Verdet constant is proportional to both the wavelength of the light and its dispersion in the medium.

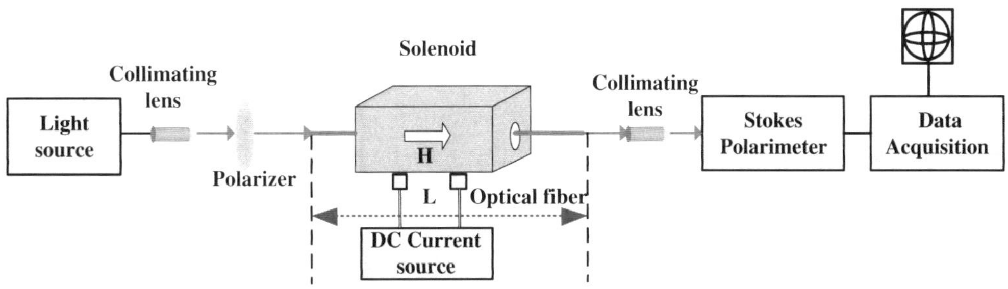

Figure 1. Magneto-optical effect measurement system.

A magneto-optical effect measurement system, which is based on the Stokes polarization parameters method, is set up shown in figure 1. A light wave from the laser source (Fabry-Perot Benchtop Laser Source, Thorlabs) is passed through a collimating lens (F220FC-B, Thorlabs) and a polarizer (LPVIS050-MP2, Thorlabs) for a linearly polarized light. And then the polarized light is launched into the fiber, which is clamped on the bare fiber launch system (MAX350D, Thorlabs). To observe Faraday effect, the optical fiber, length about 1m, is placed in a 30.4 cm long copper-wire coil (coil diameter: 1.8 mm per turn, inner radius: 12.8 mm, outer radius: 74.8 mm, the number of layer: 38, the number of turns: 5890). The magnetic field is produced by using a solenoid with a direct-current (DC). Its direction is parallel to the axial direction of the fiber. Maximum allowable current and magnetic fields strength of on the coil axis is 6 A, 140 mT, respectively. After passing through the optical fiber, the collimated light arrives at the Stokes polarimeter. And then the states of polarization can be displayed on the Poincare sphere model.

3. Results and Discussion

The Verdet constant are measured at different wavelengths (660, 808, 980, 1310, 1550 nm), which are listed in Table 1.

Table 1. Verdet constant of PbS-doped fiber and SMF at different wavelengths.

Wavelength(nm)	SMF	PbS-doped fiber
	Verdet Constant (rad/Tm)	
660	2.41	3.17
808	1.85	2.42
980	1.28	1.77
1310	0.73	0.86
1550	0.61	0.65

According to table 1, the Verdet constant for PbS-doped fiber and SMF at different wavelengths are plotted, and then polynomial fitted to the wavelength dependent relationship, as shown in figure 2. It indicates that the Verdet constant of PbS-doped fiber is larger than that of SMF at different wavelengths.

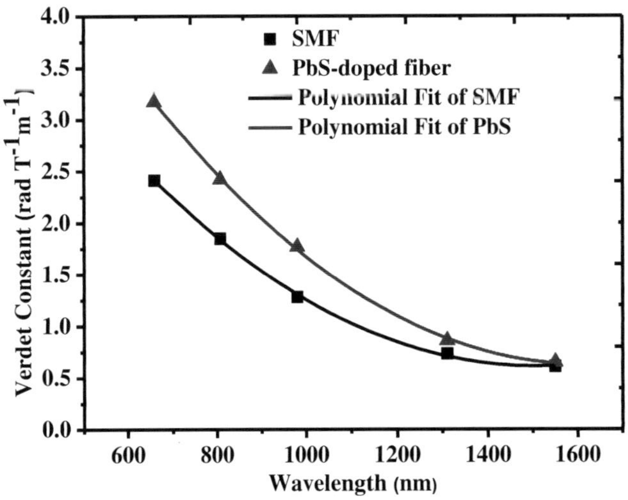

Figure 2. Verdet constant of PbS-doped fiber and SMF at different wavelengths.

And the wavelength dependence effect of the Verdet constant has been demonstrated. Furthermore, for PbS-doped fiber, its Verdet constant at 660 nm is 3.17 rad/Tm, and that of SMF is 2.41 rad/Tm, 31.5% larger than that of SMF. In summary, the larger Verdet constant of the PbS-doped fiber than that of SMF is ascribed primarily to the PbS material doped in the optical fiber.

4. Conclusion

We have been experimentally measured the Verdet constants of the PbS-doped fiber and SMF by the Stokes polarimetry and found that the Verdet constant of the PbS-doped fiber is higher than that of SMF at different wavelengths. And the increment of Verdet constant of PbS-doped fiber can reach to about 31.5% at 660 nm.

Acknowledgments

This work is supported by the National Science Foundation (NSF) (61475095, 61475096, 61520106014); Science and Technology Commission of Shanghai Municipality (STCSM) (15220721500).

References

[1] Zu P *et al* 2011 *Opt. Lett.* **36** 1425-27
[2] Zheleznov D S *et al* 2012 *J. Opt. Soc. Am. B* **29** 786-92
[3] Correa N *et al* 2014 *J. Phys.: Conf. Series* **511** 012026
[4] Dong Y H *et al* 2014 *Appl. Surf. Sci.* **320** 372-78
[5] Villora E G *et al* 2011 *Appl. Phys. Lett.* **99** 011111-3
[6] Y. S. Didosyan Y S *et al* 2001 *Int. J. Appl. Electrom.* **13** 277-83
[7] Dong Y H *et al* 2015 *Opt. Mater. Express* **5** 712-19
[8] Ruan Y *et al* 2005 *Opt. Commun.* **252** 39-45
[9] Huang Y *et al* 2016 *Opt. Express* **24** 18709-17

IOP Conf. Series: Journal of Physics: Conf. Series **844** (2017) 012020 doi:10.1088/1742-6596/844/1/012020

Efficient absorption by monolayer graphene in microring resonator

Ziyu Liu

Qian Xuesen Laboratory of Space Technology, China Academy of Space Technology, Beijing, China,

100094, Email: liuziyu@qxslab.cn

Abstract. We demonstrate a graphene photodetector based on a SOI microring resonator. The absorption of the graphene was enhanced by the resonating, so to improve the detection efficiency by 5 times. The impact of graphene's position and length on the absorption rate was also been studied. The frequency selective effect shows the potential on spectrum detection and WDM photoelectric conversion.

1. Introduction

At present, photo detector based on the two-dimensional materials provides a promising way to realize the efficient detection of optical. Graphene has emerged as an attractive photoelectrical material because of its ultra-broadband light absorption, high carrier mobility and gate-tunability of the optical absorption. However, there are still some basic scientific problems to be solved, such as dark current, limited bandwidth and low detection. Graphene-based photo detectors on Si waveguides [5-8] have been proposed to enhance the absorption of the graphene, however, the enhancement was determined by the effective length of detection.

In this paper, we demonstrate a graphene photodetector based on a SOI microring resonator. The optical wave was coupled to the ring through a bus waveguide and resonating in the ring, graphene was located on the looped optical waveguide. The absorption of the graphene was enhanced by the resonating, so to improve the detection efficiency.

2. The influence of graphene for micro ring

The schematic of the photo detector is shown in Fig. 1.The microring resonator with base width of 400nm was fabricated on a SOI substrate with 220-nm-thick silicon layer with a 3um buried oxide layer. The monolayer graphene was prepared by mechanical exfoliation on a stack of polymers on the ring resonater. The source and drain contacts on the graphene are prepared on both sides of the ring waveguide, doping for Photovoltaic effect. A gate was prepared to tunning the Fermi level of the graphene.

The transport characters of the graphene microring detector were calculated using three-dimensional (3D) FDTD method. Figure 2(a) shows an add-drop racetrack ring photo detector model with a radiu of 3.1um, the coupling gap of the ring is 100 nm, a straight waveguide was defined in the racetrack ring with an 2D rectangle graphene on it. The TE mode incident into the straight waveguide from in port and coupled to the ring, then resonate in the ring then transmitted to the drop port. The incident field which didn't conform to the resonance condition almost transmitted to the through port.

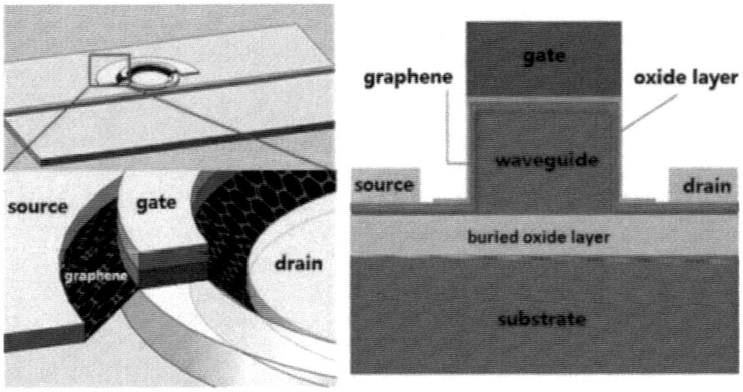

Fig. 1 The schematic of the graphene ring detecter.

2.1. Comparison of graphene on the straight waveguide and on the ring

The optical parameters of monolayer graphene refer to the article [9]. In order to compare the graphene's absorption rate between the ring resonator photodetector and traditional straight waveguide photodetector, ring resonators with graphene on the add port and pass port with the same length also being calculated as references as shown in Fig. 2(a). In order to characterize the graphene's absorption rate, the sum of the transmittance of four ports (input, pass, drop and add) was shown in Fig. 2(b), which reflects the total absorption of the device. A valley at 1.542nm in micro ring without graphene represents the loss in the SOI micro ring. The ring with graphene on the add port represents a bordband absorption, while the ring with graphene on the pass port shows an enhanced absorption at resonance peak of the ring resonator, showing the absorption effect of graphene before and after the micro ring frequency selective filter.

Compared above, the micro ring with graphene on the racetrack looped waveguide represents obvious absorption at ring's resonating wavelength than that with graphene on the drop port. With the ring resonator loss as reference, the absorption at the resonating wavelength was increased by about 5 times than the graphene on the straight waveguide.

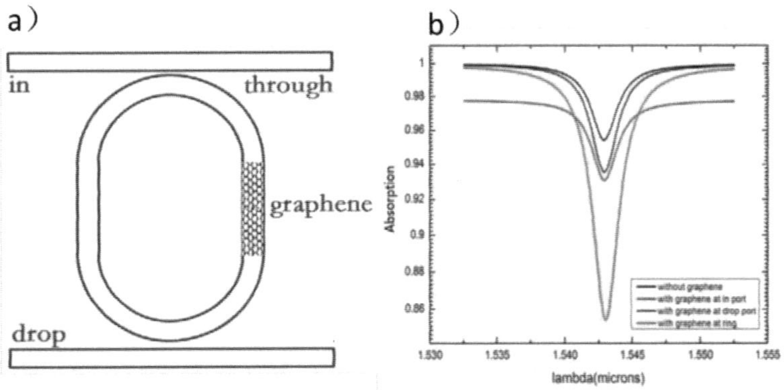

Fig. 2 (a)The schematic of the photodetector in FDTD solution (b)The obsorption of photodetector with graphene on the loop ring and straight waveguide.

2.2. Comparison of graphene on the different position of the ring

We also studied the impact of graphene's position on the absorption rate. Graphene was located on the different position of the ring, seen in Fig.3 (a), covering a 20 um long straight part of the racetrack ring. The absorption rate was then calculated as shown in Fig.3 (b), indicating that the absorption of graphene on the right side is higher than that on the left side. The intensity of the light field in the right part is stronger than that in the left part, which causes the difference on absorption.

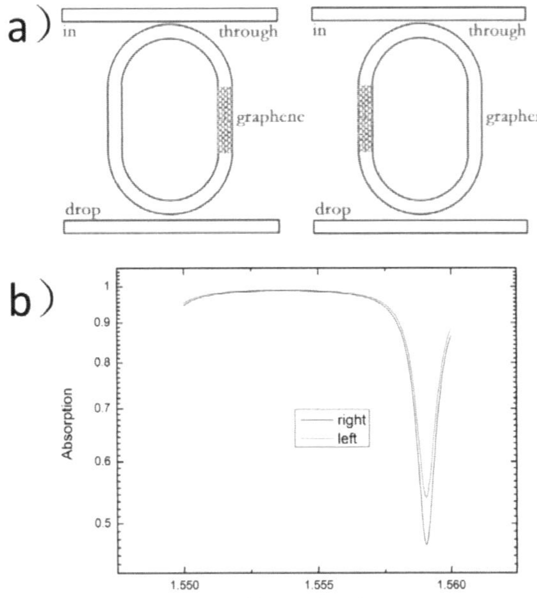

Fig. 3 (a)The schematic of the photodetector with graphene at different position of the loop in FDTD solution(b)The obsorption of photodetector with graphene at different position of the loop.

2.3. The impact of graphene's length on the absorption rate

The impact of graphene's length on the absorption rate was also been studied. A race track microring model with straight part of 20 um on both sides was built for the calculation convenient of the graphene's length, the graphene cover both straight parts with length of L, seen in Fig. 4(a). Fig. 4(b) shows the absorption rate of the ring resonator with different covering length of graphene, the covering length L varies from 0um to 40um. From the Fig. 4(b) we can see that with the increase of graphene cover length, the absorption first increases then decreases.

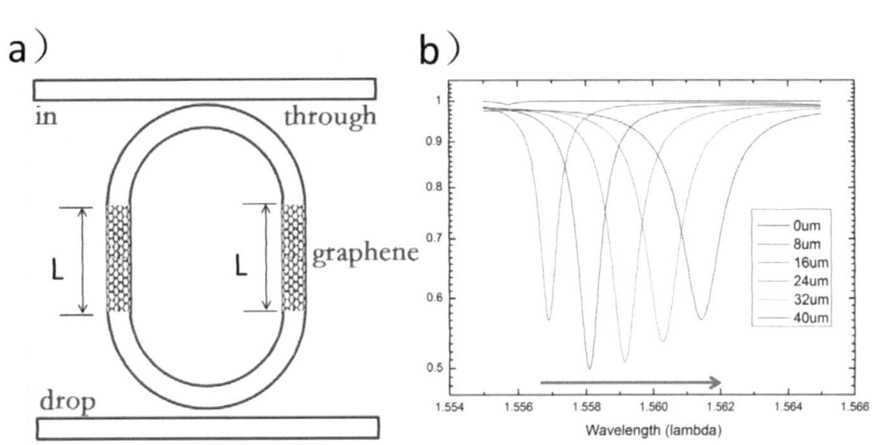

Fig. 4 (a)The schematic of the photodetector with graphene at different position of the loop in FDTD solution(b)The obsorption of photodetector with graphene at different position of the loop.

We deduced that there is a trade-off relationship between the absorption of graphene and resonance of micro ring. When graphene cover length rise to more than 16 um, the resonance of light in the microring damping so violently that it cannot enhance the light field as the growth of the graphene coverage length, the decay of the resonating absorption leads to the absorption attenuation.

In conclusion, we presented a graphene photodetector based on microring resonator. The absorption enhancement was achieved through located the graphene on parts of light field resonance. As a result, graphene detection efficiency increased 5 times than traditional waveguide mode. Location is a factor that influence efficiency of graphene detection, the absorption is related to the optical field distribution of micro rings. There is a trade-off relationship between the absorption of graphene and resonance of micro ring when the cover length of graphene increases. The frequency selective effect shows the potential on spectrum detection and WDM photoelectric conversion.

Acknowledgment
This work is supported by the independent innovation foundation of the Qian Xuesen Laboratory of Space Technology and the independent innovation foundation of CASC.

References

[1] Lee E J, Balasubramanian K, Weitz R T, Burghard M and Kern K 2008 *Nature nanotechnology* **3** 486-90
[2] Park J, Ahn Y H and Ruiz Vargas C 2009 *Nano Lett.* **9** 1742–1746
[3] Xia F, Mueller T, Lin Y M, Valdes-Garcia A and Avouris P 2009 *Nature nanotechnology* **4** 839-43
[4] Mueller T, Xia F and Avouris P 2010 *Nature Photonics* **4** 297-301
[5] Gan X, Shiue R-J, Gao Y, Meric I, Heinz T F, Shepard K, Hone J, Assefa S and Englund D 2013 *Nature Photonics* **7** 883-7
[6] Pospischil A, Humer M, Furchi M M, Bachmann D, Guider R, Fromherz T and Mueller T 2013 *Nature Photonics* **7** 892-6
[7] Wang X, Cheng Z, Xu K, Tsang H K and Xu J-B 2013 *Nature Photonics* **7** 888-91
[8] Youngblood N, Anugrah Y, Ma R, Koester S J and Li M 2014 *Nano letters* **14** 2741-6
[9] Liu M, Yin X, Ulin-Avila E, Geng B, Zentgraf T, Ju L, Wang F and Zhang X 2011 *Nature* **474** 64-7

IOP Conf. Series: Journal of Physics: Conf. Series **844** (2017) 012021 doi:10.1088/1742-6596/844/1/012021

Diode-pumped Tm:YAP/YVO_4 intracavity Raman laser

Jiaqun Zhao[1,4*], Ping Cheng[2], Feng Xu,[3] Xiaofeng Zhou,[1] Guodong Wang[1]

[1]College of Science, Hohai University, Nanjing, China 211100,
[2]College of Computer and Information, Hohai University, Nanjing, China 211100
[3]College of Science, Nanjing University of Aeronautics and Astronautics, Nanjing, China 210016
[4]College of Science, Harbin Engineering University, Harbin, China 150001
*Corresponding author: zhaojq@hhu.edu.cn

Abstract. The laser performance based on YVO_4 Raman conversion in a diode-pumped actively Q-switched Tm:YAP laser is demonstrated for the first time. With an incident diode power of 10.9 W and a pulse repetition rate of 1 kHz, the average output powers for the first Stokes laser at 2.4 μm is about 270 mW.

1. Introduction

Stimulated Raman scattering (SRS) process in crystals is one of the most important nonlinear frequency conversion methods for extending the spectral coverage of the present lasers. The commonly known Raman crystals include $Ba(NO_3)_2$, YVO_4, $GdVO_4$, diamond, and variety of tungstate crystals[1-6].

In this study, we present a diode-pumped actively Q-switched intracavity Raman laser at 2.4 μm. YVO_4 crystal as Raman gain medium and Tm:YAP crystal as laser gain medium are utilized in the intracavity Raman laser. To the best of our knowledge, this is the first time YVO_4 Raman conversion in an actively Q-switched Tm:YAP laser has been demonstrated.

2. Experimental setup and results

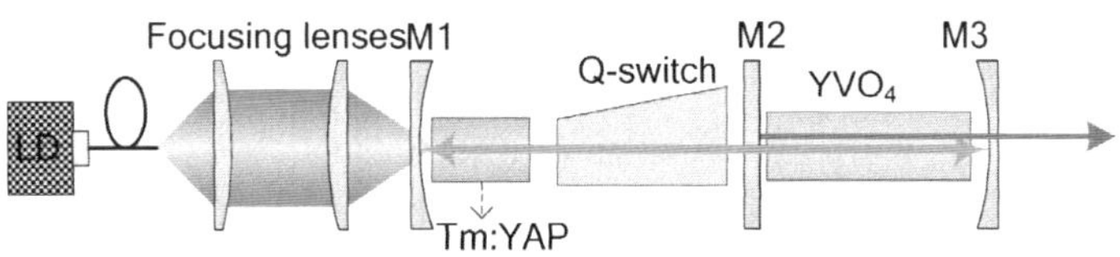

Figure 1. Configuration of the actively Q-switched Tm:YAP/YVO_4 intracavity Raman laser

Figure 1 shows the experimental configuration of the intracavity Tm:YAP/YVO_4 Raman laser. The pump source is a 20 W fiber-coupled continuous-wave 794 nm laser diode with a 400 μm fiber core diameter and a numerical aperture of 0.22. A focus lens system with a pair of plane-convex lenses is used to focus the pump beam into the Tm:YAP crystal. The a-cut Tm:YAP crystal is doped with 3.0 at.% Tm^{3+}, and the laser crystal has dimensions of 3 mm×3 mm×10 mm. Both end faces of the crystal are antireflection coated for the fundamental laser wavelength at 1.99 μm and the pump wavelength at 794 nm. The crystal is wrapped in indium foil and held in a copper heat sink whose temperature is maintained at 16 ℃. The YVO_4 crystal with dimensions of 4 mm×4 mm×30 mm is antireflection coated in the range of 1800-2550 nm (R<3%). It is wrapped with indium foil and mounted on a water-cooled copper heat sink maintained at a temperature of 18 °C. A water-cooled acousto-optic

IOP Conf. Series: Journal of Physics: Conf. Series **844** (2017) 012021 doi:10.1088/1742-6596/844/1/012021

modulator is used for Q-switched operation.

The actively Q-switched Tm:YAP/YVO_4 intracavity Raman laser comprises a fundamental laser cavity and a Raman cavity. The fundamental laser cavity is consisted of a flat mirror M1 and a concave output coupler M3 with a curvature radius of 300 mm. One surface of M1 is antireflection coated at 794 nm (R<0.2%), the other surface is coated for high reflectivity at 2 μm (R>99.8%). The other mirror M3, shared with the Raman cavity as output coupler of the Raman laser, is coated for high reflectivity at 2 μm (R>99.5%), and partial transmission at 2.4 μm (T=10%). One surface of the flat intracavity mirror M2 is antireflection coated at 2 μm (R<1%), the other surface is antireflection coated at 2 μm (R<1%) and high reflectivity at 2.4 μm (R>99.8%). Mirrors M2 and M3 compose the Raman Cavity. The physical length of fundamental laser cavity is about 145 mm, and the Raman laser cavity length is about 40 mm.

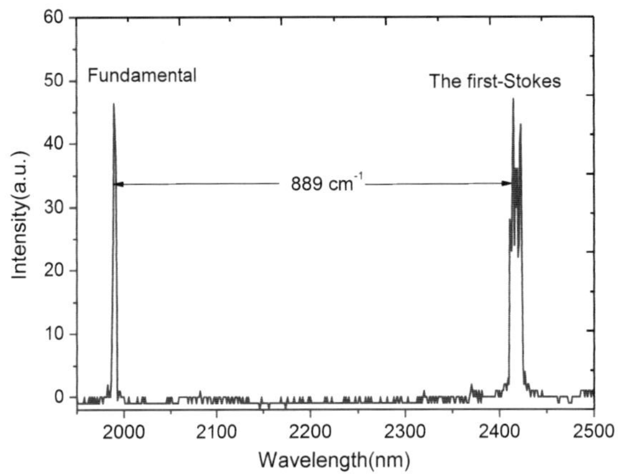

Figure 2. Optical spectrum of the actively Q-switched Tm:YAP/YVO_4 intracavity Raman laser.

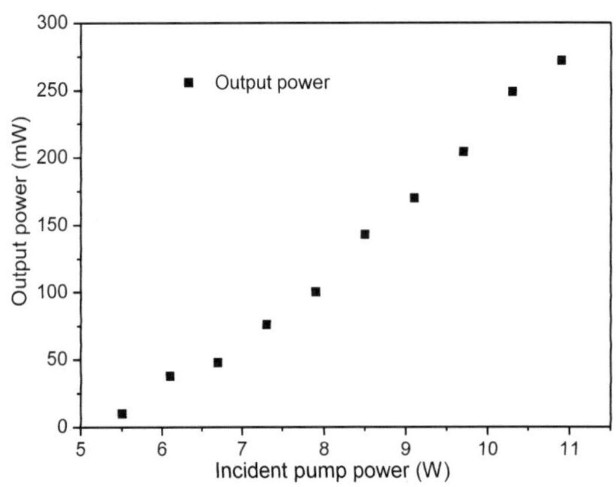

Figure 3. Raman output power versus incident diode pump power

The spectral information of the intracavity Raman laser in Q-switched operation is monitored by a Zolix spectrograph. The Raman laser output powers are measured by a power meter (Coherent, PM10). The operation of the intracavity Raman laser is investigated at a pulse repletion rate of 1 kHz. Figure 2 shows the spectral information of the intracavity Raman laser. The central wavelengths of the fundamental laser and first-Stokes laser are 1990 nm and 2418 nm. We can see that the frequency shift of the fundamental laser and the first Stokes component is 889 cm^{-1}.

Figure 3 shows the average output powers of the first Stokes laser as function of the incident pump power at 794 nm. It can be seen that the threshold powers of the Raman laser at a pulse repetition rate of 1 kHz are approximately 5.0 W. The fundamental and Raman output powers increase almost linearly with the incident pump power from 5.5 to 10.9 W. With an incident diode power of 10.9 W and a pulse repetition rate of 1 kHz, the average output powers for the first Stokes laser at 2.4 μm is about 270 mW.

3. Conclusion

The characteristics of diode-pumped actively Q-switched Tm:YAP/YVO_4 intracavity Raman laser has been demonstrated. With a pump power of 10.9 W, the average output powers of 272 mW for the first Stokes laser of 2.4 μm are obtained at a PRR of 1 kHz.

This research work is in part supported by the Natural Science Foundation of China (61378027, 61301199), and by the Fundamental Research Funds for the Central Universities (2016B02014, 2016B12114).

References

[1] A. Agnesi, E. Caracciolo, L. Carrà, F. Pirzio, and G. Reali 2012 *Appl. Phys. B* ***107*** 691

[2] A. A. Kaminskii, K. Ueda, H. J. Eichler, Y. Kuwano, H. Kouta, S. N. Bagaev, T. H. Chyba, J. C. Barnes, G. M. A. Gad, T. Murai, and J. R. Lu 2001 *Opt. Commun.* **194** 201

[3] A. J. Lee, H. M. Pask, and D. J. Spence 2012 *Opt. Lett.* **37** 3840

[4] R. P. Mildren and A. Sabella 2009 *Opt. Lett.* **34** 2811

[5]L. I. Ivleva, T. T. Basiev, I. S. Voronina, P. G. Zverev, V. V. Osiko, and N. M. Polozkov 2003 *Opt. Mater.* **23** 439

[6]I. V. Mochalov 1997 *Opt. Eng.* **36** 1660

Distributed Feedback Laser Based on Single Crystal Perovskite

Shang Sun, Shumin Xiao[#], Qinghai Song[#]

State Key Laboratory of Tunable Laser Technology, Ministry of Industry and Information Technology Key Lab of Micro-Nano Optoelectronic Information System, Shenzhen Graduate School, Harbin Institute of Technology, Shenzhen, 518055, China

qinghai.song@hit.edu.cn

Abstract. We demonstrate a single crystal perovskite based, with grating-structured photoresist on top, highly polarized distributed feedback laser. A lower laser threshold than the Fabry–Perot mode lasers from the same single crystal $CH_3NH_3PbBr_3$ microplate was obtained. Single crystal $CH_3NH_3PbBr_3$ microplates was synthesized with one-step solution processed precipitation method. Once the photoresist on top of the microplate was patterned with electron beam, the device was realized. This one-step fabrication process utilized the advantage of single crystal to the greatest extend. The ultra-low defect density in single crystalline microplate offer an opportunity for lower threshold lasing action compare with poly-crystal perovskite films. In the experiment, the lasing action based on the distributed feedback grating design was found with lower threshold and higher intensity than the Fabry-Perot mode lasers supported by the flat facets of the same microplate.

1. Introduction

Organic-inorganic halide perovskite materials ($CH_3NH_3PbX_3$) have been studied as efficient photovoltaic due to the ultralow defect density and long carrier diffusion. Besides, these key characteristics are also important for making lead halide perovskites widely applied in photonic devices, like light emitting diodes (LEDs). The most important is the electronic band gap can be tuned continuously by varying the ratio of halide from Cl, Br, I, which helps to tune the emitting color from the devices. Recently, lots of studies focus on the perovskite based amplified emission applications, with both single crystal nanostructure and poly-crystal films. Utilizing the natural regular boundaries of single crystal halide perovskites, polygon microcavities, microrods, spherical cavities and microcrystal networks have been studied to realize lasing actions. There are also several reports these days about amplified spontaneous emission based on thin halide perovskite films, made by solution-processed methods. Distributed feedback (DFB) cavity is a structure with optical feedback can be provided by Bragg scattering from an interference grating. The grating structure can be fabricated either in the active medium or in the neighborhood of the resonant cavity lead to a periodic altered refractive index distribution. A DFB cavity induced laser usually possess lots of advantages such as low threshold, inexpensive manufacturability, high quality factor, mirror-free and so on. Jia *et al.* realized a metal-clad, second order DFB perovskite laser on a silicon-based grating [1]. By nanoimprinting a corrugated structure onto a polymer template, followed by the subsequent evaporation of a conformal perovskite layer, Saliba *et al.* realized a perovskite a DFB cavity [2].

IOP Conf. Series: Journal of Physics: Conf. Series **844** (2017) 012022 doi:10.1088/1742-6596/844/1/012022

Brenner *et al.* demonstrated solution-processed $CH_3NH_3PbI_3$ perovskite distributed feedback lasers fabricated in ambient atmosphere directly on top of low cost polymer gratings [3]. However, in all the DFB cavities mentioned above, poly-crystal perovskite films were used as the active material [4, 5], in which high defect density and scattering were induced, lead to higher thresholds. Here, by simply structuring the photoresist spinned onto a single crystal $CH_3NH_3PbBr_3$ perovskite microplate, a DFB laser with lower threshold than the Fabry-Perot (FP) mode lasers for the same microplate was obtained.

2. Results and Discussion

First we synthesized $CH_3NH_3PbBr_3$ perovskite microplates with one-step solution processed precipitation method. Then, 350 nm ZEP520 film was spin-coated onto the glass substrate with perovskite microplates on it. After 24 hours, the sample was exposed to electron beam in E-beam writer (Raith E-line, 30kV) and developed in ND510 solution for 60 seconds at 0°C to form the ZEP grating structures. Dried under clean nitrogen flow, the sample finally fabricated.

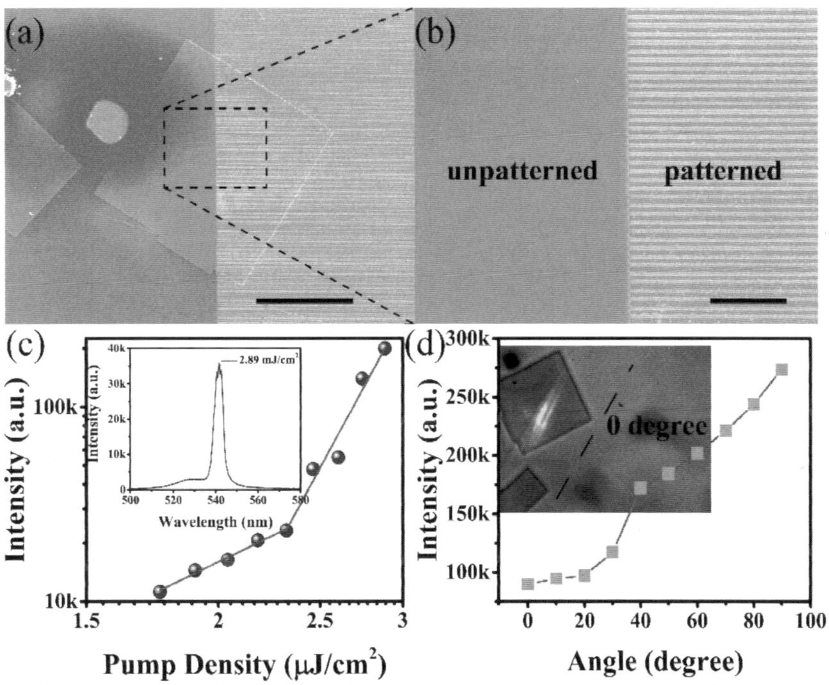

Figure 1. (a)(b) SEM images for the sample. The scale bars are 30 μm and 3 μm, respectively. (c) The output intensity as a function of pumping density. The inset shows the spectrum of the laser. (d) The polarization of the emission. The inset is the corresponding fluorescent microscope image.

As illustrated in Figure 1(a), the scanning electron microscope (SEM) image shows one perovskite microplate was divided into two parts, the left part unpatterned and the right part patterned with gratings. A high-resolution SEM image is illustrated in Figure 2(b) which shows that the period and the width of ZEP tooth were 310 nm and 120 nm. Then the sample was placed onto a three-dimensional translation stage under a homemade microscope and optically excited by a frequency doubled Ti: Sapphire laser (400 nm, 1 kHz repetition rate, and 100 fs pulse duration). The relative position between the laser spot and the sample was controlled by the translation stage. The emitted light and the fluorescent image were collected by the same objected and detected with spectrometer and CCD camera, respectively. For a specific relative pump position, a lasing action can be excited in the grating area, as shown in Figure 1(c). The emission spectrum was a broad peak for low pump density. Once the pump density was above 2.3 μJ/cm2, sharp peaks emerged around 541.6 nm (see the inset in Figure 1(c)). From the inset in Figure 1(d), the fluorescent microscope image is showed. It can be seen, different from common Fabry–Perot modes or whispering-gallery like modes lasers, bright spots locating at the edges of single crystal microplates, a bright line was formed perpendicular to the grating orientation, indicating a DFB laser. Figure 1(d) illustrates the polarization of the emitted laser, which demonstrates strong TE polarization property, closely relating to the grating orientation.

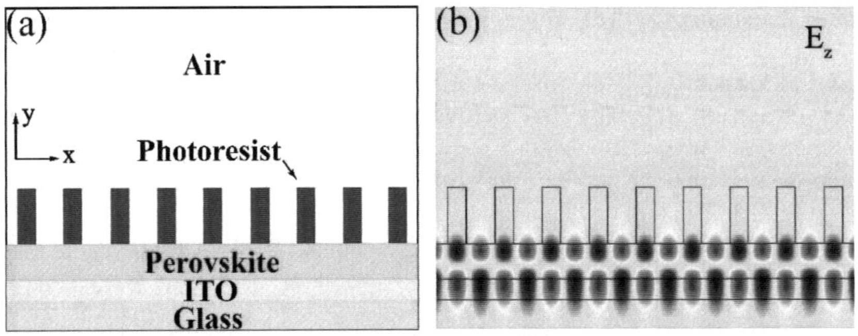

Figure 2. (a) The side-view of the single crystal perovkite based DFB structure. (d) The field pattern of high Q mode at 541 nm in x-y plane.

To further understand the lasing actions, we have also studied the resonant modes by performing full two-dimensional numerical calculation in finite element method software (COMSOL Multiphysics 4.3a). The thickness of the microplate and ITO layer are 225 nm and 135 nm, respectively. A high Q resonance can be found around 541 nm, which is consistent with the experiment results. The field pattern in the transverse plane (x-y plane, see Figure 2(b)) shows that the field distribution is a TE mode in z-direction.

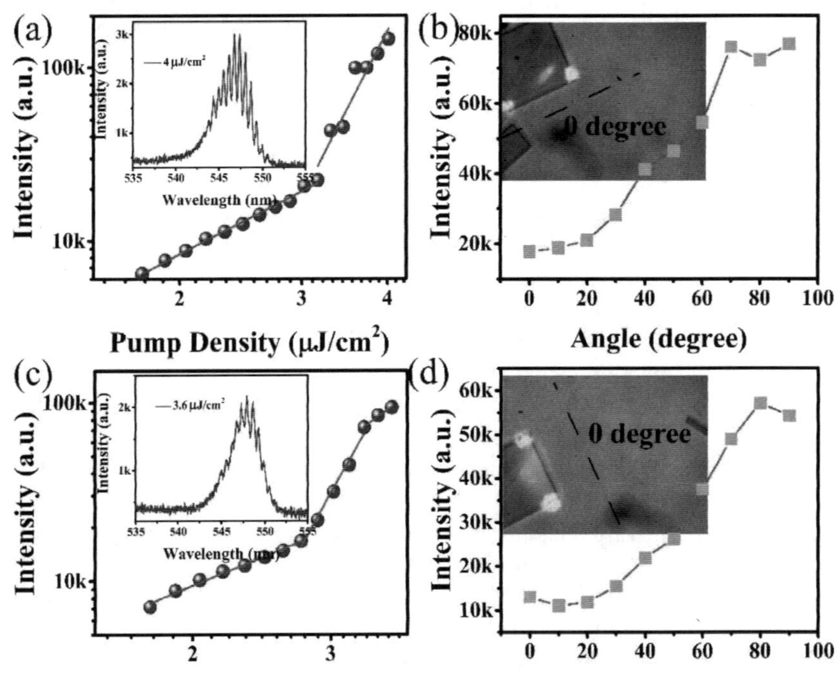

Figure 3. (a) and (c) The output intensity as a function of pumping density for two FP lasers. The insets illustrate the spectra of the lasers. (b) and (d) The polarizations of the emissions. The insets are the corresponding fluorescent microscope images.

Comparisons with the FP lasers supported by the single crystal microplate edges have been done in the experiments. The data in Figure 3(a)(b) and Figure 3(c)(d) corresponded to the two FP lasers for the perpendicular orientations of the microplate. From the curve shown in Figure 3(a) and (c), thresholds pump densities of 3.17μJ/cm^2 and 2.75μJ/cm^2 can be found, which are much higher than the one for the DFB laser supported by the grating structure. The spectra in the insets in Figure 3(a) and (c) indicate the FP modes of the two lasers. Two bright spots located at the opposite side facets of the microplate for all the two lasers as shown in the fluorescent microscope images in Figure 3(b) and (d), which also proved the nature of FP modes. From Figure 3(b) and (d), it can be found that both the two FP lasers were dominated by TE polarization.

3. Conclusions

In this study, a highly polarized distributed feedback laser based on single crystal perovskite has been realized with polymer grating fabricated on it. A much lower threshold pump density was illustrated than the common FP mode lasers supported by the single crystal microplate. This simple one-step fabricated DFB laser based on single crystal perovskite utilized the advantage of single crystal to the greatest extend. We believe our finding will be essential for the practical applications of lead-halide perovskite microlasers.

References

1. Jia Y, Kerner R A, Grede A J, Brigeman A N, Rand B P, Giebink N C 2016 *Nano Lett.* **16** 4624
2. Saliba M, Wood S M, Patel J B, Nayak P K, Huang J, Alexander-Webber J A, Wenger B, Stranks S D, Hörantner M T, Wang J T W, Nicholas R J, Herz L M, Johnston M B, Morris S M, Snaith H J, Riede M K 2016 *Adv. Mater.* **28** 923
3. Brenner P, Stulz M, Kapp D, Abzieher T, Paetzold U W, Quintilla A, Howard I A, Kalt H, Lemmer U 2016 *Appl. Phys. Lett.* **109** 141106
4. Feng Y, Deng H, He J J 2016 *Progress in Electromagnetic Research Symposium* 1951
5. Chen S, Chong W K, Lee J, Roh K, Sari E, Mathews N, Sum T C, Nurmikko 2016 *CLEO: Science and Innovations. Optical Society of America* SM2F. 6.

IOP Conf. Series: Journal of Physics: Conf. Series **844** (2017) 012023 doi:10.1088/1742-6596/844/1/012023

Transmittance characteristics of the plasmonic graphene ribbon with a wing

Lin Yuan1,2), Xin Yan1,2), Yueke Wang1,2)*, Tian Sang1,2) and Guofeng Yang1,2)

1Optical Information Science and Technology Department, Jiangnan University, Wuxi, Jiangsu 214122, China

2Optoelectronic Engineering and Technology Research Center, Jiangnan University, Wuxi, Jiangsu 214122, China

ykwang@jiangnan.edu.cn

Abstract. We investigate the transmittance characteristics of graphene ribbons with a wing numerically by Finite Element Method. By conducting the dispersion relation of edge graphene plasmons (EGPs) modes and analyzing the mode distributions, it is believed that the transmission dips originate from the resonances of three kinds of EGPs modes respectively, including the symmetrical EGPs, anti-symmetrical EGPs, and EGPs of a semi-infinite sheet. By changing the width and length of the wing, the conclusion that transmission dips originate from EGPs modes is further approved. Thanks to the tunable permittivity of graphene by gate voltages, the transmittance dips can be easily tuned.

Recently, the research of graphene attracts more and more attention due to mechanical, electrical, optical, and thermal properties1). Especially, graphene is the most suitable candidate for establishing the platform of surface plasmons (SPs) at infrared frequencies2). Compared with the counterpart in the visible frequencies, which propagates along the noble metal surface, graphene SPs(GPs) show attractive properties, including high effective refractive index, strong confinement and tunable optical properties by gate voltage3). Many novel GPs phenomena have been studied such as cloaking4, 5), Talbot effect6), superlens7), negative refraction8) and so on. Also, tunable plasmonic devices in infrared frequencies are designed, including polarization converters9, absorbers10), filters11), and switchers12).

With the development of the graphene manufacturing technique, many researchers pay more attention to the research on GPs propagating along the graphene nanoribbons, which are called as Edge GPs (EGPs) modes13, 14). EGPs modes are firstly observed experimentally by a scattering type SNOM in a patterned graphene nano-ribbon on the Al2O3 substrates15). The EGPs modes, which can possess deep-wavelength confinement, pave a new way to build compact plasmonic waveguide devices, including splitters16, resonators17, 18), demultiplexers19) and so on. Optical circuit model20) and plasmon wave function21) are also proposed to explain the physical meaning. Besides, other interesting optical phenomena of graphene nanoribbons structures are reported, such as plasmonic bistability22), plasmon induced transparency23), and sensors24).

In this paper, we propose a graphene ribbons structure, which is composed of an input graphene ribbon with a wing. Based on simulation results, we obtain the transmittance spectra in the infrared region ranging from 7μm-8.5μm, which include three transmission dips. By analyzing the mode distribution and dispersion, we believe the transmission dips come from the resonances of symmetrical EGPs (SEGPs), anti-symmetrical EGPs (AEGPs), and EGPs of a semi-infinite sheet (SIEGPs)13). By changing geometry parameters of the wing, the transmission dips are further proved to originate from the resonances of three kinds of EGPs mode. Besides, we also discuss the tuning of transmittance dips by the gate voltage.

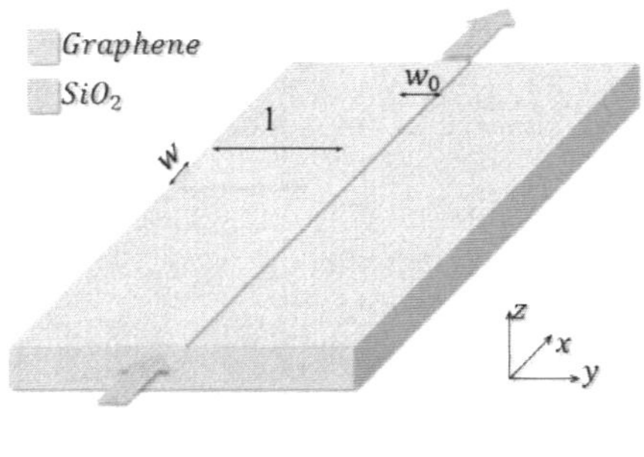

Fig. 1 Basic configuration of the graphene ribbons structure with a wing

The proposed structure consists of a graphene ribbon waveguide on the SiO2 substrate with a wing shown as Fig. 1. The thickness of the graphene structures is 1nm. The refractive index of SiO2 substrate and medium above the graphene are 1.5 and 1, respectively. The length and width of the wing are l and w. And the width of the input ribbon waveguide are w0 (Here, we choose w0 as 20nm, and incident ribbon waveguide only supports SEGPs mode ranging from 7μm to 8.5μm). Commercial software COMSOL Multiphysics based on Finite Element Method(FEM) is adopted to solve the Maxwell equations.

The surface conductivity of graphene is σg, which is obtained by the Kubo formula25):

$$\sigma_g = \frac{ie^2 E_f}{\pi\hbar^2(\omega+i\tau^{-1})} + \frac{ie^2}{4\pi\hbar}\ln\left[\frac{2E_f-(\omega+i\tau^{-1})\hbar}{2E_f+(\omega+i\tau^{-1})\hbar}\right] + \frac{ie^2 k_B T}{\pi\hbar^2(\omega+i\tau^{-1})}\ln\left[\exp\left(-\frac{E_f}{k_B T}\right)+1\right] \quad (1)$$

Here, temperature T is 300 K. It is obvious that σg depends on chemical potential Ef, momentum relaxation time τ, and photon frequency ω. Ef can be tuned by the gate voltage which is applied on graphene. Besides, e, $\hbar$, kB represent the electron charge, reduced Plank's constant and Boltzman's constant, respectively. The permittivity of graphene is governed by:

$$\varepsilon_g = 1 + \frac{i\sigma_g \eta_0}{k_0 \Delta} \quad (2)$$

η0 (≈377Ω) is the impedance of air, τ is chosen as 0.5 ps. Δ(=1nm) is the thickness of graphene. The permittivity of graphene can be electrically tuned by gate voltage, and therefore leads to a voltage-tuned value based on Eqs. (1) and (2).

Before investigating the transmittance characteristics, we first calculate dispersion relations and modes distributions of EGPs supported by graphene shown in Fig. 2. Figs. 2(a) and (b) show the real part of effective refractive index Re(neff) of EGPs with wavelength, when Ef =0.2eV. As discussed in Refs. 13 and 14, the graphene ribbon with finite width can support two EGPs modes, including SEGPs and AEGPs mode. EGPs propagation directional component of the electric field is symmetrical for SEGPs, and anti-symmetrical for AEGPs. The semi-infinite graphene sheet can support only one EGPs (SIEGPs) mode. It is found that Re(neff) decreases with wavelength for the three EGPs modes. For SEGPs mode and AEGPs mode, Re(neff) decreases and increases with the width of the ribbon increasing, respectively. For 20 nm width graphene ribbon, the ribbon only supports SEGPs mode for the wavelength region (7μm-8.5μm) and AEGPs mode is cut off. And under the same wavelength, the

Re(neff) of SEGPs is always larger than that of SIEGPs; that of SIEGPs is always larger than the Re(neff) of AEGPs. Besides, Fig. 2(c) shows that Re(neff) decreases with Ef increasing for all of the edge modes. Figs. 2(d)-(f) plot EGPs propagation directional component of the electric field of SESPs, AESPs and SIESPs respectively, which can depict the mode distributions.

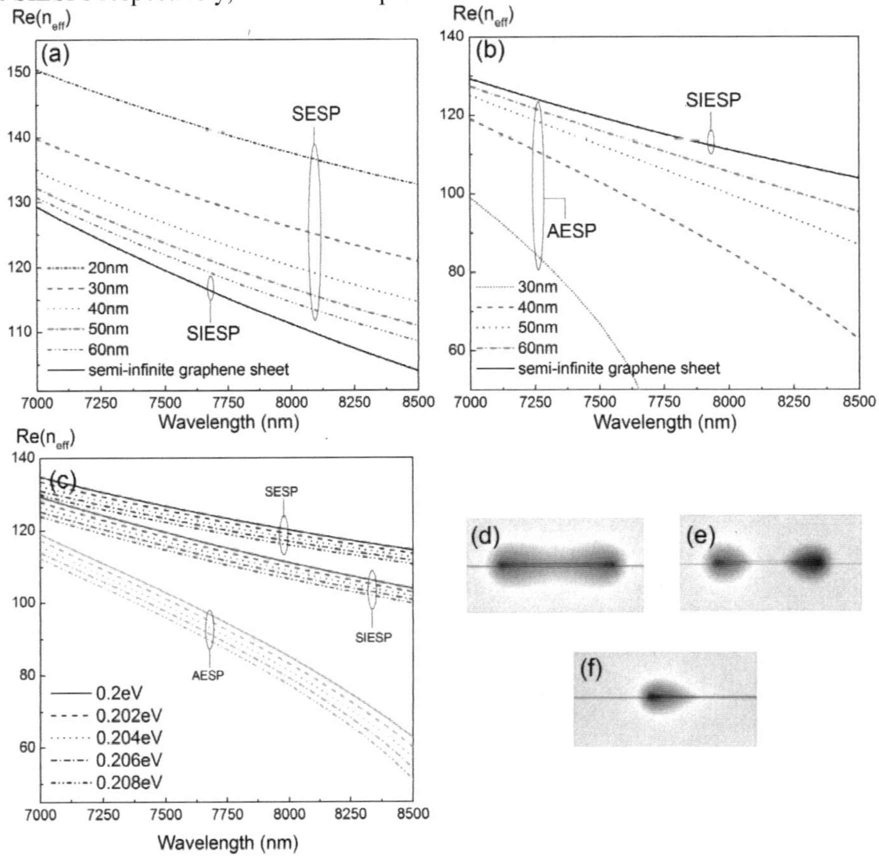

Fig.2. (a) and (b) Real part of effective refractive index Re(n_{eff}) of three ESPs modes *vs* wavelengths under different width *w* of ribbons. (c) Re(n_{eff}) of three ESPs modes vs wavelengths under different Femi levels. The color plots for the EGPs propagation directional component of the electric field of (d) SESPs mode, (e) AESPs mode, and (f) SIESPs mode.

When SEGPs mode propagates along the input ribbon along y-direction, EGPs can be excited in the wing. The wing can be assumed to be an F-P resonator for the EGPs, when the length L of the resonator is satisfied as:

$$\mathrm{Re}(n_{eff})\frac{2\pi}{\lambda}L+\theta=m\pi \tag{3}$$

the resonator is in F-P resonance, most energy of electromagnetic field is coupled into the wing, and the transmission reaches the minimum. Here, Re(neff) is the real part of the EGPs mode, m is the resonance order, L is the circumference length of EGPs resonator, and θ is the additional phase (here, we ignore θ for simplicity). The circumference length L of resonator is l for SESPs and AESPs mode; and w+2l for SIEGPs mode. The calculated transmission spectrum of the ribbon structure for different l (changing from 100nm to 104 nm) is shown in Fig. 3(a). Other parameters here are assumed to be w = 40nm, and Ef = 0.2eV. There are three groups of transmission dips (I, II and III) in the infrared region (ranging from 7μm to 8.5μm), the resonance wavelengths of which red-shift with the length l of the wing increasing. Figs. 3 (b), (d), and (d) plot the y component of the electric field of group I, II, and III, respectively, which shows the edge modes hardly propagate the wing. Concentration of electric field along the edges of wing proves that EGPs modes are excited in the wing and in resonance. Thus, we deduce transmission dips I, II, and III originate from the ring resonance of SIEGPs, third order resonance of SESPs, and second order resonance of AESPs respectively, based on Figs. 3(b)-(d). Symmetrical distribution of Ey in Fig. 3(c) illustrates dips II come from resonance of SESPs and anti-symmetrical distribution of Ey in Fig. 3(d) illustrates dips III come from resonance of AESPs, which are in accordance with Figs. 2(d) and (e), respectively. When increasing l, L of the resonator for all of three modes is lengthened. So, increasing l will produce the red-shift of the resonance wavelengths for the three transmission dips based on Eq. (3), shown in Fig. 3(a). The reason we draw the conclusion transmission dips I come from the resonance of SIEGPs will be further discussed below.

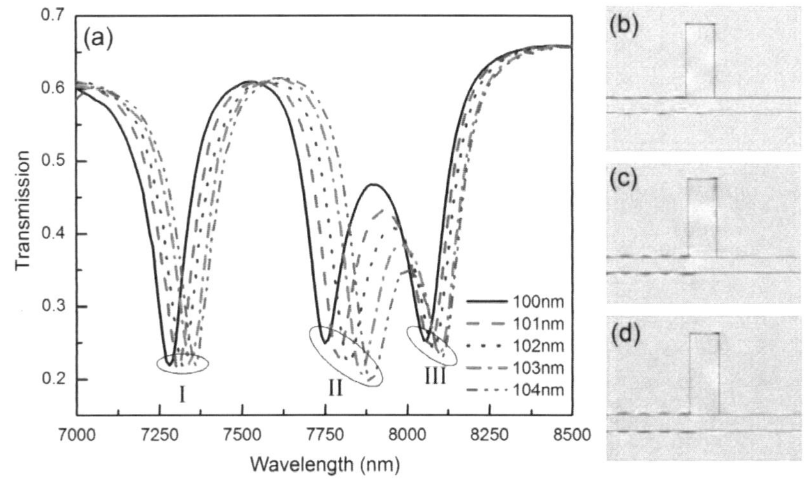

Fig.3. (a) Transmission spectra of the ribbon structure for different length *l* of the wing, when w= 40 nm and E_f = 0.2eV. *y* component of the electric field of resonant wavelengths (b) group I, (c) group II, (d) group III.

To further investigate the transmittance characteristics, we calculate transmission spectrum of ribbon structure under different width w of wing (ranging from 20nm to 60nm) in Fig. 4. Other parameters here are assumed to be l = 100nm, and Ef = 0.2eV. It is found the transmission dips II blue-shift, changing from 8.39 μm to 7.61 μm with increasing w. It is because the Re(neff) of SEGPs decreases with increasing w (shown in Fig. 2(a)), thus transmission dips II blue-shift with w based on Eq. (3), if we ignore θ. So we further confirm that the transmission dips II originate from the third resonance of the SEGPs. To the contrary, the transmission dips I change from 7.28 μm to 7.45 μm when w ranging from 40 nm to 60nm. It is because that though the Re(neff) of SIEGPs mode is independent on w, the circumference length L(=w+2l) for SIEGPs mode increases a little with w, so the transmission dips I red-shifts a little (about 170 nm), based on Eq. (3). Besides when w is 20 nm and 30 nm, there is no dip for the resonance of SIEGPs mode. It is because the width of the wing is so narrow that it can not support the SIEGPs mode in the region from 7μm to 8.5μm. The transmission dips III also red-shift changing from 7.43 μm to 8.45 μm when w is ranging from 30 nm to 50nm. We believe this group of transmission dips originates from the second order resonance of AEGPs mode. It is because Re(neff) of SEGPs increases with w(shown in Fig. 2(b)), and the resonance wavelength red-shifts based on Eq. (3). When w = 20 nm, the wing can not support AEGPs; and w = 60 nm, transmission dip of second order AEGPs moves to the wavelength larger than 8.5μm. Thus, there is no transmission dip III when w = 20 nm and 60 nm, in Fig. 4.

IOP Conf. Series: Journal of Physics: Conf. Series **844** (2017) 012023 doi:10.1088/1742-6596/844/1/012023

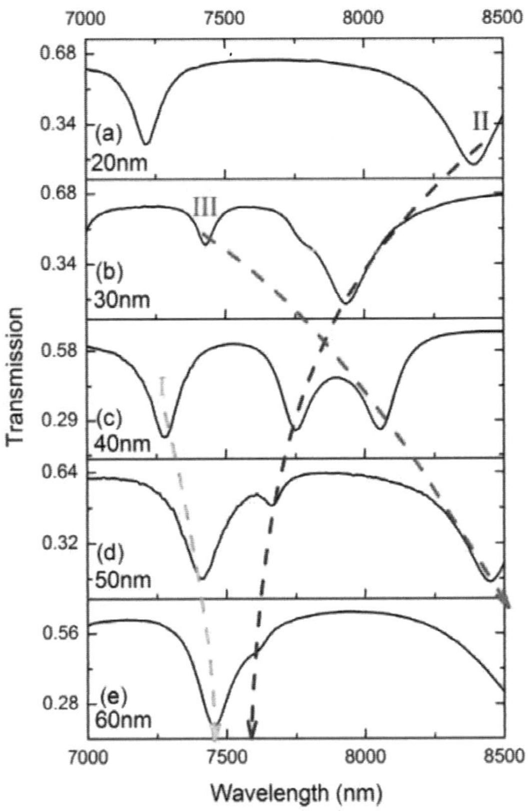

Fig.4. Evolution of the transmission spectrum with w, when l = 100nm, and E_f = 0.2eV.

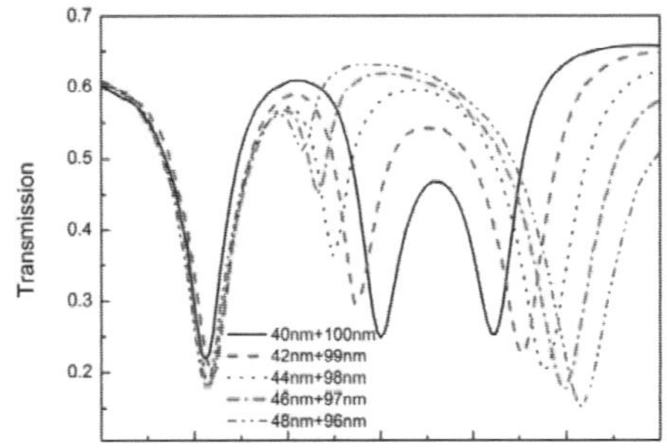

Fig.5. Evolution of the transmission spectrum with the fixed circumference 240 nm of the wing, (w and l) changing from (40 nm and 100nm) to (48 nm and 96nm), when E_f = 0.2eV.

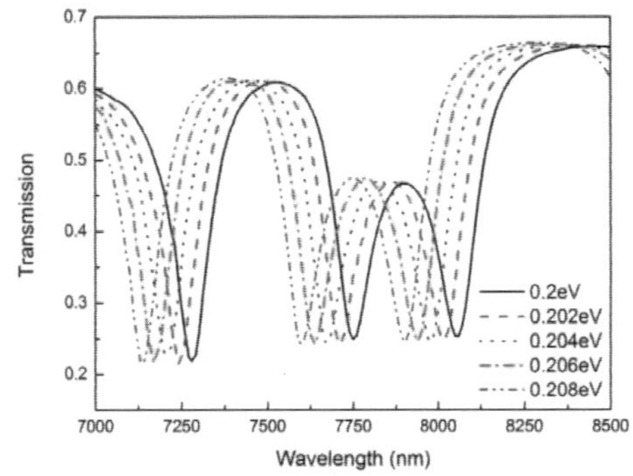

Fig.6. Evolution of transmission spectra for different Fermi level, when l = 100nm and w = 40nm.

To further verify the transmittance dips I come from the ring resonance of SIEGPs mode, we calculate the evolution of transmission spectrum when fixing the circumference (w+2l) of the wing to be 240 nm, shown in Fig. 5. It is found the wavelengths of the dips I are unchanged (7.28μm), when width w and length l of the wing varying from (40 nm and 100nm) to (48 nm and 96nm). It is because Re(neff) and L(w+2l) are unchanged for SIEGPs mode, so does the resonance wavelength of dips I based on Eq. (3). Thus, the conclusion, that transmittance dips I originate from the ring resonance of SIEGPs mode, is approved. But wavelengths of the dips II blue-shift and III red-shift as (w and l) changing from (40 nm and 100nm) to (48 nm and 96nm). It is because that Re(neff) of SEGPs and L decreases with w, so the resonance wavelength of dips II blue-shifts based on Eq. (3). Besides, Re(neff) of AEGPs increases with w, and the decrease of L is less, so the resonance wavelength of dips III red-shifts based on Eq. (3).

The transmission spectra under different Fermi level of graphene Ef(ranging from 0.2eV to 0.208eV) are shown in Fig.6. Other parameters here are assumed to be l = 100nm, and w= 40nm. It is found that the three groups of transmission dips I, II, and III all blue-shift with Ef. It is because Re(neff) of all the three EGPs modes decrease with increasing Ef shown in Fig.2(c). So the blue-shift happens when increasing Ef, based on Eq. (3). It is believed that the simulation results pave a way to design tunable multiple windows of plasmonic filter in infrared region.

We have proposed a plasmonic graphene ribbon with a wing, and investigate the transmittance characteristics by FEM. We believe the transmission dips originate from the resonances of SEGPs mode, AEGPs mode and SIEGPs mode, by studying the dispersion relations and mode distributions of the three kinds of EGPs modes. The change regulations of the three transmission dips with the geometry of the wing further confirm our findings. Besides, the transmission dips can be easily tuned by gate voltage, which provides a smart method to design tunable infrared plasmonic filters.

References

[1] A. K. Geim, and K. S. Novoselov, Nat. Mater. 6, 183(2007)
[2] M. Jablan, H. Buljan; and M. Soljacic, Plasmonics in graphene at infrared frequencies, Phys. Rev. B 80, 245435 (2009)
[3] F. H. L. Koppens, D. E. Chang, and F. J. García de Abajo, Nano Lett. 11(8), 3370–3377 (2011).
[4] V. Ashkan and N. Engheta, Science 332, 1291-1294 (2011)
[5] P. Y. Chen and A. Alù, ACS Nano 5(7), 5855–5863 (2011)
[6] Y. Fan, B. Wang, K. Wang, H. Long, and P. X. Lu, Opt. Lett.39, 3371-3373 (2011)
[7] P. N. Li and T. Taubner, ACS Nano 6, 10107–10114 (2012)
[8] K. V. Sreekanth, A. De Luca, and G. Strangi K. V. Sreekanth, A. De Luca, and G. Strangi, Appl. Phys. Lett. 103, 023107 (2013)
[9] H. Cheng, S. Q. Chen, P. Yu, J. X. Li, L. Deng, and J. G. Tian. Opt. Lett. 38(9), 1567-1569 (2013)
[10] Z. Fang, Y. Wang, A. E. Schlather, Z. Liu, P. M. Ajayan, F. J. García de Abajo, P. Nordlander, X. Zhu, N. J. Halas, Nano Lett. 14 (1), 299-304 (2014)
[11] L. Zhang, J. Yang, X. Fu, and M. Zhang, Appl. Phys. Lett. 103(16), 163114 (2013).
[12] W. –S. Chang, J. B. Lassiter, P. Swanglap, H. Sobhani, S.Khatua, P. Nordlander, N. J. Halas, and S. Link, Nano. Lett. 12, 977-4982 (2012)
[13] A. Y. Nikitin, F. Guinea, F. J. Garcia-Vidal, and L. Martin-Moreno, Phys. Rev. B 84, 161407 (2011).
[14] H. W. Hou, J. H. Teng, T. Palacios, S. J. Chua, Opt. Commun. 370, 226–230 (2016).
[15] Z. Fei, M. D. Goldflam, J. S. Wu, S. Dai, M. Wagner, A. S. McLeod, M. K. Liu, K. W. Post, S. Zhu ,G. C. A. M. Janssen, M. M. Fogler, D. N. Basov, Nano Lett. 15, 8217-8276 (2015).
[16] X. L. Zhu, W. Yan, N. A. Mortensen, and S. S. X, Opt. Express 21(3), 3481-3491 (2013)
[17] L. Zhang, J. Z. Yang, X. L. Fu, and M. Zhang, Appl. Phys. Lett. 103, 163114 (2013)
[18] G. Rosolen, and B. Maes, J. Opt. Soc. Am. B 31(5), 1092-1102 (2014)
[19] E. Forati, and G. W. Hanson, Appl. Phys Lett. 103, 133104(2013)
[20] A. Khavasi, and B. Rejaei, IEEE J QUANTUM ELECT., 50(6), 397-403 (2014)
[21] I. Silveiro, J. M. Ortega and F. J. García de Abajo, New J. Phys. 17, 083013 (2015)
[22] T. Christensen, W. Yan, A. P. Jauho, M. Wubs, and N. A. Mortensen, Phys. Rev. B 92, 121407 (2015)
[23] .X. Han, T. Wang, X. M. Li, S. Y. Xiao, and Y. J. Zhu, Opt. Express 23(25), 31945-32955 (2015)
[24] J. Wu, C. H. Zhou, J. J. Yu, H. C. Cao, S. B. Li, and W. Jia, Opt. Laser Technol. 59, 99–103 (2014)
[25] G. W. Hanson, J. Appl. Phys. 104(8), 084314 (2008)

Zero Cross-Talk Regimes in Dually Modulated Mutually-Coupled Nano-lasers

H Han[1,2], M J Zhang[1], Y C Wang[1], K A Shore[2]

[1]Taiyuan University of Technology, College of Physics and Optoelectronics, Key Laboratory of Advanced Transducersand Intelligent Control System, Ministry of Education, Taiyuan 030024, China.
[2]School of Electronic Engineering, Bangor University, Wales, LL57 1UT, UK.

e-mail: hanhong@tyut.edu.cn; k.a.shore@bangor.ac.uk

Abstract. The modulation properties of dually modulated mutually-coupled nano-lasers have been analyzed using rate equations which include the Purcell cavity-enhanced spontaneous emission factor F and the spontaneous emission coupling factor β. Analysis of the dynamical response of modulated mutually-coupled nano-lasers reveals the existence of regimes of zero cross-talk wherein the response of one nano-laser is not impacted by the dynamics of the other nano-laser. The availability of zero-cross talk regimes is seen to offer opportunities for exploitation in photonic integrated circuits.

1. Introduction

Mutually coupled lasers have been investigated for many decades [1]. Activity on mutually coupled semiconductor lasers also has long antecedents [2], [3] with significant effort having been given to identifying regimes of synchronization and instabilities [4]-[7]. Optical injection is well-known as a means for enhancing the modulation bandwidth of semiconductor lasers [8] and in recent work modulation bandwidth enhancement in mutually-coupled monolithically integrated laser diodes has been reported [9]. Semiconductor nano-lasers [10], [11] are of interest not least for their potential for inclusion in photonic integrated circuits. Fabricating nano-lasers is rather challenging and, as far as we are aware, no experimental studies of their dynamical properties have been published. In this context, it is considered appropriate to explore nano-laser dynamics using a generic model rather than limiting the analysis to any specific nano-laser structure. It is foreseen that such analysis may offer perspectives for future experimental exploration of nano-lasers as they become available.

In early work, the impact of Purcell enhanced spontaneous emission on the modulation performance of nano-LEDs and nano-lasers [12] was examined. In addition to [13], [14] a number of recent investigations of the dynamical performance of nanolasers have been undertaken. The behaviour of optically pumped nanolasers has been studied including the role of the spontaneous emission factor, β, in achieving single mode operation of nanolasers [12]. Ding et al. explored how the dynamics of electrically pumped nanolasers are impacted by F and β [15]. Theoretical work has also been reported on the control of dynamical instability in such lasers [16].

Previous investigations of the influence of F and β shows that modulation bandwidth of up to 60 GHz can be achieved for metal clad nano-lasers [17]. In later work on the effect of optical injection in nanolasers, it has been identified the modulation bandwidth of nanolaser can approach 90 GHz [18].

The theme of the present paper is the impact of direct current modulation on the dynamics of mutually-coupled nano-lasers. The paper is structured as follows. The nano-laser dynamical model is instructed in section 2. Results given in section 3 delineate the main dynamical behavior which arises when mutually-coupled nanolasers are subject to different modulation frequencies, wherein modulation responses are calculated. Finally, in section 4, conclusions are drawn based on the results obtained.

2. Nano-laser Dynamical Model

A schematic diagram of mutually coupled nano-lasers where they are modulated with different modulation frequencies is shown in Figure 1.

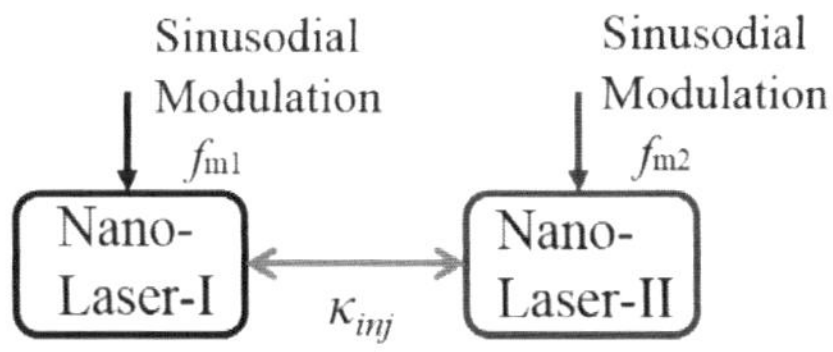

Fig. 1 Schematic diagram of modulated mutually-coupled semiconductor nano-lasers.

It is underlined that the Purcell factor and the spontaneous emission coupling factor impact the spontaneous emission rate as shown in Eqs. (1) and (2) below. Specifically it is pointed out that for Purcell factors greater than unity an effective reduction in the carrier lifetime will result. Similarly an increase of the spontaneous emission coupling factor towards unity also causes an effective reduction of the carrier lifetime. In contrast, the phase Eq. (3) is dependent on the laser gain and hence is not affected by the enhanced spontaneous emission.

$$\frac{dS_{I,II}(t)}{dt}=\Gamma\left[\frac{F\beta N_{I,II}(t)}{\tau_n}+G_n\left(N_{I,II}(t)-N_{th}\right)S_{I,II}(t)\right]+2\frac{\kappa_{inj}}{\tau_{in}}\sqrt{S_{I,II}(t)S_{II,I}\left(t-\tau_{inj}\right)}\cos\left(\theta_{I,II}(t)\right) \tag{1}$$

$$\frac{dN_{I,II}(t)}{dt}=\frac{I_{I,II}}{eV_b}-\frac{N_{I,II}(t)}{\tau_n}\left(F\beta+(1-\beta)\right)-G_n\left(N_{I,II}(t)-N_0\right)S_{I,II}(t) \tag{2}$$

$$\frac{d\phi_{I,II}(t)}{dt}=\frac{\alpha}{2}\Gamma G_n\left(N_{I,II}(t)-N_{th}\right)\pm\Delta\omega-\frac{\kappa_{inj}}{\tau_{in}}\sqrt{\frac{S_{II,I}\left(t-\tau_{inj}\right)}{S_{I,II}(t)}}\sin\left(\theta_{I,II}(t)\right) \tag{3}$$

$$\theta_{I,II}(t)=\pm\Delta\omega t+\omega_{II,I}\tau_{inj}+\phi_{I,II}(t)-\phi_{II,I}\left(t-\tau_{inj}\right) \tag{4}$$

$$I_{I,II}(t)=I_{dc}\left(1+h_m\sin\left(2\pi f_{m1,m2}t\right)\right) \tag{5}$$

In the rate equations including the modulation, the subscripts 'I' and 'II' represent laser I and laser II respectively. $S(t)$is the photon density and $N(t)$ is the carrier density,$\varphi(t)$is the phase of the laser, $\theta(t)$ is the phase of injection laser. $\Gamma$is the confinement factor; τ_n and τ_p are the radiative carrier lifetime and photon lifetime respectively. G_n is the differential gain that takes into account the effect of group velocity, $N_0$is the transparency carrier density, ε is the gain saturation factor and α is the linewidth enhancement factor. $I_{dc}=jI_{th}$ is the dc bias current, where j is the normalized injection current; I_{th} is the threshold current ($I_{th}=(F\beta+(1-\beta))N_{th}V_a/\tau_n$), V_a is the volume of the active region e is the electron charge and N_{th} ($N_{th}=N_0+1/\Gamma g_n\tau_p$) is the threshold carrier density. $\Delta\omega$ is the angular frequency detuning between laser I and laser II. $\tau_{inj}=D/c$ is the injection delay, where D is the distance between laser I and laser II, c is the speed of light in free space.$\tau_{in}=2nL/c$ is the round-trip time in of the laser cavity, where L is the cavity length and n is group refractive index. The mutually-coupled optical injection into the laser I and laser II is controlled by the injection fraction, κ_{inj}, which is related to the injection parameter [20]. Sinusoidal direct current modulation of the lasers included in Eq. (2) is characterised by a modulation frequency,

IOP Conf. Series: Journal of Physics: Conf. Series **844** (2017) 012024 doi:10.1088/1742-6596/844/1/012024

f_{m1} or f_{m2}, for the laser I and laser II, and the corresponding depth of modulation h_m. The values of the nano-lasers device parameters used in the simulations are provided in ref. [22].

3. Spectral Response

3.1. Spectral Characteristics

Figure 2 shows the response when nano-laser I is modulated at the relatively high frequency of 50 GHz whilst nano-laser II is modulated at 10 GHz. This is an example of a bi-directionally isolated zero cross-talk state where no frequency component derived from one nano-laser appears in the other nano-laser. As shown in Figure 3 a non-reciprocal zero-cross state can be accessed by changing a combination of such experimentally-accessible parameter as the injection coupling, the depth of modulation or the laser bias current.

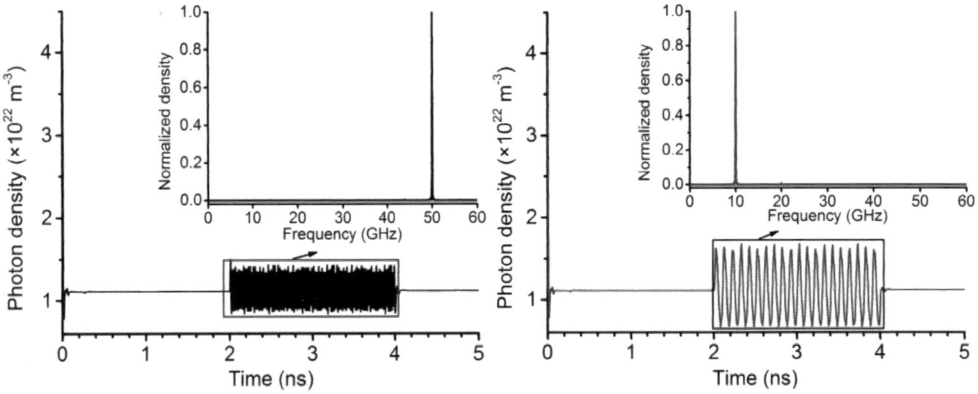

(a) Nano-laser I (b) Nano-laser II

Fig. 2 Photon density time series and FFT (insets) at $I_I= I_{II}= 2I_{th}$, h_m=0.6 and κ_{inj}=0.1×10^{-3}. (a) nano-laser I with f_{m1}=50 GHz, (b) nano-laser II with f_{m2}=10 GHz.

From Figure 3(a) it is seen that nano-laser I, which is subject to the higher modulation frequency, exhibits spectral features of nano-laser II at harmonics of 10 GHz, viz., 20 GHz, 30 GHz and 40 GHz. However it is apparent from Figure 3(b) that no spectral component from nano-laser I appears in the spectrum nano-laser II. That is, one laser displays the signature of the modulation of the other but not vice versa. The practical significance of the zero cross-talk regime is that it permits independent modulation of the coupled lasers.

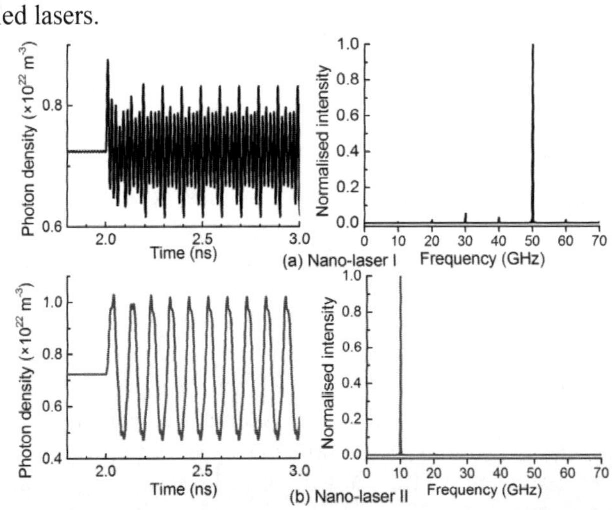

Fig. 3 Photon density time series and FFT at $I_I= I_{II}= 4I_{th}$, h_m=0.4 and κ_{inj}=0.3×10^{-3}. (a) laser-I with f_{m1}=50 GHz, (b) laser-II with f_{m2}=10 GHz.

IOP Conf. Series: Journal of Physics: Conf. Series **844** (2017) 012024 doi:10.1088/1742-6596/844/1/012024

This zero cross-talk behaviour can also be found for relative lower modulation frequency, such as f_{m1}=7 GHz and f_{m2}=3 GHz. The results are shown in the Fig. 4, where the bias currents are twice threshold and the modulation depth is 0.6. We notice that even for such modulation depths as 0.1, bi-directionally isolated zero cross-talk can also be observed if injection strength is $\kappa_{inj}=0.1\times10^{-3}$. With other parameters unchanged, if the injection strength increases to say $\kappa_{inj}=0.3\times10^{-3}$, the bi-directionally isolated behavior requires a larger modulation depth: greater than or equal to 0.2.

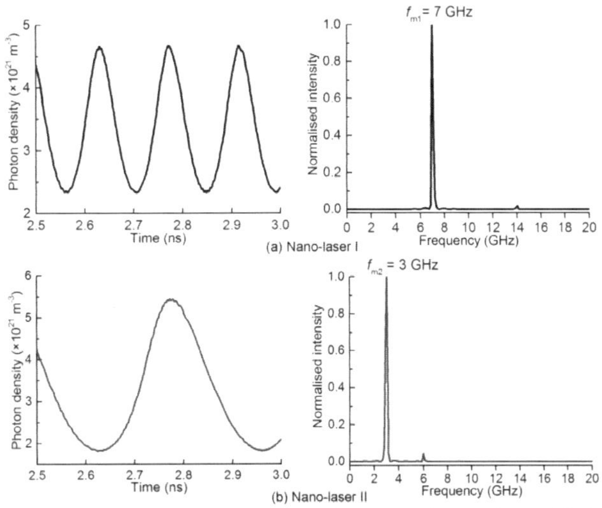

Fig. 4 Photon density time series and FFT at $I_I = I_{II} = 2I_{th}$, h_m=0.6 and $\kappa_{inj} = 0.1\times10^{-3}$. (a) nano-laser I with f_{m1} 7 GHz, (b) nano-laser II with f_{m2} 3 GHz.

3.2. Modulation Response

Having identified two general classes of zero cross-talk modulation regimes, attention is now focused on the direct current modulation response of dually modulated nano-lasers in these regimes.

The modulation response index, η defines as:

$$\eta = \frac{S_{max}(t) - S_{min}(t)}{S_{mean}(t)} \tag{6}$$

where $S(t)$ is photon density of nano-laser in modulation. By altering modulation frequency, f_{m1} or f_{m2}, the corresponding response index values are obtained.

As a function of the laser bias current when the modulation frequency of nano-laser I is varied from 10 GHz to 50 GHz while the modulation frequency for nano-laser II is maintained as 10 GHz. The results given in Figure 5(a) show that the modulation index for nano-laser I increases with increasing bias current but then saturates. The same qualitative behavior is seen for nano-laser II in Figure 4(b) albeit with the different modulation frequency leading to a different modulation index. The salient feature of Figure 5(b) is that the response curves for different modulation frequencies are over-layered thereby explicitly demonstrating the consequences of zero cross-talk.

In fact, strong driving bias currents or deep modulation depth or weak injection coupling strength makes each of the modulated mutually-coupled nanolasers independent of each other. However non-linear cross talk will occur if use is made of strong injection coupled power or a very low modulation depth or relatively small bias currents. A more detailed discussion of these aspects is given in [22].

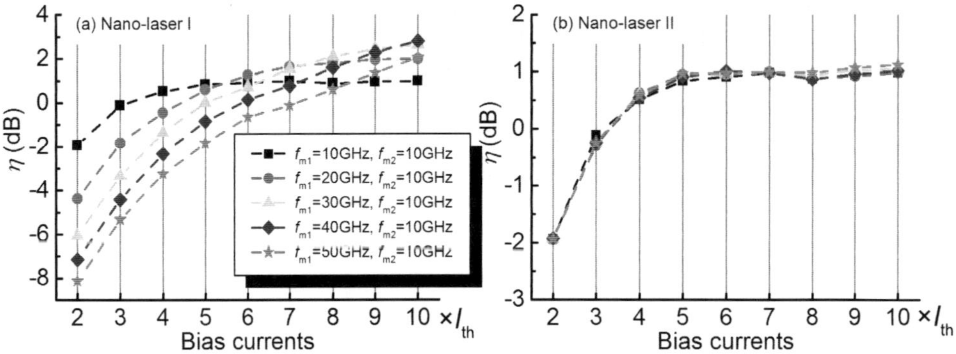

Fig. 5 Modulation response vs. bias currents at h_m=0.6 and $\kappa_{\text{inj}}=0.1\times10^{-3}$. (a) laser-I with f_{m1} from 10 GHz to 50 GHz; (b) laser-II with 10 GHz.

4. Conclusion

Theoretical analysis shows that dually modulated mutually coupled nano-lasers can display a range of interactions depending on the bias current, injection coupling and the modulation depth. Zero cross-talk regimes have been identified for which the modulation response of the lasers have been calculated. The facility to individually address nano-lasers is expected to find ready applications.

Acknowledgments
This work is supported by The Sêr Cymru National Research Network in Advanced Engineering Materials, the International Science & Technology Cooperation Program of China (2014DFA50870), The National Natural Science Foundation of China (61527819, 61475111, 61601319), The Natural Science Foundation of Shanxi Province(201601D202043).

References

[1] Spencer M B and Lamb W E 1972 *Phys. Rev. A*, **5**(2) 893-896
[2] Hohl A, Gavriledes A, Ereux T and Kovanis V 1997 *Phys. Rev. Lett.* **78** 4745-4748
[3] Hohl A, Gavriledes A, Erneux T and Kovanis V 1999 *Phys. Rev. A* **59** 3941-3949
[4] Mulet J, Mirasso C, Heil T and Fisher I 2004 *J. Optics B: Quant. semiclassical Optic* **6**, 97–105
[5] Yanchuk S, Schneider K R and Recke L 2004 *Phys. Rev. E* **69** 056221
[6] Kumar P, Prasad A and Ghosh R 2009 *J. Phys. B: At. Mol. Opt. Phys.* **42**(14) 145401–145407
[7] Hong Y 2015 *IEEE J. Sel. Topics Quantum Electron.* **21**(6) 1801007
[8] Lau E K, Wong L H and Wu M C 2009 *IEEE J. Sel. Topics Quantum Electron.* **15**(3) 618–633
[9] Sun C, Liu D, Xiong B, Luo Y, Wang J, Hao Z, Han Y, Wang L and Li H 2015 *IEEE J. Sel. Sel. Topics Quantum Electron.* **21**(6) 1802008
[10] Ding K, Hill M T, Liu Z C, Yin L J, Veldhoven P J Vand Ning C Z 2013 *Opt. Express,* **21**(4) 4728–4733
[11] Maslov A V and Ning C Z 2007 *Proc. SPIE.* **6468** 646801
[12] Gao H, Fu A, Andrews S C and Yang P 2013 *Proc. Natl. Acad. Sci.* **110**(3) 865-869
[13] Han H and Shore K A,2016 *IEEE J. Quantum Electron.* **52**(11) 2000306
[14] Lorke M, Suhr T, Gregersen N and M φ rk J 2013 *Phys. Rev. B* **87** 205310
[15] Ding K and Ning C Z 2012 *Light: Sci. Appl.* **1**(7) 20
[16] Kumar P and Grillot F 2013 *Eur. Phys. J. Special Top.* **222**(3-4) 813-820
[17] Sattar Z A and Shore K A 2015 *J. Lightw. Technol.* **33**(14) 3028 – 3033
[18] Sattar Z A, Kamel N A and Shore K A 2016 *IEEE J. Quantum Electron.* **52**(2) 1200108
[19] Ding K, Liu Z, Yin L, Wang H, Liu R, Hill M T, Marell M J H, Veldhoven P J V, Notzel

IOP Conf. Series: Journal of Physics: Conf. Series **844** (2017) 012024 doi:10.1088/1742-6596/844/1/012024

R and Ning C Z 2011 *Appl. Phys. Lett.* **98**(23) 231108
[20] Sattar Z A and Shore K A 2015 *IEEE J. Sel. Topic Quantum Electron.* **21**(6) 1800106
[21] Sattar Z A and Shore K A 2016 *IEEE J. Quantum Electron.* **52**(4) 1100108
[22] Han H and Shore K A 2017 *IEEE J. Quantum Electron.* **53**(2) 2000208

IOP Conf. Series: Journal of Physics: Conf. Series **844** (2017) 012025 doi:10.1088/1742-6596/844/1/012025

Broadband Second-Harmonic Generation with Group-Velocity Matching in 5mol%MgO-doped Aperiodic Poled Lithium Niobate

Jian Jiang, Jian-dong Zhang, Kai Wang, Xuan Xiao, and Zu-xing Zhang
School of Opto-electronic Engineering, Nanjing University of Posts and Telecommunications, Nanjing China 210023

E-mail: jiangjian@njupt.edu.cn

Abstract In this paper, we design an aperiodic poled lithium niobate by using genetic algorithm to increase the acceptance bandwidth of second harmonic generation in 5mol% MgO-doped with group-velocity matching, and an acceptance bandwidth of 243.3nm is obtained finally, which is enhanced to be 4.5 times of that in periodic structure.

1. Introduction

Broadband quasi-phase matching (QPM) is widely used in many applications, such as multi-wavelength, second harmonic generation (SHG), ultrashort pulse SHG and optical communications [1-2]. To date, several approaches have been proposed to increase the QPM acceptance bandwidth of SHG. One method is to realize broadband SHG under the conditions that the QPM and group-velocity matching (GVM) of the fundamental (FM) and second harmonic waves are satisfied simultaneously [3-5]. For instance, Yu *et al.* obtained a bandwidth of 52nm at the communication band in 5mol% MgO-doped periodically lithium niobate (MgO: PPLN) with appropriate temperature, FM wavelengths and QPM grating period [6]. Another way is to compensate the phase-mismatch of SHG by designing a proper structure of nonlinear crystal, which was confirmed by several studies. For example, Dang generated a wide flattop SHG bandwidth of 5.5nm in MgO-doped aperiodic lithium niobate (MgO: APPLN) by simulated annealing algorithm [7]; Bostani designed a special apodized chirped PPLN with which SHG spectra over 30nm could attain [8]; Vyunishev analyzed the theory of SHG under nonlinear Raman-Nath diffraction and designed a chirped 2D nonlinear optical superlattice to supports broadband multiple SHG [9]. However, the requirement of QPM bandwidth is nearly a hundred nanometers with the nonlinear crystal of 10mm for frequency doubling of tens femtosecond ultrashort pulse, and the methods mentioned above are difficult to meet the requirement.

In this paper, we propose an efficient method to realize wider acceptance bandwidth of SH with APPLN in the telecommunication band. Firstly, conditions of GVM and QPM in 5mol% MgO: PPLN are analyzed, and simulation shows that the bandwidth is about 53.9nm for 10mm crystal. An aperiodic structure of lithium niobate was selected to increase the acceptance bandwidth of SH. The APPLN has 3333 uniform domains, whose polarization directions are optimized designed by genetic algorithm to obtain the desired profile of nonlinear coefficient in 1550nm. And then we present a method of optimizing the aperiodic poled crystals by adjusting the positions and quantities of the fundamental wavelengths appropriately to broaden the acceptance bandwidth of SH. The simulation results show that the maximum bandwidth of SH is about 243nm for type-I (o+o→e) interaction of QPM around the wavelength of zero group-velocity dispersion, increased by 189nm with the optimal structure of 5mol% MgO: APPLN.

2. Theory

With small signal approximation, the QPM conversion efficiency of SHG is proportional to $\sin c^2(\Delta k_{QPM}L/2)$, here, $\sin c(x) \equiv \sin x / x$ [10], and Δk_{QPM} is the wave vector mismatching which defined as:

$$\Delta k_{QPM} = \Delta k(\lambda) - k_m = k_{2\omega} - k_{\omega} = \frac{4\pi}{\lambda}(n_{2\omega} - n_{\omega}) - k_m \quad \text{.....(1)}$$

here, λ is the wavelength of FM, $\Delta k(\lambda)$ is the phase mismatch between FM and SH waves, k_m is the reciprocal vector provided by poled nonlinear crystals, k_ω and $k_{2\omega}$ represent the wave vectors of the FM and the SH waves respectively, L represents the length of nonlinear crystal, n_ω and $n_{2\omega}$ are the index of refraction of the FM and SH waves.

The wavelength derivative of phase mismatch is given by [3]:

$$\frac{d\Delta k(\lambda)}{d\lambda} = \frac{4\pi c}{\lambda}\delta_{\omega,2\omega} \quad \text{.....(2)}$$

here, $\delta_{\omega,2\omega} = 1/v_\omega - 1/v_{2\omega}$ is defined as group-velocity (GV) mismatch with v_ω and $v_{2\omega}$ represent the group velocity of the FM and the SH waves respectively. The GV mismatch $\delta_{\omega,2\omega} = 0$ can be achieved when $v_\omega = v_{2\omega}$ (when the GVM condition is satisfied, we call the corresponding wavelength the GVM point), which means the GVM condition is satisfied at the corresponding FM wavelength.

From the theory of Fourier transformation, crystals with an aperiodic optical structure can supply reciprocal vectors flexibly and match more optical parametric processes simultaneously. Therefore, it can increase the acceptance bandwidth of SH effectively. With the small signal and the slowly varying approximation, the conversion efficiency η of SHG is proportional to [11]:

$$d_{reff}^2(\lambda) = \left| \frac{1}{L}\int_0^L d(z)e^{i\Delta k(\lambda)z}dz \right|^2 \quad \text{.....(3)}$$

here, $d_{reff}(\lambda)$ is called effective nonlinear coefficient, and $d(z)$ represents the orientation of each domain taking binary values of 1or -1, the length of each block is supposed to be ΔL, and the crystal is divided into N blocks , then $d_{reff}(\lambda)$ can be expanded as:

$$\begin{aligned} d_{reff}(\lambda) &= \left| \frac{1}{L}\int_0^L d(z)e^{i\Delta k(\lambda)z}dz \right| = \frac{1}{L}\left| \sum_{q=0}^{N-1} d(z)\int_{z_q}^{z_{q+1}} d(z)e^{i\Delta k(\lambda)z}dz \right| \\ &= \frac{1}{L}\left| \sum_{q=0}^{N} d(z)\frac{e^{i\Delta k(\lambda)z_{q+1}} - e^{i\Delta k(\lambda)z_q}}{i\Delta k(\lambda)} \right| = \frac{1}{L\Delta k(\lambda)}\left| \sum_{q=0}^{N} d(z)(e^{i\Delta k(\lambda)z_{q+1}} - e^{i\Delta k(\lambda)z_q}) \right| \end{aligned} \quad \text{.....(4)}$$

Here, $q = 0,1,2........N$, and the position of each block is located between z_q and z_{q+1}, the desired structure of aperiodic crystal will be get by optimizing $d(z)$. And then take domain structure functions into the equation (4), the harmonic conversion efficiency curve can be obtained.

3. Simulations and results

In this paper, a 5mol% MgO-doped $LiNO_3$ (5mol% MgO: LN) crystal is used with the temperature set at 20°C.The Sellmeier equation of 5mol% MgO: LN is given by [12]. As is shown in Fig.1, the GVM wavelength is located at 1.560μm for type-I (o+o→e) QPM after analyzing the dispersion property of the 5mol% MgO: LN.

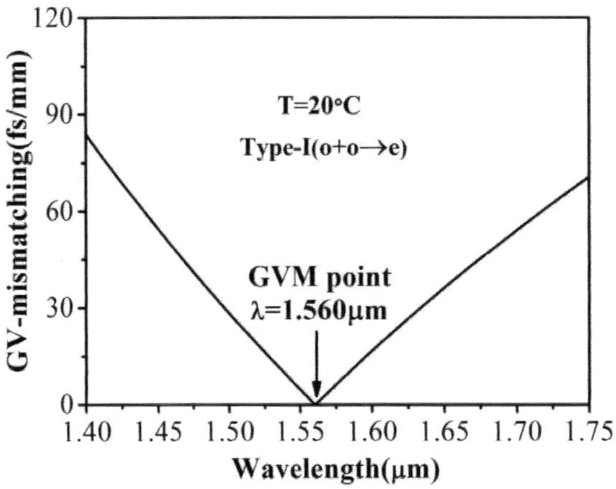

Fig.1. GV-mismatching as a function of fundamental wavelength for type-I QPM in 5mol% MgO: LN.

The schematic diagram of APPLN is illustrated in Fig.2. The length of APPLN is about 10mm which is divided into 3333 uniform domains with the domain width chosen as 3μm. The arrows which represent the direction of polarization are optimized by genetic algorithm[13]. Then the conversion efficiency will be obtained after substituting the optimal structure into Eq. (3).

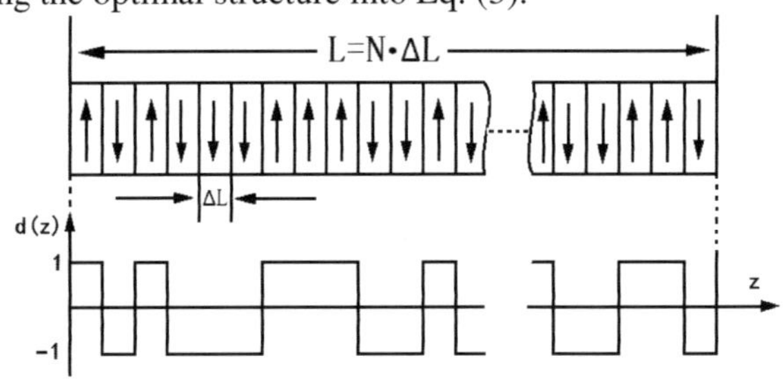

Fig.2. Schematic diagram of 5mol% MgO: CAPPLN.

Subsequently, broadband QPM SHG's property is investigated with the aperiodic structure we designed at the waveband around 1.560μm. Firstly, we set the FM wavelength at 1.560μm, 1.720μm, the optimal structure will be get after genetic operations, such as selection, crossover and mutation procedure implemented. Then substitute the optimal structure into Eq (3). The normalized conversion efficiency curve for the optimized 5mol% MgO: APPLN and the phase mismatch $\Delta k(\lambda)$ as a function of FM wavelength is shown in Figs.3. Here the acceptance bandwidth of SH is defined as the FM wavelength range of the full width at half maximum (FWHM) of the normalized conversion efficiency [14], as it can be seen from the blue line in Fig.3(a), two QPM SHG peaks with bandwidth of 52.2nm and 6.1nm can be acquired at the position of the predesigned FM wavelength, however, another QPM SHG peak at 1.432μm with acceptance bandwidth of 3.8nm can also be achieved without predesigned wavelength at the position, for this position has almost the same value as 1.720μm which is shown in Fig.3(b). Besides, as shown in the red line in Fig.3(a), two broadband QPM SHG peaks with the same bandwidth of blue line in Fig.3(a) can be obtained with the absence of the broader QPM SHG peak at the GVM wavelength with FM wave of 1.720μm selected only. Consequently, the following simulations are carried out with the the GVM wavelength of 1.560μm fixed, and the predesigned wavelength on the right side is chosen as an example. Fig.4 shows the simulation results as the right wavelength shifts to the GVM wavelength. We find that bandwidth of the two broadband QPM SHG peaks become wider and wider as the wavelength close to the GVM point. If the interval between the right FM wavelength and the 1.560μm small enough, those three broadband QPM SHG peaks would overlap to a single broader bandwidth peak but with a significant ripples on the top which will gets smaller as the interval becomes closer, and when the right FM wavelength moves to 1.600μm, a wide flattop QPM SHG bandwidth of 98.1nm is obtained as shown in Fig.4.

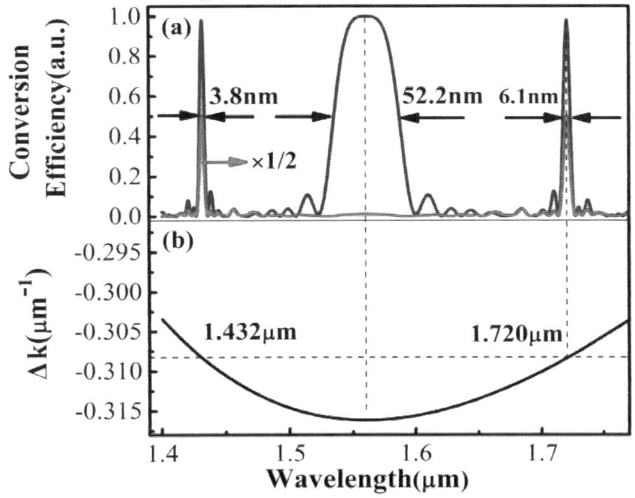

Fig. 3. Conversion efficiency and Phase-mismatch curve.

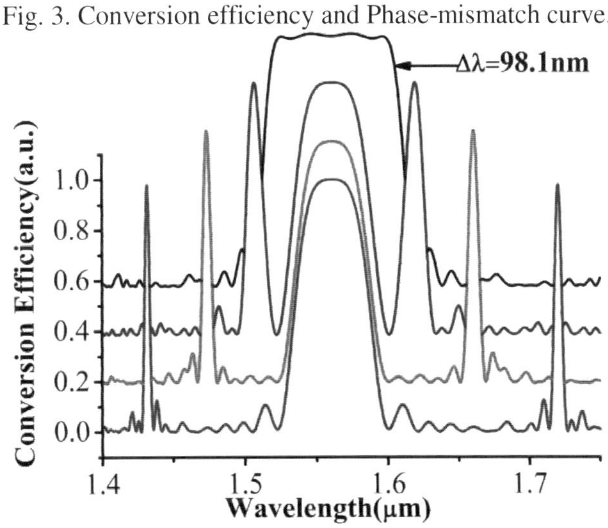

Fig. 4. Conversion efficiency curve as the predesigned wavelength shifts.

In order to further broaden the acceptance bandwidth of SH, simulations are performed with 1.560μm, 1.600μm fixed and one more FM wave added every 5nm on right side. Fig. 5(a)-(f) shows the broadband SHG obtained with the number of 1, 7, 13, 18, 22 and 24 FM waves added, respectively. It is shown that the acceptance bandwidth of SH would broaden with the number of FM wavelength increased and reaches the maximum of 243.3nm with 18 wavelengths added. However, bandwidth of SH will become narrow when further wavelength added.

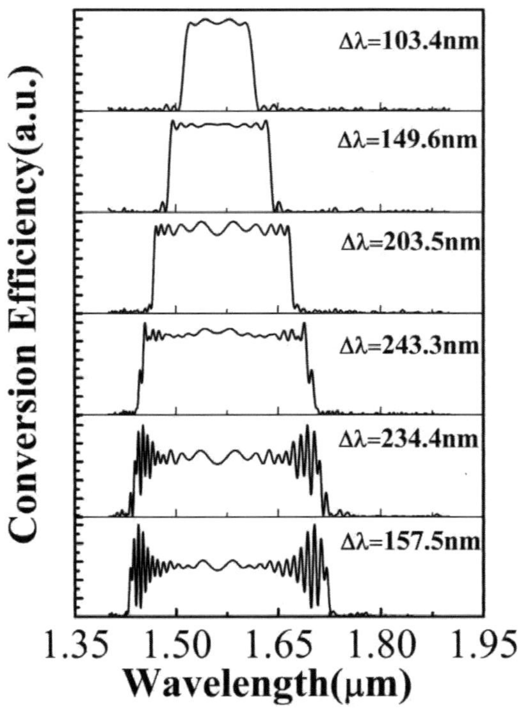

Fig.5. (a)-(f) SHG bandwidth with 1.5601μm and 1.600μm fixed and the number of 1,7,13,18,22,24 FM wavelengths are added every 5nm on the right side.

Acknowledgment

This work was supported by the Natural Science Foundation of the Higher Education Institutions of Jiangsu Province, China (Grant No. 14KJB140010) and the Nanjing University of Posts and Telecommunications Foundation, China (Grant No. NY213029), the Jiangsu Specially Appointed Professor Project (Grant No. RK002STP14001) and the Six Talent Peaks Project in Jiangsu Province (2015-XCL-023).

Reference

[1] B.Q. Chen , C. Zhang, C.Y. Hu, R.J. Liu, Z.Y. Li 2015 *Phys. Rev. Lett.* **115** 083902
[2] L.Y. Zhang, Y.J. Liu, J.J.Huang, S.Z. Pu, Z.Y. Yang 2014 *J. Opt. Soc. Am. B* **31** 1202
[3] Kwang-jo Lee, Choon Sup Yoon and Fabian Rotermund 2005 *Jpn. J. Appl. Phys.* **44** 1264
[4] J.F. Zhang, Y.P. Chen, F. Lu, W.J. Lu, W.R. Dang, X.F. Chen, Y.X. Xia 2007 *Appl. Opt.* **46** 7792
[5]Sung Hak Bae, In Hyeong Baek, Sun Young Choi, Won Bae Cho, Fabian Rotermund, Choon Sup Yoon 2010 *Opt. Commu.* **283** 1894
[6] Nan E Y, Jung H R, Myoungsik C , Sunao K, Takunori T 2002 *Opt. Lett.* **27** 1046
[7] W.R. Dang, Y.P. Chen, X.F. Chen 2012 *Photon. Technol. Lett.* **24** 347
[8] Bostani A, Ahlawat M, Tehranchi A, Morandotti R, Kashyap R 2015 Opt. Express **23** 5183
[9]Andrey M. Vyunishev, Vasily G. ArkhipkinR, Anatoly S. Chirkin2015 *J. Opt. Soc. Am. B* **32** 2411
[10] Martin M. Fejer, G. A. Magel, Dieter H. Jundt, Robert L. Byer 1992 *J. Quant. Electron.* **28** 2631
[11] X.F. Chen and F. Wu 2004 *Phys. Rev. A* **69** 0117018
[12] Q. Gayer,Z. Sacks, E. Galun, A. Arie 2008 *Appl. Phys. B* **91** 343
[13]J.Y. Lai, Y.J. Liu, H.Y. Wu, Y.H. Chen, S.D. Yang 2010 *Opt. Express* 18 5328
[14]J. Jiang, J.H. Chang, S.J. Feng, L. Wei, Q.H. Mao 2010 *Opt. Express* **18** 4740

IOP Conf. Series: Journal of Physics: Conf. Series **844** (2017) 012026 doi:10.1088/1742-6596/844/1/012026

Plasmonic welding of hybrid Au-ZnO nanostructure

Z Y Chen[1], H B Yang[1], P Ghosh[1], Q Li[1] and M Qiu[1]

[1]State Key Laboratory of Modern Optical Instrumentation, College of Optical Science and Engineering, Zhejiang University, Hangzhou, 310027, China

E-mail: qiangli@zju.edu.cn

Abstract. We report that 532 nm CW laser can be used to obtain non-wetting metal-semiconductor (Au-ZnO) Schottky heterojunctions by plasmonic welding. Single crystal Au and n-type ZnO nanowires are placed on gold (Au) and titanium (Ti) electrodes, respectively, and the junction welding is realized. The current-voltage (I-V) characteristic curve of the single Schottky rectifier is also measured.

1. Introduction

One-dimensional metal/semiconductor nano-materials have received widespread attention for their potential applications in micro/nano-electronics and -photonics. Metal/semiconductor Schottky junction with nonlinear electrical response can be applied in integrated circuits, including high-speed switches [1] and rectifier units [2]. The condition of Schottky junction formation is a large bias between the work function of the metal and the fermi level of the semiconductor. Different sizes and doping concentration can lead to different barrier heights.

The metal/semiconductor Schottky junction are generally achieved with two different methods: (1) chemical growth methods (including chemical vapor deposition [3], thermal vapor deposition [4, 5,6], self-assembly [7], and liquid phase method [8].) and (2) nano-welding technology (including chemical [9], electrical joule heat [10, 11], thermal annealing [12], and photothermal welding [13]). Chemical growth methods have disadvantages such as complex operations and strict reaction conditions, including high vacuum and high temperature. In contrast, nanoscale welding method provides possibilities for building complex structures, reducing barrier height, enhancing mechanical strength, and increasing conductivity.

2. Experiment

2.1. Experimental setup

The length of n-type ZnO nanowire ranges from 30 μm to 100 μm, and its diameter is about 150 nm. The length of single-crystal Au nanowire ranges from 20 μm to 30 μm, and its diameter is about 200 nm. Figure 1 is the schematic of the experimental set-up for photothermal nano-welding. A Nd: YAG laser is used as the light source. The operating wavelength is 532 nm. An attenuator is placed next to the laser to control the laser intensity. To regulate the exposure time, a shutter is used. Here, the exposure time is set as 4 ms. For uniform heating, circularly polarized light is used. A CCD is used to monitor the nanomanipulation by two fiber probes.

IOP Conf. Series: Journal of Physics: Conf. Series **844** (2017) 012026 doi:10.1088/1742-6596/844/1/012026

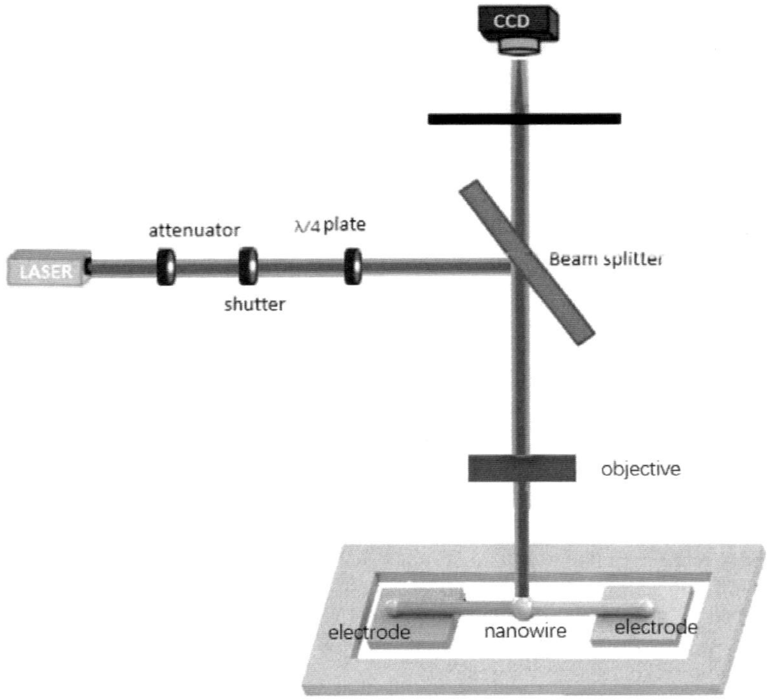

Figure 1. Schematic of experimental set-up for photothermal nano-welding.

2.2. Results and analysis

A single crustal Au nanowire is placed on top of a n-type ZnO nanowire using a nano-manipulator. The circularly polarized laser beam of wavelength 532 nm is focused on the top of the nanowire junction using a microscope objective (100 X). The scanning electron microscope (SEM) images are used to observe the quality of the welded junction. Figure 2 shows the welded junction of the single crystal Au and n-type ZnO nanowires.

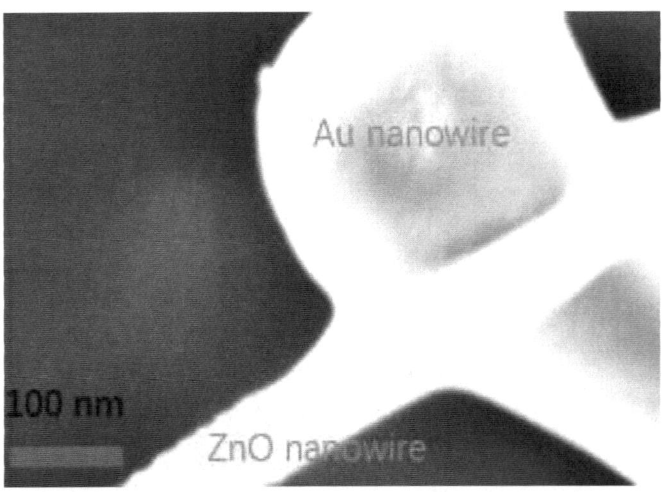

Figure 2. SEM image of the welded junction of the single crystal Au and n-type ZnO nanowires.

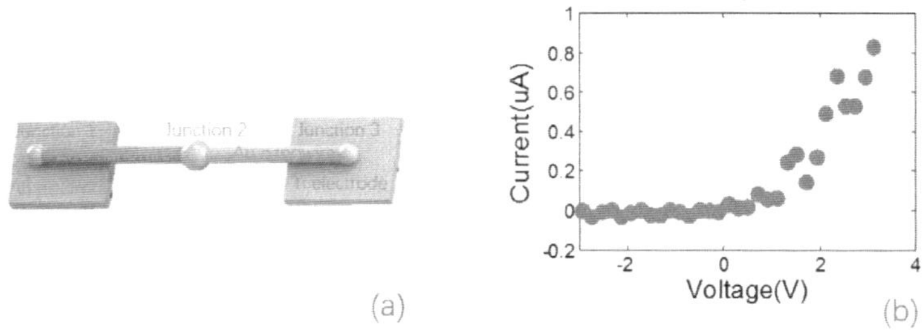

Figure 3. (a) The schematic diagram of the single crystal Au and n-type ZnO nanowires placed on gold and titanium electrodes, respectively. (b) The I-V characteristic curve of the single Schottky rectifier.

We studied electrical property of this structure towards further probing the quality of the welded junction. In order to obtain the characteristic current-voltage (I-V) curve, the single crystal Au and n-type ZnO nanowires were placed on Au and Ti electrodes, respectively. We welded three junctions: (1) single crystal Au nanowire on Au electrode, (2) single crystal Au and n-type ZnO nanowires, and (3) n-type ZnO nanowire on Ti electrode (see figure 3. a). Junction 1 and 3 are ohmic contacts because the barrier height, which is the difference between work function of Ti (4.1 eV) and Fermi level of ZnO (4.3 eV), is less than 1.0 eV. On the other hand, junction 2 is Schottky contact due to the large barrier height between single-crystal Au (work function is 5.2 eV) and n-type ZnO nanowires. The I-V characteristic curve (see figure 3.b) shows that the current remains about zero when the applied bias voltage is below +0.7 V, and it starts increasing beyond this voltage. The maximum current is about 1 A at bias voltage 3 V. This is typical behavior of single Schottky rectifier. It can be concluded that good quality junction is realized by plasmonic welding.

3. Conclusion
Our experimental results demonstrate the feasibility of the nano-scale device fabrication of metal-semiconductor (Au-ZnO) Schottky heterojunctions by photothermal welding technique. It has great significance in large-scale integration applications such as high-speed switches and rectifier units.

Acknowledgment
This work is supported by the National Natural Science Foundation of China (grant nos 6142503, 61575177 and 61235007).

References
[1] Hara E, Machida S and Ikeda M 1981 A high speed optoelectronic matrix switch using heterojunction switching photodiodes *IEEE. J. Quant.* **17** 1539-46
[2] Polyakov A Y, Milnes A G, Govorkov A V, Druzhinina L V and Tunitskaya I V 1995 Band offsets in heterojunctions of InGaAsSb and AlGaAsSb *Solid-Stat. Electron* **38** 525–9
[3] Jung S W and Park W I 2002 Time-resolved and time-integrated photoluminescence in ZnO epilayers grown on Al_2O_3(0001) by metalorganic vapor phase epitaxy *Appl.Phys.Lett.* **80** 1924-6
[4] Hu J, Bo Y, Zhan J, Li C and Golberg D 2007 Mg_3N_2-Ga: Nanoscale Semiconductor-Liquid Metal Heterojunctions inside Graphitic Carbon Nanotubes *Adv. Mat.* **19** 1342–6
[5] Hu J, Bo Y, Zhan J, Xu F, Sekiguchi T and Golberg D 2004 Growth of Single-Crystalline Cubic GaN Nanotubes with Rectangular Cross-Sections *Adv.Mat.* **16** 1465–8
[6] Hu J, Bo Y and Golberg D 2005 Sn-Catalyzed Thermal Evaporation Synthesis of Tetrapod-Branched ZnSe Nanorod Architectures *Small* **1** 95-9

[7] Yang J, Levina L, Sargent E H and Kelley S O 2006 Heterogeneous deposition of noble metals on semiconductor nanoparticles in organic or aqueous solvents *J. Mater. Chem* 16 4025-8
[8] Wen B, Liu C and Liu Y 2005 Bamboo-shaped ag-doped TiO_2 nanowires with heterojunctions. Inorganic *Chemistry* **44** 6503-5
[9] Lee J, Lee P, Lee H B, Hong S, Lee I and Yeo J 2014 Room-temperature nanosoldering of a very long metal nanowire network by conducting-polymer-assisted joining for a flexible touch-panel application *Adv. Funct. Mater.* **23** 4171-6
[10] Peng Y, Cullis T and Inkson B 2009 Bottom-up nanoconstruction by the welding of individual metallic nanoobjects using nanoscale solder *Nano. Lett.* **9** 91-6
[11] Song T B, Chen Y, Chung C H, Yang Y M, Bob B and Duan H S 2014 Nanoscale joule heating and electromigration enhanced ripening of silver nanowire contacts *Acs Nano* 8 2804-11
[12] Gu Z, Ye H, Gracias D H and Gracias D 2005 The bonding of nanowire assemblies using adhesive and solder *Jom* **57** 60-4
[13] Li Q, Liu G, Yang H, Wang W, Luo S and Dai S 2016 Optically controlled local nanosoldering of metal nanowires *Appl. Phys. Lett.* **108** 193101

IOP Conf. Series: Journal of Physics: Conf. Series **844** (2017) 012027 doi:10.1088/1742-6596/844/1/012027

Enhanced Nonlinear Effect of Lithium Niobate Based Periodic Nano-antenna Array

X L Pei[1], S A Bai[1], J Y Tian[1], P Ghosh[1], Q Li[1] and M Qiu[1]

[1]State Key Laboratory of Modern Optical Instrumentation, College of Optical Science and Engineering, Zhejiang University, Hangzhou, TN 310027, China

E-mail: qiangli@zju.edu.cn

Abstract. We report nonlinear properties of lithium niobate based periodic nano-antenna array. The resonances of this nano-antenna array can be engineered by tuning the geometrical parameters. The nonlinear effect gets enhanced when the electric and magnetic resonances overlap.

1. Introduction

It is well known that nano-antenna can precisely control electromagnetic waves owing to its subwavelength structure. Research on nano-antenna has become a new focus because of its diverse applications, including thermal absorption [1-3] and biomedicine [4-6]. Furthermore, the strong field confined to a small region due to electromagnetic resonances in nano-antenna provides a new way to study nonlinear effects. However, conventional metallic nano-antenna manipulates electromagnetic field with their surface plasmon resonance which is difficult to tune, and it also manifests large ohmic losses. In addition, these metals, which are typically used to fabricate the nano-antenna, are also not ideal nonlinear materials. In this regard, dielectric nano-antenna adequately overcomes these drawbacks. Unlike metal, dielectric nano-antenna has low ohmic losses [7-9] and higher intrinsic nonlinearity. Particularly, lithium niobate is an important candidate for dielectric nano-antenna because of its multifunctional photoelectric and highly nonlinear characteristics specifically, second order nonlinear susceptibility.

In dielectric nano-antenna, electric and magnetic resonances occur when the incident light wavelength satisfies certain condition corresponding to the geometry of the nano-structure [10]. These resonances lead to strong electric field confinement in the nano-antenna [11,12]. In the lithium niobate based dielectric nano-antenna, different kind of electromagnetic resonances can be induced in different wavelength regime. The wavelength position of electric and magnetic resonances can be changed by manipulating the shape and size of the nano-antenna [13] as the resonance-wavelength is related to the geometry of the nano-antenna. The nonlinear effects can be further enhanced by strong electric field which can be achieved by overlapping the electric and magnetic resonances. This overlapping can be obtained by manipulating the shape and size of the nano-antenna structure [14].

2. Multipole

When electromagnetic wave is incident on a non-magnetic nanoparticle, polarization current will be generated. This polarization current can be derived using Poynting theorem when the surrounding medium is nonmagnetic ($\boldsymbol{\mu_r = \varepsilon_r = 1}$). Total extinction power can be calculated following the equation given by [10]

$$P_{ext} = \mathrm{Im}(\frac{\omega}{2}\int_V \vec{E}_0^*(\vec{r})\vec{P}(\vec{r})dV'), \tag{1}$$

where P_{ext} is the total extinction power, $\vec{E_0}$ is the electric field of the incident electromagnetic wave, $\vec{r}$ refers to the position, and $\vec{P}(\vec{r})$ is the polarization vector induced in the nanoparticle by the incident electromagnetic wave. $\vec{P}(\vec{r})$ can be expressed as

$$\vec{P}(\vec{r}) = \int_V \vec{P}(\vec{r}')\delta(\vec{r}-\vec{r}')dV'. \tag{2}$$

The extinction power can be approximated to be the scattering power as the absorption is negligible. The total extinction power (P_{sca}) can be given by

$$P_{sca} \square P_{sca}^P + P_{sca}^m + P_{sca}^Q + P_{sca}^M, \tag{3}$$

where P_{sca}^P, P_{sca}^m, P_{sca}^Q, and P_{sca}^M are electric dipole (ED), magnetic dipole (MD), electric quadrupole (EQ), and magnetic quadrupole (MQ), respectively. Only the first four electromagnetic multipole terms are considered as the higher order terms are negligible. P_{sca} can also be written as

$$P_{sca} \square \frac{c^2k_0^2Z_0}{12\pi}\left|\vec{p}\right|^2 + \frac{c^2k_0^2Z_0}{12\pi}\left|\vec{m}\right|^2 + \frac{c^2k_0^2Z_0}{12\pi}\left|Q_{\alpha'\beta'}\right|^2 + \frac{c^2k_0^2Z_0}{12\pi}\left|M_{\alpha'\beta'}\right|^2, \tag{4}$$

where

$$\vec{p} = \varepsilon_0(1-\varepsilon_r)\int_V \vec{E}(\vec{r}')dV',$$

$$\vec{m} = \frac{i\omega\varepsilon_0(1-\varepsilon_r)}{2c}\int_V \left|\vec{r}\times\vec{E}(\vec{r}')\right|dV',$$

$$Q_{\alpha'\beta'} = \frac{\varepsilon_0(1-\varepsilon_r)}{2}\int_V \left[r_{\alpha'}E_{\beta'} + r_{\beta'}E_{\alpha'} - \frac{2}{3}(\vec{r}.\vec{E}(\vec{r}'))\delta_{\alpha'\beta'}\right]dV',$$

$$M_{\alpha'\beta'} = \frac{i\omega\varepsilon_0(1-\varepsilon_r)}{3c}\int_V [(\vec{r}\times\vec{E}(\vec{r}'))_{\alpha'}r_{\beta'} + (\vec{r}\times\vec{E}(\vec{r}'))_{\beta'}r_{\alpha'}]dV',$$

c is the speed of light in vacuum, k_0 is the wavenumber of incident light in vacuum, and Z_0 is the free-space impedance. α' and β' refer to the two of the three coordinates (x, y, z). The electromagnetic resonances, which are calculated using equation (4), are tuned by varying the size and shape of the nano-antenna.

3. Simulation

3.1. Structure of the nano-antenna

In order to analyse the characteristics of the lithium niobate based periodic nano-antenna array, following simulations are done. The inverted-cone antenna was modelled by the finite-difference time-domain method (FDTD Solutions). The schematic of the nano-antenna structure is shown in figure 1. In this case, both substrate and the nano-antenna are lithium niobate. Thus, we can fabricate the structure directly on the lithium niobate bulk which reduce the complexity of the process. The xz- or yz- cross-section of each unit is a trapezoid. The radius of the top surface is larger than that of the bottom surface for each unit. The slope, θ, of the side wall is about 80 degree. The simulation shows that the overlapping will be better if the structure has higher slope, but the difficulty of the fabrication increases sharply as the slope is increased. The periodicity of the nano-antenna array is T (=2 × r_{top} + 200 nm), where r_{top} is the radius of the top surface. The whole system are illuminated by plane-wave with the –z direction. The boundary conditions of the z direction is perfectly matched layers. The mesh size used for the simulation was 10 nm.

IOP Conf. Series: Journal of Physics: Conf. Series **844** (2017) 012027 doi:10.1088/1742-6596/844/1/012027

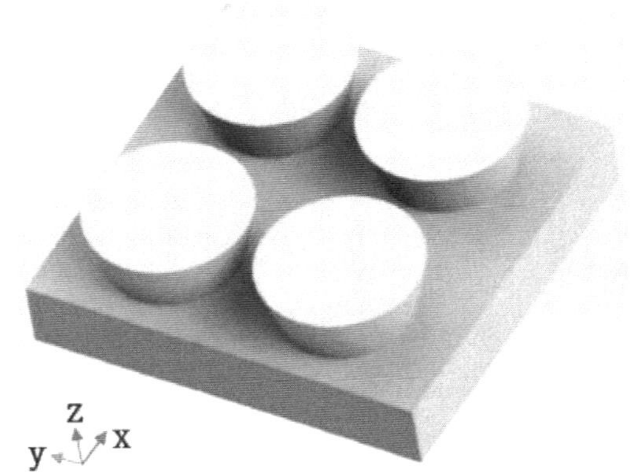

Figure 1. Schematic diagram of nano-antenna array.

3.2. Results and discussion

Figure 2 (a) shows the transmission spectrum of the periodic nano-antenna array. The radius of the top and bottom surfaces are 320 nm and 250 nm, respectively. The height of the nano-antenna is 430 nm. In the transmission spectrum (figure 2 (b)), the three marked valleys from right to left correspond to magnetic dipole, electric dipole, and magnetic quadrupole resonances, respectively. The magnetic quadrupole mode is mixed with the higher order modes and could not be identified in the transmission spectrum.

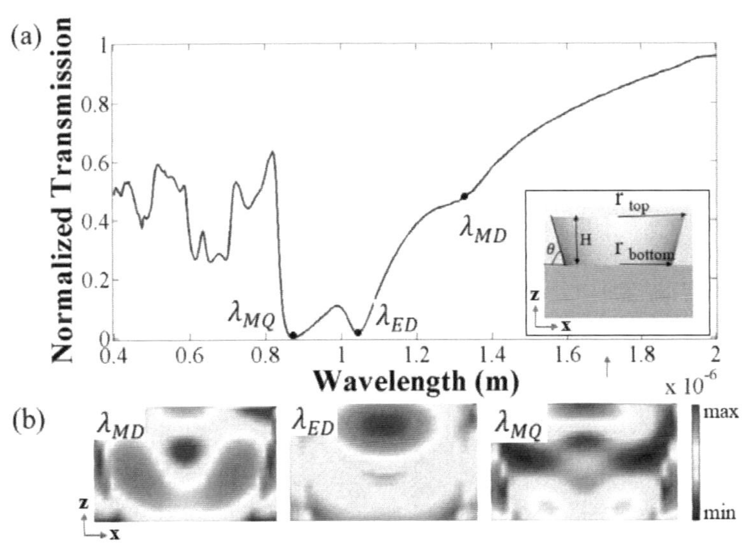

Figure 2. (a) Simulated transmission spectrum of the periodic nano-antenna array. (b) 2D electric field distribution patterns corresponding to the three modes: magnetic dipole (MD), electric dipole (ED), and magnetic quadrupole (MQ) resonance.

For the simulations of nonlinear effects, the incident laser beam wavelength is chosen according to the resonant wavelength obtained from the transmission spectrum of the nano-antenna array. The second order nonlinear susceptibility of lithium niobate is 3×10^{-11} m/V [8]. From the simulation, we get the power of the second harmonic generation (SHG) (see figure 4 (a)).

Furthermore, the electric dipole and magnetic quadrupole resonances can be overlapped by varying the geometrical parameters of the nano-antenna (figure 3). Thereafter, we derive the SHG signal strength again (figure 4 (b)) and compare it with previous results. It demonstrates that the nonlinear effect gets enhanced when the two resonances are overlapped.

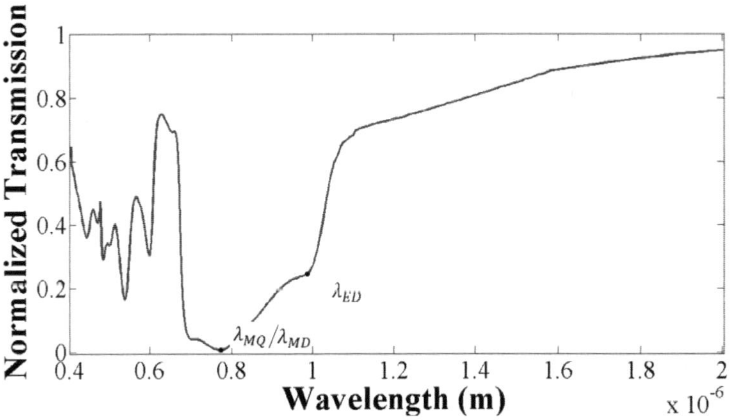

Figure 3. (a) Simulated transmission spectrum of the periodic nano-antenna array with r_{top} = 240 nm, r_{bottom} = 170 nm, *H*=30 nm, *and* θ = 80°.

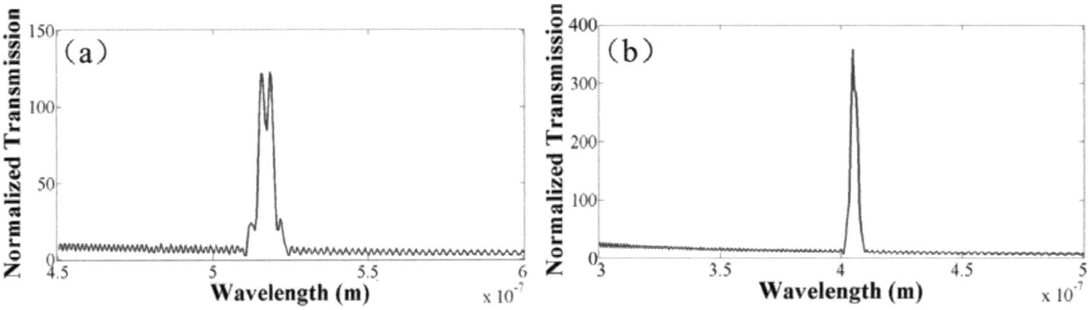

Figure 4. Simulated transmission spectrum showing the peak corresponding to SHG with r_{bottom} = 250 nm (a) and 170 nm (b).

4. Conclusion

We report nonlinear properties of lithium niobate based periodic nano-antenna array. The resonances of this nano-antenna array can be engineered by tuning the geometrical parameters. The electric field confined to the nano-antenna gets enhanced when resonance occur. Intensity of the field gets further enhanced if two of the resonances are overlapped. Therefore, the nonlinear effect can be enhanced by tuning the geometrical parameters of the nano-antenna.

5. Acknowledgment

This work is supported by the National Natural Science Foundation of China (grant nos. 61425023, 61575177, and 61235007).

6. Reference

[1] Liu X, Fan K, Shadrivov I V and Padilla W J 2017 *Opt. Express* **25** 191-201
[2] Callewaert F, Chen S, Butun S and Aydin K 2016 *J. Opt.* **18** 075006
[3] Luo M, Shen S, Ye Y, Liu Y, Zhou Y and Chen L 2016 *Plasmonics* 1-5
[4] Zhao X, Xue J, Mu Z, Huang Y, Lu M, and Gu Z 2015 *Biosens. Bioelectron.* **72** 268-74
[5] Liu Y, Lu Z, Lin X, Zhu H, Hasi W, Zhang M and Lou X 2016 *RSC Adv.* **6**, 58387-93
[6] Petrella R A, Schoenbach K H and Xiao S 2016 *IEEE Trans. Plasma Sci.* **44** 708-14
[7] Biagioni P, Huang J S and Hecht B 2011 *Rep. Prog. Phys.* **75** 0034-4885
[8] Krasnok A E, Miroshnichenko A E, Belov P A and Kivshar Y S 2012 *Opt. Express* **20**, 20599-604
[9] Filonov D S, Krasnok A E, Slobozhanyuk A P, Kapitanova P V, Nenasheva E A and Kivshar Y

IOP Conf. Series: Journal of Physics: Conf. Series **844** (2017) 012027 doi:10.1088/1742-6596/844/1/012027

S 2012 *Appl. Phys. Lett.* **100** 245403-1896
[10] Sikdar D, Cheng W and Premaratne M 2015 *J. Phys.* **117** 083101
[11] Smirnova D A, Khanikaev A B, Smirnov L A and Kivshar Y S 2016 *ACS Photonics* **3** 1468–76
[12] Camacho-Morales R, Rahmani M, Kruk S, Wang L, Xu L and Smirnova D A 2016 *Nano Lett.* **16** 7191-7
[13] Staude I, Miroshnichenko A E, Decker M, Fofang N T, Liu S and Gonzales E 2013 *ACS Nano* **7** 7824-32
[14] Tian J Y, Li Q, Yang Y Q and Qiu M 2016 *Nanoscale* **8** 4047-53

Geometric-Phase Polarization Fan-out Grating Fabricated with Deep-UV Interference Lithography

Chenhao Wan[1,2], David Lombardo[2], Andrew Sarangan[2], and Qiwen Zhan[2*]
[1] School of Optical and Electronic Information, Huazhong University of Science and Technology, Wuhan, China 430074
[2] Department of Electro-Optics and Photonics, University of Dayton, Dayton, USA 45469
*E-mail: qzhan1@udayton.edu

Abstract We report the design, fabrication and testing of a highly efficient polarization fan-out grating for coherent beam combining working at 1550 nm. The grating design exploits the geometric-phase effect. Deep-UV interference lithography is used to fabricate the designed grating. Such a polarization fan-out grating demonstrates several advantages that are ideal for laser beam combining.

1. Introduction

Coherent beam combining offers a path to higher radiance of laser sources by combining the power produced from a number of lower-power gain elements. Dammann gratings are commonly utilized as a fan-out element for beam combining because their binary phase feature facilitates the design and fabrication. However, the power diffracted to undesired higher orders plagues the splitting efficiency and deteriorates the overall efficiency of the laser combining system. Multilevel and continuous phase gratings can considerably increase the fan-out efficiency at the expense of fabrication with high complexity and cost.

Recently optical elements taking advantage of the geometric-phase shift have attracted interest and spur numerous applications. Contrary to adjusting the optical path length, the geometric-phase shift arises from the evolution of a lightwave through an anisotropic parameter space [1]. The geometric-phase holograms (GPHs) can be made extremely thin with an arbitrary and continuous phase profile. Two leading approaches have been adopted to fabricate GPHs including metasurfaces consisting of carefully aligned V- or rod-shaped nanoparticles and liquid crystals assisted by the photo-alignment technique. However, transmissive metasurfaces are generally low in efficiency and photo-alignment liquid crystals are not highly resistant to laser induced damage. Therefore fan-out gratings made with these two techniques are not quite suitable for high-power laser beam combining purposes. To surmount this hurdle, we fabricate the geometric-phase fan-out grating on silicon with deep-UV interference lithography. Although the interference lithography is limited in its use due to the inability to create arbitrary geometries, it is sufficient to fabricate geometric-phase fan-out gratings that consist of blocks of binary subwavelength structures with various orientations. The intrinsic nature of the interference lithography also makes the fabrication of large-area gratings an inexpensive task.

2. Grating design and fabrication

2.1. Geometric-phase fan-out grating

The geometric phase was first described by S. Pancharatnam in 1956, and applied to quantum mechanics by M. Berry in 1984 [2-3]. The geometric-phase shift can be generated by inhomogeneous and anisotropic optical elements and imparted on two of the three outgoing waves: the primary wave and the conjugate wave [1]. By controlling the polarization state of the incident wave and the form birefringence of the material, it is possible to direct all the power to the primary wave or the conjugate wave. The schematic of a 16-phase-level 1-by-5 beam splitter is shown in Fig. 1. The grating consists of 16 periods and each period is constituted by eight segments. Each segment is formed by binary subwavelength structures with different widths (40.8 μm, 44.1 μm, 34.2 μm, 30.9 μm) and orientations (123.75°, 33.75°, 22.5°, 0°)

as listed in Fig. 1. Therefore the geometric-phase shifts provided by each segment are 1.375π, $0.375\ \pi$, $0.25\ \pi$ and 0 respectively. The etching depth of the binary structures, according to the rigorous coupled wave analysis, is 666 nm for the pitch of 314 nm to maximize the power distributed to the primary wave.

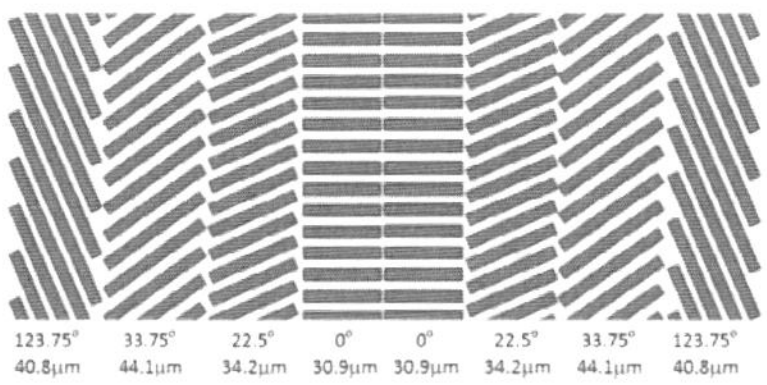

Fig. 1. Schematic diagram of the 1-by-5 geometric-phase fan-out grating (one period).

2.2. Deep-UV interference lithography

The deep-UV interference lithography system consists of a diode-pumped continuous-wave YAG laser that is quadrupled to 266nm. The laser beam is expanded, collimated and shaped to produce a uniform illumination to cover a 4-inch diameter at the Lloyd's mirror stage as shown in Fig. 2 [4]. Since the mirror and sample are both physically mounted on the Lloyd's mirror stage, the effect of vibration between the two beams is reduced considerably, allowing for a sharp pattern even after very long exposure times.

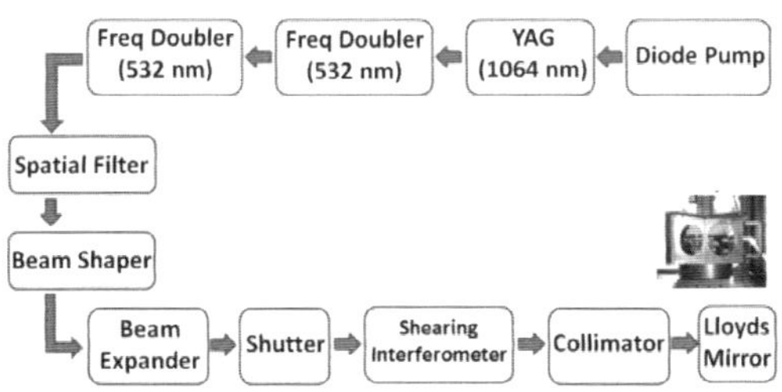

Fig. 2. The optical system of the deep-UV interference lithography.

The period of the fringes created by the two beam interference is given by:

$$\Lambda = \frac{\lambda}{2\sin\theta}\ , \tag{1}$$

where λ is the wavelength (266 nm) and θ is the incident angle. The incident angle is controlled by the rotation of the stage and a 25-degree incident angle renders a period of 314 nm. The size of the grating is 4.8 mm by 4.8 mm. The grating contains four orientations of subwavelength structures and each orientation is accomplished with one exposure. A grating of a larger size can be fabricated just by expanding the laser beam. The scanning electron microscopy (SEM) images of the grating are presented in Fig. 3 showing great fabrication quality. The gaps between adjacent segments have negligible effects on the far-field intensity distribution.

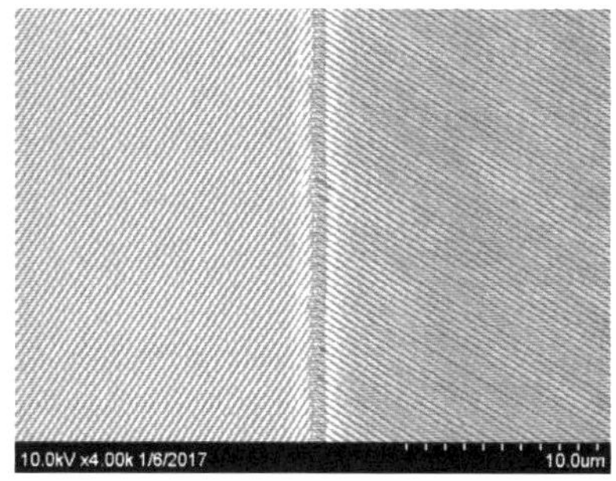

Fig. 3. SEM images of the geometric-phase fan-out grating.

3. Grating test

The grating is tested with a left-handed circular polarization (LHCP) laser beam at 1550nm. The far-field intensity distribution is measured by an infrared camera. The experimental setup is illustrated in Fig. 4.

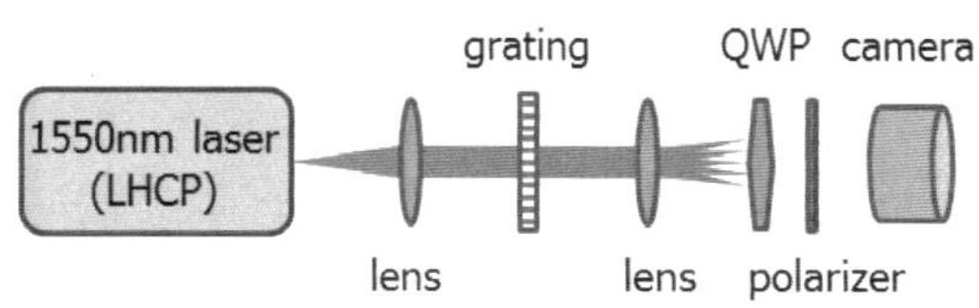

Fig. 4. Schematic of grating test setup. QWP: quarter-wave plate.

The transmitted field of the grating splits into five individual beams and their polarization states are all right-handed circularly polarized (RHCP) verified by a quarter-wave plate (QWP) and a linear polarizer. The far-field intensity distribution is shown in Fig. 5.

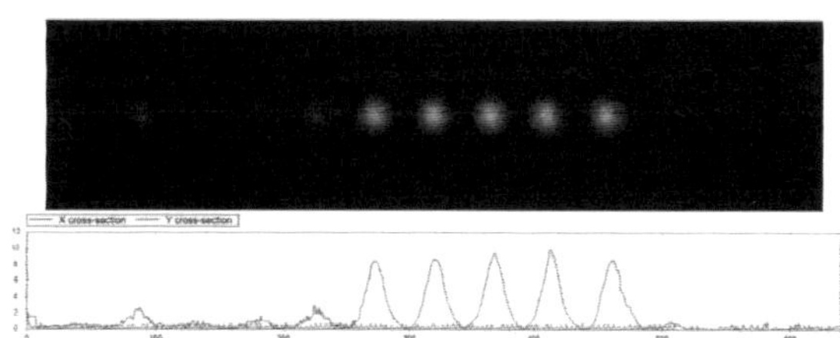

Fig. 5. The far-field intensity distribution captured by the

IOP Conf. Series: Journal of Physics: Conf. Series **844** (2017) 012028 doi:10.1088/1742-6596/844/1/012028

camera.

The five desired orders are theoretically to have identical power and the experimental result gives a power ratio of $\eta_{-2}:\eta_{-1}:\eta_0:\eta_1:\eta_2$ =1 : 1.01 : 1.1 : 1.15 : 1.01.

Conclusions

We design and fabricate a geometric-phase polarization fan-out grating with deep-UV interference lithography. The grating splits one LHCP laser beam into five RHCP laser beams of almost identical power. The grating may find application to coherent beam combining by coupled laser resonators.

Acknowledgment

Chenhao Wan acknowledges the support by the National Natural Science Foundation of China (NSFC) (61505062).

Reference

[1] M. J. Escuti, J. Kim and M. W. Kudenov, "Controlling Light with Geometric-Phase Holograms," Opt. and Photon. News, 27(2), 22-29 (2016).

[2] S. Pancharatnam, "Generalized theory of interference, and its applications. Part i. Coherent pencils," Proc. Indian Acad. Sci. A 44, 247–262 (1956).

[3] M. Berry, "Quantal phase factors accompanying adiabatic changes," Proc. R. Soc. London Ser. A 392, 45–57 (1984).

[4] D. Lombardo, P. Shah, P. Guo, and A. Sarangan, "Deep-UV interference lithography combined with masked contact lithography for pixel wiregrid patterns," In SPIE Advanced Lithography, International Society for Optics and Photonics (2016).

IOP Conf. Series: Journal of Physics: Conf. Series **844** (2017) 012029 doi:10.1088/1742-6596/844/1/012029

Theoretical Study on the Dual-wavelength Second Harmonic Generation with Transverse Gradient Nested Optical Superlattice

Xuan Xiao, Jian Jiang, Yun-qing Lu, Jian-dong Zhang, Kai Wang, and Zu-xing Zhang
School of Opto-electronic Engineering, Nanjing University of Posts and Telecommunications, Nanjing China 210023

E-mail: jiangjian@njupt.edu.cn

Abstract Dual-wavelength second harmonic generation with transverse gradient nested optical superlattice is investigated. The ratio of SHG of two fundamental wavelengths can be controlled by adjusting the position of incident light. This study shows that the efficiency of SHG is related to the temperature and coherent length of the nested structure.

1. Introduction

With the development of nonlinear optical technology, researchers are not satisfied with the periodic structure of crystals to achieve a single quasi-phase matching. Lots of optical superlattices have been reported to achieve multiple quasi-phase matching in a specific domain reversal structure on a single crystal, such as dual-periodic, quasi-periodic, non-periodic and aperiodic[1-4]. As the regularity of the domain structure is weakened, optical superlattice can provide abundant reciprocal lattice vectors, which makes the multiple quasi-phase matching process more flexible and efficient. Although the structures mentioned above can achieve multiple wavelengths second harmonic generation (SHG) simultaneously, however, the multi-wavelength frequency doubling scheme does not involve the efficiency controlled and the output ratio of each fundamental wave. In this letter, we propose transverse gradient nested optical superlattice structure to obtain the dual-wavelength frequency doubling simultaneously. The ratio of SHG of two fundamental wavelengths can be controlled by adjusting the position of incident light. And the study shows that the efficiency of SHG is related to the temperature and coherent length of the nested structure.

2. Structrue and theoretical analysis

The schematic diagram of a nested transverse gradient optical superlattice is shown in Fig. 1. The black and white region is the positive and negative domain, respectively. The periodic structure of superlattice is constituted with triangular positive domain, trapezoidal negative and positive domain, and triangular negative domain. The domain of optical superlattice is transverse gradient nested by coherence length of (e point) and 2 (a point). The coherence length of the fundamental wavelength is calculated as:

$$l_c = \lambda/4\left(n_{2\omega} - n_{\omega}\right)\ldots\ldots(1)$$

With the small signal and the slowly varying approximation, the conversion efficiency of SHG is proportional to, d_{reff}^2 , and the relatively effective nonlinear coefficient d_{reff} is defined as [5]:

$$d_{reff}(\lambda) = \left| \frac{1}{L} \int_0^L d(z) e^{i\Delta k(\lambda) z} dz \right| \quad \text{(2)}$$

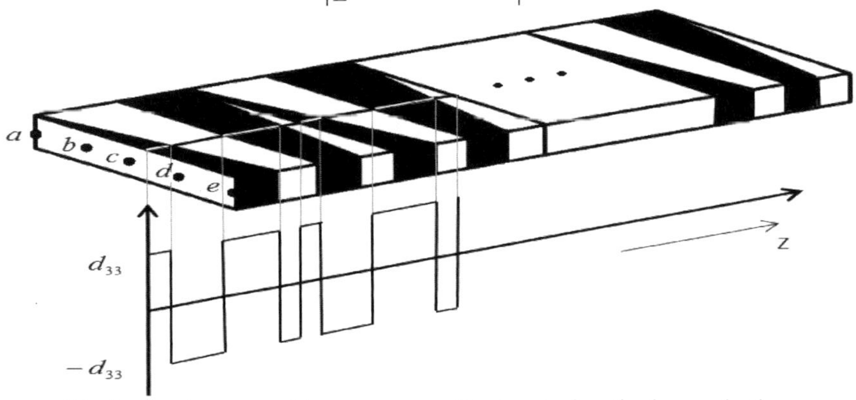

Fig.1 The structure of transverse gradient nested optical superlattice

3. Simulations and results

In this paper, the crystal used is 1% MgO-doped LiNbO3, and we discuss the output of dual-wavelength frequency doubling type0 (e + e→e) in crystal. When the temperature of crystal is controlled at 114oC, for the wavelengths of 1.06um and 0.86um, the coherence length is 3.3148um and 1.6574um, respectively.

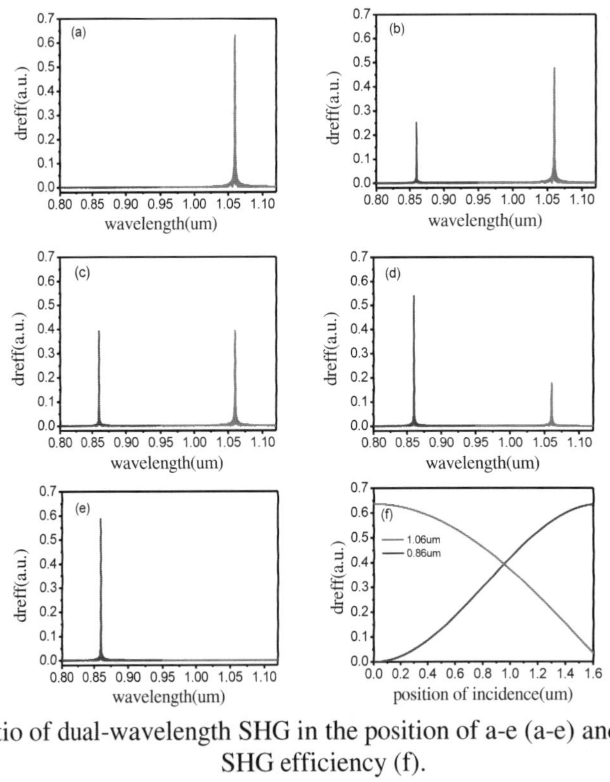

Fig.2 The output ratio of dual-wavelength SHG in the position of a-e (a-e) and the evolution graph of SHG efficiency (f).

Fig.2 shows the different intensity ratio of SHG efficiency spectrum with the altered crystal positions of a, b, c, d, e and SHG efficiency evolution curves of the two fundamental wave when the position of incident light changed.

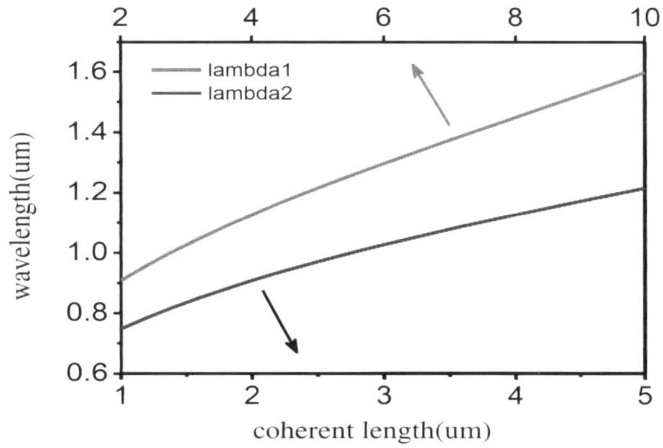

Fig.3 Fundamental wavelengths increase with rising coherence length of optical superlattices

The influence of the domain length error on the conversion efficiency is shown in Fig.5. When the unit domain length, four times of coherence length, is 6.6296um, the conversion efficiency d_{reff} are 0.395 for both fundamental waves with the incidence position of point c. It can be seen that the conversion efficiency is decreased with the increase of the domain error, and the tolerance of domain length are 11.6nm and 20.3nm for 0.86um and 1.06um, respectively.

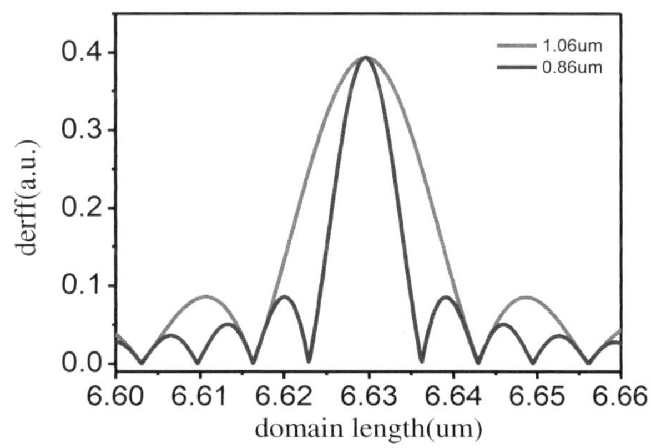

Fig. 4. Conversion efficiency curve via domain length.

Acknowledgment

This work was supported by the Natural Science Foundation of the Higher Education Institutions of Jiangsu Province, China (Grant No. 14KJB140010) and the Nanjing University of Posts and Telecommunications Foundation, China (Grant No. NY213029), the Jiangsu Specially Appointed Professor Project (Grant No. RK002STP14001) and the Six Talent Peaks Project in Jiangsu Province (2015-XCL-023).

Reference

[1] Imeshev G, Fejer M M and Galvanauskas A, 2001 Generation of dual-wavelength pulses by frequency doubling with quasi-phase-matching gratings, Opt. Lett. 26(5), 268-70.
[2] Zhu S N, Zhu Y Y and Qin Y Q, 1997 Experimental realization of second harmonic generation in a fibonacci optical superlattice of LiTaO3, Phys. Lett. 78(14), 2752-2755.
[3] Nemati M, Kheradmand R and Rezaei M, 2012 Multiple quasi-phase-matching in a one-dimensional aperiodically poled optical superlattice, Opt. Lett. 45(17), 3119-3130.
[4] Chen X, Wu F and Zeng X, 2004 Multiple quasi-phase-matching in a nonperiodic domain-inverted optical superlattice, Phys. Rev. A. 69(1), 13818-013818.
[5] Gayer O, Sacks Z and Galun E, 2008 Temperature and wavelength dependent refractive index equations for MgO-doped congruent and stoichiometric LiNbO3, Appl. Phys. B., 91(2): 343-348.

IOP Conf. Series: Journal of Physics: Conf. Series **844** (2017) 012030 doi:10.1088/1742-6596/844/1/012030

SERS enhancement dependence on the diameter of Au nanoparticles

Yifei Yang
Shanghai Jiaotong University

Abstract Series of Au colloidal solutions with different diameters were synthesized by the chemical reaction method. The influence of Au nanoparticles with different size on SERS of R6G was investigated. Experiments indicate that the enhancement factor grows in direct proportion to size of Au nanoparticles within limit.

1. Introduction

Surface-Enhanced Raman Scattering (SERS) has been a useful tool to study the vibrational mode and structure information for molecules. It is called "fingerprint of molecules" and its sensitivity has reached to a single molecule detecting level [1–3]. The enhancement factor (Ef) is an important parameter of SERS spectra which depends on the metal surface roughness, the size of metal particles used for the SERS substrate, and the wavelength chosen to stimulate SRES. Only suitable surface roughness or size of metal particles could produce stronger Raman signal [4–5]. Yabing Du studied the enhancement factor of different surface roughness for Cu and Ag substrate, respectively, and result indicated that the maximum Ef in visible excitation can be obtained for silver with an average surface roughness of 100 nm [6] and for copper with 50 nm surface roughness in red-visible excitation [7].The construction of substrate can also have an influence in the effect of SERS. Conventional methods make use of metal colloidal,metal electrode and island film to prepare SERS substrates. In recent years, a few kinds of metal substrates embellished by organic molecule and oxide has been fabricated to get a more stable and effective SERS system. Sanchez-Cortes research group synthesized metal nanoparticles embellished by calixarene molecules with different substituent group, and used Ag nanoparticles embellished as SERS substrate to detect four PAHs such as purine, benzanthracene, benzophenanthrene and coronene. The minimum concentrate which can be detected has reached 10-10 mol/L. Li fabricated Au nanoparticles substrate with thickness-adjusted SiO2 shell. Zhang realized quantitative determination of nitrite ion in food sample with Au@SiO2 particles. Although these metal substrates have more effective enhancement, they are usually complex to synthesize.

Compared with other SRES substrates, metal colloidal is easy to fabricated and has a comparable enhancement factor. The size of particles in metal colloidal is easy to control. There are many reports that metal nanoparticles with only one diameter are used as SERS substrate. But to our knowledge ,the study of substrates for Au nanoparticles with different diameters is rare to be seen. The research may help us to learn more about the physical mechanism of SERS.In this paper a series of Au nanoparticles with different diameters was synthesized by the chemical reaction method. The influence of diameter of Au nanopartcles was investigated by detecting SRES spectra of R6G.

2.Experimental

2.1 synthesis of colloidal Au nanoparticles with different diameters

Colloidal Au nanoparticles were prepared using Frens method[8].By changing the amount of reductive agent,the size of Au nanoparticles can be controlled.The standard procedure is as follows:50 ml HAuCl4 aqueous solution (1.0× 10-4 g/ml) was heated to boiling. x ml C6H5Na3O7 (1.0×10-2g/ml) was added by a micro-pipette. Boiling was continued for 15 min. A purple-red Au colloidal solution was obtained after cooling at ambient. The analytical grade HAuCl4 and C6H5Na3O7 were purchased from Shanghai chemical reagent corporation and Sinopharm Chemical Reagent Co. Ltd. in China, respectively. Five samples were obtained by adding 5 different volumes of C6H5Na3O7， which are shown in table 1. TEM images of Au nanoparticles are shown in Fig 1,which were made in Tecnai G2 SpiritBiotwin. The average diameters obtained are 70nm,60nm,48nm,24nm,12nm, which reduce with the increasing concentrate of reductive agent. When the size of Au nanoparticles grows up, the shape are irregular and the ellipticity become larger.

simple	a	b	c	d	e
x	0.21	0.3	0.5	0.75	1.5
Diameter	70nm	60nm	48nm	24nm	14nm

Table.1. amount of adding reductive agent

The linear absorption spectra of solutions are shown in Fig. 2, which were obtained using a UV–vis–NIR spectrophotometer (type: Varian Cary 5000). For the five Au colloids prepared by chemical methods, the surface plasma resonance absorption peaks are located between 519nm and 539nm.And absorption peak shifts to longer wavelength as the diameter increases. The relation between absorption peak and diameter of Au nanoparticles is fitted linearly as shown in Fig.3.

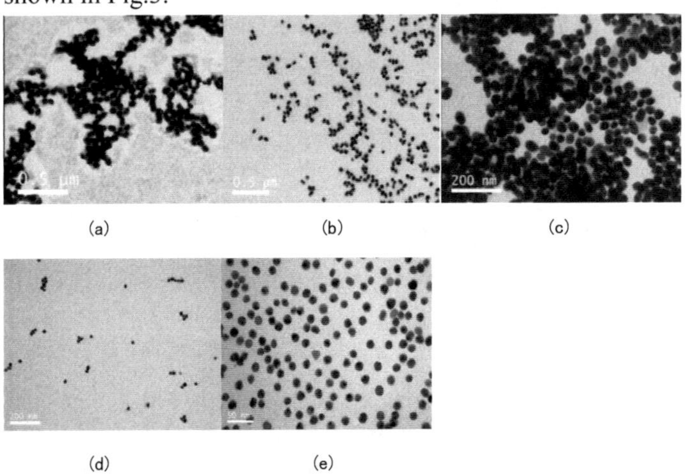

Fig.1.TEM images of Au nanoparticles fabricated by adding (a)0.21ml (b)0.3ml (c)0.5ml (d)0.75ml (e)1.5ml C6H5Na3O7 solutions

IOP Conf. Series: Journal of Physics: Conf. Series **844** (2017) 012030 doi:10.1088/1742-6596/844/1/012030

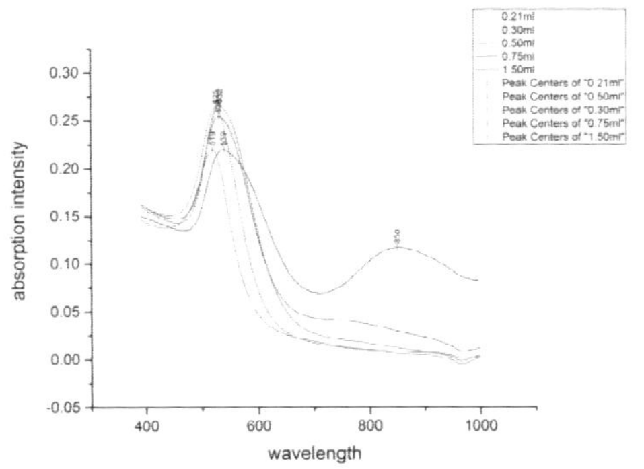

Fig.2.The linear absorption spectra of solutions

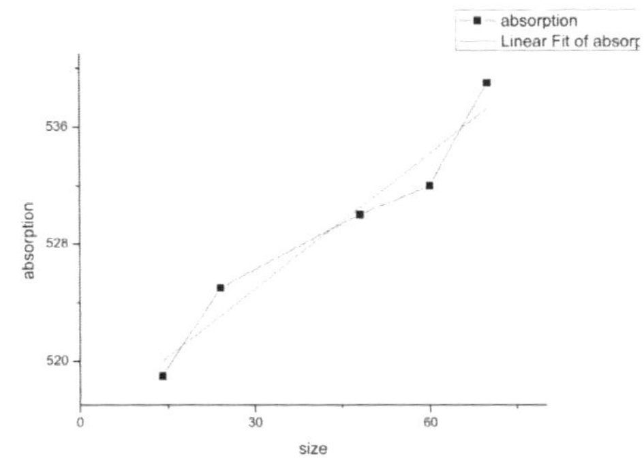

Fig.3.The relation between absorption peak and diameter of Au nanoparticles

2 .2 detection of SERS spectra

5 Au colloidal samples with different average diameters (14nm、24nm、48nm、60nm、70nm) were prepared. Every sample was added with equal volume of R6G aqueous solution and dispersed by ultrasonic agitation to ensure a homogeneous distribution of Au nanoparticles. The concentrate of R6G is 10-4 M. After 4 hours,load the samples with capillary tube and detect the SERS spectra. The Raman spectra was made in Dispersive Raman Microscope Senterra R200-L.

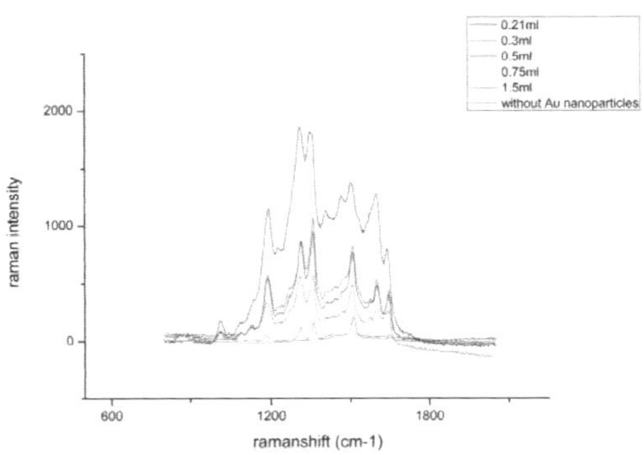

Fig.4.SERS spectra of R6G solution (10^{-4}M) doped with Au nanoparticles

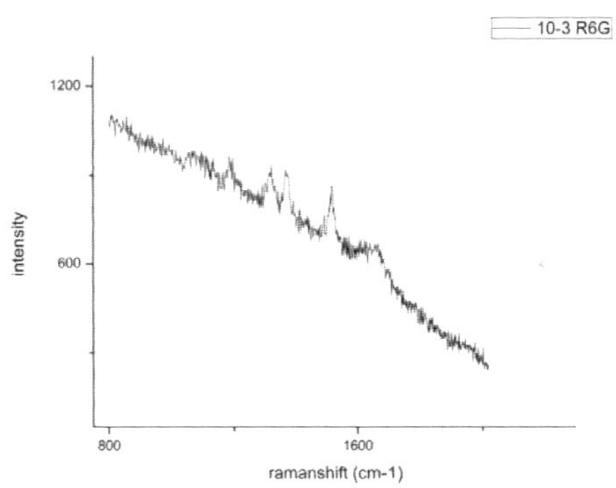

Fig.5. Normal raman spectra of R6G solution (10^{-3}M)

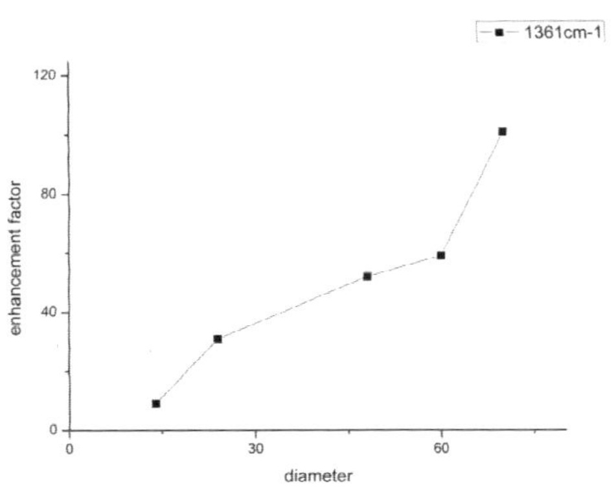

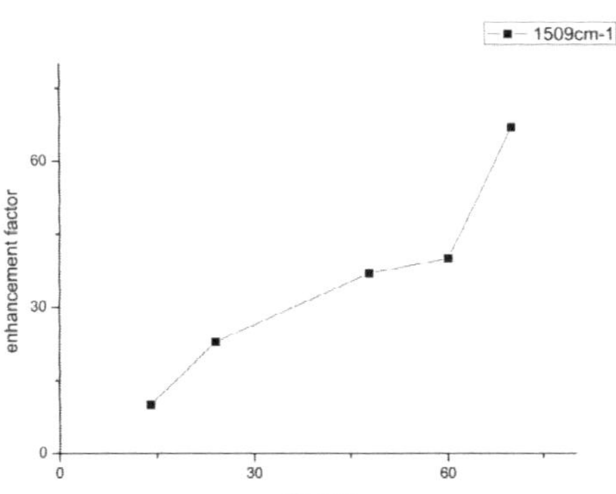

(a) (b)

Fig.6.enhancement factor of mode (a)1361cm^{-1} and (b)1509cm^{-1}

3. Result and discussion

The result of SERS detection is shown in Fig.4. To the pure R6G aqueous solution, intensity of ramanshift is nearly hard to recognize because of the low concentrate(10-4M) while there are obvious raman peaks for the samples doped Au nanoparticles with different diameters. To calculate the enhancement factor exactly, we also make a detection of pure solution with a higher concentrate which is shown in Fig.5. The relation between raman peaks and molecule vibration mode is shown in table.2. The raman peak 1308cm belongs to the flexural vibration of Cx-H while 1361cm^{-1}, 1509cm^{-1} and 1647cm^{-1} belong to stretching vibration of Cx-Cx. Formula used to calculate enhancement factor is as follows:

$$Ef= (Is/In) \times (Cn/Cs)$$

In is the normal raman intensity of peak while Is is the SERS intensity. Cn and Cs are the concentrate of molecule for normal and SERS spectra, respectively. We calculate the Ef of mode 1361 and 1509,respectively. The result is shown in Fig.6.On the whole,Ef increases as the diameter of Au nanoparticles becomes larger for both 1361 and 1509. As we known, the physical motion of SERS is attribute to surface plasma resonance of metal and local field enhancement. After the SERS of single molecule was successfully observed, a new mechanism for SERS, so-called “hot spot”or “hot site”, had been put forward [1].“Hot spot”is a very special place where a huge enhancement up to 1014 could be produced. Combining the experiments and related theoretic analysis, it was understood that the “hot spot” is the one between two very closely placed solid metal nano-particles. The theoretic calculation showed that the enhancement could reach up to 1010 or more, when a molecule is placed at a gap with 1 nm distance surface-to-surface between two silver nano-particles. And this enhancement would also dramatically decrease with the increase of gap width, for instance, when the distance surface-to-surface increases from 1 to 5.5 nm, the enhancement decreases more than three orders of magnitude [9]. When R6g solution is added,the distance of Au nanoparticles will reduce to form Au nanaoparticle aggregates after absorbing the R6G molecules. Au particles in these aggregates could become “hot-spot” if they were close enough.The intensity of electric field excited from“hot spot”is larger than that from only one nanoparticle. Compared with small Au nanoparticles in colloidal, the Au colloidal with large nanoparticles is more unstable dispersion system. So there may be more hot spots in the solution with larger nanoparticles. Besides this, the local field excited by periodic arrangement Au nanopartcles will become larger as the increasing of particle size in a certain range[10]. This two point may cause for the influence of Au nanoparticle size on SERS. On the other hand, from the result,the Ef of different raman peaks is different. We speculate there is chemical enhancement besides electric enhancement. The charge transfer model is selective to the vibration mode of molecules. But it is difficult to obtain how much share from the chemical enhancement at this moment, because the experimental data obtained up to now are not enough to analyze this problem.

Reference

[1] S. Nie, S.R. Emory, Probing single molecules and single nanoparticles by surfaceenhanced Raman scattering, Science 275 (1997) 1102–1106.

[2] K. Kneipp, Y. Wang, H. Kneipp, et al., Single molecule detection using surfaceenhanced Raman scattering, Phys. Rev. Lett. 78 (9) (1997) 1667–1670.
[3] Y.J. Mo, G. Mattei, M. Pagannone, et al., Study of photochemical reaction from C70 adsorbed on silver film by surface enhanced Raman scattering, Appl. Phys. Lett. 66 (1995) 2591–2593.
[4] G.F. Li, H. Li, Y.J. Mo, et al., Surface enhanced resonance Raman spectroscopy of rhodamine 6G adsorbed on silver electrode in lithium battery, Chem. Phys. Lett. 330 (2000) 249–254.
[5] Y.J. Mo, I. Mo ¨rke, P. Wachter, Surface enhanced Raman scattering of pyridine on silver surface of different roughness, Surf. Sci. 133 (1983) L452–L458.
[6] S.W. Joo, Adsorption of bipyridine compounds on old nanoparticle surfaces investigated by UV–vis absorbance spectroscopy and surface enhanced Raman scattering, Spectrosc. Lett. 39 (2006) 85–96.
[7] Y.J. Mo, I. Mo ¨rke, P. wachter, The influence of surface roughness on the Raman scattering of pyridine on copper and silver surface, Solid State Commun. 50 (9) (1984) 829–832.
[8] Frens G . Nat. Phys. Sci. ,1973, 241: 20
[9] H. Xu, J. Aizpurua, M. Ka ¨ll, P. Apell, Electromagnetic contributions to singlemolecule sensitivity in surface-enhanced Raman scattering, Phys. Rev. E 62 (2000) 4318–4324.
[10] Wang Tao, Zhang Zhaoshun, Liao Fan, Cai Qian, Li Yanqing,* Lee Shuit-Tong,* Shao Mingwang,* The Effect of Dielectric Constants on Noble Metal/Semiconductor SERS Enhancement: FDTD Simulation and Experiment Validation of Ag/Ge and Ag/Si Substrates, Sci. Rep. 2014, 4.

IOP Conf. Series: Journal of Physics: Conf. Series **844** (2017) 012031 doi:10.1088/1742-6596/844/1/012031

Simulation of the effect of incline incident angle in DMD Maskless Lithography

L W Liang, J Y Zhou *, L L Xiang, B Wang, K H Wen and L Lei

School of Physics & Optoelectronic Engineering, Guangdong University of Technology, Guangzhou 510006, China

*Zhou Jin-yun, E-mail: zhoujinyun1962@163.com

Abstract. The aim of this study is to provide a simulation method for investigation of the intensity fluctuation caused by the inclined incident angle in DMD (digital micromirror device) maskless lithography. The simulation consists of eight main processes involving the simplification of the DMD aperture function and light propagation utilizing the non-parallel angular spectrum method. These processes provide a possibility of co-simulation in the spatial frequency domain, which combines the microlens array and DMD in the maskless lithography system. The simulation provided the spot shape and illumination distribution. These two parameters are crucial in determining the exposure dose in the existing maskless lithography system.

1. Introduction

Lithography is a central topic in integrated circuit manufacturing, and maskless lithography has drawn much attention in recent years. In maskless lithography, a digital micromirror device (DMD) plays the decisive role as a spatial light modulator. Maskless lithography involves the use of a DMD as a virtual mask in place of a physical mask. A DMD is composed of millions of micrometer-scale square mirrors. The micromirrors are tilted at $\pm 12°$ along the diagonal line, which correspond to "on" and "off" states. However, the boundary of the micrometers results in diffraction. The stray diffracted light leads to disturbance once it accumulates to the exposure threshold of the photoresist.

In conventional photolithography using a photomask, various optical proximity correction (OPC) have been used to compensate for imaging errors such as edge placement and corner-rounding errors in order to achieve desired features in a pattern. Image errors such as the edge placement error and the corner-rounding effect also occur in maskless lithography[1].It is necessary to consider the pattern quality in maskless lithography as in conventional photolithography resulting from the diffraction effect. J Hur et al. derived by approximating the intensity of the beam to the Gaussian distribution[2]. It would be better if Hur's compensation is based on specific illumination distribution instead of Gaussian distribution. Furthermore, the incident light has tilt angle between DMD pixels in DMD maskless lithography system[3][4]. The illumination distribution may be different from normal incidence. Thus, the diffraction effect in non-parallel region should be studied.

In addition, illumination distribution is a crucial parameter in non-linear maskless lithography. Non-linear lithography is a potential sub-diffraction technology for resolution enhancement with physical method. J S Wei et al.[5] and J Koch et al[6]. broke the optical resolving limit in the aspect of physics by nonlinear effect. V Lyubin et al.[7] promote the non-linear lithography in the aspect of material by developing the nonlinear photoresists with transition dissolution curve for maskless photolithography. Research on the light source may drive the non-linear technology in the aspect of optics. For the reason that, slight intensity fluctuation within a light spot directly result in the change of spot size in non-line lithography system, which means edge placement error and the change of critical dimension.

Lastly, designer used to reduce edge placement error though increasing the redundancy. Redundancy means illumination accumulation both the target spot and the diffraction light. Redundancy tolerance should be under concerned.

The present paper extends the method to analyze the diffraction problem in the DMD lithography system. We primarily focus on light propagation in the non-parallel region. Although the angular spectrum had been applied extensively in the field of ultrasound and hologram, it was not commonly used in the diffraction problem of DMD. This algorithm can be used to obtain the illumination distribution in any of the near field, Fresnel field, and Fraunhofer field. This simulation method provides a possibility to substitute the diffraction light with a plane wave as an incident beam with specific diffraction light in the MLA design. This model of calculation is under MATLAB's software environment, which can process dynamic data exchange between Zemax flexibly.[8] Thus, this model may be suitable for co-simulation between a DMD and an MLA. In addition, the aperture function used in DMD was simplified for minimizing the computational cost. The aperture function was classified into four types according to whether the spatial frequencies fx and fy are zero or not. The aperture function, which contains zero spatial frequency, can be simplified.

2. Formula in simulation

In recent years, researchers have built some models to fit the diffraction effect of DMD lithography system. J P Rice et al.[9] considered a micromirror array that acts as a blazed grating. H Ryoo et al.[10] further hypothesized that a micromirror array acts as two perpendicular gratings. They utilized the two-dimensional grating equation to calculate the optimum incident angle. Furthermore, they suggested that the tilted incident angle(θ) contributed to the shift of the envelopes. When the incident angle and groove tilt were arranged to meet a blazed condition, the majority of the energy was concentrated at the blazed order. X. Zheng et al.[11] calculated the efficiency of different diffraction orders by means of Fourier transformation. X. Chen et al.[12] focused on the dispersion effect of high diffraction orders.

The above studies were based on the grating equation or Fraunhofer diffraction model. However, most of them only focused on the diffraction efficiency. These methods share a common constraint in calculating the diffraction efficiency in the non-parallel condition or in the near field. In addition to diffraction efficiency, the illumination distribution is important. These two parameters provide a possibility of co-simulation between a microlens array (MLA) and a DMD. Conventional microlens array optimization presupposes that the light source is a plane wave. However, after modulation by the DMD, the light source is no longer a plane wave, because of the diffraction effect of DMD pixels. J Li et al.[13] compared the numerical computational accuracy of the Kirchhoff formula, Rayleigh-Sommerfeld formula, and angular spectrum transmission formula. They suggested that the angular spectrum transmission formula was the easiest for achieving sufficient sampling. K. Matsushima et al.[14] proposed a Jacobian matrix enable coordinate transformation in angular spectrum transmission.

2.1. Angular spectrum

For comprehension convenience, the basic concept of angular spectrum is introduced briefly in 2.1. the angular spectrum theory is a more rigorous method to study light propagation in comparison with the Fraunhofer diffraction model and Fresnel diffraction model. In particular, the diffraction distance is relatively small such that the image plane is not in the Fresnel field.

The basic concept of the angular spectrum method is to describe a complex wave field as an equivalence of an infinite number of plane waves in all directions[15]. The method aims at determining the field $E_x(x,y,z)$ as a function of $E_x(x,y,0)$ without approximation. The z co-ordinate is considered as the propagation direction.

$$G_z(f_x,f_y) = G_0(f_x,f_y)\exp\left[j\frac{2\pi}{\lambda}z\sqrt{1-(\lambda f_x)^2-(\lambda f_y)^2}\right] \quad (1)$$

Equation (1) defines the angular spectrum transfer function. In the propagating wave model, corresponding to spatial frequencies with $(f_x^2+f_y^2)<1/\lambda^2$, the frequency spectrum acquires a phase delay due to the propagation in the z direction. Therefore, this component is called the propagating component.

2.2. Co-ordinate transform

According to the Huygens-Fresnel principle, every point on the wavefront becomes a source of a spherical wave. These secondary waves reconstruct the form of the wave at subsequent times. As shown in Fig. 1, only frequency spectra within the dashed line can reach the imaging plane.

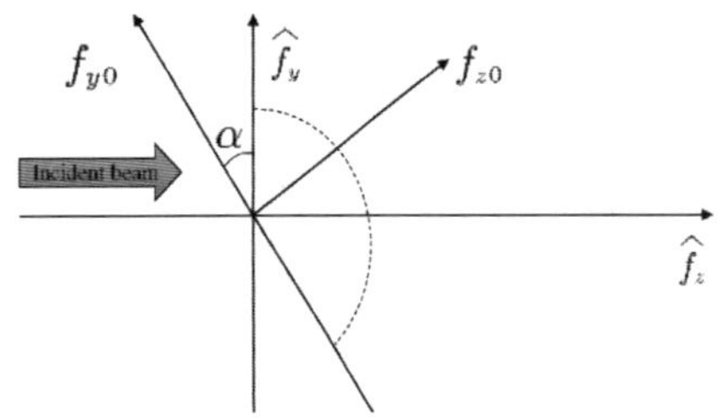

Fig. 1 Schematic of co-ordinate rotation.

The rotational matrix about the x-axis, A, is given by

$$A=\begin{bmatrix}\alpha_1 & \alpha_2 & \alpha_3\\ \alpha_4 & \alpha_5 & \alpha_6\\ \alpha_7 & \alpha_8 & \alpha_9\end{bmatrix}=\begin{bmatrix}1 & 0 & 0\\ 0 & \cos\alpha & -\sin\alpha\\ 0 & \sin\alpha & \cos\alpha\end{bmatrix} \tag{2}$$

$$\begin{bmatrix}f_{x0}\\ f_{x1}\\ f_{x2}\end{bmatrix}=\begin{bmatrix}\alpha_1 & \alpha_2 & \alpha_3\\ \alpha_4 & \alpha_5 & \alpha_6\\ \alpha_7 & \alpha_8 & \alpha_9\end{bmatrix}\begin{bmatrix}\hat{f}_x\\ \hat{f}_y\\ \hat{f}_z\end{bmatrix} \tag{3}$$

Because the z direction is the propagation direction, the x-y plane is observed as the cross-section. We only pay attention to the x and y direction. Equation(3) can be partially expressed as

$$\begin{cases} f_{x0}=\alpha\left(\hat{f}_x,\hat{f}_y\right)=a_1\hat{f}_x+a_2\hat{f}_y+a_3\hat{f}_z\\ f_{y0}=\beta\left(\hat{f}_x,\hat{f}_y\right)=a_4\hat{f}_x+a_5\hat{f}_y+a_6\hat{f}_z \end{cases}. \tag{4}$$

The nonlinearity of the rotation transformation was offset by the Jacobian matrix $df_x df_y$, which was replaced by $\left|J\left(\hat{f}_x,\hat{f}_y\right)\right|d\hat{f}_x d\hat{f}_y$ to give[16]

$$J\left(\hat{f}_x,\hat{f}_y\right)=\begin{bmatrix}\dfrac{\partial\alpha}{\partial\hat{f}_x} & \dfrac{\partial\alpha}{\partial\hat{f}_y}\\ \dfrac{\partial\beta}{\partial\hat{f}_x} & \dfrac{\partial\beta}{\partial\hat{f}_y}\end{bmatrix} \tag{5}$$

$$=(\alpha_2\alpha_6-\alpha_3\alpha_5)\frac{\hat{f}_x}{\hat{f}_z}+(\alpha_3\alpha_4-\alpha_1\alpha_6)\frac{\hat{f}_y}{\hat{f}_z}+(\alpha_1\alpha_5-\alpha_2\alpha_4)$$

In the mirror co-ordinate system, an incident beam is given by

$$E_1(x,y,0)=E_0(x,y,0)\exp\left(-j\frac{2\pi}{\lambda}\sin\theta\right) \tag{6}$$

According to the shifting property in the Fourier transform,

$$\mathcal{F}\{E_1(x,y,0)\}=G\left(f_{x0},f_{y0}+\frac{\sin\theta}{\lambda}\right). \tag{7}$$

2.3. *Simulation procedure*

The illumination distribution can be calculated as a function of distance z. The basic idea and steps are as follows:

Step 1. An inclined wave is propagated to an intermediate plane that is parallel to the imaging plane.
Step 2. Waves on the intermediate plane are decomposed into plane waves by Fourier transformation.
Step 3. Co-ordinate translation and rotation in the spatial domain by means of shifting theorem.

Step 4. Co-ordinate rotation in the Fourier domain.
Step 5. Nonlinearity compensation with a Jacobian Matrix.
Step 6. Spectrum filtering for efficacious propagation.
Step 7. Propagation calculation with angular spectrum method.
Step 8. Spectrum restoration and inverse Fourier transformation.

2.4. *Simplification of aperture function of DMD pixel*

The center of the micromirror should coincide with the origin of the co-ordinate system. Further, it is tilted by ±12° along the diagonal line. However, the computational cost for such a model is high because all four quadrants are occupied, as shown in Fig. 2. To simplify the analytical solution, we initially move the DMD to the first quadrant, and then return it in the frequency domain based on the shifting property. The spatial spectrum of a single micromirror is given by

$$T(f_x,f_y) = \int_0^a \exp(-j2\pi x_0 f_x)\exp(-j2\pi y_0 f_y)dx_0 dy_0 \tag{8}$$

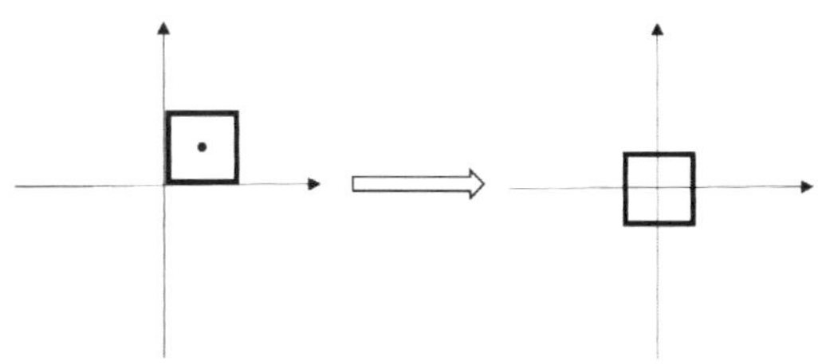

Fig. 2 Rectangular light beam in the spatial domain

To deal with a possible case in which a denominator in equation (8) becomes zero, a brief discussion is made on different classification criteria.

1)When $f_x \neq 0, f_y \neq 0$
$$T(f_x,f_y) = \frac{-[\exp(-j2\pi a f_x) - 1][\exp(-j2\pi a f_y) - 1]}{4\pi^2 f_x f_y} \tag{9}$$

2) When $f_x = 0, f_y \neq 0$
$$T(f_x,f_y) = \frac{-ja(1-\exp[-j2\pi a f_y])}{2\pi f_y} \tag{10}$$

3) When $f_x \neq 0, f_y = 0$
$$T(f_x,f_y) = \frac{-ja(1-\exp[-j2\pi a f_x])}{2\pi f_x} \tag{11}$$

4) When $f_x = 0, f_y = 0$
$$T(f_x,f_y) = a^2 \tag{12}$$

3. Examples using this algorithm for analysis

As analysed in Section 1, the diffraction effect is affected by the incident wavelength λ, image distance z, incident angle θ, and pixel size determined by x0 and y0. The rotation angle about the x-axis also affects the diffraction effect. In the DMD system, this rotation angle is a constant ($\alpha = 45°$).

In the following examples, we defined the control group as the incident angle $\theta_i = 12°$, incident wavelength $\lambda = 405nm$,and the DMD pixel size(x_0 and y_0) is 7.25 μm.

3.1. *Influence of propagation distance z*

For the purpose of observing the diffraction effect that changes along with the propagation distance, we obtained the simulation results for z = 0.01 mm, z = 1 mm, and z = 130 mm; these parameters were chosen from the definition of near field, Fresnel field, and Fraunhofer field, respectively.

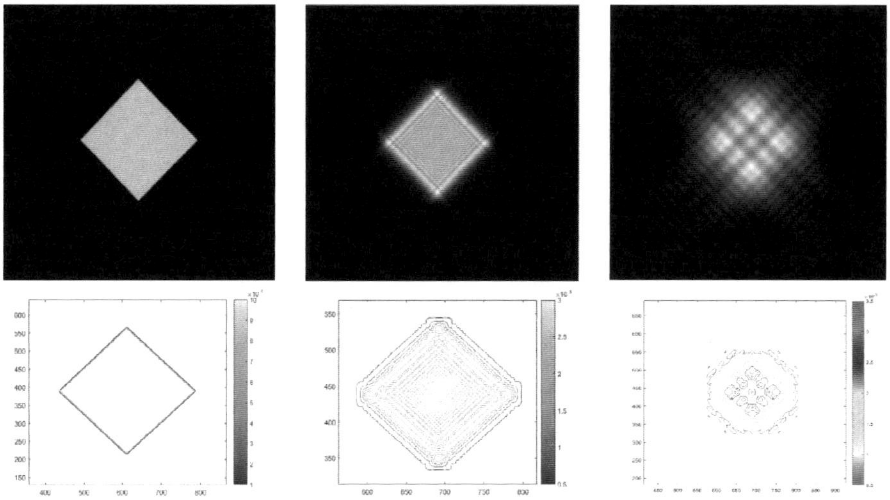

z = 0.01 mm z = 1 mm z = 130 mm

Fig. 3 Illuminance distribution and its contour lines obtained at different distances

Fig. 3 demonstrates that the diffraction effect becomes more obvious as the imaging distance increases. The illumination area is not flat, and there are peaks in the middle. Four tails arise around the pixel. As the propagation distance increases, the light uniformity decreases.

3.2. *Influence of pixel size*

Commercial DMDs have four types of standard sizes. To investigate the influence of pixel size, we obtained the simulation results under experimental conditions of a_1 = 7.25 μm, a_2 = 10.8 μm, a_3 = 13.68 μm, and a_4 = 16 μm. The parameter a is the pixel size corresponding to x0 and y0.

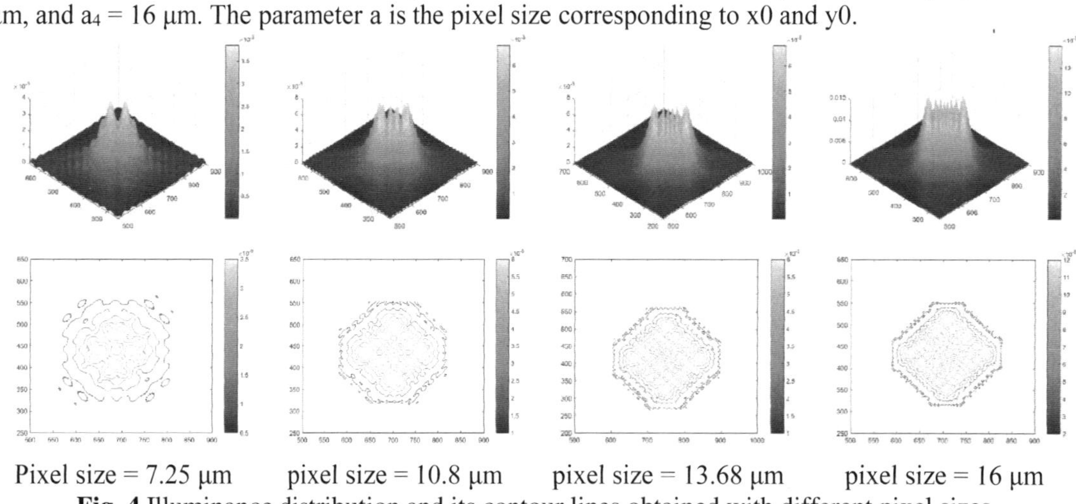

Pixel size = 7.25 μm pixel size = 10.8 μm pixel size = 13.68 μm pixel size = 16 μm

Fig. 4 Illuminance distribution and its contour lines obtained with different pixel sizes

Fig. 4 demonstrates that the number of peaks increases as the pixel size increases. Meanwhile, peaks appear denser and less undulatory. When a_1 = 7.25 μm, there are two peaks. When a_2 = 10.8 μm, there are four peaks. When a_3 = 13.68 μm, there are six peaks. When a_4 = 16 μm, there are eight peaks.

3.3. *Influence of incident angle θ*

The incident angle is affected by the optical structure owing to the space limitation. To investigate the influence of incident angle θ, we obtained the simulation results under the experimental conditions of θ_1 = 12°, θ_2 = 18°, θ_3 = 24°, θ_4 = 30°.

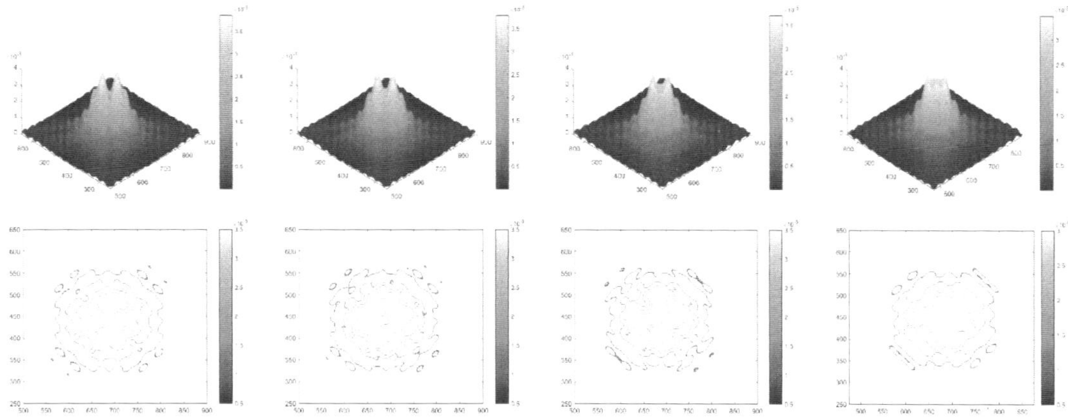

$\theta = 12°$ $\theta = 18°$ $\theta = 24°$ $\theta = 30°$

Fig. 5 Illuminance distribution and its contour lines obtained for different incident angles θ

Neighboring peaks move closer to each other as the incident angle θ increases. The peaks are merged when the incident angle θ is greater than 24°. Fig. 5 (d) demonstrates the light intensity when θ = 30°. Three distinct color steps represent intensity steps. The spot appears as a clip in the center.

3.4. Influence of wavelength λ

There are three types of typical wavelengths in conventional lithography. To investigate the influence of the wavelength, we obtained the simulation result under experimental conditions of $\lambda_i = 0.365$ μm, $\lambda_h = 0.405$ μm, $\lambda_g = 0.436$ μm.

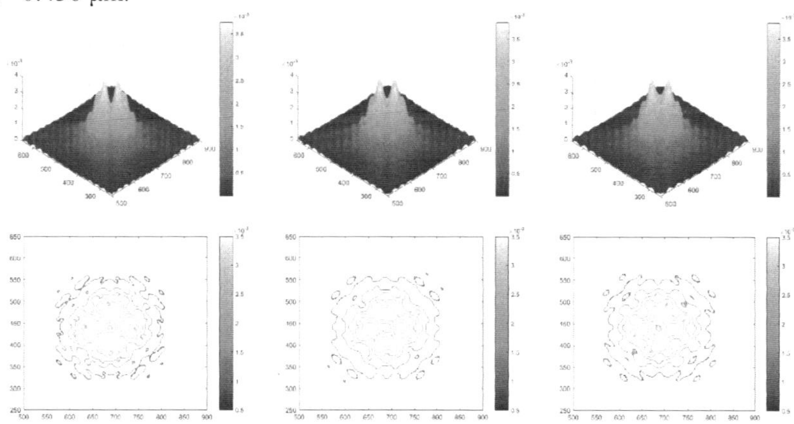

$\lambda = 365$ nm $\lambda = 405$ nm $\lambda = 436$ nm

Fig. 6 Illuminance distribution and its contour lines obtained at different wavelengths λ

Fig. 6 demonstrates that the incident wavelength has less influence on the spot shape in comparison with other independent variables. It mainly contributes to the slope of peaks. However, the influence is small and seems nonlinear.

Figures 3 to 6 show the common characteristics of illumination distribution:

(1) Illumination is not a flat-topped mode. (2) Sawteeth encircle the depression center. (3) There are four main peaks at the corner. (4) Tails are distributed along the horizontal and normal directions.

4. Calculation of diffraction disturbance

Diffraction may lead to a disturbance between the pixels. Stray light must be controlled within threshold level, particularly in a pulse width modulation resolution enhancement system.[17] This system obtains a smaller spot by controlling the incident intensity, which is based on the nonlinear absorption effect.

IOP Conf. Series: Journal of Physics: Conf. Series **844** (2017) 012031 doi:10.1088/1742-6596/844/1/012031

Furthermore, this system is sensitive to the light intensity. Superfluous illumination causes pixel deformation and blots.

A new criterion is defined to describe the disturbance in a DMD lithography system:

$$P_d = \frac{1}{4}\left(1 - \frac{I_e}{I_r}\right) \tag{13}$$

where I_r is the total illumination reaching the imaging plane and I_e is the effective illumination at the center. The effective illumination is different from diffraction efficiency. The criterion of the diffraction efficiency is that the ratio of the power diffracting into a designated direction to the power incident at the diffractive element. In contrast, the disturbance percentage quantifies how much optical power is diffracted into a designated direction compared to the optical power reaching the imaging plane. The power lost in propagation is negligible with respect to the disturbance.

It should be noticed that, there is a $\frac{1}{4}$ in the beginning of the formula (18). That's because diffraction occurs in four direction. P_d only indicates the disturbance from a certain direction. We defined P_d from the perspective of the photoresist, instead of the DMD pixel. Because in scanning maskless lithography system, the pattern contribute by multi-exposure, which means the actual exposure does is the sum of the actived DMD pixels and the disturbances from adjacent pixels passing over a certain point of photoresist. For example, in Fig 7, "A" is a certain point in photoresist, while 1~4 is 4 pixels passing over "A" in sequence. Pixel 1 and 3 is actived, which means they will be in "on state" when they pass over A. Pixel 2 and 4 is not active when they pass over A. The total exposure dose of "A" in photoresist is the sum illumination contribute by pixel 1 and 3, and the disturbance caused by acitved pixels adjacent pixel 1~4. Fig 8 shows the disturbance caused by actived adjacent pixels as an example. Fig 8 reveals that the disturbance is not always come from all 4 direction. The disturbance in Fig 8 is 3 Pd in such situation. In consideration of practical situation mentioned above, we defined the new criteria Pd, which indicates the disturbance percentage of a certain pixel in one direction. The total disturbance T_d in an area of photoresist is

$$T_d = \sum_1^m n \times P_d \tag{14}$$

Where n is the number of adjacent actived pixels in a particular moment, m is the number of pixel passing over the area "A" through the scanning process.

It should be noticed that, illumination accumulates at the same moment should be superposed in the form of complex amplitude. But the illumination accumulates at different time should be superposed in the form of light intensity. That is because the data is in the form of plural, which implies it contains the phase information. The superposition of complex amplitude refers to coherent addition in the same moment while the superposition of light intensity refers to incoherent addition in different time.

Fig.7 take 4 pixels passing over a certain point "A" on the photoresist as an example of considering the total exposure does from the perspective of photoresist.

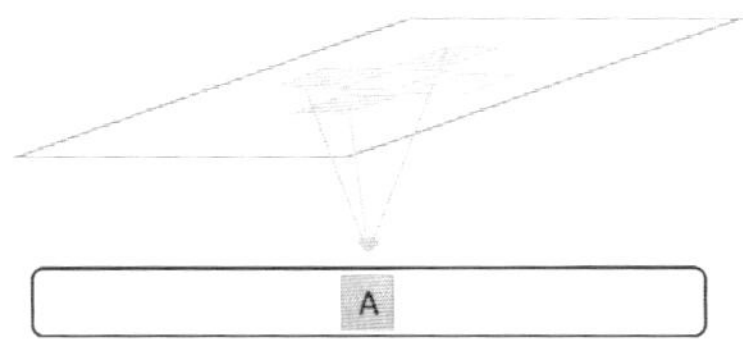

Fig.8 disturbance caused by 3 adjacent acitved pixels

The simulation parameters are set as $\lambda = 0.405$ μm, z = 1 mm, x0 = y0 = 7.25, incident angle $\theta = 12°$ and rotation angle $\alpha = 45°$.The simulation result of P_d is 7%.

5. Discussion

This simulation method can be used to determine the exposure doses, scanning path, and incident light source characteristics in maskless lithography. This method is based on a half analytical method. The simulation has three application ranges. The first application range relates to non-linear sub-diffraction lithography, which requires precise control of the spot shape and illumination distribution. These are key parameters to determine the resolution and error tolerance. Thus, this study may contribute to further study in non-linear sub-diffraction lithography, from spot-scanning to plane-scanning. The second application range deals with the scanning path. The aerial image of DMD stretches along the horizontal and normal direction. This extension may lead to a disturbance between the neighboring pixels. This disturbance accumulates further in the scanning procedure. In the maskless lithography system, the pixel field source gives multiple exposures to the imaging point, instead of a single, static exposure during scanning. Thus, the scanning procedure results in disturbance accumulation. On the other hand, the exposure repetition number relates to the scanning path. In the above condition, the degree of disturbance is 7%. We assume that the exposure threshold is 50% and that only one diffracting pixel exists near the non-exposed area. The number of acceptable repeated exposures is less than 7. Otherwise, stray light accumulates to the exposure threshold of the photoresist, resulting in the error spot. The third application involves co-simulation. The conventional optimization of a microlens array assumes that the incident beam is a plane wave. In fact, the aerial image of DMD has light intensity in the form of sawteeth on the edge. These sawteeth may result from interference within a pixel. Reducing the degree of coherence may make a more uniform intensity pattern, but reduce the resolution at the same time.[18] Hence, the MLA should co-simulate with DMD pixel so as to balance the resolution and the uniformity of the aerial image as they are projected onto the photoresist.

6. Conclusion

In this study, the non-parallel spectrum method was applied for the diffraction analysis of DMD. This algorithm can be utilized in the non-parallel region, whereas the previous simulation methods cannot. The simulation procedure involves eight main steps was presented. The process of computing was simplified for the DMD. A critical point was calculated for determining the simulation range. This model can be utilized in practical application with various incident angles, propagation distances, and pixel sizes. As an example of using this model, we obtained the diffraction efficiency of the control group. We also obtained the qualitative relationship between illumination distribution and the wavelength λ, image distance z, incident angle θ, and pixel size x0 and y0 as another example using this algorithm. We summarized the common characteristics of illumination distribution and the shape of the light spot according to the simulation results for perceptual knowledge. These simulation results provide a reference for the scanning path and exposure doses in maskless lithography.

Acknowledgments

This work was supported by the National Natural Science Foundation of China (61475037, 11304044, 61405039) and Science and Technology Research Projects of Guangdong Province (2016B090918124).

References

[1] Levenson, M. D. "Using Destructive Optical Interference in Semiconductor Lithography." *Optics & Photonics News* 17.**4**(2006): 30-35.
[2] Hur, Jung Yu, and M. S. Seo. "Optical Proximity Corrections for Digital Micromirror Device-based MasklessLithography." *Journal of the Optical Society of Korea* 16.**3**(2012):221-227.
[3] Shi, Zhaojun, and Y. Gao. "Pixel-based partially coherent image method for maskless lithography system." *Optik - International Journal for Light and Electron Optics* 124.**18**(2013):3292-3295.
[4] Ryoo, Hoonchul, W. K. Dong, and J. W. Hahn. "Analysis of the line pattern width and exposure efficiency in maskless lithography using a digital micromirror device." *Microelectronic Engineering* 88.**10**(2011):3145-3149.
[5] Wei, Jingsong, and R. Wang. "Maskless direct laser writing with visible light: Breaking through the optical resolving limit with cooperative manipulations of nonlinear reverse saturation absorption and thermal diffusion." *Journal of Applied Physics* 115.**12**(2014):123102-123102-7
[6] Koch, J., et al. "Maskless nonlinear lithography with femtosecond laser pulses." *Applied Physics A Materials Science & Processing* 82.**1**(2005):23-26.
[7] Lyubin, V., et al. "Nonlinear photoresists for maskless photolithography on the basis of Ag-doped As2S3 glassy films." *Applied Physics Letters* 92.**1**(2008):011118-011118-3.
[8] Zhang,Wen Jing,W.G.Liu, and Z.J.Liu. "Dynamic data exchange between Zemax and Matlab." *Journal of Appliled Optics* (2008)
[9] J. P. Rice et al., "DMD diffraction measurements to support design of projectors for test and evaluation of multispectral and hyperspectral imaging sensors," *Proc. SPIE* **7210**(3), 351-7 (2009).
[10] H. Ryoo, D. W. Kang, and J. W. Hahn, "Analysis of the effective reflectance of digital micromirrordevices and process parameters for maskless lithography," *Microelectron. Eng.***88**(3), 235-9 (2010).
[11] X. Zheng et al., "Diffraction analysis of digital micromirror device in maskless photolithography system," *J. Micro Nanolithoghapy*. MEMS MOEMS **13**(4), 353-5 (2014).
[12] X. Chen et al., "Diffraction of digital micromirror device gratings and its effect on properties of tunable fiber lasers," *Appl. Opt.* **51**(30), 7214-20 (2012).
[13] J. Li, Z. Peng, and Y. Fu, "Diffraction transfer function and its calculation of classic diffraction formula," *Opt. Commun.* **280**(2), 243-8 (2007).
[14] K. Matsushima, H. Schimmel, and F. Wyrowski, "Fast calculation method for optical diffraction on tilted planes by use of the angular spectrum of plane waves," *J. Opt. Soc. Am. A* **20**(9), 1755-62 (2003)
[15] K. Khare, *Fourier Optics and Computational Imaging*, Wiley, UK (2015).
[16] S. Yonghuan, *Practical Mathematics Handbook*, Science Press, Beijing (1995).
[17] H. Kuo and Y. Huang, "Resolution enhancement using pulse width modulation in digital micromirror device-based point-array scanning pattern exposure," *Opt. Laser Eng.* **79**, 55-60 (2016)
[18] Joseph W. Goodman, *Introduction to Fourier Optics, Tird Edition*, Publishing House ofElectronics industry, Beijing (2011)

IOP Conf. Series: Journal of Physics: Conf. Series **844** (2017) 012032 doi:10.1088/1742-6596/844/1/012032

Strong coupling of phonon polaritons in h-BN nanowires with graphene plasmons

Y Zhou[1] and Y Q Cao[2]

[1]School of Science, Hangzhou Dianzi University, Hangzhou 310018, China
[2] College of Physics Science and Technology, Yangzhou University, Yangzhou 225002, China

E-mail: yzhou@hdu.edu.cn

Abstract. The phonon polaritons in hexagonal boron nitride (h-BN) can couple with graphene plasmons, forming new hybrid modes which are called plasmon-phonon polaritons. In this paper, we point out that such coupling can occur within the strong coupling regime, where the dispersion relations of the graphene plasmons and the phonon polaritons avoid crossing with each other. We have investigated graphene-wrapped h-BN nanowires, and found that such strong couplings exist between the graphene plasmons and the fundamental as well as the high-order phonon polariton modes, even if the graphene layer is far away from the h-BN. This is interesting because the high-order phonon polariton modes are mostly confined within the bulk h-BN with little fields outside the nanowire, i.e., the interactions of these modes with the graphene plasmons should be rather weak; however, our calculations disagree with such judgements. Due to the dispersive permittivity of h-BN, the graphene plasmons can couple simultaneously with two branches of phonon polariton modes under carefully chosen electron doping levels. Our investigations may open the way to explore the plasmon-phonon polaritons in the strong coupling regime.

1. Introduction

The phonon resonances in the material may lead to negative permittivity components in the dielectric tensor, creating so-called hyperbolic materials [1]. Hexagonal boron nitride (h-BN) is such a material, which has two active infrared phonon modes: out-of-plane A_{2u} phonon mode, enabling type I hyperbolic response; and in-plane E_{1u} phonon mode, enabling type II hyperbolic responses [2,3]. Compared with artificial hyperbolic metamaterials, which often suffer from ohmic losses and low-yield fabrication processes, these naturally formed hyperbolic materials may be more useful for applications in infrared photonic devices, such as superlens and waveguides [3-5].

Despite that h-BN supports low-loss surface phonon polariton modes in the mid-infrared region, its phonon resonances are intrinsic properties of the material; thus, they are not tunable by external fields. Recently, h-BN graphene heterojunctions have been proposed which support plasmon-phonon polaritons [6,7]. These hybrid modes possess low propagation losses while maintaining tunable through the fermi level of graphene, providing a promising platform for low-loss mid-infrared nanophotonic applications. In this paper, we investigate the forming of these hybrid modes in the strong coupling regime, which is often ignored by other researchers, but indeed can occur once the graphene is separated from the bulk h-BN with a moderate gap. Our investigations may open the way to explore the plasmon-phonon polaritons in the strong coupling regime.

2. Phonon polaritons in h-BN nanowires

We analytically calculated the transverse magnetic (TM) polarized phonon polariton modes in h-BN nanowires. Naturally grown h-BN nanowires have lattice structures which can be seen as constructed by rolling up the planar boron nitride sheets into a cylinder. Due to the existence of the in-plane E_{1u} phonon mode, the effective permittivity within the boron nitride plane is negative, which makes h-BN similar to metallic materials. After the rolling, the nanowire of course possesses negative permittivities along the axial and azimuthal directions.

2.1. Theoretical model

H-BN is an anisotropic material. In the upper Reststrahlen band, the nanowire has positive permittivity along the radial direction, but negative permittivities along the axial and azimuthal directions. Thus, in order to find the phonon polariton modes, h-BN nanowires must be treated as anisotropic waveguides. Under the cylindrical coordinates, the permittivities along the axial, radial and azimuthal directions are respectively denoted as ε_z, ε_r and ε_θ. The TM-polarized modes have electric field component E_z and E_r, and magnetic field component H_θ in the cylindrical coordinates; from the mode electric field components, we can expect that the dispersion relations depend on both the radial and axial permittivities. After matching the boundary conditions, the dispersion relations of the TM-polarized phonon polariton modes are found to be determined by the following equations:

$$0 = \left[\alpha_1 k_r^{\,1} \frac{\mathrm{I}_1(k_r^{\,1}R)}{\mathrm{I}_0(k_r^{\,1}R)} + \alpha_2 k_r^{\,2} \frac{\mathrm{K}_1(k_r^{\,2}R)}{\mathrm{K}_0(k_r^{\,2}R)} \right] \tag{1}$$

where $\alpha_i = k_0 \varepsilon_r^{\,i} / (k_z^2 - \varepsilon_r^{\,i} k_0^2)$, $k_r^{\,i} = \sqrt{\varepsilon_z^{\,i}(k_z^2 - \varepsilon_r^{\,i} k_0^2) / \varepsilon_r^{\,i}}$ (i = 1 or 2, respectively corresponding to the region inside and outside the nanowire). I_0 and K_0 are the modified Bessel functions of the first and the second kind, respectively. Using Equation 1, we can simultaneously determine the real and the imaginary parts of the mode propagation constant accurately.

2.2. Phonon polaritons in h-BN nanowires

After tedious calculations, we have derived the TM-polarized phonon polariton modes in a h-BN nanowire with a radius of 100 nm, as shown in Figure 1. As one can see, multiple modes are found. Both the real and the imaginary parts of the mode propagation constants are derived, as shown by the solid and dashed curves in Figure 1(a), respectively. Only the first three modes are plotted, TM_0, TM_1 and TM_2, denoted by black red and blue curves, respectively. The subscripts indicate the number of the nodes of the field component E_z along the radial direction. The pink line in Figure 1(a) is the light line; thus, the phonon polariton modes in the h-BN nanowire are extremely dispersive, which of course caused by the permittivities of h-BN in the upper Reststrahlen band.

Due to the fact that h-BN is a natural hyperbolic material, the h-BN nanowire supports multiple TM-polarized phonon polariton modes; all of them are propagative modes, with the fields deeply confined within the bulk material. The normalized electric field distributions of the TM_0, TM_1 and TM_2 modes are shown in Figures 1(b), 1(c) and 1(d), respectively. The TM_0 mode has most of the electric fields confined around the surface, while the TM_1 and TM_2 modes have electric fields that are mostly confined within the bulk material. Such different field distributions may lead to the conclusion that the TM_0 mode may have strong couplings with the surroundings, while the high-order ones do not. However, our calculations strongly disagree with such conclusion, and show that the high-order phonon polariton modes can have ultra-strong couplings with other fundamental excitations. The physical reason for this is quite simple: high-order phonon polaritons have mode profiles with mostly fields within the material, while other excitations such as plasmons may possess extensive field distributions which may cause large field overlaps with the high-order phonon polariton modes; thus, strong couplings can still exist. In the followings, we analytically investigate such couplings within the framework of electrodynamics.

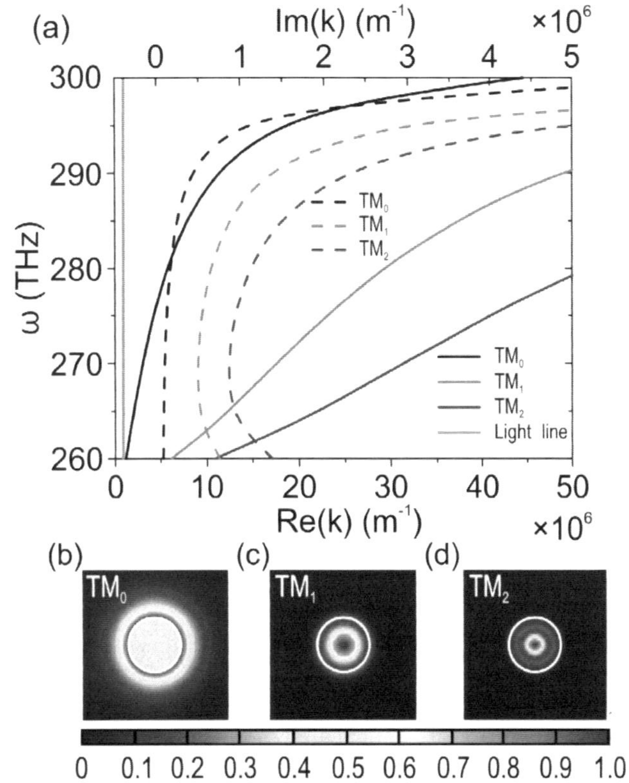

Figure 1. (a) Phonon polariton modes in the h-BN nanowire with a radius of 100nm. The normalized electric field distributions of the (b) TM_0, (c) TM_1, and (d) TM_2 modes.

3. Plasmon-phonon polaritons in the strong coupling regime

In the upper Reststrahlen band, h-BN supports TM-polarized phonon polariton modes. Since graphene plasmons are also TM-polarized, couplings between these two kinds of polaritons are possible. The graphene-wrapped h-BN nanowire is investigated in this paper as an analytic model for deriving the plasmon-phonon polaritons in the strong coupling regime, as schematically shown in Figure 2.

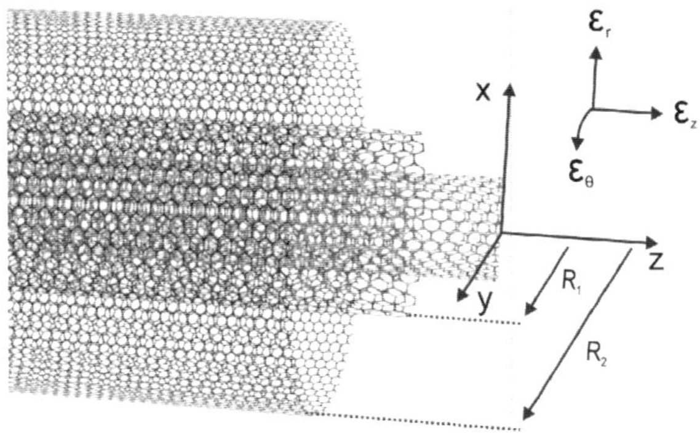

Figure 2. Schematic of the graphene-wrapped h-BN nanowire, where the propagation is along the z-axis. Graphene sheet is also rolled up into a cylinder, indicated by the large black tube with a radius of R_2, and the h-BN nanowire has a radius of R_1.

H-BN nanowire forms the core of the system, which has a radius of R_I; graphene layer is rolled up in to a cylinder, which has a radius of R_2. In this paper, we have $R_2 \geq R_I$. The permittivities along the radial, azimuthal and axial directions are respectively denoted as ε_r, ε_θ and ε_z. Given the lattice structures, we have $\varepsilon_\theta = \varepsilon_z = \varepsilon_{//}$ and $\varepsilon_r = \varepsilon_\perp$, where $\varepsilon_{//}$ and $\varepsilon_\perp$ are the permittivities parallel and perpendicular to the c-axis of h-BN [2-4]. In the calculations, the permittivities of the bulk h-BN were used; the TM-polarized plasmon-phonon polaritons were determined using the waveguide theory. After matching the boundary conditions, we have derived a determinant:

$$\begin{vmatrix} \mathrm{I}_0(a) & -\mathrm{I}_0(b) & -\mathrm{K}_0(b) & 0 \\ 0 & \mathrm{I}_0(c) & \mathrm{K}_0(c) & -\mathrm{K}_0(d) \\ -i\beta_1 \mathrm{I'}_0(a) & i\beta_2 \mathrm{I'}_0(b) & i\beta_2 \mathrm{K'}_0(b) & 0 \\ 0 & i\beta_2 \mathrm{I'}_0(c) - \gamma \mathrm{I}_0(c) & i\beta_2 \mathrm{K'}_0(c) - \gamma \mathrm{K}_0(c) & -i\beta_3 \mathrm{K'}_0(d) - \gamma \mathrm{K}_0(d) \end{vmatrix} = 0 \quad (2)$$

where $\gamma = Z_0\sigma / 2$ and $\beta_i = \alpha k_{ri}$ (i = 1, 2, 3. The h-BN core, the space outside graphene and the volume between them are respectively denoted by i = 1, i = 3, and i = 2.). Z_0 is the vacuum impedance and σ is the graphene conductivity. Letters *a*, *b*, *c and d* respectively denote $k_{rI}R_I$, $k_{r2}R_I$, $k_{r2}R_2$ and $k_{r3}R_2$. Other letters have the same meanings as in Equation (1). In this paper, graphene has been treated as a curved conductive surface. Using Equation 2, the dispersion relations can be determined. We chose R_I = 100 nm, R_2 = 200 nm. The fermi level of graphene can be varied, modifying the coupling strength between the phonon polaritons and the graphene plasmons.

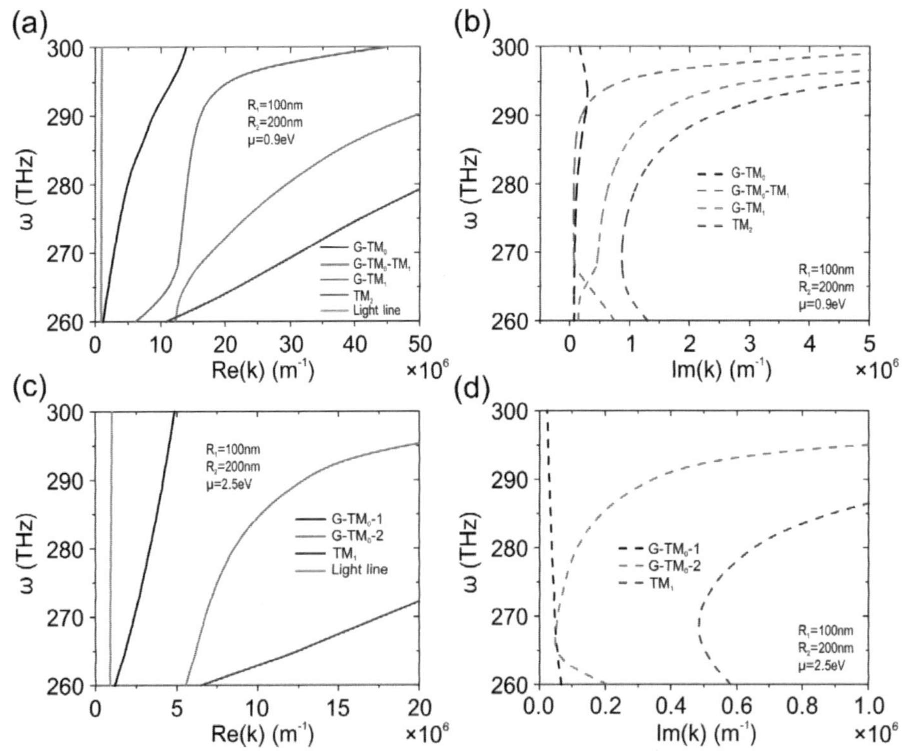

Figure 3. (a) The dispersion relations of the plasmon-phonon polariton modes when μ = 0.9 eV. (b) The corresponding imaginary parts of the modes in (a). (c) The dispersion relations of the plasmon-phonon polariton modes when μ = 2.5 eV. (d) The corresponding imaginary parts of the modes in (c).

Figures 3(a) and 3(b) respectively show the real and the imaginary parts of the propagation constants of the hybrid plasmon-phonon polaritons modes under the condition of $\mu = 0.9$ eV; Figures 3(c) and 3(d) respectively show the real and the imaginary parts of the propagation constants of the hybrid plasmon-phonon polaritons modes under the condition of $\mu = 2.5$ eV. With $\mu = 0.9$ eV, the dispersion relation of the graphene plasmons, denoted as G, simultaneously across with the TM_0 and TM_1 phonon polariton modes in h-BN nanowire, as shown in Figure 3(a), indicating two couplings: one is near 295 THz, forming the G-TM_0 plasmon-phonon polariton mode, denoted by the black curve; the other is near 265 THz, forming the G-TM_1 plasmon-phonon polariton mode, denoted by the red curve. The green curve in Figure 3(a) originates from the graphene plasmons; however, due to the mode coupling, it shows three distinct sections: the first one is below 265 THz; the second one is from 265 THz to 295 THz; the third one is above 295 THz. The first and the third sections are phonon-like, while the second section is plasmon-like. We come to these conclusions based on the slopes of the dispersion curves in three sections. The corresponding imaginary parts of the propagation constants (denoted as Im(k)) are plotted in Figure 3(b). At the coupling frequencies, their imaginary parts across with each other, showing totally different phenomena from Figure 3(a). With $\mu = 2.5$ eV, the dispersion relation of the graphene plasmons, denoted as G, only across with the TM_0 phonon polariton mode, causing only one coupling near 265 THz, as shown in Figures 3(c) and 3(d), forming two hybrid plasmon-phonon polariton modes, respectively denoted as G-TM_0-1 and G-TM_0-2. One can see that the high-order phonon polariton modes, like TM_1, are still important, since they can have strong couplings with other excitations, for example graphene plasmons. TM_2 and even higher order phonon polariton modes may also have such strong couplings, depending on the configuration of the composite systems.

4. Conclusion

In our paper, we show that the plasmon-phonon polaritons in graphene and h-BN composite structures can be formed in the strong coupling regime. The high-order phonon polariton modes in h-BN nanowires are still important due to the possibility of their strong couplings with other fundamental excitations, for example, plasmons. Due to the dispersive permittivity of h-BN, such strong coupling can occur simultaneously in two branches of phonon polaritons, depending on the parameters of the graphene. Our investigations may lead to the exploration of plasmon-phonon polaritons in graphene and h-BN composite structures within the strong coupling regime.

Acknowledgments

This work is supported by the Zhejiang Provincial Natural Science Foundation (Grant No. LQ17A040003) and the National Natural Science Foundation of China (Grant No. 11647117).

References

[1] Low T, Chaves A, Caldwell J D, Kumar A, Fang N X, Avouris P, Heinz T F, Guinea F, Martin-Moreno L and Koppens F 2017 *Nat. Mater.* **16** 182-194
[2] Kumar A, Low T, Fung K H, Avouris P and Fang N X 2015 *Nano Lett.* **15** 3172-3180
[3] Zhao Z W, Wu H W and Zhou Y 2016 *Opt. Express* **24** 22930-22942
[4] Xu X G, Ghamsari B G, Jiang J H, Gilburd L, Andreev G O, Zhi C, Bando Y, Golberg D, Berini P and Walker G C 2014 *Nat. Commun.* **5** 4782
[5] Li P, Lewin M, Kretinin A V, Caldwell J D, Novoselov K S, Taniguchi T, Watanabe K, Gaussmann F and Taubner T 2015 *Nat. Commun.* **6** 7507
[6] Dai S, Ma Q, Liu M K, Andersen T, Fei Z, Goldflam M D, Wagner M, K Watanabe, Taniguchi T, Thiemens M, Keilmann F, Janssen G C A M, Zhu S E, Jarillo-Herrero P, Fogler M M, Basov D N 2015 *Nat. Nanotech.* **10** 682-686
[7] Woessner A, Lundeberg M B, Gao Y, Principi A, Alonso-González P, Carrega M, Watanabe K, Taniguchi T, Vignale G, Polini M, Hone J, Hillenbrand R and Koppens F H L 2015 *Nat. Mater.* **14** 421-425

IOP Conf. Series: Journal of Physics: Conf. Series **844** (2017) 012033 doi:10.1088/1742-6596/844/1/012033

Dual-color super-resolution imaging by fluorescence emission depletion (STED) microscopy

Wen-sheng Wang[1], Cui-fang Kuang[1], Shao-cong Liu[1], Shi-yi Sun[1], Xu Liu[1]

[1]State Key Laboratory of Modern Optical Instrumentation, College of Optical Science and Engineering, Zhejiang University, Hangzhou, 310027, cfkuang@zju.edu.cn

Abstract We design and build a stimulated emission depletion (STED) microscopy which is capable of super-resolution imaging in two colors. The enhancement in the lateral resolution is demonstrated experimentally through fluorescent beads and biological samples. The dual-color modality enables the system to be suitable for further biological investigations.

1. Introduction

Limited by the diffraction, the spatial resolution of far-field optical microscopy is restricted to about half the wavelength of the illumination beams, which prevents its further application in the investigation of the interior of living cells. In order to overcome this bottleneck, several techniques have been proposed since 1990s, such as stimulated emission depletion microscopy(STED), stochastic optical reconstruction microscopy(STORM), photoactivated localization microscopy(PALM) and structured illumination microscopy(SIM). STED nanoscopy is typically based on a confocal scanning microscope, in which the focused excitation beam is overlaid by a STED beam featuring an intensity distribution of doughnut shape[1-3]. In addition, the wavelength of the STED beam is tuned to the red edge of the fluorophore's emission spectrum to stimulate the de-excitation of the excited molecules as well as to avoid the re-excitation of the fluorophores. Hence, the spontaneous emission is spatially confined to the central subregion of the excitation beam with a subdiffraction scale through the de-excitation of the STED beam. Furthermore, in multi-color STED instrument, the individual molecules are labelled with fluorophores of different emission spectrum and detected through respective color channel[4,5]. Here we propose and implement a dual-color STED system employing both pulsed and continuous wave (CW) laser beams, with one color of pulsed modality and the other of CW. We present a thorough characterization of this system by imaging fluorescent beads and its biological viability is also demonstrated.

2. Dual-color STED Setup

A schematic is shown in figure.1. For the green-emission fluorophore, two continuous wave laser beams are employed, a 488nm excitation laser (Thorlabs MCLS1) and 592nm STED beam (MPB Communications VFL-P-1000-592-OEM1). For the STED beam, the laser output is guide into a polarizing beam splitter (PBS) and two wave plates to modulate the polarization of the beams, guaranteeing the circular polarization before guided into the objective lens. Next, the depletion beam is sent through a Vortex Phase Plate (RPC Photonics, VPP-1A) to imprint the phase pattern necessary for forming the doughnut shape with a zero intensity at the center. On the other hand, two pulsed lasers are employed for the red-emitting fluorophore, 640nm laser diode (Becker&Hickl GmbH, BDL-SMN-640, 80MHz) for excitation and 750nm femtosecond Ti:Sapphire laser (Coherent, Chameleon Ultra I) for STED. The femtosecond laser pulse is firstly pre-stretched to about 250ps using a glass rod and a 100-meters-long polarization-maintaining single mode fiber (OZ OPTICS, PMJ-A3AHPC). The power of the pulsed STED beam is controlled by a half-wave plate and a PBS. Also, the output STED laser from the fiber is transferred to two wave plates and the Vortex Phase Plate to get the desired polarization and phase characteristics. In addition, the pulse synchronization of the excitation and STED beam is operated through a manual optical delay stage together with an electronic timing device. Next, the spatial co-alignment of the four beams is accomplished through three dichroic mirrors and the coaligned beams are finally guided into the galvo-mirror to realize the scanning through the samples. The fluorescent signal is imaged onto a multimode fiber with a core diameter of 62.5 um （Thorlabs, M31L01） and detected separately by two avalanche photo diodes after filtering by their respective band-pass filter.

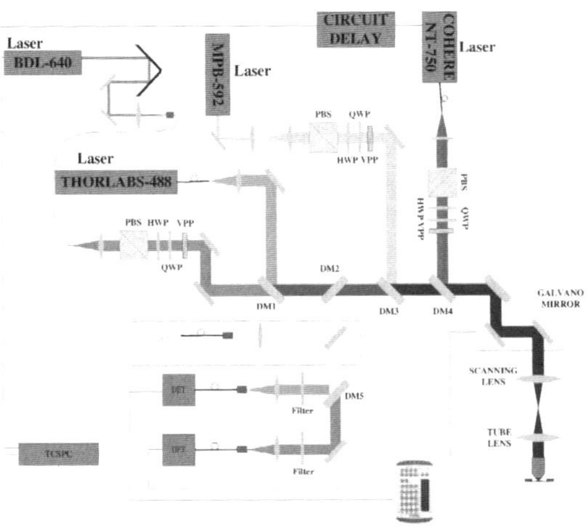

Fig. 1. Schematic of the proposed STED system.

3.Results and Conclusion

To verify the STED imaging capabilities elementarily, we test the dark red fluorescent beads (Thermofisher, 200nm, respectively). Figure.2 present the confocal and STED images of the fluorescent beads in 5 * 5 μm area with an excitation power of 20μW and depletion power of 60mW measured at the entrance pupil of the objective lens. The results of isolated beads shows an FWHM of about 200nm, which is the actual size of these beads. In addition, microtubes stained with Star635p is also observed under same experimental conditions (Figure 3). Contrary to the confocal image, some details are clearly discerned in the STED image, indicating an evident improvement in the spatial resolution.

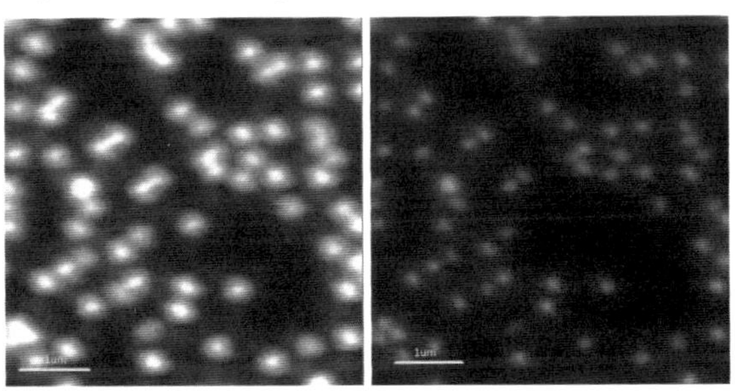

Fig. 2. Imaging of 200nm fluorescent beads. Confocal (left) and STED (right). The images are recorded with an excitation power of 20μW and depletion power of 60mW.

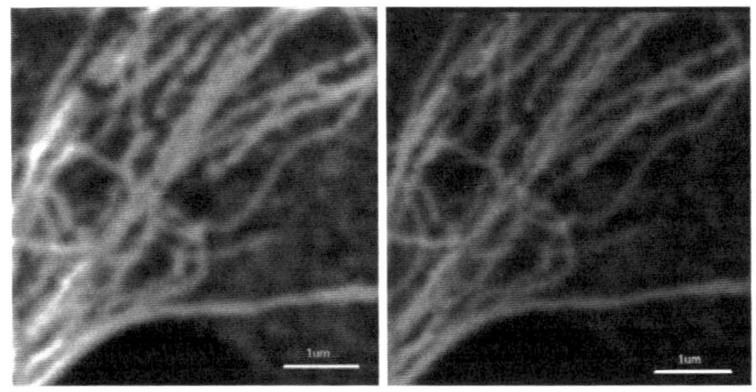

Fig. 3. Imaging of microtubes stained with Star635p. Experimental parameters is same with the Fig.2.

Acknowledgment

National Basic Research Program of China (973Program) (2015CB352003); The National Key Research and Development Program of China(2016YFF0101400); National Natural Science Foundation of China (NSFC) (61335003, 61377013, 61378051, 61427818); Natural Science Foundation of Zhejiang province (LR16F050001); and the Fundamental Research Funds for the Central Universities.

Reference

[1] Willig K I, Harke B, Medda R, et al. STED microscopy with continuous wave beams.[J]. Nature Methods, 2007, 4(11):915-918

[2] Eggeling C, Hell S W. STED Fluorescence Nanoscopy[J]. Springer, 2014, 14:3-25.

[3] Harke B, Keller J, Ullal C K, et al. Resolution scaling in STED microscopy[J]. Optics Express, 2008, 16(6):4154-4162.

[4] Han K Y, Ha T. Dual-color three-dimensional STED microscopy with a single high-repetition-rate laser[J]. Optics Letters, 2015, 40(11):2653.

[5] Osseforth C, Moffitt J R, Schermelleh L, et al. Simultaneous dual-color 3D STED microscopy[J]. Optics Express, 2014, 22(6):7028-7039.

A Novel WA-BPM Based on the Generalized Multistep Scheme in the Propagation Direction in the Waveguide

Yang Ji[1,2,*], Hong Chen[2], and Hongwu Tang[2]
[1]College of Science, Hohai University, Nanjing, China, 210098

[2]State Key Laboratory of Hydrology-Water Resources and Hydraulic Engineering, Hohai University, Nanjing, China, 210098

Corresponding author's e-mail: jiyang@hhu.edu.cn

Abstract. A highly accurate wide-angle scheme, based on the generalized mutistep scheme in the propagation direction, is developed for the finite difference beam propagation method (FD-BPM). Comparing with the previously presented method, the simulation shows that our method results in a more accurate solution, and the step size can be much larger

1. Introduction

The beam propagation method (BPM) has been one of the most popular techniques for modeling and simulation of optical guided-wave device. The original BPMs are derived from paraxial approximation of the scalar wave equation and hence can only perform the simulation of scalar wave propagation along waveguide axis in weakly guiding structures. To get rid of the paraxial restriction, several wide-angle (WA) BPMs have been developed in recent years. The original wide-angle algorithm was based on padé approximants given by Hadley[1,2]. The resulting formulas were derived by using the padé recurrence scheme to be approximated to the square root of the characteristic matrix. This formulation is, however, based on the Crank-Nicholson(CN) scheme with a truncation error of $O(\Delta x^2)$ in the transverse direction. Yamauchi *et al.* [3,4] proposed an improved wide-angle BPM based on a generalized Douglas(GD) scheme for variable coefficients. In their algorithm, the second derivative of function E with respect to x was improved by replacing the Crank-Nicholson (CN) scheme with the Douglas scheme. The truncation error of their algorithm was deduced to $O(\Delta x^4)$ in the transverse direction. Yevick *et al.*[5] introduced a wide-angle scheme using complex padé approximants. They rotated the branch-cut of standard padé expansion from the negative real axis into the lower half-plane. So the conventional real padé approximant is replaced by the modified padé approximant form with complex constants to evaluate the square-root operator in the exponential of the propagation operator. However, all of the above efforts were directed to the improvements in the transverse direction, i.e, x axis, while in the propagation direction, i.e, z axis, the operator was still approximated by the conventional CN scheme. This approximant have severely restricted the accuracy of the BPMs. To overcome this limitation, Chiou and Chang[6] proposed an improved WA-BPM in which they performed the padé approximants in the longitudinal directions as well as in the transverse direction. Although their method gives a more accurate result, it has to factor a polynomial of nth order in both the numerator and the denominator in the propagation equation in each step. So the efficient is limited.

In this letter we propose a novel wide-angle beam propagation method based on the generalized multistep scheme in the propagation direction. The exponential of the square root operator is approximated by a continued-multiplication algorithm which is generalizable to arbitrarily high order in the longitudinal direction. Combining with the previously presented algorithm, the simulation results show that the accuracy of our method is much higher and the longitudinal step size can be much larger.

2. Formulation

We consider a slowly varying waveguiding structure characterized by the refractive index $n(x, y, z)$, where x and y are the transverse direction and z is the propagation direction. By using the slowly varying envelope approximation formalism, the harmonic Maxwell equations can be reduced to the governing equation in

$$\frac{\partial^2\phi}{\partial z^2} - 2jk_0n_0\frac{\partial\phi}{\partial z} + k_0^2n_0^2\mathrm{P}\phi = 0 \tag{1}$$

in which ψ is a column vector of the transverse components of the electromagnetic field, k_0 is the wave number in free space, n_0 is the reference refractive index to be appropriately chosen, and P is the differential operator [5]. Hadley had solve Eq. (1) using the recurrence formula of the continued fraction expansion[2]. Here we solve Eq. (1) as a quadratic equation with respect to $\partial\phi/\partial z$ and obtain the solution of one-way propagation scheme:

$$\frac{\partial\phi}{\partial z} = -jk_0n_0\left(\sqrt{\mathrm{P}+1}-1\right)\phi \tag{2}$$

Eq. (2) is often discretized directly with the standard Crank-Nicolson scheme, resulting in the midpoint scheme:

$$\frac{\phi^{l+1}-\phi^l}{\Delta z} = -jk_0n_0\left(\sqrt{\mathrm{P}+1}-1\right)\frac{\phi^{l+1}+\phi^l}{2} \tag{3}$$

This treatment will lead to the conventional CN scheme. Considering Eq. (2) as a first-order differential equation, the more accurate solution can be given as exponential form:

$$\phi^{l+1} = \exp\left[-jk_0n_0\left(\sqrt{\mathrm{P}+1}-1\right)\Delta z\right]\phi^l \tag{4}$$

where Δz is the step size between the preceding field and the current field, and the superscript l indicates the position along the z axis. By using the standard padé approximate [7] to the square root of the transverse operator in Eq. (4), it becomes

$$\phi^{l+1} = \exp\left(-jk_0n_0\Delta z\sum_{k=1}^{m}\frac{a_{k,m}\mathrm{P}}{1+b_{k,m}\mathrm{P}}\right)\phi^l \tag{5}$$

or denoted as

$$\phi^{l+1} = \prod_{k=1}^{m}\exp\left(-jk_0n_0\Delta z\frac{a_{k,m}\mathrm{P}}{1+b_{k,m}\mathrm{P}}\right)\phi^l \tag{6}$$

Where

$$a_{k,m} = \frac{\sin^2(k\theta)}{m+0.5}, \qquad b_{k,m} = \cos^2(k\theta)$$

for $\theta = \pi/(2m+1)$. By introducing the parameter α, we rewritten the exponential of the square root operator as the following form:

$$\exp\left(\frac{-jk_0n_0\Delta z a_{k,m}P}{1+b_{k,m}P}\right) = \frac{\exp\left[(1-\alpha)\dfrac{-jk_0n_0\Delta z a_{k,m}\mathrm{P}}{1+b_{k,m}P}\right]}{\exp\left(-\alpha\dfrac{-jk_0n_0\Delta z a_{k,m}\mathrm{P}}{1+b_{k,m}P}\right)} \tag{7}$$

Here the original definition for the Euler's Number e is considered:

$$\lim_{x\to\infty}(1+\frac{1}{x})^x = e \tag{8}$$

Substituting Eq. (8) into the numerator of Eq. (7), the numerator can be expressed as

$$\exp\left[(1-\alpha)\frac{-jk_0n_0\Delta za_{k,m}\mathrm{P}}{1+b_{k,m}\mathrm{P}}\right]=\left(\lim_{x\to\infty}(1+\frac{1}{x})^x\right)^{\left[\frac{(\alpha-1)jk_0n_0\Delta za_{k,m}\mathrm{P}}{1+b_{k,m}\mathrm{P}}\right]} \tag{9}$$

Now we introduce a positive integer r $(r=1,2,3\cdots)$ and a function $f_r=r\left(1+b_{k,m}\mathrm{P}\right)/\left[(\alpha-1)jk_0n_0\Delta za_{k,m}\mathrm{P}\right]$. Note that, when r approaches ∞, f_r approaches ∞ too. So we can replace the x in Eq.(9) with f_r, and rewrite Eq. (9) as follows:

$$\begin{aligned}&\exp\left[(1-\alpha)\frac{-jk_0n_0\Delta za_{k,m}\mathrm{P}}{1+b_{k,m}\mathrm{P}}\right]\\&=\left(\lim_{r\to\infty}\left(1+\frac{1}{\frac{r\left(1+b_{k,m}\mathrm{P}\right)}{(\alpha-1)jk_0n_0\Delta za_{k,m}\mathrm{P}}}\right)^{\frac{r\left(1+b_{k,m}\mathrm{P}\right)}{(\alpha-1)jk_0n_0\Delta za_{k,m}\mathrm{P}}}\right)^{\frac{(\alpha-1)jk_0n_0\Delta za_{k,m}\mathrm{P}}{1+b_{k,m}\mathrm{P}}}\\&=\lim_{r\to\infty}\left(1+\frac{(\alpha-1)jk_0n_0\Delta za_{k,m}\mathrm{P}}{r\left(1+b_{k,m}\mathrm{P}\right)}\right)^r\end{aligned} \tag{10}$$

Similarly the denominator in Eq. (7) can be rewritten as

$$\exp\left(-\alpha\frac{-jk_0n_0\Delta za_{k,m}\mathrm{P}}{1+b_{k,m}\mathrm{P}}\right)=\lim_{r\to\infty}\left(1+\frac{\alpha jk_0n_0\Delta za_{k,m}\mathrm{P}}{r\left(1+b_{k,m}\mathrm{P}\right)}\right)^r \tag{11}$$

Then substituting Eq. (10) and Eq. (11) into Eq. (7), and using the basic limit theorem that is, the quotient of limits is a limit of the quotient, we obtain

$$\psi(z+\Delta z)\approx\prod_{k=1}^{m}\lim_{r\to\infty}\left(\frac{1+\frac{1-\alpha}{r}(-jk_0n_0\Delta z)\frac{a_{k,m}\mathrm{P}}{1+b_{k,m}\mathrm{P}}}{1+\frac{-\alpha}{r}(-jk_0n_0\Delta z)\frac{a_{k,m}\mathrm{P}}{1+b_{k,m}\mathrm{P}}}\right)^r\psi(z) \tag{12}$$

This is the exact solution for Eq.(6). It can be seen that when r is large enough, the formalism can be generalizable to arbitrarily high order, but it will lead a concomitant increase in numerical complexity. In actual simulation, we can choose a proper value for r, considering the calculation efficiency. Finally, assuming $r=R$ and $m=M$, we obtain the following generalized multi-step algorithm as

$$\psi(z+\Delta z)\approx\prod_{k=1}^{M}\left(\frac{1+\frac{1-\alpha}{R}(-jk_0n_0\Delta z)\frac{a_{k,M}\mathrm{P}}{1+b_{k,M}\mathrm{P}}}{1+\frac{-\alpha}{R}(-jk_0n_0\Delta z)\frac{a_{k,M}\mathrm{P}}{1+b_{k,M}\mathrm{P}}}\right)^R\psi(z) \tag{13}$$

and reduced to

$$\psi(z+\Delta z)=\prod_{k=1}^{MR}\frac{1+\left(b_{k,M}-\frac{1-\alpha}{R}a_{k,M}jk_0n_0\Delta z\right)\mathrm{P}}{1+\left(b_{k,M}+\frac{\alpha}{R}a_{k,M}jk_0n_0\Delta z\right)\mathrm{P}}\psi(z) \tag{14}$$

where R is a positive integer which is the order of the continued-multiplication approximant of the exponential function in the propagation direction, M is the order of the padé approximant of the square root in the transverse direction, and α $(1/2\le\alpha\le1)$ is a weighted factor employed to ensure the convergence.

The case with α =1/2 corresponds to the Crank-Nicholson scheme and the case with α =1 corresponds to the backward Euler method. Note that when R=1 and α=1/2, Eq. (14) is deduced to the frequently used Hadley's wide-angle BPM with the CN implicit scheme:

$$\psi(z+\Delta z)=\prod_{k=1}^{M}\frac{1+\left(b_{k,M}-\frac{1}{2}a_{k,M}jk_0n_0\Delta z\right)\mathrm{P}}{1+\left(b_{k,M}+\frac{1}{2}a_{k,M}jk_0n_0\Delta z\right)\mathrm{P}}\psi(z) \tag{15}$$

The conventional Hadley's algorithm is the simplest case of our derived algorithm. From the derivation above, it can be proved that the higher the value of R is, the more approximate to the exact one the simulation result becomes. A higher value of R not only leads to a more accurate result but also means that the longitudinal step Δz can be much larger.

In the analysis of a wide-angle waveguide structure by the semi-vector or full-vector BPM, the z-varying structure is discretized by piecewise z-invariant segments. The discontinuousness of the refract index at the longitudinal interface will excite the evanescent modes which respond to the negative eigenvalues of the operator $\mathrm{P}+1$ of the eigenmode equation. Assuming that p is one of the eigenvalues of P, we have $\left|\exp\left[-jk_0n_0\left(\sqrt{p+1}-1\right)\Delta z\right]\right|<1$, for $p<-1$. This means that for the exact one-way beam propagation, the evanescent mode are always damped. But the conventional Hadley's scheme with CN scheme can not suppress the evanescent modes. It can be shown that the advancing ψ in spatial domain by one longitudinal step corresponds to the amplification factor g_z given by

$$g_z=\left[1+\left(b_{k,M}-\frac{1-\alpha}{R}a_{k,M}jk_0n_0\Delta z\right)\mathrm{P}\right]\Big/\left[1+\left(b_{k,M}+\frac{1-\alpha}{R}a_{k,M}jk_0n_0\Delta z\right)\mathrm{P}\right] \tag{16}$$

It can be proved that the finite-difference propagation scheme are stable if $|g_z|\le1$. For the case of α =1/2, the formula is unconditionally stable and is ideally power conserving, since $|g_z|$=1. Although the Crank-Nicholson method corresponding to α =1/2 has the desirable power-conserving property for propagating modes in a straight waveguide, it incorrectly treats the evanescent modes as propagating modes. The evanescent modes are often excited in the analysis of semi-vector or full-vector BPM for three-dimension structures. If they are not suppressed efficiently, they will accumulate and give rise to a large error in the final solution. Our wide-angle method of Eq. (14) provide the desired damping for the evanescent modes, since $|g_z|<1$ for $1/2<\alpha<1$.

3. Simulation and discussions

We demonstrate the accuracy and usefulness of our method in a tilted slab waveguide shown in Fig.1. The refractive indices of core and cladding are n_1=1.469 and n_2=1.460, respectively. The core width is w=3 μm. The transverse sampling width is Δx =0.1 μm, and the tilt angle is $\theta = 30^\circ$. The wavelength is λ =1.55 μm. The reference refractive index is n_0=($n_1 + n_2$)/2.

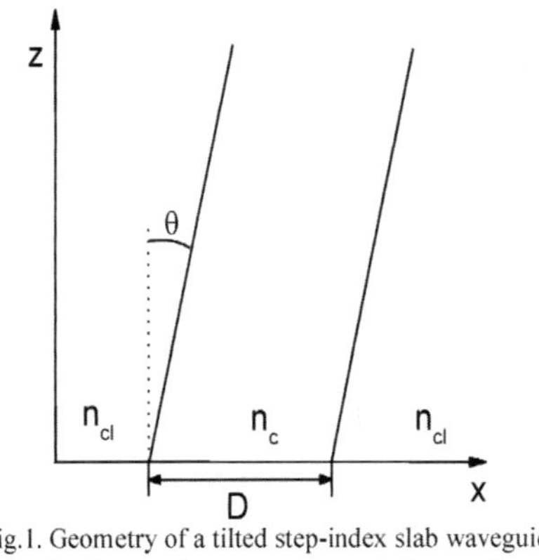

Fig.1. Geometry of a tilted step-index slab waveguide

For boundaries of numerical window, the transparent boundary condition (TBC) is used to suppress the reflections at the edges of the finite computational window. The (3,3) padé approximant operator is employed to the square root of the transverse operator. The second derivative in transverse direction is evaluated by the generalized Douglas(GD) scheme[3]. To validate the accuracy of the present method, we calculate the mode-mismatch loss η, which take into account the dissipation of the initial power as well as the deformation of the propagation field. In a two-dimensional waveguide, η is defined as

$$\eta = -10\log\left[\left|\int E_0 E^*\right|^2 \Big/ \left(\left|E_0\right|^2 dx\right)^2\right][dB] \tag{17}$$

where E is the propagating field and E_0 is the incident field of the fundamental mode.

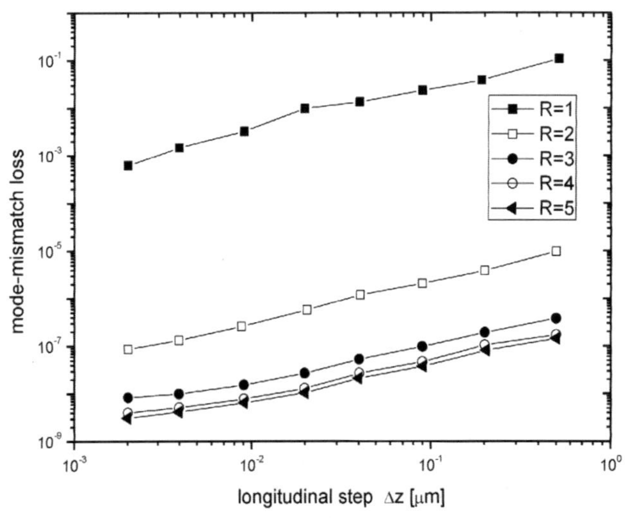

Fig.2. Mode-mismatch loss as a function of step size Δz in the longitudinal direction using the same weighted factor α =1/2. The tilt angle is set at $\theta = 30^\circ$ and the order of padé is M=3.

Fig.2 shows the mode-mismatch loss η evaluated at a propagation distance of 20 μm as a function of the step size in propagation direction. The curve denoted as R=1 corresponds to the conventional Hadley's method. It can be seen that comparing with the conventional method, the accuracy of our method is greatly enhanced when the higher-order value of R is adopted. The presented formulation is generalizable to arbitrarily high order, with a concomitant increase in numerical complexity. Not that,

IOP Conf. Series: Journal of Physics: Conf. Series **844** (2017) 012034 doi:10.1088/1742-6596/844/1/012034

although the simulation results in a more accurate solution when a higher value R is used, the reduction of η descends progressively with the growth of R. In our simulation, the order of the generalized multistep scheme is set to be 4. It is also shown in Fig.2 that, for a fixed η, the step size of our methods can be much larger than that of the conventional method. This means that we could apply it to simulate the large-size photonic integrated devices.

Conclusion

In conclusion, we have proposed an efficient and accurate wide-angle beam propagation method based on the generalized multi-step scheme in the propagation direction. Both two-dimension and three-dimension cases were demonstrated. Compared with the conventional method, our method show higher accuracy and the step size can be much larger.

References

[1] G. D. Hadley, "Wide-angle beam propagation using padé approximant operators," Opt. Lett. **17**, 1426-1428 (1992).
[2] G. D. Hadley, "Multistep method for wide-angle beam propagation," Opt. Lett. **17**, 1742-1745 (1992).
[3] Yamauchi, J. Shibayama, and H. Nakano, "Wide-angle propagating beam analysis based on the generalized Douglas scheme for variable coefficients," Opt. Lett. **20**, 7-9 (1995).
[4] J. Yamauchi, J. Shibayama, M. Sekiguchi, and H. Nakano, "Improved multistep method for wide-angle beam propagation," IEEE Photon. Technol. Lett. **8**, 1361-1363 (1996).
[5] D. Yevick, "The application of complex padé approximants to vector field propagation," IEEE Photon. Technol. Lett. **12**, 1636-1638 (2000).
[6] Y.P.Chiou and H.C.Chang, "Efficient beam-propagation method based on padé approximants in the propagation direction," Opt. Lett. **22**, 949-951 (1997).
[7] G. A. Baker, Essentials of padé approximants (Academic, New York, 1975)
[8] A.Behie and P. K. W. Vinsome, "Block iterative methods for fully implicit reservoir simulation," Soc. Pet. Eng. J. **22,** 658-668 (1982)
[9] H. A. Van Der Dorst, "Bi-CGSTAB: a fast and smoothly converging variant of Bi-CG for the solution of nonsymmetric linear systems," SIAM J. Sci. Statist. Comput. **13**, 631-644 (1992)
[10] W. P. Hung, C. L. Xu, W. Lui, and K, Yokoyama, "The perfectly matched layer (PML) boundary condition for the beam propagation method," IEEE Photon. Technol. Lett. **8** 649-651 (1996)
[11] Y. Y. Lu and J. Zhu, "Propagation modes in optical waveguides terminated by perfectly matched layer," IEEE Photon. Technol. Lett. **17** 2601-2603 (2005)

IOP Conf. Series: Journal of Physics: Conf. Series **844** (2017) 012035 doi:10.1088/1742-6596/844/1/012035

Luminescent AIZS-GO nanocomposites as fluorescent probe for detecting copper (II) ion

Liu Yongfeng[1], Deng Ming[1,2]*, Tang Xiaosheng[1], Zhu Tao[1]

[1]Key Laboratory of Optoelectronic Technology & Systems (Ministry of Education), Chongqing University, Chongqing, China, 400044

[2]Guangxi Colleges and Universities Key Laboratory of Optoelectronic Information Processing, Guilin University of Electronic Technology, Guilin, China, 541000

Abstract

A selective copper (II) sensor based on graphene oxide (GO) modified Ag-In-Zn-S (AIZS) quantum dots (QDs) was proposed. The fluorescence of the water-soluble AIZS QDs was quenched after addition of Cu^{2+} ion, which was triggered by the aggregation of the QDs in the form of R-COO-(Cu^{2+})-OOC-R, as verified by the TEM and FTIR. The fluorescence probe can be applied to detect Cu^{2+} ion in a wide concentration range of 0-850 μM based on the degree of fluorescence quenching, with its limitation of detection (LOD) and R^2 of0.18μM and 0.99, respectively. Moreover, the nanosensor can distinguish Cu^{2+} ion from other ions, offering great potentials in copper (II) determination in drinking water.

1. Introduction

Copper element is one of crucial transition metallic elements to human health and many living tissues. It is physiologically essential in various aspects such as bone formation, cellular respiration, and connective tissue development and serves as the significant catalytic cofactor for hemoglobin, elastin, and collagen [1, 2]. In the recent decades, quantum dots (QDs), zero-dimensional nanomaterials, are ever-increasingly attractive to scientists, and have already become a hot issue [3-5]. Compared with traditional organic dyes and lanthanide probes, such materials possess broad absorption spectra, tunable wavelength, big Stokes shift, fueled by the quantum confinement, and provide a new method to the quantification of heavy metals [6-9]. Up to now, several types of functionalized QDs have been used for selective ion recognition such as silver, copper, cadmium, and mercury [10-16]. Phase transfer by thiol-derivative molecules into water-soluble QDs always renders low quantum yield (QY) (~1%), which exerts detrimental influence on the sensitivity of these developments [17]. In addition, the time response of theses sensors is relatively long, which is unfavorable in practical applications.

In the work, water-soluble AIZS QDs transferred by GO were synthesized and used as fluorescent probe for copper (II) detection. The AIZS-GO nanocomposites have a QY as high as (~25%), which is much higher than the thiol-derivative modified water-soluble QDs mentioned above. Its pH value effect and selectivity toward copper were discussed, and quenching mechanism was experimentally verified by the Fourier transform infrared (FTIR) spectra and TEM images. The fluorescent probe has a wide detection span of 0 - 850μM, detection of limitation of 0.18μM, and the fast time response of one minute, which is superior to that of the probes mentioned above. This offers great potentials in copper (II) detection in drinking water.

2. Experimental

2.1. Copper (II) studies

The quenching effect of Cu^{2+} ion on the fluorescence of GO AIZS QDs was performed as following procedures. Stock solution of copper (II) (50 mM) was prepared by the dissolution of copper sulfate in deionized water. 2 mL as-prepared water-soluble AIZS was added to a 10 mL volumetric flask with the following addition of some Tris-HCl buffer solution to adjust the solution pH value at 7.4, and then 3 μL stock solution. The mixture was finally diluted to 5.0 mL by deionized water and the final concentration of Cu^{2+} ion and AIZS QDs are 30 μM and 0.5 mg/mL, respectively. The fluorescence intensity was measured under the excitation wavelength of 365 nm at the room temperature at different response time using a fluorescence spectrophotometer with an accelerating voltage of 200 kV. The slit widths of excitation and emission are 10 nm and 10 nm, respectively.

2 mL as-prepared water-soluble AIZS QDs, Tris-HCl buffer solution and different amount of Stock solution of copper (II) were sequentially added into a 10 mL volumetric flask. After thoroughly shaking for one minute, the fluorescence was recorded by a fluorescence spectrophotometer in the same operation conditions as the above.

3. Results and discussion

3.1. Characteristics and optical properties of AIZS-GO QDs

The absorption and fluorescence spectra were shown in Fig.1 (A). We can see that the absorption spectra of AIZS (black curve) and AIZS-GO (red curve) QDs are almost identical in shape. The PL spectrum of AIZS-GO QDs (purple curve) keeps nearly the same as AIZS QDs (blue curve) after phase transfer, and there is no wavelength shift in PL spectrum but only a small difference is PL intensity, which is mainly due to the fact that the surface traps were exposed to the aqueous solution and the existence of blackish GO. The full width at half maximum (FWHM) fluorescence spectra of AIZS and AIZS-GO QDs were both about 100nm and symmetric, indicating that they were homogeneous. Therefore, we can draw a conclusion that phase transfer by GO did not affect its optical properties. The size and morphology of AIZS-GO QDs were evaluated by TEM as shown in Fig.1 (B). From Fig.1 (B) and (D), we can clearly observe that the average diameter is about 5.2 nm and the QDs are monodisperse. All the characterizations indicate that AIZS-GO QDs were successfully synthesized.

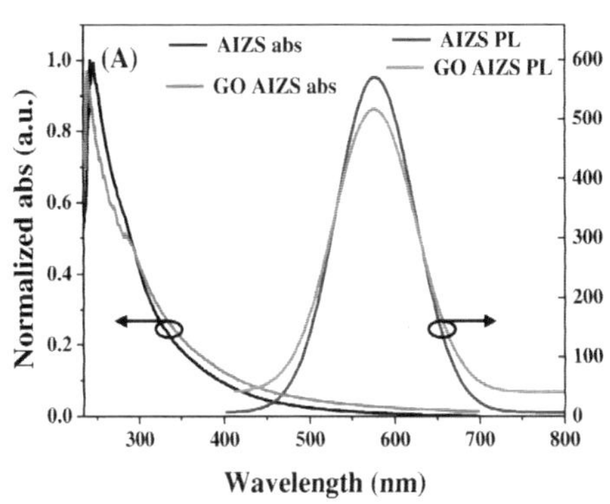

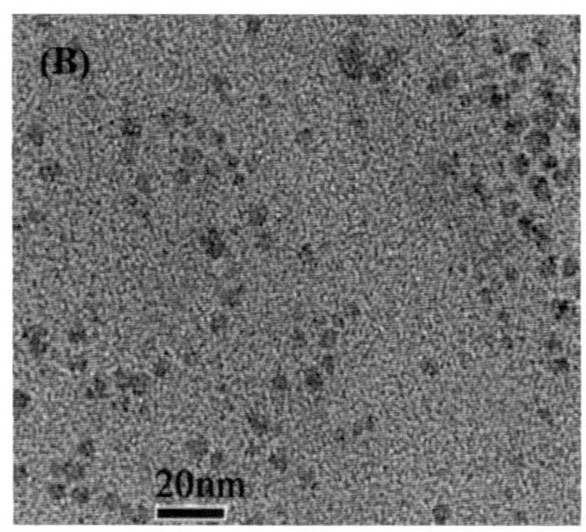

Fig.1. (A) Absorbance and PL spectrum of AIZS and AIZS-GO QDs; (B) TEM image of AIZS-GO QDs。

3.2. Copper (II) studies

Time response of the AIZS-GO QDs in 10 μM Cu^{2+} ion was studied (data not shown). It is clear that the PL keeps stable after one minute of the addition of copper (II), which indicates that the AIZS-GO QDs can be used as a fast time response sensor. Time response (1 minute) is much faster than reported in [18] (~10 minutes) and [19] (~20 minutes). Therefore, the PL in the following experiments was recorded at one minute after the addition of copper (II).

The PL dependence on Cu^{2+} ion was analyzed by adding different Cu^{2+} ion concentration into probe solution, as shown in Fig.2 (A). We can observe that the PL intensity decreased after addition of copper (II) with an obvious red shift triggered by increasing the size scale of QDs, and when copper (II) concentration was larger than 850 μM, there was no significant quenching anymore in the PL intensity in that binding rate between carboxyl groups and Cu^{2+} ion decreased with the increase of copper (II) concentration, and gradually saturated.

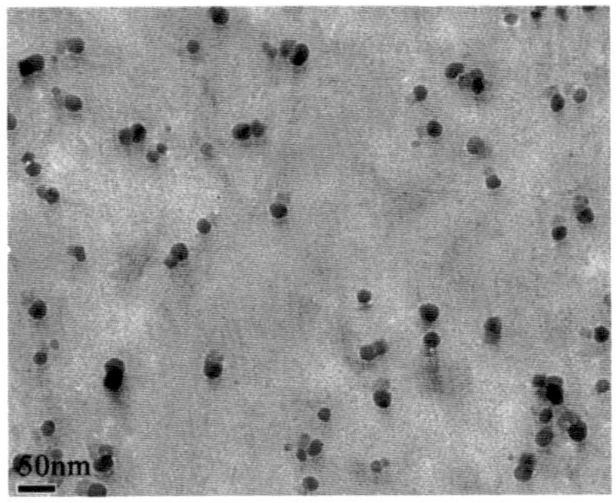

Fig. 2 TEM image of the QDs in the presence of 850μM Cu^{2+}.

Upon the addition of Cu^{2+} ion into the solution, the Cu^{2+} ion was drawn to the QDs physically due to the negatively charged carboxyl groups on GO, and then chemical reaction occurred between carboxyl groups and the added Cu^{2+} ion. As a result, the carboxylate copper complex in the form of R-COO-(Cu^{2+})-OOC-R was formed, aggregating the QDs, and the size scale of AIZS-GO QDs ion became larger, an indication of the quenching of fluorescence [20]. This quenching mechanism was experimentally proved by the TEM image as shown in Fig. 2. It can be seen that the average diameter of AIZS-GO QDs in the presence of 850 μM Cu^{2+} ion was 21.7 nm, which is larger than that in solutions without Cu^{2+} ion, verifying the quenching mechanism.

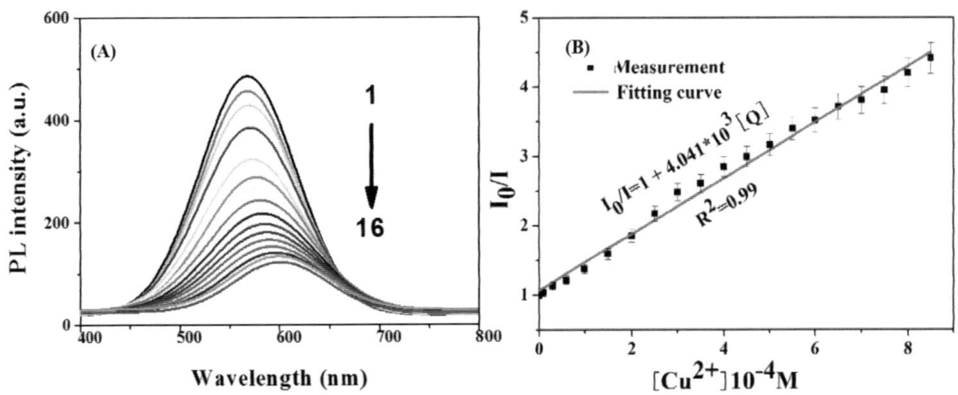

Fig. 3 (A) The fluorescence spectra of the QDs after one minute of the addition of different Cu^{2+} concentrations. The Cu^{2+} concentrations from 1 to 16 were 0, 10, 30, 60, 100, 150, 200, 250, 300, 350, 400, 500, 550, 650, 750, 850 μM, respectively; (B) the calibration curve for the PL intensity of the QDs with the function of Cu^{2+} ion, extracted from peak PL intensity in Fig. 3 (A).

The relationship between I_0/I and Cu^{2+} ion concentration is shown in Fig. 3 (A). The PL dependence on Cu^{2+} ion concentration can be best expressed by the following Stern-Volmer equation:

$$I_0/I = 1 + K_{SV}\,[Q]$$

Where I_0 and I are the fluorescence intensity of quantum dots in solution without and with a given concentration of copper(Ⅱ) respectively. Herein, K_{SV} is the Stern-Volmer constant (slope), 4.041×10^3 /M in this study. As shown in Fig. 3 (B), a good linear relationship between I_0/I and Cu^{2+} ion concentration is clearly observed with the R^2 of 0.99 and a broad linear span of up to 850μM which is larger than ones reported by others [21-23]. The limitation of detection (LOD) is defined by $3\sigma/K_{SV}$, where the σ and K_{SV} are standard deviation (n=10) and the sensitivity of the calibration curve (slope), is 0.18 μM.

3.3. The selectivity of AIZS-GO QDs

The selectivity toward Cu^{2+} ion was evaluated by measuring PL intensities of QDs in the presence of six other kinds of heavy metal ions of 500 μM in the same experimental condition as the Cu^{2+} ion as shown in Fig. 4. It can be seen that in the presence of other 500 μM metal ions, the PL intensity of the QDs probe had slight changes, possibly due to the relatively weak binding ability between the carboxyl groups and other metal ions, which indicates that other cations exert small influence on determination for Cu^{2+} ion, making the solution sensor in this paper a selective and practical probe.

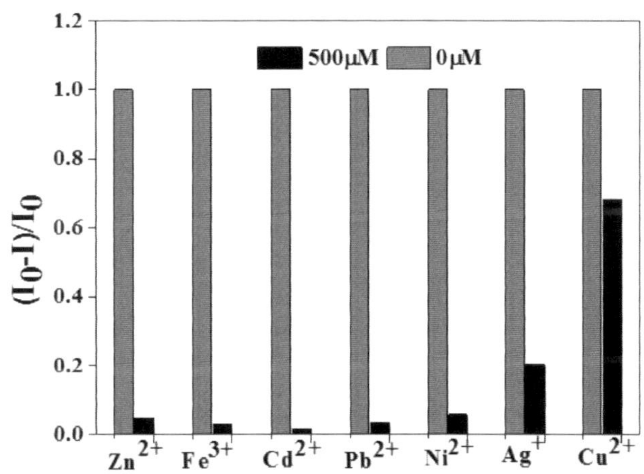

Fig. 4 Relative PL intensity of the QDs at pH=7.4 in the absence (red column) and presence (black column) of 500 μM solution containing a specified metal cation.

4. Conclusions

In this work, cadmium-free AIZS QDs in aqueous solution were successfully used as a fluorescent probe for Cu^{2+} ion detection in the samples. In the presence of Cu^{2+} ion, the PL quenched due to the quantum dots aggregation caused by carboxylate copper complex formation through reaction between carboxyl group and Cu^{2+} ion, and the relationship between I_0/I and Cu^{2+} ion concentration can be well described by the Stern-Volmer equation. The fluorescent probe possesses good selectivity toward Cu^{2+} ion and a wide linear range of 0-850μM with short time response within one minute, which offers great potentials in copper (Ⅱ) detection in drinking water.

Acknowledgement

This work is supported by the Natural Science Foundation of China under Grants (No.61475028, 61475029 and 61575190, the Chongqing Outstanding Youth Project under Grant cstc2014jcyjjq40002) and the Fundamental Research Funds for the Central Universities under Grant 106112015CDJXY120010.

References

[1] L.R. McDowell, "Minerals in Animal and Human Nutrition," 2nd ed., Elsevier: New York, 2003.

[2] P. A. Balch, "Prescription for Nutritional Healing," 4th ed., Penguin Group: New York, 2006. 17–22 (2001).

[3] B.K. Chen, H.Z. Zhong, Z.A. Tan, "Highly Emissive and Color-Tunable CuInS2-Based Colloidal Semiconductor Nanocrystals: Off-Stoichiometry Effects and Improved Electroluminescence Performance," Adv. Funct. Mater. 22(10), 2081–2088(2012).

[4] T.T. Zhao, X.Q. Shen, Z.P. Guan, Q.H. Xu, "Gold nanorods as dual photo-sensitizing and imaging agents for two-photon photodynamic therapy," Nanoscale 4(24), 7712–7719 (2012).

[5] F. Huang, J.C. Zhou, Y.H. Wang, "Formation of $AgGaS_2$ nano-pyramids from Ag_2S nanospheres through intermediate Ag_2S–$AgGaS_2$ heterostructures and $AgGaS_2$ sensitized Mn^{2+} emission," Nanoscale, 6(4), 2340–2344 (2014).

[6] X.H. Gao, L.L. Yang, S.M. Nie, "In vivo molecular and cellular imaging with quantum dots," Current Opinion in Biotechnology 16(1), 63–72 (2005).

[7] A.M. Smith, X.H. Gao, S.M. Nie, "Quantum dot nanocrystals for in vivo molecular and cellular imaging," Photochemistry and. Photobiology 80(3), 377–385 (2004).

[8] B.J. Marcel, M. Mario, G. Peter, W. Shimon, A.P. Alivisatos, "Semiconductor Nanocrystals as Fluorescent Biological Labels," Science 281(25), 2013–2016 (1998).

[9] X. Michalet, F.F. Pinaud, L.A. Bentolila, J.M. Tsay, S. Weiss, "Quantum Dots for Live Cells, in Vivo Imaging, and Diagnostics," Science 307(28), 538–544 (2005).

[10] J.G. Liang, X.P. Ai, Z.K. He, "Functionalized CdSe quantum dots as selective silver ion chemodosimeter," Analyst 129(7), 619-622 (2004).

[11] K.M, Gattas-Asfura, R.M. Leblanc, "Peptide-coated CdS quantum dots for the optical detection of copper(II) and silver(I)," Chem. Commun. 2003(21), 2684 – 2685 (2003).

[12] M.J. Ruedas-Rama, X. Wang, E. A. H. Hall, "A multi-ion particle sensor, Chem. Commun.2007(15)," 1544–1546 (2007).

[13] A. Touceda-Varela, E.I. Stevenson, J.A. Galve-Gasion, D. T. F. Dryden, "Selective turn-on fluorescence detection of cyanide in water using hydrophobic CdSe quantum dots," Chem. Commun., 2008(17), 1998–2000 (2008).

[14] J. Chen, A. Zheng, Y. Gao, C. He, G. Wu, Y. Chen, X. Kai, X. Zhu, "Functionalized CdS quantum dots-based luminescence probe for detection of heavy and transition metal ions in aqueous solution," Spectrochim. Acta, Part A 69(3'), 1044–1052 (2008).

[15] C. Dong, H. Qian, N. Fang, J.J. Ren, "Study of Fluorescence Quenching and Dialysis Process of CdTe Quantum Dots, Using Ensemble Techniques and Fluorescence Correlation Spectroscopy," Phys. Chem. B 110(23), 11069–11075 (2006).

[16] H.B. Li, Y. Zhang, X. Q. Wang, "L-Carnitine capped quantum dots as luminescent probes for cadmium ions," Sens. Actuators, B 127, 593–597 (2007).

[17] Z.Q. Liu, S.P. Liu, P.F. Yin, Y.Q. He, "Fluorescence enhancement of CdTe/CdS quantum dots by coupling of glyphosate and its application for sensitive detection of copper ion," Analytica Chimica Acta 745, 78-84 (2012).

[18] Y.L. Ding, S.Z. Shen, H.D. Sun, K.N. Sun, F.T. Liu, "Synthesis of L-glutathione-capped-ZnSe quantum dots for the sensitive and selective determination of copper ion in aqueous solutions," Sensors and Actuators B 203, 35–43 (2014).

[19] C. Yang-Hsiang, J.X, Chen, J.D. Batteas, "Ultrasensitive copper(II) detection using plasmon-enhanced and photo-brightened luminescence of CdSe quantum dots," Anal. Chem. 82(9), 3671-3678 (2010).

[20] A.M. Munro, I.J-L, Plante, M.S. Ng, D.S. Ginger, "Quantitative Study of the Effects of Surface Ligand Concentration on CdSe Nanocrystal Photoluminescence," J. Phys. Chem. C 111(17), 6220–6227 (2007).

[21] Y.F. Chen and Z. Rosenzweig, "Luminescent CdS Quantum Dots as Selective Ion Probe," Anal. Chem. 74(19), 5132-5138 (2002).

[22] G.X. Liang, H.Y. Liu, J.J. Zhu, "Ultrasensitive Cu^{2+} sensing by near-infrared-emitting CdSeTe alloyed quantum dots," Talanta 80(5),2172–2176 (2010).

[23] S. Kawakubo, H. Kato and M. Iwatsuki, "Catalytic spectrofluorimetric determination of copper using aerial oxidation of ascorbic acid in the presence of o-Phenylenediamine," Analyst 119(9), 2119-2122 (1994).

IOP Conf. Series: Journal of Physics: Conf. Series **844** (2017) 012036 doi:10.1088/1742-6596/844/1/012036

All-fiber power sensor based on silicon-germanium core fiber F-P cavity

Tengjiao Wang, Na Chen*, Ziwen Zhao, Fufei Pang, Zhenyi Chen and Tingyun Wang

Key Laboratory of Specialty Fiber Optics and Optical Access Networks, Shanghai University, 149 Yanchang Road, Shanghai 200072, China

*E-mail: na.chen@shu.edu.cn

Abstract. An all-optical power sensor based on silicon-germanium (SiGe) core fiber Fabry-Perot (F-P) cavity is proposed. The F-P sensor was formed by fusion splicing SiGe core fiber with conventional single-mode fiber (SMF). The 980 nm laser can be absorbed efficiently by the silicon-germanium material, resulting in the temperature increase inside the cavity, which induces the drifts of the reflection spectrum. The injected power has a very good linear relationship with the drifts experimentally, and the power sensitivity is up to 1.7 nm/mW, which may have great potential for all-fiber power detector.

1. Introduction

Recently, due to the excellent material advantage of semiconductor core fiber, its development has been widely concerned. Especially semiconductor core fiber takes advantages of high nonlinearity characteristics, high Raman gain and medium-infrared transparency in optical properties [1-2], so its presence has already given optical fiber new features and enormous possibilities for fiber-based devices and sensors. Since J. Ballato, et al. firstly proposed molten core method to prepare silicon core fiber in 2008 [3], the research on semiconductor core fiber has got rapid development. Various kinds of semiconductor core fiber have been prepared, such as germanium core fiber, zinc-selenide core fiber and so on [4-5]. Further characteristic exploration of semiconductor core fiber provides the basis for the realization of all-optical device. For example, silicon core fiber (SCF) has a thermo-optic coefficient (TOC) approximately 10 times larger than that of fused silica, resulting in higher temperature sensitivity [6]. And the large thermal diffusivity of the SCF leads to high response speed, which is more than 60 times larger than that of fused silica [7]. Especially, SiGe alloy has the greater TOC compared with silicon. Thus, SiGe core fiber can potentially provide better performance for optical fiber sensing technology. Currently, most of cavity medium of F-P sensing devices is either silica or air, the temperature sensitivity of fiber FPI is only 14 pm/℃ owing to the limitation of fiber sensor materials [8]. Such a low sensitivity cannot meet the actual requirements in some special cases. Therefore, the SiGe core fiber has been introduced to improve the performance of sensor.

In this paper, we present an all-fiber optical power sensor based on the silicon-germanium core F-P cavity, and demonstrate its power sensitivity is up to 1.7 nm/mW when 980 nm laser irradiating on the SiGe core cavity.

IOP Conf. Series: Journal of Physics: Conf. Series **844** (2017) 012036 doi:10.1088/1742-6596/844/1/012036

2. Principle

The schematic diagram of SiGe core F-P sensor is shown in Fig. 1. The silicon-germanium material of F-P cavity has a strong absorption between 500 nm and 1000 nm and could transform it into internal energy high efficiently inside the cavity [9]. This photothermal effect causes temperature distribution on F-P cavity, which leads to interference spectrum variation. In this scheme of Fig. 1, 980 nm laser is injected into the SiGe core formed F-P cavity. Meanwhile, broadband light with the central wavelength of 1550 nm is also guided into the cavity to monitor reflection spectrum variation, in order to characterize its response to the injected laser power. Different injected laser power corresponds to varying degrees of interference wavelength drift.

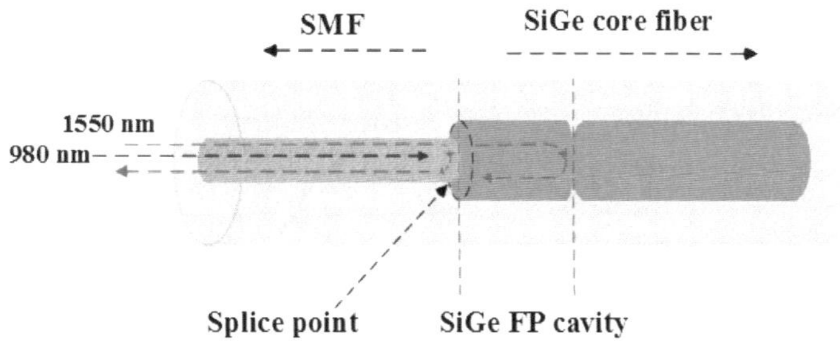

Fig. 1. The schematic diagram of SiGe core fiber F-P cavity.

According to the principle of temperature sensing of F-P cavity, the resonant cavity length and the refractive index of the cavity will change with the temperature inside the cavity. The optical path difference between the two reflection beams will also change, resulting in changes in the phase difference between the interference light beams. From the reflection spectrum, we can intuitively observe the drift of interference wavelength. For fiber F-P cavity with cavity length of L and dielectric refractive index of n, the relationship of the two parameters and the temperature change can be expressed [8]:

$$\Delta L = L \cdot \alpha \cdot \Delta T, \quad (1)$$

$$\Delta n = n \cdot \varepsilon \cdot \Delta T, \quad (2)$$

where ΔL is the cavity length change, α is the thermal expansion coefficient of the medium, Δn is the index change, ε is the thermo-optical coefficient of the medium, and ΔT is temperature change. As can be seen from the above equations, the phase difference of F-P cavity is proportional to the temperature, and it is mainly determined by the thermo-optical coefficient and the thermal expansion coefficient of the cavity medium. In order to improve the sensing resolution of fiber-optic F-P cavity, the SiGe core material with greater thermal expansion coefficient and thermo-optical coefficient is considered.

3. Fabrication and experiments

The SiGe core fiber employed in this experiment was fabricated with rod-in-tube method with homemade fiber draw tower of our laboratory. Before the drawing process of SiGe core fiber, we mixed the germanium powder and the silicon powder in a specific ratio, then put the mixed powder into the silica tube. One end of the silica tube was collapsed beforehand to form the fiber preform. Through the melt and drawing process in the graphite furnace, we prepared the SiGe core fiber with a perfect circle core and a clean core-cladding interface. Then, to form F-P cavity inside the SiGe core fiber, the homemade SiGe core fiber was manually spliced with the conventional single mode fiber (SMF) with a commercial arc splicer. Since the melting points of the two fibers are different during the splicing process, we could introduce certain thermal gradient into the fiber core by tuning the splicing parameters such as interval, stepping rate and location aligning the electrode axis, then the SiGe core F-P cavity was formed close to the splice point.

In order to verify that the cavity medium is a silicon-germanium material, we used a confocal Raman spectrometer to measure the Raman spectrum of the fiber core. The result is shown in Fig. 2. We can find clearly the Raman vibrational peaks of 289.2 cm^{-1}, 406.8 cm^{-1} and 484.9 cm^{-1}, which correspond to the optical phonon modes of Ge-Ge, Si-Ge and Si-Si respectively. And the SiGe alloy is confirmed in the cavity. The strain in the silicon-germanium core causes the peak shifts of the three vibrational modes, and the vibration peak position is related to the proportion of the material in the alloy.

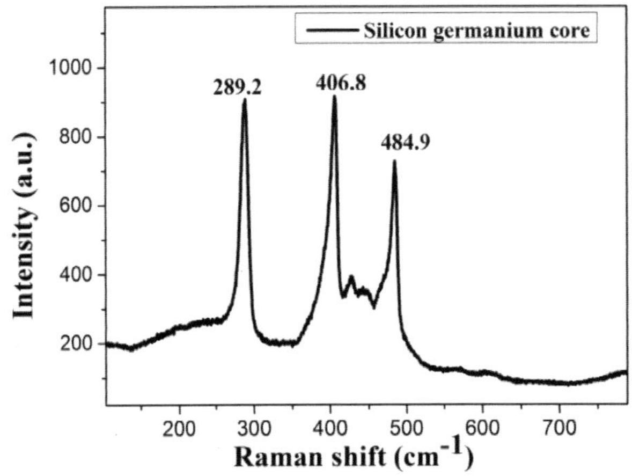

Fig. 2. Raman spectrum of SiGe core material of F-P cavity.

After that, to characterize the 980 nm laser power detecting property of the SiGe core F-P sensor, the experimental setup was built according to the schematic diagram of Fig. 3a. The incident light from 1550 nm broadband light source was propagated through the input port (port 1) of the fiber optic circulator (OC), one of the output (port 2) was connected with the 1550 nm input port of the 980/1550 nm wavelength division multiplexer (WDM), and the backward reflection was collected to the optical spectrum analyzer (OSA) through the port 3 of the OC, recording the F-P's reflection spectrum. The 980 nm laser was delivered to heat the SiGe core through the other input port of the WDM. The reflection spectrum of F-P was affected by the thermal effect induced by the laser injection. Then, we can evaluate the performance of the sensor through the resonance wavelength shift by altering the injected laser power. The inset Fig. 3b privates the microscope image of the formed F-P cavity after fusion splices of the SiGe core fiber and the SMF.

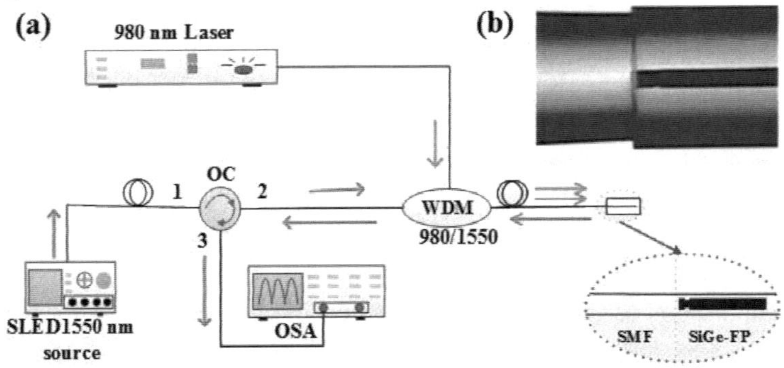

Fig. 3. Experimental setup for laser power detecting

4. Result and discussion

To characterize the laser power sensing property of SiGe core fiber F-P cavity, the optical power of 980 nm laser was successively increased from 0 mW to 6 mW. As illustrated in Fig. 4, the reflection spectrum shifts to longer wavelength regularly when the laser power increases (Fig. 4a), and vice versa (Fig. 4b).

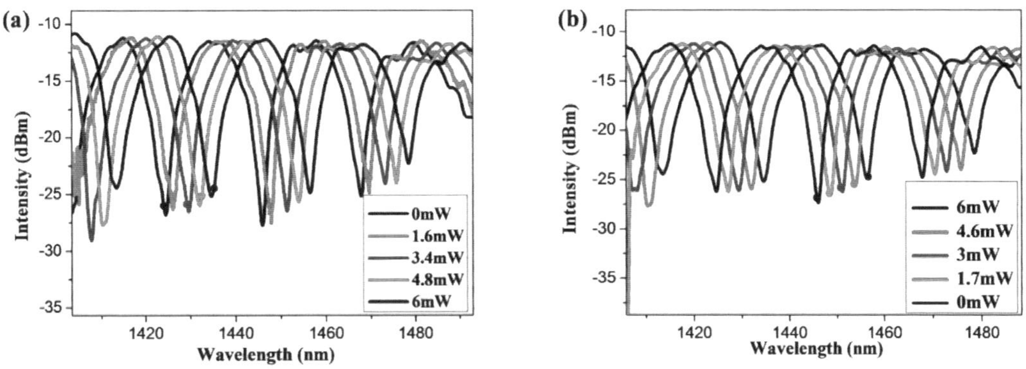

Fig. 4. Reflection spectra with respect to the injection optical power of 980 nm laser: (a) optical power increases; (b) optical power decreases.

In order to quantitatively observe the drift at different injection power, Fig. 5 gives the resonant peak wavelength of 1415.1 nm shifting with the applied 980 nm laser power. We could see that the optical power response curves are nearly linear, and the 980 nm optical power sensitivity of the F-P cavity is about 1.7 nm/mW.

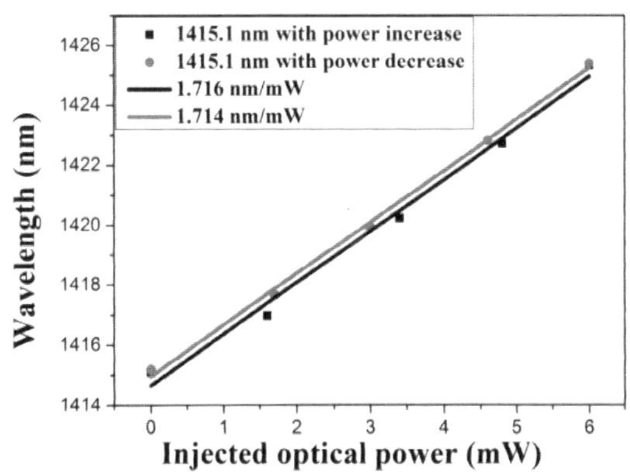

Fig. 5. Linear fitting of laser power to the reflection wavelengths of 1415.1 nm, where the black line is fitting as power increases, and the red line is fitting as power decreases.

Such phenomena support the analysis that high absorption of silicon-germanium core to the laser power leads to high temperature on the cavity, which induces the variation of F-P's cavity length, thus the F-P's reflection spectrum changes. The reflectance spectra coincide before and after laser irradiation, which proves the good stability of SiGe core fiber F-P sensor.

5. Conclusion

In conclusion, an all-fiber power sensor based on the SiGe core fiber F-P cavity was proposed and demonstrated. Firstly, SiGe core fiber F-P cavity could be prepared by splicing the homemade SiGe core fiber with the SMF. Then the presence of SiGe core material was confirmed by Raman spectroscopy. After that, based on the strong absorption of silicon-germanium core from the visible to the near IR wavelength range, we experimentally characterized the response of reflection spectrum on the injected 980 nm optical power change. Good linearity exists between the detecting power and the resonant wavelength. The optical power sensitivity is up to 1.7 nm/mW. This kind of SiGe core fiber F-P cavity may have great potential for in-fiber optical power detector.

Acknowledgment
This work is supported by the National Natural Science Foundation (Project Nos: 61575120, 61227012, 61422507, 61475095, and 61635006).

References

[1] Ballato J, Hawkins T, Foy P, Mcmillen C and Burka L 2010 Advancements in semiconductor core optical fiber *Opt. Fiber Technol.* **16** 399-408.

[2] Mehta P, Healy N, Baril N F, Badding J and Peacock A 2010 Nonlinear transmission properties of hydrogenated amorphous silicon core optical fibers *Opt. Express.* **18**, 16826-31.

[3] Ballato J, Hawkins T, Foy P, et al 2008 Silicon optical fiber *Opt. Express.* **16** 18675-83.

[4] Ballato J, Hawkins T, Foy P, Stolen R and Ellison M 2009 Glass-clad single-crystal germanium optical fiber *Opt. Express.* **17** 8029-35.

[5] Sparks J R, He R, Healy N, et al 2011 Zinc selenide optical fibers *Adv. Mater.* **23** 1647.

[6] Zhang S, Zhao Z, Chen N, Pang F, Chen Z and Liu Y 2015 Temperature characteristics of silicon core optical fiber Fabry-Perot interferometer *Opt. Lett.* **40** 1362-5.

[7] Liu G and Han M 2015 Fiber-optic gas pressure sensing with a laser-heated silicon-based Fabry-Perot interferometer *Opt. Lett.* **40** 2461.

[8] Guan B, Wu C, Fu H and Qureshi K et al 2011 High-pressure and high-temperature characteristics of a Fabry-Perot interferometer based on photonic crystal fiber *Opt. Lett.* **36** 412-414.

[9] Wang H, Liu X and Zhang M 2013 Absorption Coefficients of Crystalline Silicon at Wavelengths from 500nm to 1000nm *Int. J. Thermophys* **34** 213-225.

IOP Conf. Series: Journal of Physics: Conf. Series **844** (2017) 012037 doi:10.1088/1742-6596/844/1/012037

Tunable mid-infrared absorption of metamaterial integrated with graphene

X B Chen, Q Li, P Ghosh, Q Min

State Key Laboratory of Modern Optical Instrumentation, College of Optical Science and Engineering, Zhejiang University, Hangzhou, 310027, China

E-mail: qiangli@zju.edu.cn

Abstract. A metal-isolator-metal (MIM) metamaterial absorber with embedded graphene shows tunable at mid-infrared frequency regime. We report tunable absorption spectra of the MIM structure for different Fermi levels of graphene.

1. Introduction

Metamaterials refer to artificially structured materials with arrays of subwavelength elementary particles (known as meta-atoms) [1]. Much work has been done, especially in creating negative-refraction materials [2-3]. Recently, the metamaterial absorbers based on metal-isolator-metal (MIM) structure have been reported. To realize the tunability of MIMs, several methods have been implemented, including altering the capacitance of the resonators and changing the property of the surrounding media [4-6]. However, those methods are limited in the modulation range and are complicated to implement.

Graphene has drawn enormous attention because of its electrically tunable conductivity from terahertz to mid-IR frequency regime [7]. The coupling of graphene and photo/plasmonic structures could achieve dynamic tunable devices.

In this article, we report the design of a hybrid graphene/MIM device which offers an efficient manner of coupling free-space light into nanoscale plasmonic modes. The electronic control of that coupling process is attained by tuning the Fermi level of graphene. A schematic diagram of the graphene/MIM device is shown in figure 1. The MIM structure consists of three layers: (1) a thick gold layer (thickness H = 100 nm) at the bottom, (2) a 35 nm thick alumina (Al_2O_3) spacer, and (3) a periodic array of gold nanostructures (width d = 1200 nm, thickness W = 30 nm) at the top. A single layer graphene (thickness t = 0.34 nm) is sandwiched between the gold particles and alumina layer to form the graphene/MIM hybrid structure.

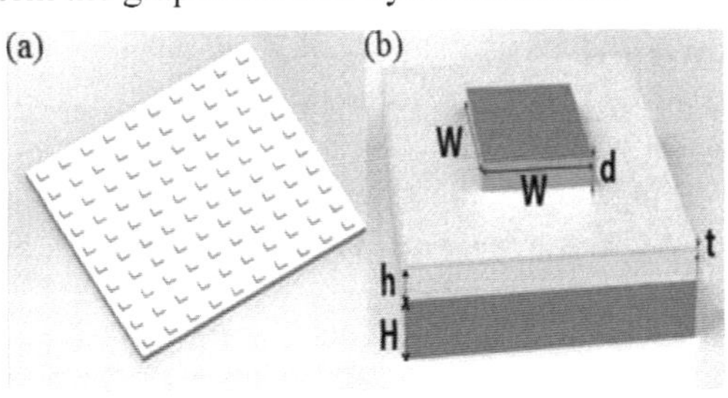

Figure 1. (a) Geometry of graphene/MIM device. (b) Unit cell of absorber. H, d, t, and h represent thickness of gold layer at bottom, gold particle, graphene, and Al_2O_3. W is width of the gold particle. The periodicity of the gold nanostructure array is 1500 nm.

IOP Conf. Series: Journal of Physics: Conf. Series **844** (2017) 012037 doi:10.1088/1742-6596/844/1/012037

2. Result and Discussion

The simulation for graphene/MIM nanostructure is performed with commercial COMSOL MULTIPHYSIS software based on the two-dimensional finite elements method. The absorption spectrum is derived using the formula: $A(\lambda) = 1-R(\lambda)-T(\lambda)$. Figure 2 (a) shows the simulated absorption spectra, $A(\lambda)$, for different Fermi level of graphene ranging from 200 meV to 400 meV. The resonant frequency changes as the Fermi level is varied. The variation of resonance frequency and corresponding absorption with different Fermi levels is listed in table 1. The peak absorption is 89% (at 5.52 μm) when the Fermi level of graphene is 200 meV. The absorption moves to 87% (at 5.62 μm) when its Fermi level becomes 400 meV. The minimum absorption also changes with Fermi level of graphene. It ranges from 2.7% (at 5.54 μm) to 2.0% (at 5.62 μm) as Fermi level changes from 200 meV to 400 meV. Figure 2 (b)-(g) illustrate the magnetic field for both maximum and minimum absorption corresponding to different Fermi level of graphene.

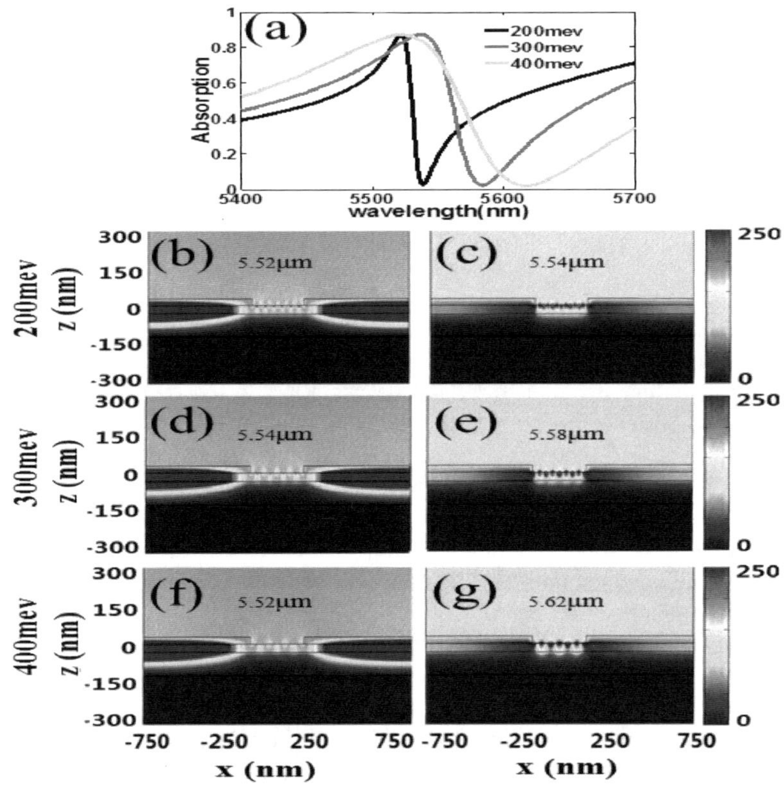

Figure 2. (a) Simulated absorption spectra for different Fermi levels. (b), (d) and (f) represent the magnetic field when the structure reaches maximum absorption. (c), (e) and (g) are also the magnetic field when it reaches minimum absorption.

IOP Conf. Series: Journal of Physics: Conf. Series **844** (2017) 012037 doi:10.1088/1742-6596/844/1/012037

Table 1. The variation of resonance frequency and corresponding absorption with different Fermi levels

Fermi level (meV)	Maximum Efficiency	Absorption Wavelength	Minimum Efficiency	Minimum Efficiency
200	89%	5.52μm	2.7%	5.54μm
300	87%	5.54μm	2.2%	5.58μm
400	87%	5.52μm	2.0%	5.62μm

In order to understand the absorption properties of these graphene/MIM nanostructure, the electromagnetic field distributions for the resonant modes are thoroughly investigated. Maximum absorption is observed due to the excitation of localized magnetic and electric dipole resonances resulting from the coupling of accumulated charges at the sides of gold particle with their image charges in the gold surface. These strong resonances effectively trap light energy and provide sufficient time to dissipate it by the Ohmic losses within the metals and graphene. [1] It can be noted that Al_2O_3 is lossless in this wavelength regime. There also exists another resonance probably because of standing waves localized on the graphene layer between two gold particles. Here the graphene layer can be considered as a plasmonic metamaterial where the plasmons can be excited at the air-graphene-Al_2O_3 interface. The dispersion of transverse magnetic (TM) plasmons in graphene layer between air and Al_2O_3 with the permittivities ε_1 and ε_2 is described by the dispersion equation [8]

$$\frac{\varepsilon_1}{[q^2(\omega)-\varepsilon_1]^{1/2}}+\frac{\varepsilon_2}{[q^2(\omega)-\varepsilon_2]^{1/2}}+iQ_s(\omega)Z_0=0 \quad (1)$$

Where ε_1 and ε_2 represent permittivity of air and Al_2O_3, σ_s (ω) is the surface conductivity of graphene and $Z_0 = \sqrt{\mu_0/\varepsilon_0}$ is the free-space impedance, q (ω) represents the effective index of the graphene plasmon.

3. Conclusion

In this article, we have demonstrated that the graphene/MIM hybrid structure can be tuned by controlling the Fermi level of the graphene. This tunable optical response allows for graphene being an attractive platform for optoelectronic applications such as light modulators, selective thermal emitters.

Acknowledgment

This work is supported by the National Natural Science Foundation of China (grant nos 6142503, 61575177 and 61235007).

References

[1] Hao J, Wang J and Liu X 2010 High performance optical absorber based on a plasmonic metamateroal *Appl.Phys.Lett.*96(25),251104-251104-3
[2] Pendry J, Schurig D and Smith D 2006 Controlling electromagnetic fields *Science* 312,1780
[3] Zhang S, Genov D, Sun C and Zhang X 2008 Cloaking of Matter Waves *Phys. Rev. Lett.*100,123002
[4] Chen H 2010 Active Terahertz Metamaterial Devices IEEE 28-29
[5] Zhang S, Park Y and Li J 2009 Negative refractive index in chiral metamaterials *Phys. Rev. Lett.* 102(2) 023901
[6] He X, Zhao Z and Shi W 2015 Graphene-supported tunable near-IR metamaterials *Opt. Lett.* 40(2) 178-81

[7] Fan Y, Shen N and Koschny T 2015 Tunable Teraherz Meta-Surface with Graphene Cut-Wires *Acs Photonics* 2(1) 150106103850009
[8] Gu T, Andryieuski A and Hao Y 2016 Photonic and Plasmonic Guided Modes in Graphene–Silicon Photonic Crystals *Acs Photonics* 2(11)

IOP Conf. Series: Journal of Physics: Conf. Series **844** (2017) 012038 doi:10.1088/1742-6596/844/1/012038

THz Wave Modulator Based on the Decay of THz Surface Plasmon along an Intrinsic InSb Surface

Tao Yang[1,*], Bing Jiang[1], Ho-pui Ho[2], Yong-yuan Zhu[3], and Wei Huang[1,4]

[1]Key Laboratory for Organic Electronics and Information Displays (KLOEID) & Institute of Advanced Materials (IAM), Nanjing 210023, Jiangsu, China
[2]Department of Electronic Engineering, The Chinese University of Hong Kong, Hong Kong
[3]National Laboratory of Solid State Microstructures, Nanjing University, Nanjing 210093, China
[4] Key Laboratory of Flexible Electronics(KLOFE) & Institute of Advanced Materials(IAM), Jiangsu National Synergetic Innovation Center for Advanced Materials (SICAM), Nanjing Tech University, Nanjing; 211816, China

E-mail: iamtyang@njupt.edu.cn

Abstract. THz surface plasmon are decayed exponentially when propagating along the surface of a semiconductor. A metal razor bladeis used to couple THz surface plasmon to THz waves at different position on the surface of an indium antimonide wafer. Thus the intensity of the output THz wave is modulated.

1. Introduction

Terahertz (THz) waves are electromagnetic radiation in the frequency range of 0.1 to 10 THz. Terahertz frequencies are attractive for the communication industry because this frequency band is currently unallocated by the Federal Communications Commission [1]. THz waves have greater bandwidth than microwave and suffer less Rayleigh scattering compared with infrared radiation. However, several obvious shortcomings of traditional THz modulators have prevented further development of THz based communications, for example, operation at cryogenic temperatures [2]; achievable modulation depth of only a few percent of the signal intensity [3]; difficulty in realizing high frequency modulation [4]. In view of the difficulty of controlling a THz wave, we turn to investigating how to control surface plasmon polaritons (SPPs) with THz frequencies because THz SPPs and THz wave can couple to each other with two razor blades.

SPPs have been the subject of intense research in recent years [5]. Most of the effort has been concentrated on investigating SPPs in the visible and near-infrared domain on metal surfaces [6]. However, not only metals support electron plasma oscillations, but semiconductors can also support SPPs. Intrinsic indium antimonide (InSb) with carrier densities of 1×10^{16} cm^{-3} and plasma frequency of 1.9 THz at 300 K is an ideal material for THz modulation devices because it offers the possibility of ultrafast active modulation with extremely small fluctuations of a razor blade position. Compared with traditional optical switching THz SPP devices that required grating with complex structure [7], the new device reported in this paper only need a flat InSb wafer and two parallel razor blades with a modulated distance, therefore the whole setup is quite simple.

2. Design of the Modulator

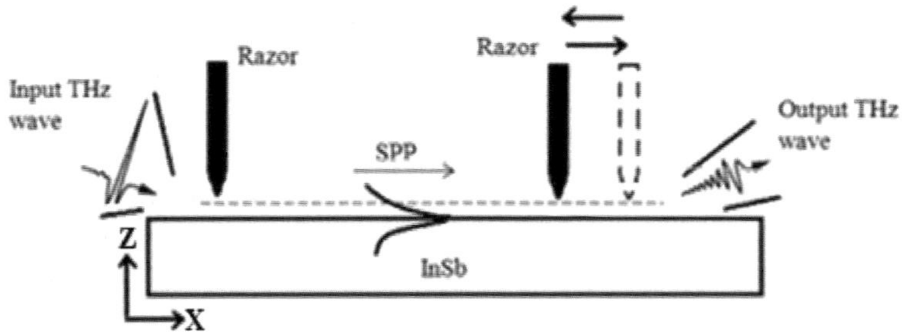

Figure 1. Schematic of the design.

Figure 1 shows the schematic of the design. Here free-space THz radiation is scattered at the aperture defined by a razor blade and the surface of the semiconductor coupling to SPPs if the frequency of the THz SPPs is smaller than the plasma frequency of the InSb semiconductor [8]. After propagating in the surface of the InSb semiconductor, the SPPs are coupled back to free-space THz radiation by another parallel razor blade. THz modulation is realized by changing the distance of the two razor blades.

3. Principles of Operation

THz surface plasmon are decayed exponentially when propagating at the interface of two media with opposite signs of permittivity. According to the Drude model [9], the permittivity of a semiconductor is related to the plasma angular frequency which is given by

$$\tilde{\varepsilon} = \varepsilon + i\varepsilon' = \varepsilon_{static}\left[1 - \frac{\omega_p^2\tau^2}{1+\omega^2\tau^2} + i\frac{\omega_p^2\tau}{\omega(1+\omega^2\tau^2)}\right] \tag{1}$$

where ε_{static} is the static permittivity; τ is the average collision time of the charge carriers; ω is the angular frequency of THz SPPs; ω_p is the plasma angular frequency, which can be expressed as

$$\omega_p = 2\pi f_p = \sqrt{\frac{Ne^2}{\varepsilon_0\varepsilon_{static}m^*}} \tag{2}$$

where f_p is the plasma frequency; N is the free carrier density; e is the fundamental charge; ε_0 is the vacuum permittivity and $m^*$is the carrier effective mass.

Thus, the propagation length of THz SPPs along the direction of a semiconductor–air can be written as [10]

$$\delta_{sp} = \frac{c}{\omega}\left(\frac{1+\varepsilon}{\varepsilon}\right)^{3/2}\frac{\varepsilon^2}{\varepsilon'} \tag{3}$$

Therefore, we can modulate the intensity of the output THz wave by putting two razor blades above the surface of the semiconductor with a modulated separation distance. Figure 2 shows the propagation lengths of THz SPPs on the surface of a flat intrinsic InSb wafer and a gold flat film. Compared with the propagation lengths on the gold film, that on the semiconductor are much short. Therefore, we can modulate the THz SPPs with a small distance fluctuation of razor blades on the semiconductor, which may lead to high modulation frequency.

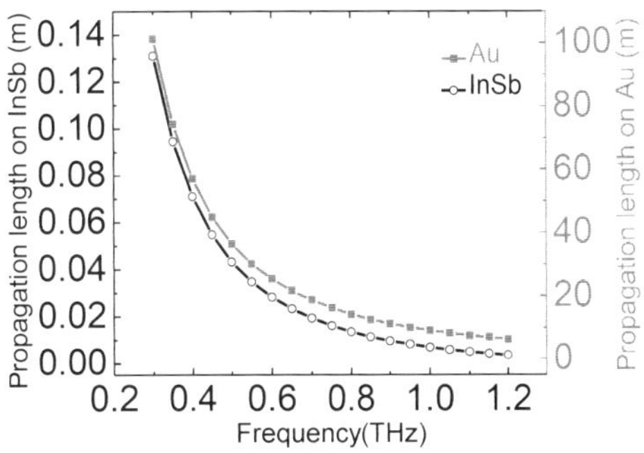

Figure 2. Relation between the frequency and propagation length of THz SPPs propagating on a InSb wafer and on a gold film.

4. Simulation Results

In order to demonstrate the propagation properties of THz SPPs in the surface of an intrinsic InSb wafer as shown in figure 1, we have performed a series of two-dimensional simulations with varying razor blade distances by using a finite element software. The modeled structure is shown in figure 1 and a minimum 0.32 μm mesh size is used in the simulations. We assume the SPP waveguide to be the intrinsic InSb semiconductor of thickness 0.45 mm. A *p*-polarized THz wave is incident at an angle of 63°. For simplicity, the THz wave used in the simulation is a monochromatic wave with a frequency of 0.5 THz because the coupled THz SPPs with this frequency have the merits of long propagation lengths and high field confinement to the surface. The size of the gap, defined as the distance between the two razor blades and the semiconductor surface, is 150 μm and the distance between the two razor blades is 3 mm, 4 mm and 5 mm respectively. The electric field distributions shown in figure 3 indicate that the intensity of the surface plasmon and its output THz wave are gradually weak when the distance between the two razor blades is increased from 3 mm to 5 mm. Our simulation results clearly show that for a THz wave with a frequency of 0.5 THz the device may act as a modulator by changing the distance of the two razor around 4 mm with 1 mm variation.

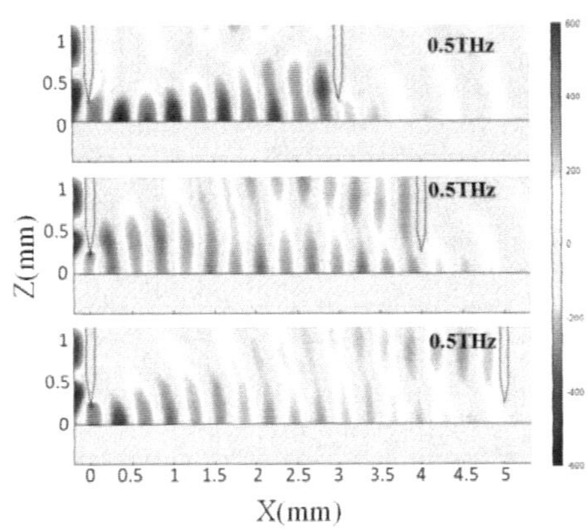

Figure 3. Spatial distribution of the electric field amplitudes of THz SPP components with frequencies of 0.5 THz on the surfaces of intrinsic InSb semiconductors.

5. Experimental Results

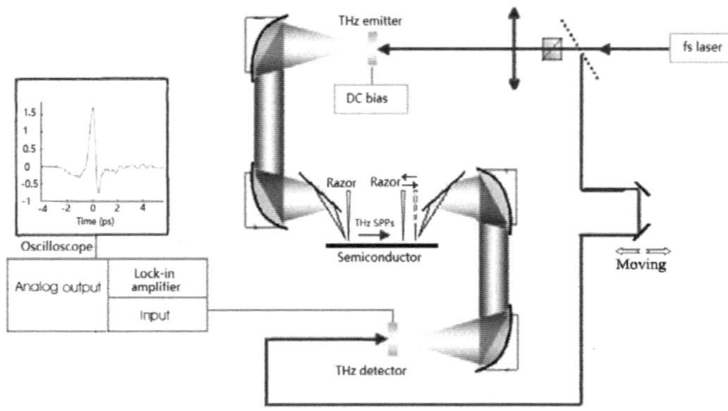

Figure 4. Schematic of the THz-TDS to excite and to detect THz SPP pulses.

We can also modulate a broadband THz wave. We demonstrate it by measuring the transmission power spectra with a free-space THz time-domain spectroscopy (THz-TDS, EKSPLA) system [11-12]. The experimental setup is shown in figure 4. In the figure, a train of ultra short optical pulses (120 fs, 800 nm) from a Ti:Sapphire laser (Coherent, Chameleon) is split into two beams. One of these beams is used to optically generate coherent and linearly polarized THz radiation from a low temperature grown gallium arsenide (LT-GaAs) surface field emitter. The thickness of GaAs substrate is 400 μm and the antenna is formed using Ti/Au metallization. The generated THz radiation is coupled into THz SPPs propagating in the surface of an InSb wafer for modulation. The other beam gates a photoconductive antenna, which detects the THz field coupled out from the THz SPPs. By varying the time delay between the two pulse trains, the THz pulse amplitude can be detected as a function of time with subpicosecond resolution as shown in figure 5 [13]. The measured THz pulses are Fourier transformed to obtain the transmission power spectra as shown in figure 6. The InSb wafer used in the experiment is 0.5 mm thick with 2 inches diameter. The size of the razor apertures and the incident angle of THz radiation are the same as those used in simulations.

The spectra in figure 5 and figure 6 show that the amplitude of the output THz wave is decreased by increasing the separation of the two razor blades, which indicates that the setup can act as a THz modulator by shifting the position of the razor blade on an intrinsic InSb wafer. The experimental results fit well with the experimental results. We also note that the transmission spectra fluctuate at some frequencies in figure 6, which is estimated to be due to water vapor absorption [14-15] and low signal to noise ratio.

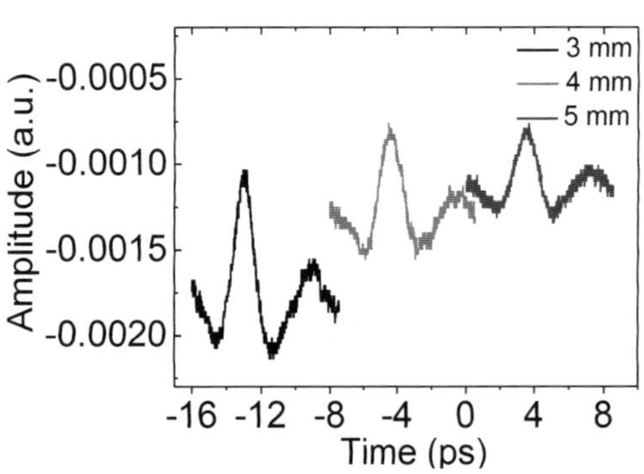

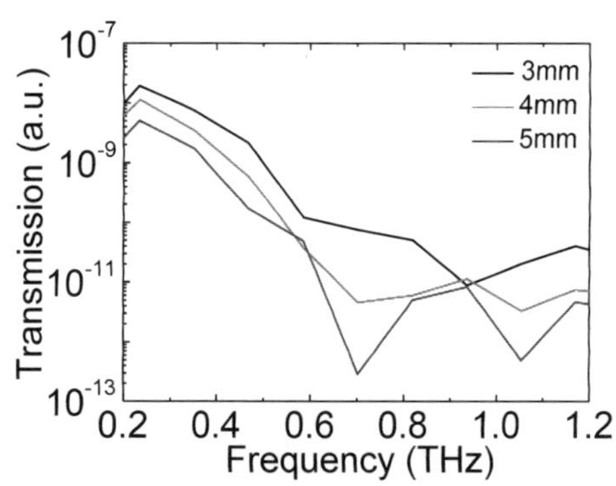

Figure 5. Time domain spectra of the output THz wave for different razor blade separation on an intrinsic InSb wafer.

Figure 6. Frequency domain spectra of the output THz wave for different razor blade separation on an intrinsic InSb wafer..

6. Conclusion

In summary, we have proposed a THz wave modulator with a flat InSb wafer and two parallel razor blades. The modulation is realized by changing the distance of the two razor blades. Compared with conventional modulators, this method of manufacturing is simpler and the modulation bandwidth is wider.

7. Acknowledgement

This work is supported by the National Natural Science Foundation of China (NSFC) (61377019), the Natural Science Foundation of Jiangsu Province (BK20151512), Open Project of the National Laboratory of Solid State Microstructures, Sponsored Foundation of Nanjing University of Posts and Telecommunications (NUPTSF) (NY217085) and Priority Academic Program Development of Jiangsu Higher Education Institutions(YX03001).

References

[1] http://www.ntia.doc.gov/osmhome/allochrt.pdf
[2] R. Kersting, G. Strasser, K. Unterrainer 2000 Terahertz phase modulator *Electron. Lett.* **36**, 1156–8
[3] T. Kleine-Ostmann, P. Dawson, K. Pierz, G. Hein and M. Koch 2004 Room-temperature operation of an electrically driven terahertz modulator *Appl. Phys. Lett.* **84** 3555–7
[4] H. T. Chen, W. J. Padilla, J. M. O. Zide, A. C. Gossard, A. J. Taylor and R. D. Averitt 2006 Active terahertz metamaterial devices *Nature* **444** 597–600
[5] M. S. Tame et al 2013 Quantum plasmonics *Nat. Phys.* **9** 329–40
[6] T. Yang and H. P. Ho 2009 Computational investigation of nanohole array-based SPR sensing using phase shift *Opt. Express* **17** 11205–16
[7] J. A. Sánchez-Gil and J. G. Rivas 2006 Thermal switching of the scattering coefficients of terahertz surface plasmon polaritons impinging on a finite array of subwavelength grooves on semiconductor surfaces *Phys. Rev. B* **73** 205410
[8] C. Janke, J. Go´mez Rivas, C. Schotsch, L. Beckmann, P. Haring Bolivar, and H. Kurz 2004 Optimization of enhanced terahertz transmission through arrays of subwavelength apertures *Phys. Rev. B* **69** 1324–32
[9] M. van Exter and D. Grischkowsky 1990 Optical and electronic properties of doped silicon from 0.1 to 2 THz *Appl. Phys. Lett.* **56** 1694–6
[10] H. Raether 1988 *Surface Plasmons on Smooth and Rough Surfaces and on Gratings* (Berlin: Springer) chapter 2
[11] D.R. Grischkowsky, S. Keiding, M. van Exter, and C. Fattinger 1990 Far-infrared time-domain spectroscopy with terahertz beams of dielectrics and semiconductors *J. Opt. Soc. Am. B* **7** 2006-15
[12] J. G. Rivas et al 2004 Time-domain measurements of surface plasmon polaritons in the terahertz frequency range *Phys. Rev. B* **69** 1124–33
[13] K. Reimann 2007 Table-top sources of ultrashort THz pulses *Rep. Prog. Phys.* **70** 1597–632
[14] W.Shi, and Y. J. Ding 2004 Direct measurement of resonant frequencies for H/sub2/O in the range of 0.2-4.2 THz by frequency-tuning monochromatic THz source *Lasers and Electro-Optics* **1** 3
[15] D. M. Slocum, E. J. Slingerland, R. H. Giles and T. M. Goyette 2003 Atmospheric absorption of terahertz radiation and water vapor continuum effects *J. Quant. Spectrosc. Radiat. Transfer* **127** 49–63

IOP Conf. Series: Journal of Physics: Conf. Series **844** (2017) 012039 doi:10.1088/1742-6596/844/1/012039

Mode/Polarization Manipulation in Silicon Photonics

Daoxin Dai, Ke Ma, and Hao Wu.

Centre for Optical and Electromagnetic Research, State Key Laboratory for Modern Optical Instrumentation, Zhejiang Provincial Key Laboratory for Sensing Technologies, Zhejiang University, Zijingang Campus, Hangzhou 310058, China.

dxdai@zju.edu.cn

Abstract. Mode-division-multiplexing and polarization-division-multiplexing are becoming very important to enhance the capacity of single-wavelength carrier. One of the keys is the realization of efficient mode/polarization manipulation. Silicon photonics provides an excellent platform to achieve high-performance on-chop mode/polarization manipulation. This paper gives a review for our recent work on the high-performance silicon photonic devices for mode/polarization manipulation.

1. Introduction

The demand for data increases exponentially and further enhancement of the information capacity is a perennial goal from the access to data-centre interconnects and to long-haul transmissions [1]-[2]. One of the most cost-effective solution is utilizing some advanced multiplexing technologies. A straightforward way is using wavelength-division-multiplexing (WDM) technology, which has been developed very successfully for long-haul optical communication network. As it is well known, the cost for WDM systems is usually very high and the wavelength management is pretty complicated. Therefore, it is very desired to improve the capacity of a single-wavelength carrier. The polarization-division-multiplexing (PDM) utilizing dual polarizations and the mode-division-multiplexing (MDM) utilizing multiple guided-modes are become the most popular technologies [2]. In order to enable the MDM and PDM technologies, one of the keys is the realization of efficient mode/ polarization manipulation. Silicon photonics provides an excellent platform to achieve high-performance on-chop mode/polarization manipulation. This paper gives a review for our recent work on the high-performance silicon photonic devices for mode/polarization manipulation.

The devices for on-chip polarization manipulation include polarizers, polarization beam splitters (PBSs), and polarization rotators (PRs). For an SOI (silicon-on-insulator) nanowire, which has an ultra-high birefringence, it possible to realize on-chip polarization-manipulation devices with ultra-compact footprints [3]-[4]. However, one should note that it is still a challenge to realize ultra-compact on-chip polarization-manipulation devices with excellent performances, e.g., low excess loss and high extinction ratio in a broad wavelength-band. In this paper, we give a review on the recent work of high-performance silicon-based on-chip polarization-manipulation devices.

The devices for on-chip mode manipulation include mode-division-multiplexers and mode converters, which work as the key component for the combination/separation/conversion of different mode-channels. Great progresses has been made on the development of on-chip mode (de)multiplexers by using multimode interferometers, adiabatic mode-evolution couplers, asymmetric Y-junctions, and asymmetric directional couplers. In this paper, we give a review recent progresses of multi-channel mode-division-multiplexer. Recent results for hybrid (de)multiplexers enabling the MDM, PDM, and WDM simultaneously are also reviewed.

2. Silicon photonic devices for on-chip polarization-manipulation

As discussed before, an SOI nanowire has the birefringence as high as 10^{-1} [3]-[4], which makes it possible to realize on-chip polarization-manipulation devices with ultra-compact footprints [3]. Various devices have been proposed theoretically and demonstrated experimentally for on-chip polarization-manipulation [3]-[4].

However, one should realize that it is still a challenge to obtain high-performance on-chip polarization-manipulation devices with ultra-compact footprints. In the following part, a review is given on the recent work of high-performance silicon-based on-chip polarization-manipulation devices, including polarizers, PBSs and PRs.

(1). Polarizers.

Fig. 1(a) shows the ultra-compact TM-pass polarizer realized with a subwavelength grating (SWG) structure based on a SOI nanowire [5]. Here this SWG waveguide is designed to work as a Bragg reflector for TE polarization, in which way the incident TE-polarized light is reflected efficiently. On the other hand, as TM polarization mode has a much lower effective index than TE polarization, a Bloch mode is supported for TM polarization. Consequently, the incident TM-polarized light goes through the structure with a low excess loss. Fig. 1(b) shows the measurement and calculation results for the transmissions of the polarizers as the period number N for the grating varies. It can be seen that the measured excess loss for the fabricated polarizers is <1 dB and is not sensitive to the period number N. For TE polarization, the transmission decreases almost linearly (in dB) as the period number N increases. When choosing N=40, the measured extinction ratio is as high as 40dB, while the polarizer is as short as ~18 μm.

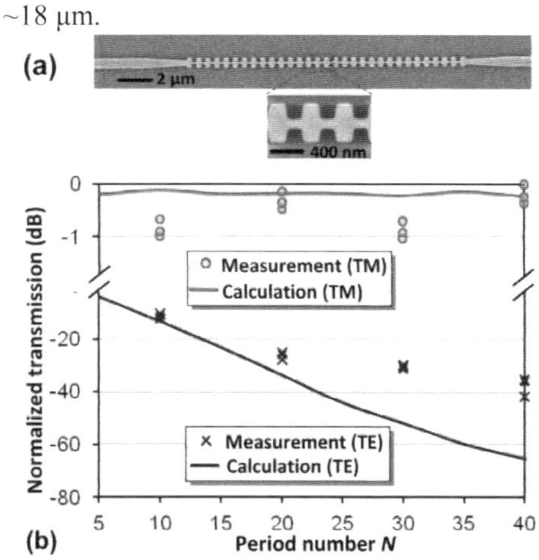

Fig. 1. (a) SEM pictures of a fabricated polarizer. (b) Measured transmissions of the SWG-based polarizers. [5]

(2). Polarization beam splitters (PBSs).

In the past years, the concept of using an ADC to realize ultra-broadband and ultra-short PBSs is becoming one of the most popular options [3], because of the design convenience and structural simplicity. An ADC can be formed by combining different types of optical waveguides. For example, when choosing a regular SOI nanowire as one of the coupling waveguides, one can introduce a special optical waveguide for the other waveguide in the coupling region, e.g., silicon HPWGs [9]-[10], silicon nano-slot waveguides [11]-[12], etc. It is also possible to form an ADC by choosing two or three non-identical waveguides, like a three-waveguide ADC [6], as well as a bent DC [7]-[8]. For a bent DC, one can make the TM-polarized light cross-coupled completely by choosing the waveguide widths and the length of the coupling region appropriately [7]-[8]. In contrast, for TE polarization, the cross-coupling is weak due to the intrinsic phase-mismatch.

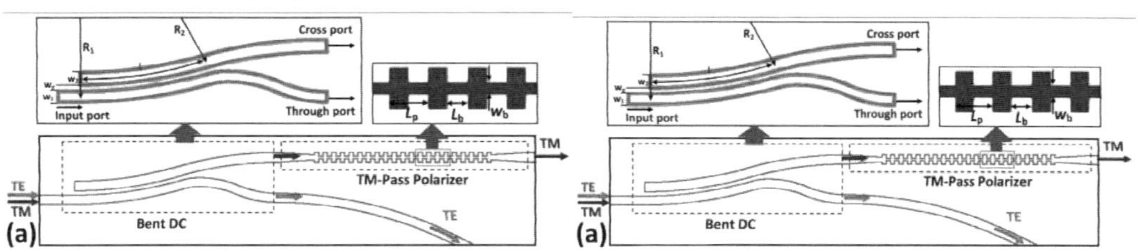

Fig. 2 A PBS consisting of a bent DC and a SWG polarizer. (a) Schematic configuration; (b) SEM pictures. [13]

One might note that the extinction ratio for those compact PBSs demonstrated previously is usually not high (e.g., <15 dB). It is still a challenge to achieve a broadband high-performance PBS with a compact footprint. In [13], a compact PBS consisting of a bent DC cascaded with a SWG TM-passed polarizer was realized for the first time, as shown in Fig. 2(a)-(b). For TE polarization, there exists some weak cross-coupling, which is filtered out with the SWG-based TM-pass polarizer cascaded at the cross port, as shown in Fig. 3(a)-(b). From the measurement results shown in Fig. 4(a)-(b), it can be seen that the extinction ratio for TE polarization is improved to be >20dB within a 70nm wavelength band when choosing N=20. For TM polarization, the extinction ratio is >20dB over a 60nm band (very similar to that of a single bent DC).

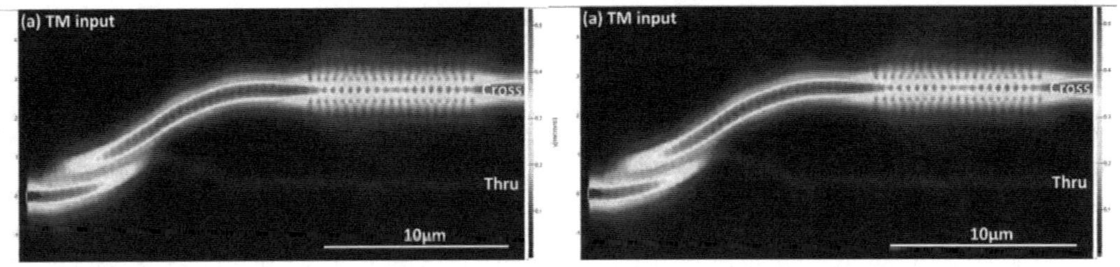

Fig. 3 Simulated light propagation in the designed PBS consisting of a bent DC and a SWG polarizer. (a) TM polarization; (b) TE polarization. [13]

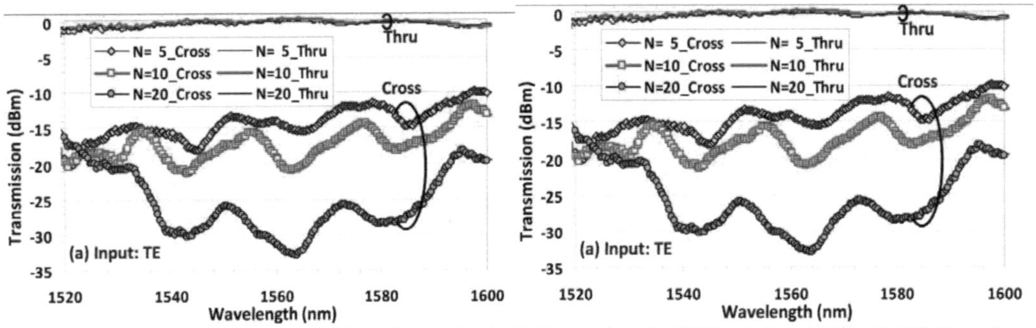

Fig. 4 Measured transmission at the cross- and through-ports for the PBS consisting of a SWG polarizer; (a) TE; (b) TM. Here the period number of the SWG structure is chosen as N=5, 10, and 20. [13]

More recently we realized a near-perfect PBS by using a cascaded structure of bent DCs [14],, which has a compact footprint as small as ~6.9×20μm^2, as shown in Fig. 5(a). Fig. 5(b) shows the measured transmissions at the through- and cross-ports when TE and TM polarization modes are launched at the input port, respectively. This PBS has a world-record bandwidth for a high ER and a low EL. For example, the measured bandwidths are as large as ~126nm, ~97nm and ~70nm for an ER of >20dB, >25dB and >30dB, respectively, while the measured ELs are <1dB and <0.5dB over a bandwidth of ~137nm and ~91nm, respectively. Meanwhile, the PBS has a large fabrication tolerance (±20nm) of the core-width variation. It can be seen that the present PBS has significant improvement compared to those best silicon-based PBSs reported previously, as shown in Table 1.

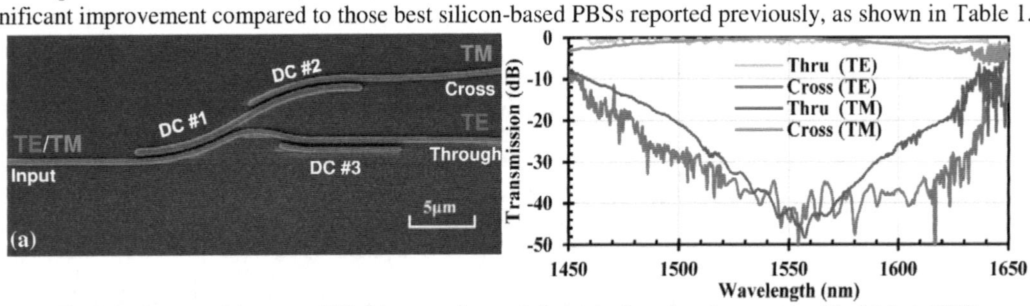

Fig. 5. (a) Structure of the present PBS; (b) measured transmission at the through-and cross-ports of the fabricated PBS.

Table 1 Comparison of demonstrated best-performed PBSs on silicon.

Structures	ER_{TE}/ ER_{TM}	EL	$BW_{ER>20dB}$	$BW_{ER>25dB}$	$BW_{ER>30dB}$	Tolerance Δw	Length
Symmetric DC **[16]**	25/34 dB	0.5 dB	125 nm	NA	NA	NA	97.4 μm
Grating **[17]**	30/30 dB	1 dB	29 nm	25 nm	21 nm	<±10 nm	27.5 μm
Bent DC **[20]**	26/33 dB	1 dB	60 nm	12 nm	NA	NA	22.5 μm
Bent DC **[21]**	33/36 dB	<1 dB	57 nm	32 nm	15 nm	±20 nm	15 μm
This work	44/42 dB	0.35 dB	126 nm	97 nm	70 nm	±20 nm	20 μm

*NA: Not available. * $BW_{ER>20dB}$, $BW_{ER>25dB}$ and $BW_{ER>30dB}$: Bandwidth for a >20dB, >25dB and >30dB ER.

(3). Polarization splitter-rotators (PSRs)

In 2011, a novel concept was proposed for realizing a PSR by using a simple structure consisting of an adiabatic taper and an ADC [15]. For this design, the fabrication is very simple and convenient because one-step etching is only required. An improved PSR with an MMI mode filter cascaded at the through port was proposed and demonstrated recently [22]. This MMI section is designed as a mode filter which is helpful to filter out the residual power of the TE_1 mode originated from the launched TM_0 mode. The measured results have shown that the launched TE_0 mode outputs from the thru-port with a low transmission loss (~0.5 dB) and the extinction ratio is >20 dB in the band from 1520 nm to 1610 nm. For the TM_0 mode, light outputs from the cross port with a loss of ~1.5 dB in a broad band from 1520 nm to 1610 nm. For this PSR, the bandwidth for an extinction ratio of >20 dB is ~50nm and the bandwidth for an extinction ratio of >15 dB is ~80 nm.

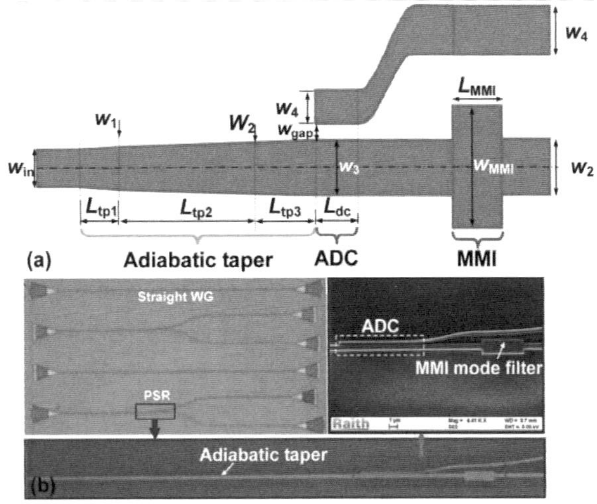

Fig. 6. An improved PSR with an MMI mode filter. (a) Schematic configuration; (b) Microscopy / SEM pictures. [22]

3. Silicon photonic devices for on-chip mode-manipulation

As the key component to combine/separate the signals carried by different mode-channels, various structures have been developed for realizing mode (de)multiplexers, including MMI couplers [24]-[25], adiabatic mode-evolution couplers [26], asymmetric Y-junctions [27]-[29], and ADCs [30]-[34]. In 2012, a 1×N mode-(de)multiplexer was proposed with cascaded ADCs, as shown in Fig. 7(a) [31]-[32]. The ADC is designed with an optimal width w_i for the i-th straight section according to the phase matching condition. Fig. 7(b) show the measured transmission responses for all output ports O_1, O_2, O_3, and O_4 when light is input from port I_1, I_2, I_3, and I_4 respectively. The (de)multiplexer shows low on-chip excess loss (<0.5dB) and low crosstalk of <−20dB in a wide wavelength-band. Because of this WDM-compatibility, it is possible to realize a kind of hybrid multiplexing technology for improving the capacity further. By using the ADC structure, it is also possible to realize a mode (de)multiplexer with dual polarizations [35]-[36].

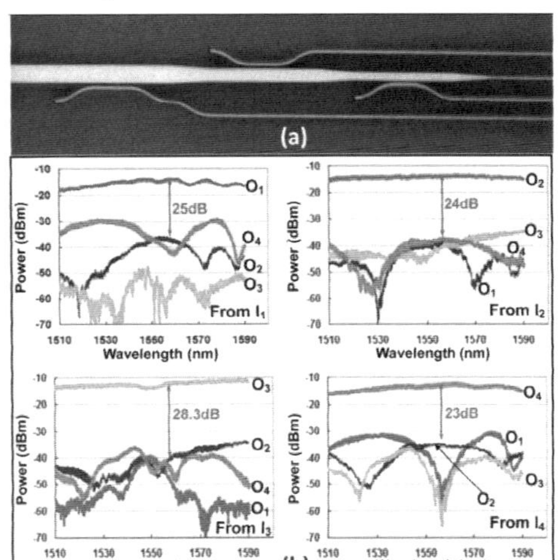

Fig. 7. (a) A four-channel mode demultiplexers based on ADCs' (b) Measured responses at output ports (O_1, O_2, O_3, and O_4) when light is input from port I_1~I_4, respectively. [32]

As demonstrated before, the mode (de)multiplexer based on cascaded ADCs works well in a broad wavelength-band. Thus it is promising to develop hybrid (de)multiplexer enabling MDM and WDM simultaneously by integrating an ADC mode (de)multiplexer [32] and wavelength-selective devices (e.g.,

AWGs). In [37], a hybrid multiplexer was realized with a four-channel mode demultiplexer and four identical AWG demultiplexers integrated on the same chip. A simplified hybrid MDM-WDM (de)multiplexer was then demonstrated by introducing bi-directional AWGs [38]. As shown in Fig. 8, the hybrid WDM-MDM multiplexer consists of a ADC-based mode demultiplexer with four channels and two bi-directional 17×17 AWG demultiplexers. It is expecting to develop high-performance hybrid WDM-MDM (de)multiplexer enabling more mode-/wavelength-channels in the near future.

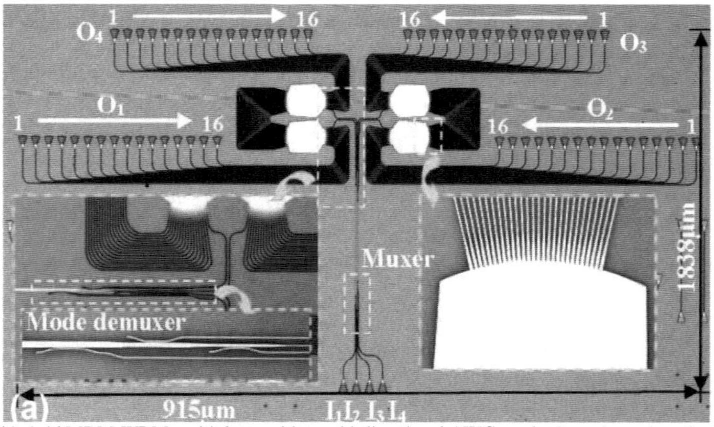

Fig. 8. A 64-channel hybrid MDM-WDM multiplexer with two bi-directional AWGs and a four-channel mode (de)multiplexer. [38]

4. Conclusion.

As a summary, we have given a brief review for the recent work on the high-performance silicon photonic devices for mode/polarization-manipulation, including PBSs, PRs, and mode converters/(de)multiplexers. It is expected to develop excellent hybrid (de)multiplexers enabling more than one (de)multiplexing technologies simultaneously, which is important for keeping the increase of the capacity of photonic networks in the future.

References

[1] N. Zhao, X. Li, G. Li, and J. M. Kahn, “Capacity limits of spatially multiplexed free-space communication,” Nature Photon. 9(12), 822-826 (2015).

[2] D. Dai and John E Bowers. "Silicon-based on-chip multiplexing technologies and devices for Peta-bit optical interconnects," Nanophoton. **3**(4-5): 283-311 (2014)

[3] D. Dai, L. Liu, S. Gao, D.-X. Xu, and S. He. “Polarization management for silicon photonic integrated circuits”, Laser Photon. Rev. **7**(3): 303-328, 2013.

[4] D. Dai, J. Bauters, and J. E Bowers, “Passive technologies for future large-scale photonic integrated circuits on silicon: polarization manipulation, light non-reciprocity, and loss reduction,” Light Sci.and Appl. **1**: 1-12 (2012).

[5] X. Guan, P. Chen, S. Chen, P. Xu, Y. Shi, and D. Dai, "Low-loss ultracompact transverse-magnetic-pass polarizer with a silicon subwavelength grating waveguide," Opt. Lett. 39, 4514-4517 (2014).

[6] D. Dai, “Silicon polarization beam splitter based on an asymmetrical evanescent coupling system with three optical waveguides,” J. Lightwave Technol. 30(20), 3281-3287 (2012).

[7] D. Dai, and J. E. Bowers, “Novel ultra-short and ultra-broadband polarization beam splitter based on a bent directional coupler,” Opt .Express 19, 18614-18620 (2011).

[8] J. Wang, D. Liang, Y. Tang, D. Dai, and J. E. Bowers, “Realization of an ultra-short silicon polarization beam splitter with an asymmetrical bent directional coupler,” Opt. Lett. 38(1), 4-6 (2013).

[9] F. Lou, D. Dai, and L. Wosinski,“Ultracompact polarization beam splitter based on a dielectric–hybrid plasmonic–dielectric coupler,” Opt. Lett. 37(16),3372-3374 (2012).

[10] J. Chee, S. Zhu, G. Q. Lo, “CMOS compatible polarization splitter using hybrid plasmonic waveguide,” Opt. Express 20(23): 25345-25355 (2012).

[11] D. Dai, Z. Wang, and J. E. Bowers, “Ultrashort broadband polarization beam splitter based on an asymmetrical directional coupler,” Opt. Lett. 36(13), 2590-2592 (2011).

[12] S., Lin, J. Hu, and K.B. Crozier, “Ultracompact, broadband slot waveguide polarization splitter,” Appl Phys. Lett 98, 151101 (2011).

[13] S. Chen, H. Wu, and D. Dai, “High extinction-ratio compact polarization beam splitter on silicon,” Electron. Lett. 52(12): 1043-1045, 2016.

[14] Hao Wu, Ying Tan, and Daoxin Dai, "Ultra-broadband high-performance polarizing beam splitter on silicon," Opt. Express 25, 6069-6075 (2017).

[15] D. Dai, and J. E. Bowers, “Novel concept for ultracompact polarization splitter-rotator based on silicon nanowires,” Opt. Express 19, 10940-10949 (2011).

[16] Z. Lu, Y. Wang, F. Zhang, N. A. F. Jaeger, and L. Chrostowski, "Wideband silicon photonic polarization beamsplitter based on point-symmetric cascaded broadband couplers," Opt. Express 23(23), 29413–29422 (2015).

[17] Y. Zhang, Y. He, J. Wu, X. Jiang, R. Liu, C. Qiu, X. Jiang, J. Yang, C. Tremblay, and Y. Su, "High-extinction-ratio silicon polarization beam splitter with tolerance to waveguide width and coupling length variations," Opt. Express 24(6), 6586–6593 (2016).

[18] D. Dai, and J. E. Bowers, "Novel ultra-short and ultra-broadband polarization beam splitter based on a bent directional coupler," Opt. Express 19(19), 18614–18620 (2011).

[19] J. Wang, D. Liang, Y. Tang, D. Dai, and J. E. Bowers, "Realization of an ultra-short silicon polarization beam splitter with an asymmetrical bent directional coupler," Opt. Lett. 38(1), 4–6 (2013).

[20] S. Chen, H. Wu, and D. Dai, "High extinction-ratio compact polarisation beam splitter on silicon," Electron. Lett. 52(12), 1043−1045 (2016).
[21] H. Wu and D. Dai, "High-Performance Polarizing Beam Splitters Based on Cascaded Bent Directional Couplers," IEEE Photon. Technol. Lett. 29(5), 474−477 (2017).
[22] D. Dai, and H. Wu, "Realization of a compact polarization splitter-rotator on silicon," Opt. Lett. 41, 2346-2349 (2016).
[23] L. Soldano, and E. Pennings. "Optical multimode interference devices based on self-imaging - principles and applications," J. Lightwave Technol. 13(4): 615-627 (1995).
[24] Y. Kawaguchi and K. Tsutsumi, "Mode multiplexing and demultiplexing devices using multimode interference couplers," Electron. Lett. 38, 1701-1702 (2002).
[25] T. Uematsu, Y. Ishizaka, Y. Kawaguchi, K. Saitoh, and M. Koshiba, "Design of a Compact Two-Mode Multi/Demultiplexer Consisting of Multimode Interference Waveguides and a Wavelength-Insensitive Phase Shifter for Mode-Division Multiplexing Transmission," J. Lightwave Technol. 3, 2421-2426 (2012).
[26] J. Xing, Z. Li, X. Xiao, J. Yu, and Y. Yu, "Two-mode multiplexer and demultiplexer based on adiabatic couplers," Opt. Lett. 38, 3468-3470 (2013).
[27] N. Riesen and J. D. Love, "Design of mode-sorting asymmetric Y-junctions," Appl. Opt. 51, 2778-2783 (2012).
[28] J. D. Love, R. W. C. Vance, and A. Joblin, "Asymmetric, adiabatic multipronged planar splitters," Opt. Quant. Electron. 28, 353-369(1996).
[29] W. Chen, P. Wang, and J. Yang, "Mode multi/demultiplexer based on cascaded asymmetric Y-junctions," Opt. Express 21, 25113-25119 (2013).
[30] M. Greenberg and M. Orenstein, "Multimode add-drop multiplexing by adiabatic linearly tapered coupling," Opt. Express 13, 9381-9387 (2005).
[31] D. Dai, "Silicon mode-(de)multiplexer for a hybrid multiplexing system to achieve ultrahigh capacity photonic networks-on-chip with a single-wavelength-carrier light," Asia Communications and Photonics Conference, Guangzhou, China, 2012, ATh3B.3.
[32] D. Dai, J. Wang and Y. Shi, "Silicon mode (de) multiplexer enabling high capacity photonic networks-on-chip with a single-wavelength-carrier light," Opt. Lett. 38, 1422-1424 (2013).
[33] Y. Ding, J. Xu, F. Da Ros, B. Huang, H. Ou, and C. Peucheret, "On-chip two-mode division multiplexing using tapered directional coupler-based mode multiplexer and demultiplexer," Opt. Express 21, 10376-10382 (2013).
[34] H. Qiu, H. Yu, T. Hu, G. Jiang, H. Shao, P. Yu, J. Yang, and X. Jiang, "Silicon mode multi/demultiplexer based on multimode grating-assisted couplers," Opt. Express 21, 17904-17911 (2013).
[35] J. Wang, S. He, and D. Dai, "On-chip silicon 8-channel hybrid (de)multiplexer enabling simultaneous mode- and polarization-division- multiplexing," Laser & Photonics Rev. 8(2), L18–L22 (2014).
[36] J. Wang, P. Chen, S. Chen, Y. Shi, and D. Dai, "Improved 8-channel silicon mode demultiplexer with grating polarizers," Opt. Express 22, 12799-12807 (2014).
[37] J. Wang, Sitao Chen, and Daoxin Dai, "Silicon hybrid demultiplexer with 64 channels for wavelength/mode-division multiplexed on-chip optical interconnects," Optics Letters 39, 6993-6996 (2014).
[38] Daoxin Dai, J. Wang, Sitao Chen, Shipeng Wang, and Sailing He. "Monolithically integrated 64-channel silicon hybrid demultiplexer enabling simultaneous wavelength and mode-division-multiplexing", Laser & Photonics Reviews, 9(3): 339-344 (2015).

Acknowledgments

This work is supported by the National Natural Science Foundation of China (NSFC) (61431166001, 11374263, and 61422510), and National Key Research and Development Plan (2016YFB0402502).

IOP Conf. Series: Journal of Physics: Conf. Series **844** (2017) 012040 doi:10.1088/1742-6596/844/1/012040

A CMOS Microdisplay Utilizing Hot Carrier Electroluminescence From Reverse-biased Si PN Junctions

Kaikai Xu[1,*], Lukas W. Snyman[2], Jean-Luc Polleux[3], and Herzl Aharoni[5]

[1]State Key Laboratory of Electronic Thin Films and Integrated Devices, University of Electronic Science and Technology of China, Chengdu, China 610054

[2]Department of Electrical Engineering, University of South Africa, Pretoria 392, South Africa 0001

[3]Université Paris-Est, ESYCOM, ESIEE Paris, Le Cnam, UPEM, Noisy-le-Grand, France 93162

[4]Department of Electrical and Computer Engineering, Ben- Gurion University, Beer-Sheva, Israel 84105

*Corresponding e-mail address: kaikaix@uestc.edu.cn

Abstract The emission of visible light by a monolithically integrated Si p-n junction under reverse bias is presented, and a fully CMOS integrated optical-type fingerprint sensor using Si light emitting devices is achieved.

1. Introduction

Various researchers have highlighted the need for small-dimension light emitters which are compatible with mainstream Si CMOS integrated circuit technology [1]. Recent progress in silicon compatible photonics is driving high density integration of photonic and electronic components manufactured by CMOS-based technology on the same platform [2].

Light emission from silicon devices has previously been realized in reverse-biased Si p-n avalanche structures [3]. It is particularly postulated that light emission occurs from these structures through phonon assisted intra-band and inter-band recombination phenomena [4]. Utilizing current density and surface engineering technology, Snyman et al. have succeeded in realizing a number of practical and utilizable Si light emitting devices (Si-LEDs) utilizing standard CMOS design and processing technology [5]. Several models are presented which postulates that enhanced light emission can be obtained in these devices through carrier energy and carrier momentum engineering [6].

The Si CMOS platform is a viable solution to integrate both, electronic and photonic devices. Significant progress has been made with the potential to move beyond application specific electronic photonic ICs and apply this concept to a more general class of ICs. Because of the ease of integration into CMOS technology, these type of devices show great potential to be integrated into CMOS based optical interconnect, optical RF connection systems and lab-on chip micro-photonic systems. Although present emission levels are sufficient in order to sustain diverse applications, a higher emission power as associated with these devices is desirable. In this paper, we report on further progresses that have been made with regard to modeling of the physical processes in Si-LEDs and in realizing increases in the optical emission power, as well with regard to higher frequency modulation capability of such devices.

2. Device description and experiments

In this research, two different device structures are hence designed and realized in order to promote and test some of the above hypotheses. A 0.35 micron RF bi-polar and TEOS oxide based process having Si-Ge layer capability was used for this purpose. With minor modifications, the process is largely CMOS compatible. Fig. 1 shows schematic diagrams of the cross sectional views of the Si-LED design.

In the first structure, p^+ and n^+ regions are created in an elongated columnar region utilizing the n-epi-layer of the process. A p^+nn^+ conduction channel device is such created. The device dimension of each sub-section is about 1 micron cube. Appropriate metal contacting is supplied to each regions. The first p^+n junction is reverse biased such that the depletion region reached through to the second p-n junction. Since the third region was n^+, the electric field in the lower doped n region is enhanced when the depletion region reaches through to the n^+ region, and assumes a almost flat E-field profile. This leads to enhanced multiplication and avalanching in the lower doped n region. Energetic electrons and holes under the influence of the high E-field undergo various excitation and recombination processes in this region. Since the middle lowly doped region is of epi material, it can be assumed that the residual defect density in this region is very low, and carrier recombination occurs mainly in a defect free environment. This device was hence nomenclated a "p^+nn^+-epi Si Av LED" device.

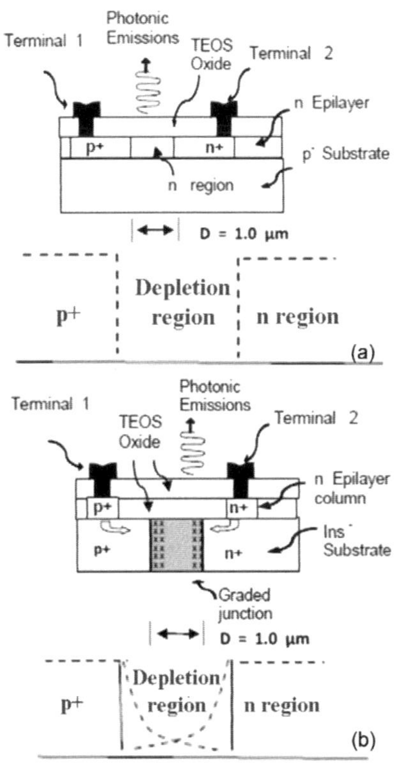

Fig. 1. Device design for controlling impurity and associated defect densities: (a) p^+nn^+ device in epi-layer; (b) Device with graded p^+ and n^+ regions in the substrate. The respective impurity doping profile distributions are indicated in each case.

In the second structure, the impurity profiles are graded with less impurity concentration in the centre middle region. The inter-junction distance was also increased to 1.0 micron. The device were generated in the substrate region below the epi-layer where dopant profiles were much deeper and also subjected to several successive thermal processing procedures. Subsequently, up to 1 micron, graded doping profiles are formed at the edges of these regions; and only some dopant compensation occurred in the center region. A non-linear Gaussian type electric field is hence maintained in the device's centre region. The device was hence nomenclated "p^+nn^+ graded junction Si Av LED".

Fig. 2 presents the *I-V* curves as observed for the various devices. The p^+nn^+ epi-layer device had the lowest operating voltage, while the higher voltage operations of the other two devices are attributed to lower E field and the larger middle traverse regions of these devices, respectively.

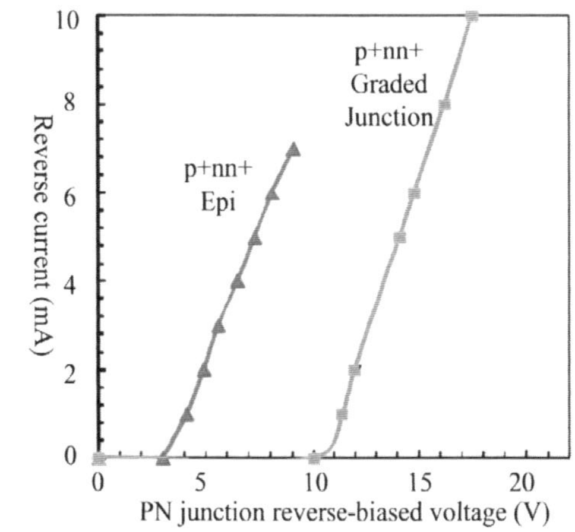

Fig. 2. Experimental results as obtained for p^+nn^+, p^+n^+ overlap and and p^+nn^+ graded junction Si LED devices: *I-V* curves as measured.

Fig. 3 gives a simple synopsis of the light emission mechanisms in silicon as they occur in a silicon reverse biased p-n junction. Electrons are accelerated in the strong electric field of the junction; the energy gained by the carriers are transferred to the lattice and electron-hole pairs are formed during the subsequent host atom ionization processes. During the subsequent energy relaxation and carrier recombination processes, light is generated in and around the junction region.

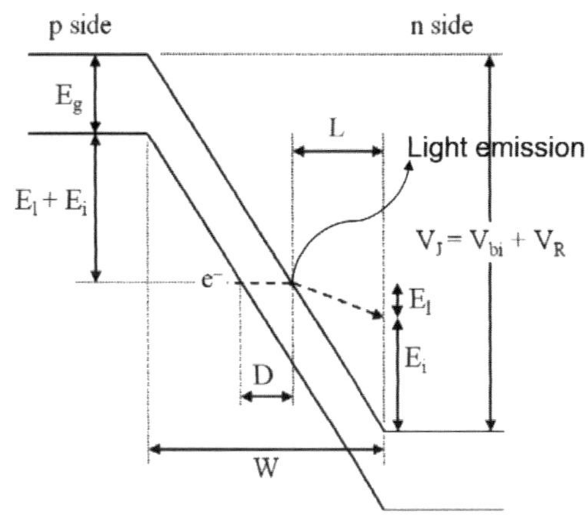

Fig. 3. The energy band diagram of the p-n junction illustrating an electron tunneling from the valence to the conduction band at the electric field peak position, leading hot-carrier induced light emission.

Fig. 4 shows the spectral power densities versus wavelength that is observed for each device. These are

IOP Conf. Series: Journal of Physics: Conf. Series **844** (2017) 012040 doi:10.1088/1742-6596/844/1/012040

measured with an Anritso MS9710B Spectrum Analyzer with a lensed fiber probe having a numerical aperture of 0.4. Logarithmic vertical scaling is used. Very evident is the predominantly 750 nm and 850 nm emissions for the p^+nn^+ epi-layer device. These emissions corresponds well with 1.8 eV and 1.5 eV emissions. These may correspond with intra-band optical emissions of transition. The 650-nm peak has been generally observed for electron relaxation in n-p junctions[7], and may be related to the threshold for pair production during ionization of the Si host atoms with electrons (1.8 eV); the secondary peak at 800 nm may be caused by transmission and interference effects due to the thin residual oxide layer on the n^+-Si surface[8].

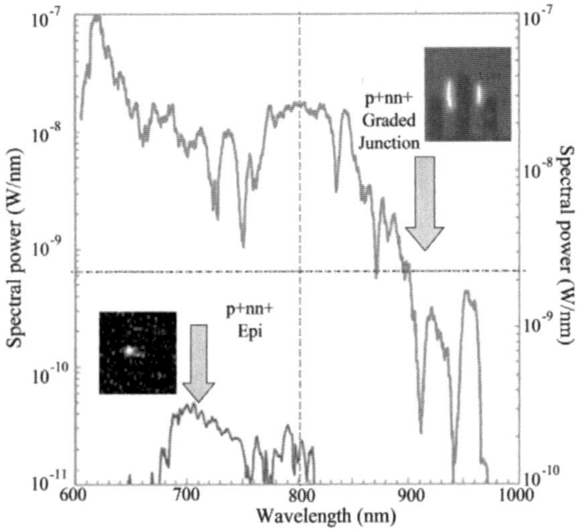

Fig. 4. Normalized electroluminescence spectra of the two types of p-n junctions.

The dominant intraband c-c photon generating process can be either a direct c-c or an indirect phonon assisted (PA) c-c transition. It was found by simulation that the direct c-c radiation dominates at lower photon energies, and the indirect PA c-c radiation dominates at higher energies[9]. A detailed theoretical study was made of the relationship between the kinetic energy of a carrier, the effective carrier temperature, and the corresponding photon distribution assuming an exponential distribution of hot-carriers in the conduction band[10]. Therefore, it can be obtained that energy distribution function $f(E_{photon})$ of photon intensity can be given by

$$f(E_{photon}) \sim \exp\left(-\frac{E_{photon}}{kT_e}\right) \sim \exp\left(-\frac{E_{photon}}{qE_x d}\right) \quad (1)$$

where E_{photon} is the energy of photon and T_e is the effective temperature of hot-carriers.

Also, Eq. (1) presents that the reverse-biased p-n junction depletion region's electric field E_x increases with the photon intensity.

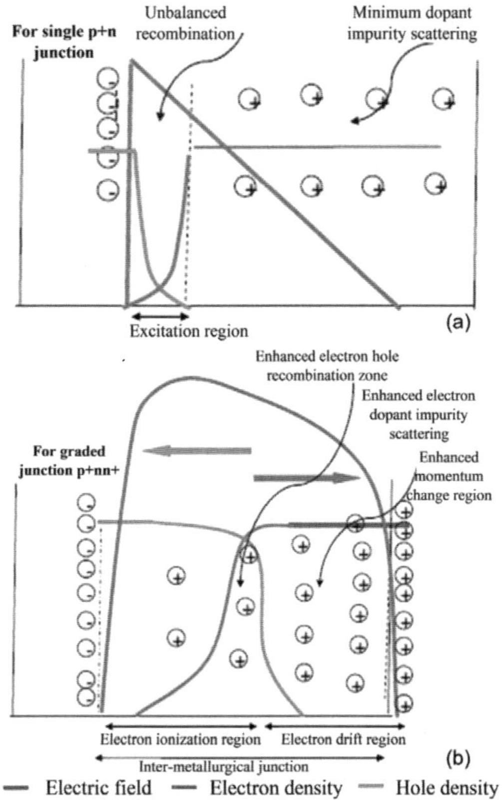

Fig. 5. Device modeling for two types of p^+n Si Av LEDs. (a) Electric field and carrier density profiles for a single junction abrupt doping profile p^+n abrupt junction Si Av LED, showing unbalanced and minimized scattering of carriers by impurity centers. (b) Electric field and carrier density profiles for a graded junction p^+nn^+ Si Av LED showing balanced and maximized scattering of diffusing electrons by positively charged dopant impurity atoms.

The electric field distribution for the two types of p-n junction-based device structures is shown in Fig. 5, and this scenario leads to the enhanced photonic emission in the visible as well as in the near infrared wavelength regions as observed in Fig. 4. Furthermore, the broad spectrum is attributed to the defects and states at or near the Si/SiO_2 interface[11].

3. Conclusion

CMOS based light sources have been shown to exhibit illumination levels that can be utilized in micro displays. In addition to the hot electron electroluminescence in Si, the spectral characteristics and the light intensity highly depend on the specific structure and materials, as well as processing technique.

Acknowledgment

This work is supported by the Natural Science Foundation of China under Contract 61674001, the Department of Science and Technology of Sichuan Province under Contract 2016JY0217, the Open Foundation of State Key Laboratory of Electronic Thin Films and Integrated Devices under Contract KFJJ201508. Finally, Prof. K. Xu would like to pay specially high appreciation to Dr. G.P. Li (UCI) and Dr. Qi Yu (UESTC) for constructive suggestions.

Reference

[1] K. Wada, "Electronics and photonics convergence on silicon CMOS platforms," *Proc. SPIE*, 5357 , 16, (2004)

[2] L. Tsybeskov, D. Lockwood, and M. Ichikawa, "Silicon Photonics: CMOS going optical," vol. 97, no. 7, *Proc. IEEE*, pp. 1161-1165, (2009)

[3] R. Newman, "Visible light from a silicon p-n junction," *Phys. Rev.*, vol. 100, no. 2, pp. 700-703, (1955)

[4] L. Snyman, H. Aharoni, M. du Plessis, J. Marais, D. van Niekerk, and A. Biber, "Planar light emitting electro-optical interfaces in standard silicon complementary metal oxide semiconductor integrated circuitry," *Opt. Eng.*, vol. 41, no. 12, pp. 3230-3240, (2002)

[5] L. Snyman, H. Aharoni, and M. du Plessis, "Two order increase in the quantum efficiency of silicon CMOS n+pn avalanche-based light emitting devices as a function of current density" *IEEE Photonic Tech. Lett.*, vol. 17, no. 10, pp. 2041-2043, (2005)

[6] L. Snyman, K. Xu, J. Polleux, K. Ogudo, and C. Viana, "Higher intensity SiAvLEDs in a RF bipolar process through carrier energy and carrier momentum engineering," *IEEE J. Quantum Electron.*, vol. 51, no. 6, 3200110, (2015)

[7] L. Snyman, H. Aharoni, M. du Plessis, and R. Gouws, "Increased efficiency of silicon light emitting diodes in a standard 1.2 micron complementary metal oxide semiconductor technology," *Opt. Eng.*, vol. 37, pp. 2133-2141, (1998)

[8] L. Heikkilä, T. Kuusela, and H. Hedman, "Electroluminescence in Si/SiO_2 layer structures," *J. Appl. Phys.*, vol. 89, no. 4, pp. 2179-2184, (2001)

[9] M. du Plessis, H. Wen, and E. Bellotti, "Temperature characteristics of hot electroluminescence in silicon," *Opt. Exp.*, vol. 23, no. 10, pp. 12605-12612, (2015)

[10] K. Xu and G. Li, "A novel way to improve the quantum efficiency of silicon light-emitting diode in a standard silicon complementary metal–oxide–semiconductor technology," *J. Appl. Phys.*, vol. 113, no. 10, 103106, (2013)

[11] J. Yuan and D. Haneman, "Visible electroluminescence from native SiO_2 on n-type Si substrates," *J. Appl. Phys.*, vol. 86, no. 4, pp. 2358-2360, (1999)

IOP Conf. Series: Journal of Physics: Conf. Series **844** (2017) 012041 doi:10.1088/1742-6596/844/1/012041

Interference of Multiple Surface Plasmon Polaritons

Dapeng Wang[1,2], Xiaocong Yuan[1,*], and Jiao Lin[1,2,*]

[1]Nanophotonics Research Centre, Shenzhen University, Shenzhen, China 518060,

E-mail: xcyuan@szu.edu.cn

[2]School of Engineering, RMIT University, Melbourne, Australia 3000,

E-mail: jiao.lin@osamember.org

Abstract Benefiting from strongly electromagnetic confinement and enhancement effects, surface plasmon polaritons (SPPs) hold great promises for tailoring light on micro and nanoscale. By contrast with previous efforts which massively concentrate on localized SPP mode, we investigated the propagating SPPs in this paper. A number of symmetrical gratings on metal surface are employed to excite multiple SPPs. Interestingly, the exotic interfering phenomena have been observed. They show good agreement with free-space interferences and take advantage of precise controllability. These findings will be promising in the applications of optical tweezers and SPP lithography.

1. Introduction

Surface plasmon polaritons (SPPs) typically originate from the interface between noble metal and dielectric layers. Through coupling incident light onto a particle or a curved surface, a very significant near-field confinement will be brought about due to the small volume of localized surface plasmon mode. As a consequence, the confinement will lead to an extreme enhancement, i.e. electric or magnetic hot spot, which intrinsically offers a promising tool for bio-sensing and detecting, such as surface enhanced Raman scattering (SERS) and surface enhanced fluorescence (SEF). Besides them, SPP as a versatile tool has also been used in a broad range of applications, e.g. nano lithography, optical trapping and integrated optoelectronics.

Since individual SPP carrying finite energy inevitably suffers from a high attenuation, the progress of SPPs has faced challenges in expanding the propagation length in order to fill the gap between electronics and photonics interconnections. Lin et al. presented a non-diffraction plasmonic beam by means of two coupling gratings[1]. The near-field profile of the singular SPP was observed experimentally and it's bounded to a tightly narrow line up to 80 μm distance on the Au surface. Wei et al. adopted a similar strategy but achieved dark channel SPP via the phase-shifted gratings[2]. With a lateral displacement at the adjacent endpoint of two gratings, the launched SPP was modulated to parallel distribution. Furthermore, they rationally tuned the intersecting angles of four metal nanoslits. The designed structure produced a self-imaging surface plasmon void array, which can be regarded as a 2D surface bottle beam array[3]. Although the above accomplishments are non-intuitive, the underlying interpretation can be attributed to the multi-beam interference phenomenon. Compared with the counterpart of free-space optical elements, two-dimensional surface polariton optics is an ideal candidate for manipulating and directing SPP beams arbitrarily.

In this paper, we have systematically investigated the interference of multiple SPP beams. Through theoretical analysis and numerical simulation, it is demonstrated that the distribution of multiple SPPs interference shows good agreement with that of free space optical cases. Nevertheless, the planar interference of two-dimensional surface waves has numerous advantages: (i) miniaturization; (ii) theoretical limitation of interference period (λ/2), (iii) precise control of phase shift, (iv) extraordinary interfering patterns. These promising potentials will facilitate the progress of SPPs in the applications of optical tweezers and SPP lithography.

2. Principle of Multiple SPP interference

A general form of N SPP-beam interference can be described as the superposition of electric field vectors of N surface waves ($\vec{E}_1, \vec{E}_2, \vec{E}_3$, ... and $\vec{E}_N$), as formulated below

$$\vec{E}_m = A_m \vec{P}_m \cos(\vec{k}_m \cdot \vec{r} - \omega \cdot t + \varphi_m) \quad (1)$$

In the electric field vector, $\vec{k}_m$ and $\vec{p}_m$ can be expressed by

$$\vec{k}_m = k(\sin\theta_m \cdot \cos\varphi_m \cdot \vec{i} + \sin\theta_m \cdot \sin\varphi_m \cdot \vec{j} - \cos\theta_m \cdot \vec{k}) \quad (2)$$

$$\vec{p}_m = \sin\phi_m \cdot \vec{i} - \cos\phi_m \cdot \vec{j} \quad (3)$$

where A_m is the amplitude, $\vec{p}_m$ is the unit polarization vector, $\vec{k}_m$ is the vector in the propagation direction, $\vec{r}_m$ is the position vector, ω is the frequency, φ_m is the initial phase, $k = 2\pi/\lambda$, λ is the wavelength, θ_m is the incident angle, ϕ_m is the azimuthal angle, respectively.

We utilized a series of gratings for the excitement of SPPs. The metal layer is selected to Ag and the effective wavelength of SPP is 613 nm for the incident 633 nm laser. Each nanoslit is the length of 10 micrometers and the width of 100 nanometers. A grating is composed of three nanoslits and the pitch is 613 nm for the maximum efficiency. When these gratings distribute as an axially symmetrically array, multi-beam propagating SPPs will take place and the interference occurs at the central area. We used the finite-different time-domain (FDTD) method to carry out the numerical simulations. As shown in the Fig. 1, three-beam and four-beam SPP interferences have been found. Specially, three-beam interference generates a triangle pattern and four-beam interference generates a square pattern.

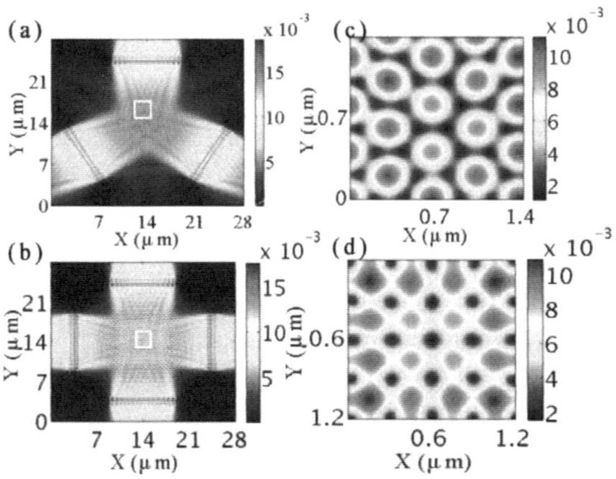

Fig. 1 Planar interference of SPP. (a) Three-beam SPP interference; (b) For-beam SPP interference; (c) and (d) represents the enlarged pattern which is highlighted by a square in (a) and (b), respectively.

Interestingly, the distribution of multiple SPPs is in accordance with those of free-space cases[4]. It's well known that the period interference is equal to $(\lambda/2)\cdot\sin\theta$, where θ is the incident angle. It means that the theoretical limitation of laser interference lithography is a half of incident light, however, in practice, it's no way to achieve such purpose. By contrast, in the case of planar interference, the period can reach to limitation as the incident angle is inherent equal to 90 degree. Moreover, up to now there is few effort about six-beam interference in free space. Because when the number of interfering beams is over four, the interference is very complicated and hard to control. A very slight offset of incident angle or phase will give rise to the resulting pattern significantly[5]. As shown in Fig. 2, we can control the phase shift by adjusting the grating location. The capability of phase controllability is important to obtain extraordinary interfering pattern, i.e. periodic honeycomb structures. It's also regarded as a 2D bottle beam.

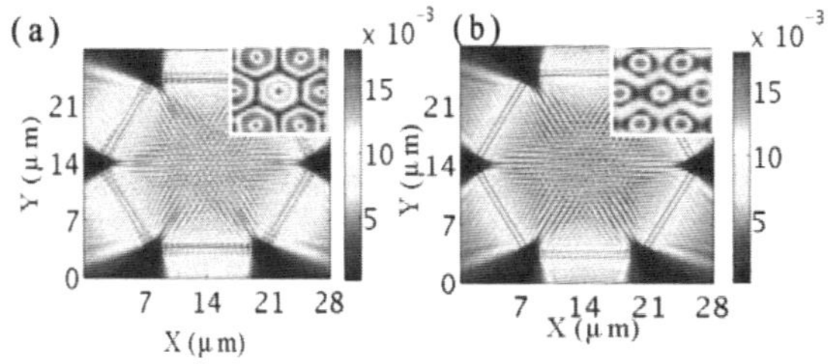

Fig. 2 The phase-shifted six-beam SPP interference. (a) no phase shift; (b) π/2 phase shift of one of six beams

3. Conclusion

We systematically investigated the multiple SPP interference. By comparing with free-space interference, the planar counterpart has numerous advantages, such as minimum limitation of period, miniaturization and controllability of phase. When the number of interfering beam is up to 6, we found the exotic interfering pattern. As a consequence, the method is promising to the applications of optical tweezers and SPP lithography.

Acknowledgment

This work is partially supported by the National Natural Science Foundation of China under Grant Nos. 61427819/61138003/61675136, the Science and Technology Innovation Commission of Shenzhen under grant No. KQTD2015071016560101.

Reference

[1] J. Lin, J. Dellinger, P. Genevet, B. Cluzel, F. de Fornel, and F. Capasso, 'Cosine-Gauss Plasmon Beam: A Localized Long-Range Nondiffracting Surface Wave', *Physical Review Letters*, 109 (2012), 093904.

[2] S. Wei, J. Lin, Q. Wang, G. Yuan, L. Du, R. Wang, L. Xu, M. Hong, C. Min, and X. Yuan, 'Singular Diffraction-Free Surface Plasmon Beams Generated by Overlapping Phase-Shifted Sources', *Optics Letters*, 38 (2013), 1182-1184.

[3] S. Wei, J. Lin, R. Wang, Q. Wang, G. Yuan, L. Du, Y. Wang, X. Luo, M. Hong, C. Min, and X. Yuan, 'Self-Imaging Generation of Plasmonic Void Arrays', *Optics Letters*, 38 (2013), 2783-2785.

[4] D. Wang, Z. Wang, Z. Zhang, Y. Yue, D. Li, and C. Maple, 'Direct Modification of Silicon Surface by Nanosecond Laser Interference Lithography', *Applied Surface Science*, 282 (2013), 67-72.

[5] D. Wang, Z. Wang, Z. Zhang, Y. Yue, D. Li, and C. Maple, 'Effects of Polarization on Four-Beam Laser Interference Lithography', *Applied Physics Letters*, 102 (2013), 081903.

IOP Conf. Series: Journal of Physics: Conf. Series **844** (2017) 012042 doi:10.1088/1742-6596/844/1/012042

Influence of Incident Polarization State on Performance of Real-time Transverse Force Sensing using Stokes Parameters in Fiber Bragg Grating

Yang Su*, Xinhao Fu, , Hua Zhou, Yong Zhu, Baofu Zhang, Guangyu Cai, and Yang Guo

Institute of Communications Engineering, PLA University of Science and Technology, Nanjing, China 210017
qieziyangyang@163.com

Abstract. Stokes parameters can be advantageously used to obtain real-time transverse force measurements with uniform FBGs. In this paper, we demonstrate here that the state of polarization (SOP) of incident light can be used to improve the performance of the sensor. The model and the simulations are presented. A 3-paddle fiber polarization controller was used to adjust the azimuth angle and ellipse angle of the light. The experimental results show the great influences of incident SOP on the sensitivity and linear range of the system, which is well agreed with the theoretical prediction. The highest sensitivity of 0.9475/kgf was obtained with good linearity.

1. Introduction

Fiber Bragg gratings (FBGs) are key components for both optical fiber telecommunications and sensing applications. While the most straightforward applications remain temperature and axial strain sensing, which changes the center wavelength of the reflected spectrum when the sensing parameters cause grating effective index or grating period variation [1–2]. In practice, the characteristics of a FBG under transverse load are quite different from axial load [3–4]. In the case of transverse load, stress induced birefringence effects will cause the unique Bragg grating condition to break down and even produce two distinct Bragg wavelengths [5,6].

However, this effect cannot be directly observed on the amplitude spectrum for limited load values (<1 N/mm)[7]. In the other hand, it has been demonstrated that the spectral response of the polarization dependent loss (PDL) [8] or the differential group delay (DGD) [9] is very sensitive to the birefringence within the grating, which can provide more effective information and therefore lead to the potential development of new types of FBG-based optical sensors [7–9] used to monitor weak transversal load. By measuring PDL or DGD peak amplitudes the amount of applied force is determined. Although the structure is simple, this design requires a complex measuring system to determine the peak values. Large amounts of data points over a wide range of spectrum must be continually analyzed in high resolution, resulting in a rather slow processing speed.

FBGs used as transversal load sensors can be applied in many domains such as structural health monitoring of composite materials [10–11] or for biomedical applications [12-13]. Therefore it is necessary to do research for real-time sensing. We have proposed a compact real-time transverse-force sensing system through Stokes parameters measurement at single wavelength [14]. The measuring system in our approach is simple and does not require processing of massive amount of spectral data,

IOP Conf. Series: Journal of Physics: Conf. Series **844** (2017) 012042 doi:10.1088/1742-6596/844/1/012042

enabling real-time transverse force monitoring. The proportional relationship and linear fit can be observed between the applied force and the Stokes parameters from the FBG.

For practical applications, it is of high importance to have the possibility to improve the sensitivity and enlarge the linear range. In this paper we put emphasis on how to improve system performance. By simulations, it is demonstrated that the state of polarization (SOP) of incident light has great influence of on sensitivity and linearity of the sensor. A 3-paddle fiber polarization controller (PC) was used to modify the polarization state of the incident light in the experiment and the experimental data confirm the simulations well. In the experiment results, a highest sensitivity of 0.9475/kgf is gained with good linearity. Therefore, a simple method to improve sensor performance by adjusting incident SOP of light using PC is proposed in this paper.

2. Principle and simulations

When the FBG is subjected to a transverse load, the difference between the effective refractive indices of the two orthogonal modes of the fiber within the loaded region will be produced. For simplicity the direction of the transverse force is assumed as y (fast axis), another direction perpendicular to y-axis is x direction (slow axis), z is along the fiber axial direction, as shown in Fig. 1.

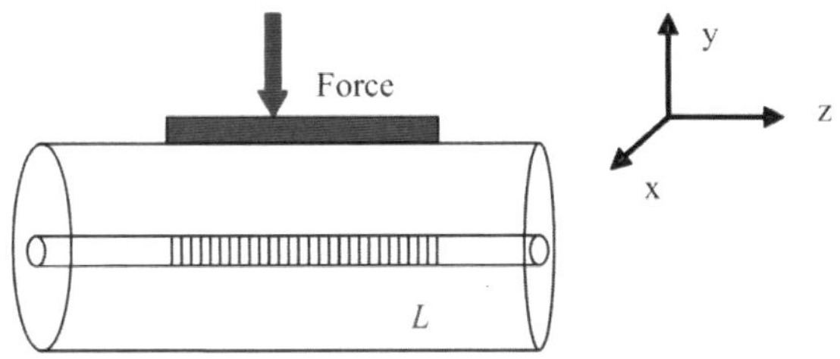

Fig. 1. Schematic showing the defined coordinate system.

The refractive index changes $\left(\Delta n_{eff}\right)_x$ and $\left(\Delta n_{eff}\right)_y$ within the loaded zone are given by[6,15]

$$\left(\Delta n_{eff}\right)_x = -\frac{n_{eff}^3}{2E}\left\{\left(p_{11}-2vp_{12}\right)\sigma_x + [(1-v)p_{12}-vp_{11}](\sigma_y+\sigma_z)\right\} \quad (1)$$

$$\left(\Delta n_{eff}\right)_y = -\frac{n_{eff}^3}{2E}\left\{\left(p_{11}-2vp_{12}\right)\sigma_y + [(1-v)p_{12}-vp_{11}](\sigma_x+\sigma_z)\right\} \quad (2)$$

Where E and v are the Young's modulus and Poisson's coefficient of the optical fiber respectively, p_{11} and p_{12} are the strain-optic coefficients. For typical optical fiber, E=74.52 GPa, v=0.17, p_{11} =0.121, and p_{12}=0.270. σ_x, σ_y and σ_z are the stress components in the grating in the x, y and z directions, respectively. They are expressed in the Hertz equations [5,6]:

$$\sigma_x = \frac{2F}{\pi DL};\ \sigma_y = -\frac{6F}{\pi DL};\ \sigma_z = u\cdot v(\sigma_x+\sigma_y) \quad (3)$$

Where D is the fiber diameter, F is the applied force, and L is the length of the region under stress, u is the correction coefficient which is related to the smoothness of the optical fiber and compression platform. The case of u = 0 denotes that the contact plane is perfectly smooth (no friction) and the fiber is in plane stress state (ends-free). In order to simplify the model of fiber stresses, we limit ourselves to this case.

The refractive indexes in the x and y-directions are then given by $n_{eff,x} = n_{eff} + (\Delta n_{eff})_x$ and $n_{eff,y} = n_{eff} + (\Delta n_{eff})_y$, respectively, with birefringence:

$$\Delta n = (\Delta n_{eff})_x - (\Delta n_{eff})_y \quad (4)$$

Due to Δn, the x and y modes undergo different couplings through the grating. The total transmitted signal is, therefore, the combination of the x and y mode signals. In a Cartesian coordinate system whose reference axes match the FBG eigenmodes, the Jones matrix of the grating is diagonal [8,9]:

$$J_{grating} = \begin{bmatrix} t_x & 0 \\ 0 & t_y \end{bmatrix} \quad (5)$$

In Eq. (5), $t_{x(y)}$ denotes the transmission coefficient of the FBG corresponding to the $x(y)$ mode [16].

The Stokes parameters represent the state of polarization. We firstly translate the Jones matrix of the grating to a 4×4 Muller matrix $M_{grating}$ and then get the Stokes vector of the transmitted signal $(S_0, S_1, S_2, S_3)_{out}^T$ by

$$\begin{pmatrix} S_0 \\ S_1 \\ S_2 \\ S_3 \end{pmatrix}_{out} = M_{grating} \cdot \begin{pmatrix} S_0 \\ S_1 \\ S_2 \\ S_3 \end{pmatrix}_{in} \quad (6)$$

Where $(S_0, S_1, S_2, S_3)_{in}^T$ represents the Stokes vector of incident light. It can be determined by two para meters, ie. ellipticity angles κ and azimuth angle θ, and therefore Stokes vector can be written as

$$(1, (\cos 2\kappa)(\cos 2\theta), (\cos 2\kappa)(\sin 2\theta), \sin 2\kappa)_{in}^T \quad (7)$$

Then the normalized Stokes parameters (s_1, s_2 and s_3) can be computed using:

$$s_i = S_i / S_0 \quad (i = 1, 2, 3) \quad (8)$$

So the relationship between transversal force and the normalized Stokes parameters are found using Eqs. (1)-(8) [19].

Simulated s_1, s_2 and s_3 spectrum evolutions for different values of birefringence are shown in [14] and the amplitude of Stokes parameters at single wavelength changes with increase of applied force. However the incident SOP will affect the response performance, which sometimes presents very low sensitivity. So how to find a solution to get a better response is important.

The changes of amplitude of Stokes parameters respect to applied force are studied. Simulated results for different κ with θ of 10 degree are shown in Fig. 2(a). And Fig.2 (b) shows simulated results for different θ with κ of 25 degree. From these results we can conclude that κ and θ has grate influence on the sensitivity and linear range of the system. And in some cases non-linearity is shown. So the choice of incident SOP is important. Maybe κ of 45° is a better choice. However we should realize that the simulations can not agree with experimental results exactly. That is because we can not control or get the accurate SOP of incident light due to the existence of fiber segment between grating and PC controller and another segment between grating and polarization detector. That is illustrated in Fig .3 in section 3. Therefore, by the directions of simulations, the experiments will give practical performance.

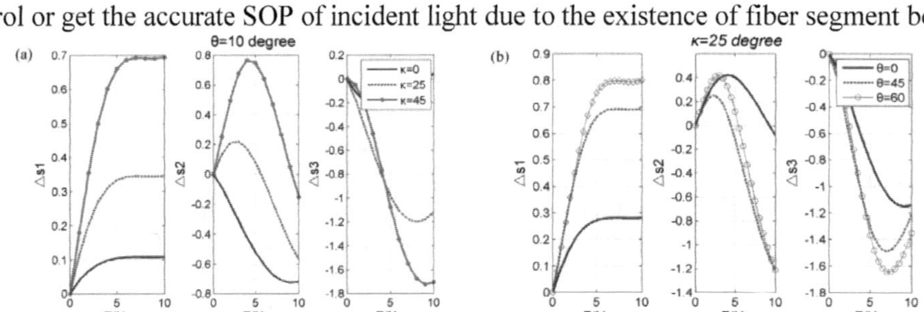

Fig. 2. (a) Simulated results for different ellipse angles κ at θ=10° (b) simulated results for different azimuth angles at κ=30° .

IOP Conf. Series: Journal of Physics: Conf. Series **844** (2017) 012042 doi:10.1088/1742-6596/844/1/012042

3. Experiment and discussion

We will now describe our experimental set-up and present some experimental results that we have obtained to validate our numerical analysis. The schematic of the experiment setup is illustrated in Fig. 3. The input light from a tunable laser source (Agilent 8163B lightwave Multimeter) was launched into FBG. A 3-paddle fiber polarization controller (PC) was used to modify the polarization state of the light. The transmitted signal from FBG was directed into a polarization detector (General Photonics's in-line polarimeter POD-15-SS-02) which measured the Stokes parameters at a certain wavelength. After A/D conversion and signal process the experimental results were displayed and recorded in a computer.

During the measurements, the FBG was not strained and the ambient temperature was kept constant. Moreover, the fiber connectors were fixed to avoid polarization instability. The length of fiber grating is 10 mm and was placed parallel to a balance reference fiber.

For the 3-paddle fiber PC used to change the incident SOP, the second circle works as a $\lambda/2$ wave plate to alter azimuth angle (θ) and the first or third circle works as a $\lambda/4$ wave plate to alter ellipse angles (κ) of incident light.

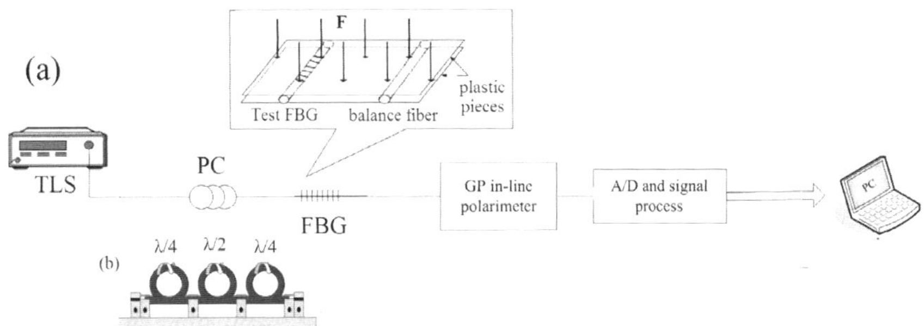

Fig. 3 (a) Experimental set-up. (b) 3-paddle fiber polarization controller.

3.1. *Real-time measurement under different incident SOP*

Firstly we applied a mobile telephone (Iphone 6 plus) to FBG and set it to vibrate mode to verify the performance of real-time measurement under different incident SOP. SOP of incident light was changed by adjusting PC for four cases randomly. The experimental results were recorded in Fig. 4(a) which depicts the evolutions of the four cases. It can be seen that all of the Stokes parameters can respond to the vibration of phone. The Stokes parameters present lowest sensitivity for the first case and highest sensitivity for the second case, which was enlarged as shown in Fig. 4(b) for example. The period is about 4.5ms, meaning a vibration frequency of about 222 Hz. So it is demonstrated that incident SOP has great influence on the performance of system.

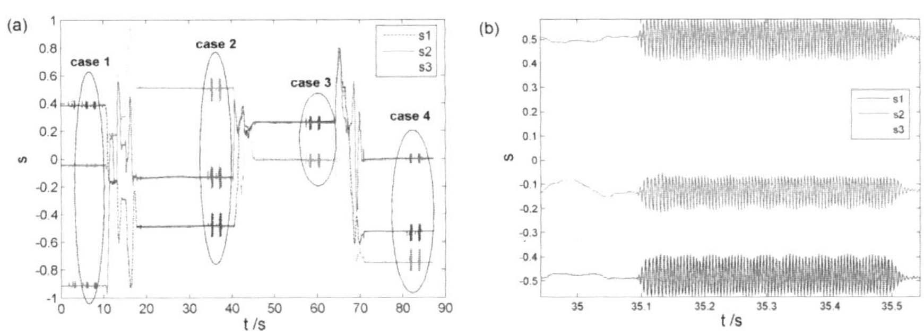

Fig.4 (a) Experimental evolutions of Stokes parameters response to a vibrated phone and (b) the enlarged responded curve for the second case.

3.2. Effect of azimuth angle and ellipse angle

Next we will adjust the 3-paddle fiber PC steadily to observe the influences.

a) Adjusting azimuth angle

The second paddle of PC was rotated from 0° to 180°, stepped by 45° to alter azimuth angle, with the first and third circle fixed. It is should be noted that these angels are just the rotated angle of the second paddle, not the values of azimuth angle.

For each case, the force to be measured was applied to the FBG by placing weights on a plastic piece against the fiber. The weight applied was increased gradually from 0 to 800g, stepped by 200g. Experimental results were recorded and saved.

Fig. 5 (a) and (b) shows the experimental evolutions of three Stokes parameters during the process of placing weights when the second paddle of PC was rotated to 90° and 135°, respectively. It can be seen that s_1, s_2 and s_3 changes as applied weight increases while presents different sensitivity and linear range.

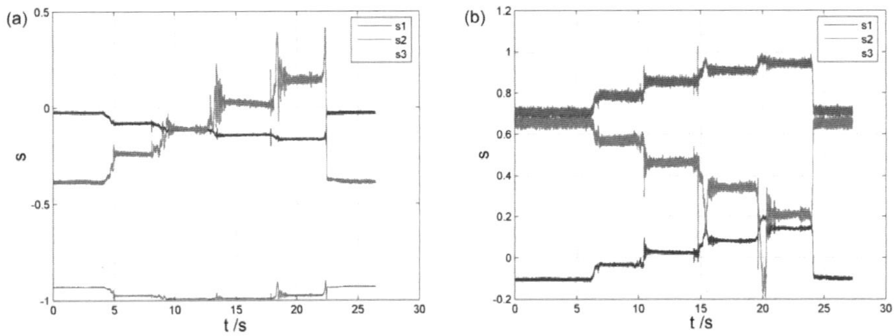

Fig.5 (a) Experimental evolutions of three Stokes parameters during the process of placing weights when the second paddle of PC ($\lambda/2$ wave plate) was rotated to (a) 90° and (b) 135°.

The amplitude of s_1, s_2 and s_3 under different amount of applied weights are recorded and used to depict the experimental performance curves. Fig. 6 (a), (b) and (c) are the amplitude changes versus weight amount under different rotated angle (that means different azimuth angles of SOP) for s_1, s_2 and s_3, respectively. It is found certainly that the difference of azimuth angle brings different characteristics. In some cases, non-linear property in range of 0-800g are presented, which also agrees with simulations, for example, rotated angel of 45° and 90° for s_2. The data with highest sensitivity are linearly fitted for each Stokes parameter, as shown in Fig. 6. The sensitivity of 0.3525/kgf, 0.5475/kgf, and -0.685/kgf are obtained for s_1, s_2 and s_3, respectively.

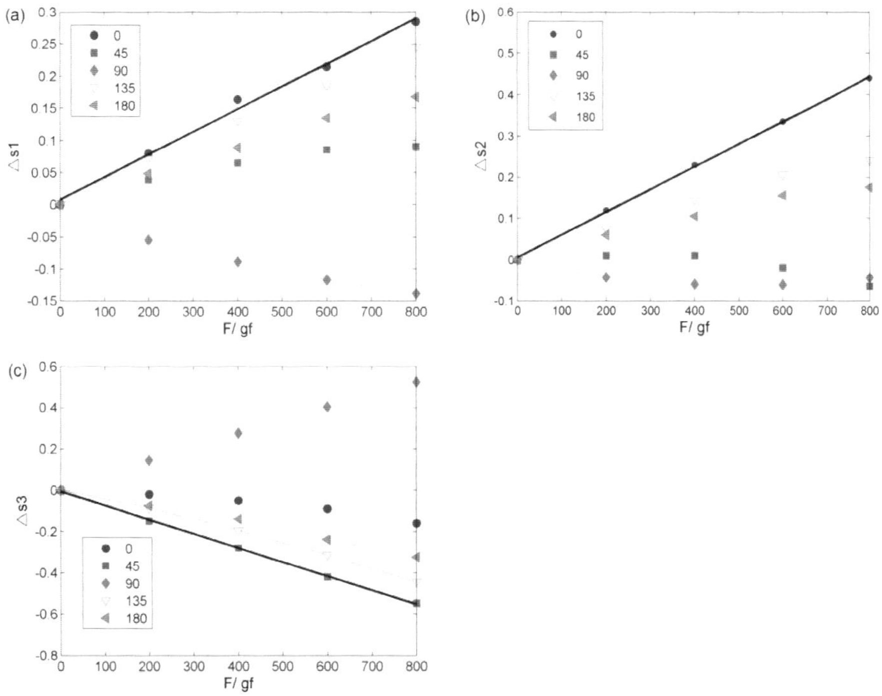

Fig. 6 The amplitude changes of Stokes parameters versus weight amount under different rotated angle of $\lambda/2$ wave plate for (a)s_1, (b)s_2 and (c)s_3..

(b) Adjusting ellipticity angle

Next the third paddle ($\lambda/4$ wave plate) of PC was rotated from 0° to 180°, stepped by 45° to alter ellipse angle, with the first and second paddles fixed.

Fig. 7 (a) and (b) depicts the experimental evolutions of three Stokes parameters during the process of placing weights when the third paddle of PC was rotated to 45° and 180°, respectively. It can be seen that for 180° case, s_3 presents very high sensitivity.

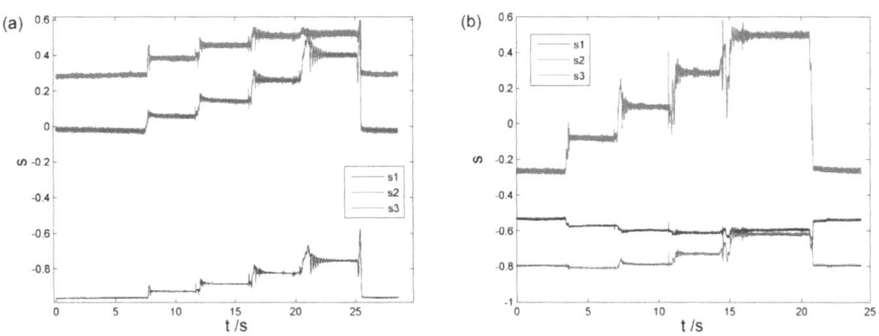

Fig. 7 Experimental evolutions of three Stokes parameters during the process of placing weights when the third paddle of PC (λ/4 wave plate) was rotated to (a) 90° and (b)180°.

Using the same method, the amplitude of s_1, s_2 and s_3 under different amount of applied weights are recorded and used to depict the experimental performance curves. Fig. 8(a), (b) and (c) are the amplitude changes versus weight amount under different rotated angles of λ/4 wave plate for s_1, s_2

and s_3, respectively. The data with highest sensitivity are linearly fitted, and the sensitivity of 0.2635/kgf, 0.52/kgf, 0.9475/kgf are obtained for s_1, s_2 and s_3, respectively.

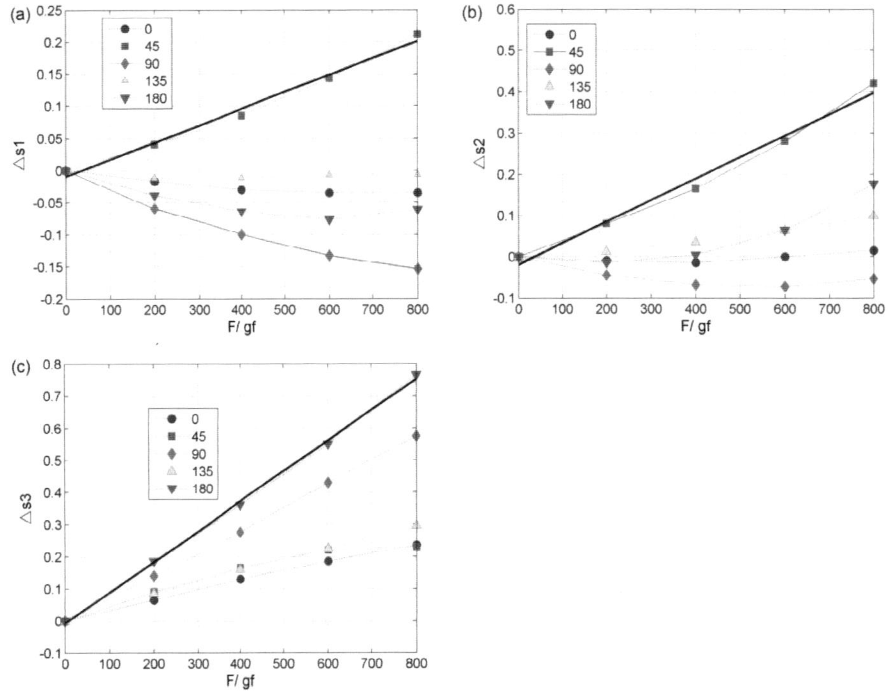

Fig. 8 The amplitude changes of Stokes parameters versus weight amount under different rotated angle of λ/4 wave plate for (a)s_1, (b)s_2 and (c)s_3.

4. Discussion and conclusion

Stokes parameters amplitude can be advantageously used to obtain transverse force measurements with uniform FBGs written into standard single mode fiber. In this paper the influences of incident SOP on the performance of sensor system are found by simulations. A 3-paddle fiber polarization controller was used in the experiment to adjust the azimuth angle and ellipse angle of the light. The experimental results show the great influences of incident SOP on the sensitivity and linear range, which is well agreed with the theoretical prediction. Although the 3-paddle fiber PC is a rough device, a satisfying response still can be found by adjusting PC steadily. In the experiments, a highest sensitivity of 0.9475/kgf was obtained with good linearity. Therefore this work offers new theories and simple method for transversal force sensing to improve the performance.

5. Acknowledgments

This work was supported by the Natural Science Foundation of Jiangsu Province (BK20161471) and Natural National Science Foundation of China (NSFC) (61371121).

6. References and Links

[1] A.D. Kersey, M.A. Davis, H.J. Patrick, M. LeBlanc, K.P. Koo, C.G.Askins, M.A.Putnam, and E.J.Friebele, "Fiber grating sensors", J. Lightwave. Technol. **15**(8), 1442–1463 (1997).

[2] K.O. Hill and G. Meltz, "Fiber Bragg grating technology fundamental and overview," J. Lightwave. Technol. **15**(8), 1263–1276 (1997).

[3] I. Abe, O. Frazão, M.W. Schiller, R. N. Nogueira, H. J. Kalinowski, and J. L. Pinto, "Bragg gratings in normal and reduced diameter high birefringence fiber optics," Meas. Sci. Technol. **17**(6), 1477(2006).

[4] Y. Wang, N. Chen, B. Yun, Z. Wang, C. Lu, and Y. Cui, "Effects of distributed birefringence on fiber Bragg grating under non-uniform transverse load," Opt. & Laser Technol. **40**(8), 1037-1040(2008).

[5] R. B. Wagreich, W. A. Altia, H. Singh, and J. S. Sirkis, "Effects of diametric load on fiber Bragg gratings in low birefringence fiber," Electron. Lett. **32**(13), 1223–1224(1996).

[6] R. Gafsi and M.A.E. Sherif, "Analysis of induced-birefringence effects on fiber Bragg gratings," Opt. Fiber Technol. **6**(3), 299-323(2000).

[7] F. Descamps, C. Caucheteur, P. Mégret, and S. Bette , "Distribution profiling of a transverse load using the DGD spectrum of chirped FBGs," Opt. Express **23**(14), 18203-18217(2015).

[8] C. Caucheteur, S. Bette, R. García-Olcina, M. Wuilpart, S. Sales, J. Capmany, and P. Mégret, "Transverse strain measurements using the birefringence effect in fiber Bragg grating," IEEE Photon. Technol. Lett. **19**(13), 966–968(2007).

[9] C. Caucheteur, S.Bette, R. García-Olcina, M. Wuilpart, S. Sales, J. Capmany, and P. Mégret, "Influence of the grating parameters on the polarization properties of fiber Bragg gratings ," J. Lightwave Technol. **27**(8), 1000–1010 (2009).

[10] Y. Qiu, Q. Wang, H. Zhao, J. Chen, and Y. Wang, "Review on composite structural health monitoring based on fiber Bragg grating sensing principle," J. Shanghai Jiaotong University **18**(2), 129–139 (2013).

[11] N. Lammens, D. Kinet, K. Chah, G. Luyckx, C. Caucheteur, J. Degrieck, and P. Mégret, "Residual strain monitoring of out-of autoclave cured parts by use of polarisation dependent loss measurements in embedded optical fiber Bragg gratings," Composites, Part A **52**, 38–44 (2013).

[12] L. Mohanty, S. C. Tjin, D. T. T. Lie, S. E. C. Panganiban, and P. K. H. Chow, "Fiber grating sensor for pressure mapping during total knee arthroplasty," Sens. Actuators, A **135**(2), 323–328 (2007).

[13] Y. Park, K. Chau, R. J. Black, and M. R. Cutkosky, "Force sensing robot fingers using embedded fiber Bragggrating sensors and shape deposition manufacturing," *in Proceedings of IEEE International Conference onRobotics and Automation* (IEEE, 2007), pp. 1510–1516.

[14] Yang Su, Yong Zhu, Baofu Zhang, and Hua Zhou, "Real-time transverse force sensing using fiber Bragg grating through direct Stokes parameters measurement", Optics Express, **23**(25),32300(2015)

[15] J. Espejo and S. D. Dyer, "Transverse-stress fiber Bragg grating sensor with high spatial resolution and temperature stability," J. Lightwave Technol. **25**(7), 1777–1785(2007).

[16] Erdogan, "Fiber grating spectra," J. Lightwave Technol. **15**(8), 1277–1294(1997

IOP Conf. Series: Journal of Physics: Conf. Series **844** (2017) 012043 doi:10.1088/1742-6596/844/1/012043

High Dynamic Range Imaging Using Multiple Exposures

Xinglin Hou [1,2,3,4], Haibo Luo[1,3,4], Peipei Zhou [1,2,3,4] and Wei Zhou[5]

[1] Shenyang Institute of Automation, Chinese Academy of Sciences, Shenyang, 110016
[2] University of Chinese Academy of Sciences, Beijing, 100049
[3] Key Laboratory of Opto-Electronic Information Processing, CAS, Shenyang 110016
[4] The Key Lab of Image Understanding and Computer Vision, Liaoning Province, Shenyang, 110016
[5] AVIC Jiangxi HONGDU Aviation Industry Group LTD, Nanchang, China

Abstract. It is challenging to capture a high-dynamic range (HDR) scene using a low-dynamic range (LDR) camera. This paper presents an approach for improving the dynamic range of cameras by using multiple exposure images of same scene taken under different exposure times. First, the camera response function (CRF) is recovered by solving a high-order polynomial in which only the ratios of the exposures are used. Then, the HDR radiance image is reconstructed by weighted summation of the each radiance maps. After that, a novel local tone mapping (TM) operator is proposed for the display of the HDR radiance image. By solving the high-order polynomial, the CRF can be recovered quickly and easily. Taken the local image feature and characteristic of histogram statics into consideration, the proposed TM operator could preserve the local details efficiently. Experimental result demonstrates the effectiveness of our method. By comparison, the method outperforms other methods in terms of imaging quality.

1. Introduction

The dynamic range of a natural scene can be as high as the order of 10^8. However, the widely used digital camera can capture a range of order from 10^2 to 10^4. A high dynamic range (HDR) scene can not be well-expressed through a single image captured by a low dynamic range (LDR) camera. It is of great significance to develop an HDR imaging method for LDR cameras.

In recent years, many HDR prototype have been designed by redesign of pixel [1]. However, because of their high price, these cameras are unacceptable for ordinary consumers. Instead the expensive hardware, an HDR image can be reconstructed by using a set of multiple exposures.

During the process of HDR reconstruction (HDRR). The recovering of the CRF is the key which determine the assignment of the scene radiance to the image gray value [2-4]. After the obtaining of the HDR image, it should be mapped to a displayable gray range using a tone mapping operator. There are some advantages of the HDRR+TM method: there is no specialized hardware required; various operations can be performed on the HDR images, such as virtual exposure; and user interactions are allowed in the TM phase to generate a tone-mapped image with the desired appearance.

The propose of HDRR is to reconstruct the HDR radiance image in which the CRF is necessary. Large number of scholars have proposed method for recovering the CRF. Debevec [5] is the pioneer who recover the CRF by minimizing a quadratic objective function. After that, a weight function is used to construct HDR radiance image. In Robertson et al.'s method [6], another object function of radiance, exposure time and the amount of light is constructed. A form of Gauss–Seidel relaxation is used to minimize the object function and CRF is constructed with a Gaussian weight function. Grossberg [7] recover the CRF from intensity mapping functions, which are determined by image histograms. In Lu et al.'s work [8], CRF is constructed through an expectation maximization like optimization algorithm using a maximum-likelihood approach based on a Bayesian framework-defined probabilistic model. Mitsunaga [9] describe the CRF using a high-order polynomial. Construction of the CRF is converted into the solution of the coefficients. The CRF is recovered by formulating an error function. In this work, the CRF is also described use a high-order polynomial. By adding a constraint, the polynomial can be solved easily using singular value decomposition.

For an HDR image, TM addresses the problem of dynamic range reduction from the scene radiance to the displayable range while preserving the image details. TM is usually applied in the luminance channel. There are two kinds of TM, local TM and global TM. Local TM methods [10-12] are based on the local spatial information; while in the global TM method [13], the same nonlinear transformation curve is used for all pixels. Reinhard et al.[10] present a local TM method based on the operation in film printing. A global TM function is initially used for the whole pixel luminance. Then a local process based on "dodging and burning" is adopted to enhance the image details. Shan et al.[11] propose a method for HDR reconstruction by using overlapping window-based linear functions. A guidance map based on local constraints is used in TM and it suppresses local high contrast and effectively preserves details. Larson et al.[13] present a global tone reproduction operator, which incorporates models for human contrast sensitivity, glare, spatial acuity, and color sensitivity to match the subjective experience. Local TM operator brightens areas around dark objects, which could produce some characteristic effects such as halos. While global TM enhances the global contrast, the local contrast may be compressed. For the two methods mentioned above, local TM provides a better result according to the perceptual evaluation introduced by Yoshida [14]. In our work, a local TM method is proposed based on histogram statics characteristic.

2. Camera Response Function Recovering

CRF is the assignment of scene radiance to the image gray value. CRF is the patent of a company and vary appreciably from one imaging system to the next. When analyze the CRF results obtained from imaging systems, we find that all CRFs are distributed like a high-order polynomial which is shown in Figure 1. We also conclude that the CRF is monotonic or at least semi-monotonic. Any curve can be fitted by a high order polynomial. Here, we modeling the CRF with a high order polynomial:

$$E = f(M) = \sum_{n=0}^{N} c_n M^n \tag{1}$$

where, M denotes pixel grey value, E denotes the exposure and n is the order of the polynomial. The minimum order of the polynomial required clearly depends on the fitting error.

Considering f_q and f_{q+1} are two images of a scene taken with different exposure time t_q and t_{q+1}. The radiance radio of a pixel can be written as:

$$\frac{E_{p,q}}{E_{p,q+1}} = \frac{I_p t_q}{I_p t_{q+1}} = \frac{t_q}{t_{q+1}} = R_{q,q+1} \tag{2}$$

where, q is the index of image, p is the index of pixel and I denotes the scene radiance. Then, we can get:

$$\frac{\sum_{n=0}^{N} c_n M_{p,q}{}^{n}}{\sum_{n=0}^{N} c_n M_{p,q+1}{}^{n}} = \frac{I_p t_q}{I_p t_{q+1}} = \frac{t_q}{t_{q+1}} = R_{q,q+1} \tag{3}$$

$$\sum_{n=0}^{N} c_n M_{p,q}{}^n - R_{q,q+1} \sum_{n=0}^{N} c_n M_{p,q+1}{}^n = 0 \tag{4}$$

In the above equation (4), M and R are known while $c_0 \ldots c_n$ are unknown for us. At least n equations are needed for the solution of $c_0 \ldots c_n$. However, the homogeneous linear equations have no solutions when the coefficient matrix is full rank. When normalized the all gray value M to the range of $[0,1]$ and suppose the largest exposure is 1, we can add another constraint to the linear equations.

$$\sum_{n=0}^{N} c_n = 1 \tag{5}$$

Then, we can get the polynomial coefficients by solving the nonhomogeneous equations. A total of $n+1$ images are needed in the process. In fact, the sampled values of a given pixel in the images can be used for the solution. However, the distribution of the sampled value in the gray range will affect the fitting result. When the sampled values locate at the right part of the gray range, the left part of fitted curve will not precise. Therefore, a screen will be implemented before fitting and removing the uneven distributed value sets. For a set of values, if more than two of them are 255 or 0, the set should be removed. After screening, all available sample set can be applied for solving the equations. We can get multiple sets of $c_0 \ldots c_n$. To make the fitting result more accurate and reduce the impact of noise, the mean of all c_0 is regarded as the final value. Figure 2 depicts the CRF of camera Canon EOS 60D.

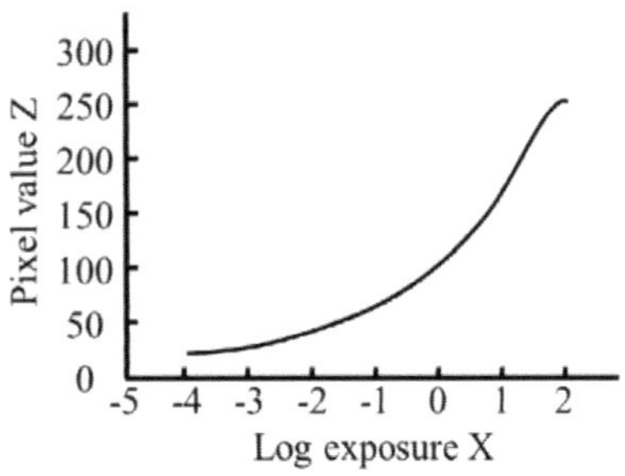

Figure 1. CRF of an imaging system

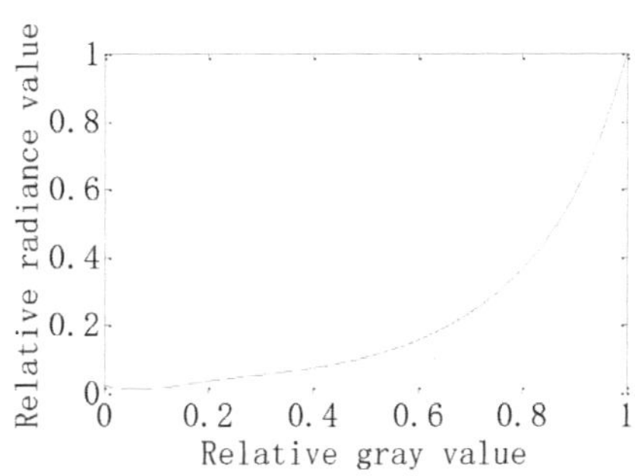

Figure 2. CRF of camera Canon EOS 60D.

3. Constructing the High Dynamic Range Radiance Map

Once the response curve f is recovered, it can be used to convert the gray values to relative radiance values. Assuming the exposure time is t_q, the radiance map can be obtained by equation (6)

$$I_q = \frac{\sum_{n=0}^{N} c_n M_q{}^n}{t_q} \tag{6}$$

For precise and robustness, all available exposure images should be used to recover high dynamic range radiance values. Here, the weighting function is used to give higher weight to exposures in which the pixel's value is closer to the middle of the response function [5]:

$$I = \frac{\sum_{q=0}^{N} \omega(I_q) I_q}{\sum_{q=0}^{N} \omega(I_q)} \tag{7}$$

where, q is the index of pixel and n denotes different radiance maps. ω is a simple hat function:

$$\omega(I)=\begin{cases} I-I_{\min} & for\ I\le\frac{1}{2}(I_{\min}+I_{\max}) \\ I_{\max}-I & for\ I>\frac{1}{2}(I_{\min}+I_{\max}) \end{cases} \tag{8}$$

A set of multiple exposure images are shown in figure 3(a), the HDR radiance map is given in figure 3(b).

(a) LDR image sequence

(b) HDR radiance map

Figure 3. A set of multiple exposure images (a) and the HDR radiance map (b)

4. Local Tone Mapping

For a RGB image, if the tone mapping is implemented in each of the RGB space, color deviation may be occurred in the result image. Here, the RGB image is firstly translated into HSV space. Then, tone mapping is implemented in V space. At last, the image will be translated back to RGB space.

Image scale space is defined as the descriptions for the image in the full scale domain absolutely. If an image is processed in different scale domain, image features in different scales will be obtained. Here, to preserve the image details in different scales during the tone mapping process, we decompose the HDR image into detail layer and base layer. Then, local tone mapping is implemented in each of the layers.

4.1. Multi-scale Decomposition

Tone mapping algorithm compress the HDR radiance map into a displayable LDR image. Here, a multiscale decomposition and histogram analysis-based method is proposed for tone mapping. A HDR scene contains both high frequency and low frequency information. Therefore, an input image is firstly decomposed into a base layer, containing only large-scale variations, and a detail layer. Then, tone mapping processing is applied to all the layers. At last, the result image is reconstructed by fusing the compressed images.

Bilateral filter is a widely used non-linear filter, where each pixel is weighted by the product of a Gaussian filtering in the spatial domain and another Gaussian filtering in the intensity domain. The bilateral filter is very effective in preserving the strong edge in the smooth intensity domain. However, it can not meet the demands of extracting image details of arbitrary scale. It can not be applied in multiscale detail extraction. In our method, an edge-preserving approach [15] based on the weighted least square (WLS) optimization framework is used for multiscale decomposition. Given an input image g , another image u can be sought by minimum the formula:

$$\sum_p\left((u_p-g_p)^2+\lambda\left(\alpha_{x,p}(g)(\frac{\partial u}{\partial x})_p^2+\alpha_{y,p}(g)(\frac{\partial u}{\partial y})_p^2\right)\right) \tag{9}$$

where the subscript p denotes the spatial location of a pixel. The goal of the data term $(u_p-g_p)^2$ is to minimize the distance between u and g , while the second term strives to achieve smoothness by

minimizing the partial derivatives of u. α_x and α_y are weights to enforce the smoothness of u. λ is the balance factor and the increase of λ will result in progressively smoother images u.
Let $u^1,\cdots,u^k$ denote progressively coarser version of g. The coarsest of these version, u^k will serve as the base layer, with the k detail layer defined as:

$$d^i = u^{i-1} - u^i \tag{10}$$

where $i = 1,\ldots,k$ and $u^0 = g$.
In figure 4, the source image is shown in the left of figure 4(a). Then, the filtered results by WLS is given with λ=1,10,20,respectively. As we can see, with the increase of λ, the result image is more and more smooth. The last image is regarded as base image which is shown in the left of figure 4(b). The rest part of figure 4(b) are detail layers.

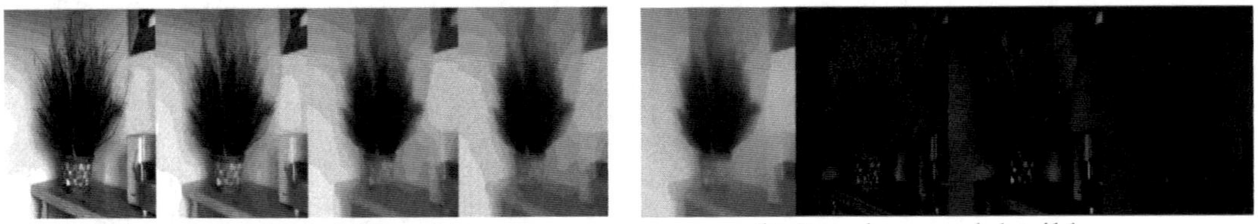

(a) Source image and filtered images (b) Base layer and detail layers

Figure 4. (a) From left to right: source image, filtered images with λ=1,10,20 and (b) From left to right: base layer, detail layers of (a).

4.2. Local Tone Mapping

For the obtained base layer and detail layers, most image details are included in the detail layers and the base layer reflects the overall brightness of the scene area. To preserve details of the image while compressing the dynamic range, nonlinear compression is only implemented for the detail layers and linear compression is applied for base layer.

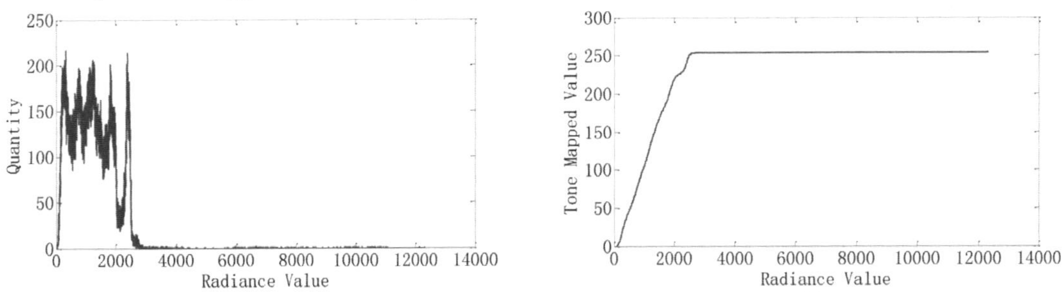

Figure 5. Histogram of an HDR radiance map (a) and tone mapping curve (b) of image in (a)

As shown in figure 5(a), for a detail layer, the main domain of the histogram locates at the left part of the radiance value range. Image details are mainly expressed by these radiance values. The higher radiance value in the right part of the histogram make less contribution to image detail. Therefore, the left part of the histogram should be slightly compressed and the right part should be seriously compressed. The compression radio should be related to the quantity of the radiance value. The larger the quantity of the value, the greater the compression ratio. Here, a cumulative distribution function (CDF) based mapping method is applied. As shown in figure 6, the quantity of the value in radiance map can be expressed by the increment in CDF of two adjacent point. Due to the fact that details of regions corresponding to radiance values with large increment are abundant, we set the compression radio inversely proportional to the increment. After linear compression of the ordinate of the CDF, we make it the tone mapping curve. For the histogram in figure 5(a), the mapping curve is shown in figure 5(b).

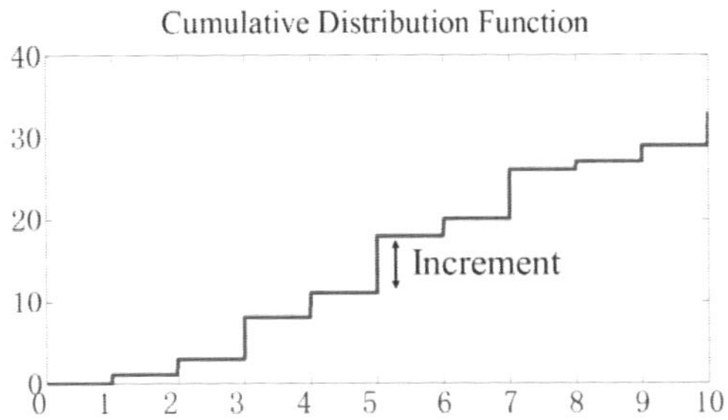

Figure 6. Cumulative distribution function (CDF), horizontal axis denoting the gray levels and vertical axis representing the number of pixels.

After the tone mapping, the result image g' can be easily recovered by simply adding up the base and detail layers:

$$g' = b' + \sum_{i=1}^{k} d^{i'} \tag{11}$$

where b' and $d^{i'}$ are compressed base layer and detail layers.

5. Experimental Results and Analysis

5.1. Local Tone Mapping

In order to demonstrate the effectiveness of the proposed method, many image sequences are captured using Canon EOS 60D camera for high dynamic range imaging. In figure 7(a), a set of multiple exposure images are captured with 1s, 1.6s, 2.5s, 4s, 5s, 6s, 8s and 10s respectively. As is shown, with the increase of the exposure time, image brightness increases. However, no image else could perfectly express the scene detail. Figure 7(b) is the linear compressed image of reconstructed HDR map. Because the dynamic range of the HDR map is much larger than 256 and the image is dark in general, there is a few details in the bright region when linear compressed. Figure 7(c) is the tone mapping result of the proposed method. Both bright and dark regions are abundant in detail. The image could perfectly express the high dynamic scene.

(a) LDR image sequence

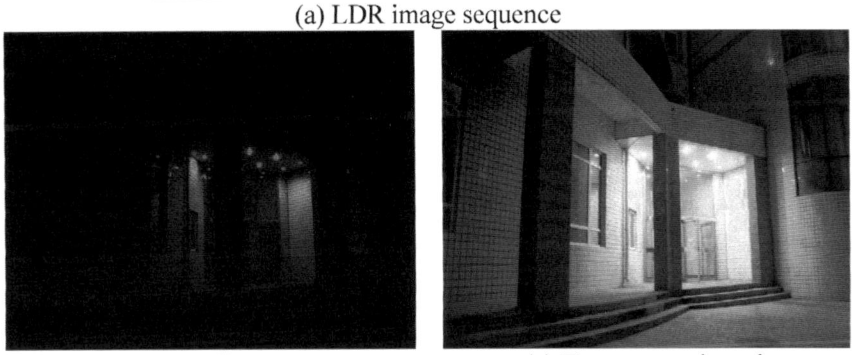

(b) HDR radiance map (c) Tone mapped result

Figure 7. Tone mapped image using the proposed method

For simulations, we compare the proposed method with previous methods [5], [16], [17] perceptually. The fundamental method [5] is chosen as baseline. Others are state-of-the-art fusion methods with the ghost removal [16], [17]. Those source codes are available on the authors' website. Free parameters of the proposed method are experimentally determined. The input image sequence of 'Desk' is shown in figure 8(a). Shutter speed settings are simulated 1/500, 1/125, 1/30, 1/8 in seconds. The HDR imaging results of each methods are shown in figure 8(b).

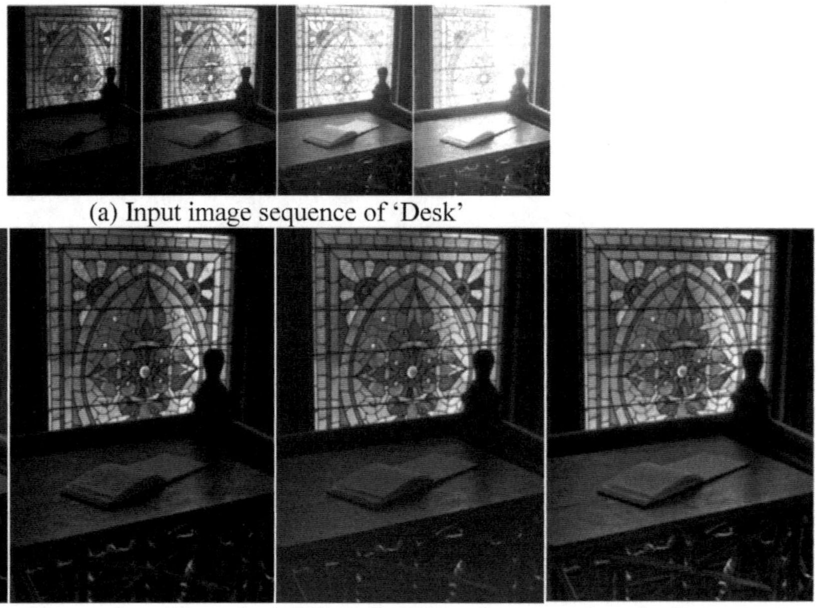

(a) Input image sequence of 'Desk'

(b) HDR imaging results with methods [5], [16], [17] and proposed method

Figure 8. HDR images of different HDR imaging methods

To better illustrate the validity of this method, an objective comparison is implemented with other state of-the-art HDR imaging methods. HDRVDP-2[18] is used for measurement. HDR-VDP-2 scores the similarity between true and fused HDR images and a perceptual quality of fused one based on the statistics is analyzed. Table 1 describes the scores of each method which indicates us that the proposed method outperforms other methods obviously.

Table 1. HDRVDP-2 scores of each method

	Ref. [5]	Ref. [16]	Ref. [17]	Proposed
Cathedral	59.94	54.24	69.18	70.21
Desk	58.84	53.61	68.26	70.33
Memorial	61.62	55.08	67.6	69.92
Night atrium	67.11	57.09	68.59	70.38
Tree	57.45	51.58	66.79	66.80

5.2. Analysis of the order of polynomial

In the proposed method, the CRF is modeled with a high-order polynomial. To recover the CRF precisely, an appropriate order should be chosen for polynomial. Here, the order n is determined by analyzing the error of the polynomial. Suppose the CRF is n-order differentiable on $[0,1]$ and (n+1)-order differentiable on $(0,1)$, then for $x \in [0,1]$, $f(x)$ an be expressed as:

$$f(x) = \frac{f(x_0)}{0!} + \frac{f'(x_0)}{1!}(x-x_0) + \frac{f''(x_0)}{2!}(x-x_0)^2 + \cdots + \frac{f^{(n)}(x_0)}{n!}(x-x_0)^n + R_n(x) \quad (12)$$

This is the famous Taylor formula. $R_n(x)$ is the error term which can be expressed by Lagrange remainder:

$$R_n(x) = f^{(n+1)}(x_0 + \theta(x - x_0))\frac{(x - x_0)^{n+1}}{(n+1)!} \tag{13}$$

Here $\theta \in [0,1]$. If $R_n(x)$ is less than an acceptable threshold, the fitted curve is accurate enough for our experiments. Here, $n = 3$ is selected for analysis. Four image is captured for CRF fitting. Supposing the sampling points are uniformly distributed, the maximum distance between any point and sampling point is 0.1. For any point $x \in [0,1]$, $f(x) < 1$. From the analysis above, we know that $R_n(x)$ is less than 0.00001. Therefore, $n = 3$ is precise enough for CRF fitting. In all our experiments, the polynomial order is set 3.

6. Conclusion

A high efficient HDR imaging method is proposed in this paper. Firstly, the CRF is modelled with a high-order polynomial. The recovery of CRF can be transformed into the solution of the polynomial coefficients. Then, HDR image can be obtained by weight summation of the relative radiance maps. After that, a local tone mapping algorithm is proposed based on the analysis of the histogram. In the result image, the local contrast of the image is preserved while compressing the dynamic range. Experimental results demonstrate that the proposed method outperforms other method obviously.

References

[1] Tocci M D, Kiser C, Tocci N and Sen P 2011 A versatile HDR video production system *ACM Trans. Graphics* **30** p 41

[2] Lin D, Huayan S, Tinghua Z and Shuai W 2016 Calibration camera response function model algorithm based on principal component analysis *Infrared and Laser Engineering* **10** pp 251-9

[3] Huo Y Q and Zhang X D 2016 Single image-based HDR imaging with CRF estimation 2016 *Int. Conf. On Communication Problem-Solving (ICCP)* IEEE pp 1-3

[4] Lin D, Huayan S, Tinghua Z and shuai W 2016 Camera Response Function Calibration Algorithm Based on Single Frame Image *Acta Optica Sinica* 07 pp 114-22

[5] Debevec P E and Malik J 2008 Recovering high dynamic range radiance maps from photographs *in ACM SIGGRAPH 2008 Classes* ACM p 31

[6] Robertson M A, Borman S, and Stevenson R L 1999 Dynamic range improvement through multiple exposures *in Proc. Int. Conf. on Image Processing (ICIP'99)* IEEE **3** pp 159–63

[7] Grossberg M D and Nayar S K 2003 Determining the camera response from images: what is knowable? *IEEE Trans. Pattern Anal. Mach. Intell.* **25(11)** pp 1455–67

[8] Lu P Y, Huang T H, Wu M S, Cheng Y T and Chuang Y Y 2009 High dynamic range image reconstruction from hand-held cameras *IEEE Conf. on Computer Vision and Pattern Recognition* IEEE pp 509–516

[9] Mitsunaga T and Nayar S K 1999 Radiometric self calibration *IEEE Computer Society Conf. on Computer Vision and Pattern Recognition* IEEE **1** pp 374-80

[10] Reinhard E, Stark M, Shirley P and Ferwerda J 2002 Photographic tone reproduction for digital images *ACM Trans. Graphics* **21(3)** pp 267–76

[11] Shan Q, Jia J and Brown M S 2010 Globally optimized linear windowed tone mapping *IEEE Trans. Visual Comput. Graphics* **16(4)** pp 663–75

[12] Kim B K, Park R H and Chang S K 2016 Tone mapping with contrast preservation and lightness correction in high dynamic range imaging *Signal, Image and Video Processing* **10(8)** pp 1425-32

[13] Larson G W, Rushmeier H and Piatko C 1997 A visibility matching tone reproduction operator for high dynamic range scenes *IEEE Trans. Visual Comput. Graphics* **3(4)** pp 291–306

[14] Yoshida A, Blanz V, Myszkowski K and Seidel H P 2005 Perceptual evaluation of tone mapping operators with real-world scenes *Int. Electronic Imaging* Society for Optics and Photonics pp 192-203

[15] ARBMAN Z, FATTAL R, LISCHINSKI D and Szeliski R 2008 Edge-preserving decompositions for multi-scale tone and detail manipulation *ACM Transactions on Graphics* **27(3)** pp 1-10
[16] Sen P, Kalantari N K, Yaesoubi M, Darabi S, Goldman D B, and Shechtman E 2012 Robust patch-based HDR reconstruction of dynamic scenes *ACM Trans. Graph.* **31(6)** p 203
[17] Kimura D, Kondo A, Yoshida T and Iwahashi M 2016 High dynamic range imaging based on camera response estimation and multi-gradients fusion 2016 *Signal and Information Processing Association Annual Summit and Conference (APSIPA)* Asia-Pacific pp 1-4
[18] R. Mantiuk, K. J. Kim, A. G. Rempel, and W. Heidrich, "HDR-VDP-2: A calibrated visual metric for visibility and quality predictions in all luminance conditions," *ACM Trans. Graph.*, vol. 30, no. 4, pp. 40:1–40:14, 2011.

IOP Conf. Series: Journal of Physics: Conf. Series **844** (2017) 012044 doi:10.1088/1742-6596/844/1/012044

Violent Interaction Detection in Video Based on Deep Learning

Peipei Zhou[1,2,3,4], Qinghai Ding[1,5], Haibo Luo[1,3,4], and Xinglin Hou[1,2,3,4]

[1] Shenyang Institute of Automation, Chinese Academy of Sciences, Shenyang, 110016
[2] University of Chinese Academy of Sciences, Beijing, 100049
[3] Key Laboratory of Opto-Electronic Information Processing, CAS, Shenyang 110016
[4] The Key Lab of Image Understanding and Computer Vision, Liaoning Province, Shenyang, 110016
[5] Space star technology co., LTD, Beijing, 100086

Abstract. Violent interaction detection is of vital importance in some video surveillance scenarios like railway stations, prisons or psychiatric centres. Existing vision-based methods are mainly based on hand-crafted features such as statistic features between motion regions, leading to a poor adaptability to another dataset. Enlightened by the development of convolutional networks on common activity recognition, we construct a FightNet to represent the complicated visual violence interaction. In this paper, a new input modality, image acceleration field is proposed to better extract the motion attributes. Firstly, each video is framed as RGB images. Secondly, optical flow field is computed using the consecutive frames and acceleration field is obtained according to the optical flow field. Thirdly, the FightNet is trained with three kinds of input modalities, i.e., RGB images for spatial networks, optical flow images and acceleration images for temporal networks. By fusing results from different inputs, we conclude whether a video tells a violent event or not. To provide researchers a common ground for comparison, we have collected a violent interaction dataset (VID), containing 2314 videos with 1077 fight ones and 1237 no-fight ones. By comparison with other algorithms, experimental results demonstrate that the proposed model for violent interaction detection shows higher accuracy and better robustness.

1. Introduction

Violent interaction (fight) behaviour makes serious effect on social and personal security and the early warning of violent activity could greatly reduce these risks. Currently, there are millions of video surveillance equipment applied in public places, such as streets, squares and railway stations. It is of vital importance to investigate the harmful abnormal contents from vast amounts of surveillance video data. For the consideration of practical application, this work focuses on the challenging task of detecting violent interaction in videos and aims to propose a new method that could automatically detect violent behaviours using computer vision techniques.

Despite the potential significance, enough attention has not been paid on violent interaction detection compared with common action recognition which have made a great progress in recent years. In the field of action recognition and also the violence detection, there are two crucial and complementary aspects: appearances and dynamics. However, extracting such information suffers from a number of complexities, including illumination variations, view point changes, camera motions and so on. Different kinds of descriptors such as GMOF and OHOF[1] or the statistics of motion regions [2,3] are extracted as the feature of a video. By feeding the features into classifiers (e.g. SVM), a model of violence detection is constructed. However, these features mainly represent the motion information estimated by optical flow images, leading to lack of appearance information. We aim to find a method that could catch both appearances and dynamics information to improve the violence detection performance.

Recently, Convolutional Networks (ConvNets) [4] have witnessed great success in classifying images of objects and video-based action recognition [5,6]. Great modelling capacity of deep ConvNets makes it possible to learn discriminative representation from raw visual data under the condition of large-scale supervised datasets. Violent interaction detection has a lot in common with general activity recognition but it differs from the common behaviours because it is random combination of multi-kinds of actions including kick, beat, push, etc. The irregular internal occlusion is the most remarkable feature and also one of the difficulties to describe the violence behaviours.

Enlightened by the design principle taken by the ConvNets on activity recognition, we make an attempt to apply the deep ConvNets to the detection of video-based violent interaction, which is impeded by two major obstacles. First, long-range temporal structure plays an important role in understanding the action dynamics in videos. We take the state-of-the-art activity network, Temporal Segment Network (TSN) [3] as the basis network for violent interaction detection, which is capable of modeling long-range temporal structure over the whole video. Based on TSN, we construct a network named FightNet fit for violent interaction detection. Second, in practice, for the sake of achieving optimal performance, a large volume of training samples are required to train the deep ConvNets. To solve this problem, we gather videos of violent interaction from datasets in the previous work [7,8,9], named Violent Interaction Dataset (VID). In addition, to prevent over-fitting of the deep ConvNets, we pre-train the model on the UCF101[8] dataset.

2. Related work

A number of researchers like Lam et al. [10] have made great effort to detect fight, violence or aggression. Some previous work [11,12] took the blood or explosion as the cues of violence, but these cues always miss alarm. Recently, Derbas et al. [13] proposed a feature which provides strong multi-modal audio and visual cues by first joining the audio and the visual features and then revealing statistically the joint multi-modal patterns. In [14], multiple kernel learning is applied to maximize the multimodality of videos based on the combined features of visual and audio. However, audio-based methods are always restricted in the real life since the absence of the audio channel.

Violent interaction detection is essentially a task of action recognition. Using computer vision technologies, the key point is to extract features that could represent the sequences during the fight. The features could be divided into two categories: hand-crafted features and learning features.

Hand-crafted features. We define the features designed by human as hand-crafted features. Among them, Space Time Interest Points (STIPs) [15] and Improved Dense Trajectories (iDTs) [16] are widely used in action recognition. Especially for violence detection, Deniz et al. [17] presented a novel method to detect violent sequences which uses extreme acceleration patterns as the main feature by applying the Radon transform to the power spectrum of consecutive frames. Recently, another features extracted from motion blobs between consecutive frames are proposed by [2] to distinguish fight and non-fight sequences. Similar work including [3] and [18] proposed a robust and intuitive approach based on motion statistical features from optical flow images. Zhang et al. [1] extract candidate violence regions using a Gaussian Model of Optical Flow (GMOF) and find out the fight regions through a linear SVM whose input vectors are constructed by Orientation Histogram of Optical Flow (OHOF). This kind of methods based on hand-crafted features are intuitive and efficient for a specific small-scale dataset, but in a big dataset their deficiencies are triggered, leading to low training speed, massive memory usage and inefficient execution.

Learning features. The features learned by deep learning networks are considered as learning features. With the enhancement of computer power brought by GPU and the collection of large-scale training set, action recognition based on deep learning makes a great progress. Designed by Simonyan et al. [19], two-stream ConvNets consist of spatial and temporal nets and exploit ImageNet dataset [20] for pre-training and exploit optical flow to explicitly capture motion information. Tran et al. [21] learned both appearance and motion features by exploring 3D ConvNets [22] trained on a large-scale supervised dataset. Recently, Varol et al. [23] employed the neural networks with long-range temporal structure (LTC) to learn the video and concluded that LTC-CNN modelled with increased temporal extents improved the accuracy of action recognition. However, these methods are limited by the computational cost so that the video could be no more than 120 frames. The temporal segment network [6] combined a sparse temporal sampling strategy and video-level supervision to learn valid information from the whole action video, achieving the state-the-of-art performance on the two complex datasets of HMDB51 (69.4%) and UCF101 (94.2%).

Although the technology on action recognition is relatively mature, little effort has been spent on the violent interaction detection using deep ConvNets. Referring to the excellent TSN that could efficiently pick up representative action features from videos, we construct a violent interaction detection ConvNet, named FightNet.

3. FightNet for Violent Interaction Detection

As the velocity magnitude of actions in violent interaction is not constant but variable randomly, i.e., there is always a nonzero acceleration field when the violent interaction occurs. Besides the basic attributes of action, we employ the acceleration feature to design FightNet fit for fight detection. As depicted in figure 1, the general flow chart of violent interaction detection is designed to model the whole video.

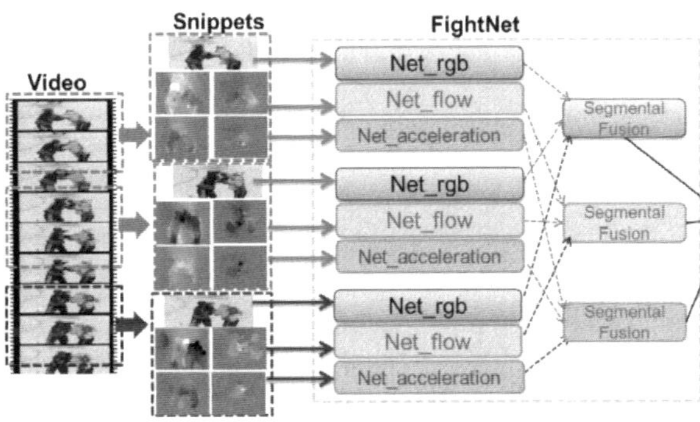

Figure 1. General flow chart of violent interaction detection, with three kinds of input modalities, i.e., RGB images for Net_rgb network, optical flow images for Net_flow network and acceleration images for Net_acceleration network.

3.1. FightNet

For a video *V*, it is divided into *K* segments $\{S_1, S_2, \cdots, S_K\}$ of equal duration and a sequence of short snippets $\{s_1, s_2, \cdots, s_K\}$ are randomly extracted in the segments. Then the *K* snippets are modelled as follows:

$$M(s_1, s_2, \cdots, s_K) = h\left(g\left(f(s_1; W), f(s_2; W), \cdots, f(s_K; W)\right)\right) \quad (1)$$

Here, each snippet s_k is randomly sampled from the corresponding segment S_k. $f(s_k; W)$ is the class score produced by FightNet with the parameter *W* operated on the snippet s_k. The fusion function g combines the outputs from different snippets. Thereon, the prediction function h provides the whole video the probability of each class, i.e., fight or non-fight. Here the widely used Softmax function is adopted to be the prediction function h. According to the standard cross-entropy loss, the loss function with respect to $G = g\left(f(s_1; W), f(s_2; W), \cdots, f(s_K; W)\right)$ is formed as equation (2):

$$L(y, G) = -\sum_{i=1}^{C} y_i \left(G_i - \log \sum_{j=1}^{C} \exp G_j \right) \quad (2)$$

where C is the number of classes, y_i is the ground truth label of class i. Here, we simplify the form of G as $G_i = g\left(f_i(s_1), f_i(s_2), \cdots, f_i(s_K)\right)$ and evenly averaging is used as the fusion function g. Depending on the fusion function g, multiple snippets are jointly used to optimize the model parameters W. In order to learn the model parameters, stochastic gradient descent (SGD) is chosen as the back-propagation algorithms and the gradients of model parameters W with respect to the loss is derived as equation (3):

$$\frac{\partial L(y,G)}{\partial W} = \frac{\partial L}{\partial G}\sum_{k=1}^{K}\frac{\partial g}{\partial f(s_k)}\frac{\partial f(s_k)}{\partial W} \tag{3}$$

Here K is the number of segments for all the videos.

3.2. Learning FightNet

As the TSN took a principle of temporal segment, the redundant frames in videos were largely reduced. Based on TSN, the FightNet makes appropriate modifications to better adapt to the detection of violent interaction.

3.2.1. FightNet Architectures. The architecture of TSN adopted the Inception of Batch Normalization (BN-Inception) [24] to make the network wider and deeper, resulting to a huge structure as deep as 34 layers. Mathematically, violent interaction detection is modeled as binary classification problem which is not so complex as the action recognition. Therefore, in terms of architectures of FightNet, the most important enhancement compared with TSN is to cut off the last inception structure which includes 6 convolution layers. This measure could largely reduce the risk of over-fitting and shorten the training time.

3.2.2. Input Data. Originally, the RGB images are applied for spatial stream and stacked optical flow fields for the temporal stream in two-stream ConvNets [19]. TSN has explored four input modalities to enhance the discriminative power, ranging from RGB images and optical flow fields to RGB difference and warped optical flow fields. Experimental results demonstrate that the former two input modalities make great contribution to the accuracy and the latter two input modalities plays little role in the improvement of accuracy. Here, we choose three modalities as the input of FightNet, i.e., RGB images, optical flow fields and acceleration fields. Considering acceleration fields as the input modality derives from the specific attributes of violent interaction that the actions are of short duration and forceful. Theoretically, acceleration fields are appropriate to learn motion features for FightNet. For two consecutive optical flow images $F(t-1)$ and $F(t)$, the difference *diff(t)* is first computed using equation (4) and the acceleration fields $Accel(t)$ is defined as equation (5):

$$diff(x,y,t) = F(x,y,t) - F(x,y,t-1) \tag{4}$$

$$Accel(x,y,t) = \begin{cases} 0, & diff(x,y,t)+128<0 \\ diff(x,y,t)+128, & 0 \le diff(x,y,t)+128 \le 255 \\ 255, & diff(x,y,t)+128>255 \end{cases} \tag{5}$$

Here, $x=\{1,2,\cdots,m\}$, $y=\{1,2,\cdots,n\}$, $t=\{1,2,\cdots,T\}$ and (x,y,t) implies the pixel position (x,y) in t-th frame. In this work, acceleration fields consist of two directions, x and y. In the acceleration image of each direction, we view the acceleration field as a general image. As shown in figure 2, RGB images are used to provide appearance information and optical flow fields and acceleration fields are the source of dynamic information.

IOP Conf. Series: Journal of Physics: Conf. Series **844** (2017) 012044 doi:10.1088/1742-6596/844/1/012044

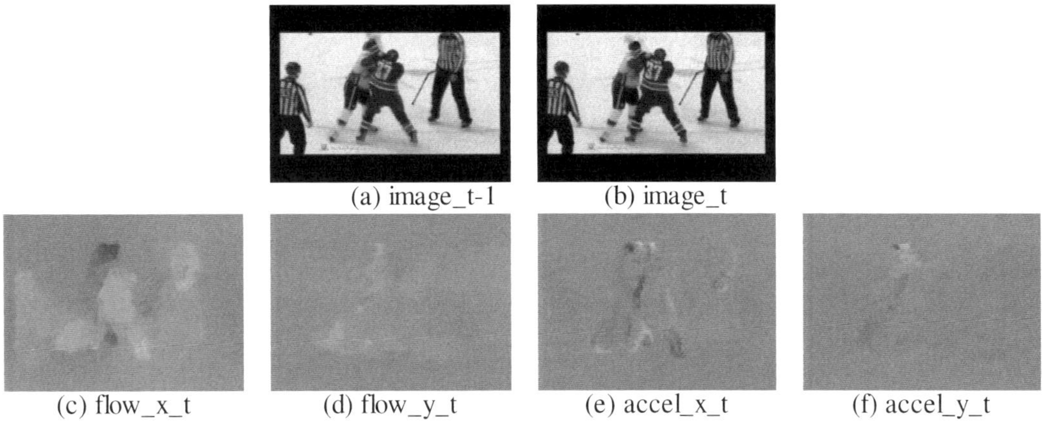

Figure 2. Examples of three input modalities, where (a) and (b) are two consecutive frames, (c) and (d) are the optical flow fields of (b) corresponding to x and y directions respectively, and (e) and (f) are the acceleration fields related to (c) and (d).

3.2.3. FightNet Training. While there exist a number of well-studied datasets for action recognition [8,9], a few significant datasets with violent actions are available. For example, the most widely used fight dataset from [7] contains only 1000 video clips which is nowhere near enough to train FightNet. Therefore, we collect a violent interaction dataset (VID) from UCF101[8], HMDB51[9] and the dataset in [7], and the total number of 2314 videos is contained, including 1077 violent interaction clips and 1237 non-violent ones. Besides, to mitigate the risk of over-fitting, we take several strategies for training FightNet as follows.
(1) Pre-training. It has been proven that pre-training is an effective way to initialize deep ConvNets when the target dataset does not have enough training samples [19]. As a deep ConvNet, FightNet takes the parameters trained on ImageNet [20] as initial values. Then RGB images extracted from UCF101 is applied to pre-train the spatial network of FightNet. As mentioned on TSN [6], we utilize RGB models to initialize the temporal networks with respect to the other two modalities, optical flow field and acceleration fields. At last, the pre-trained networks are employed on the proposed VID.
(2) Data Augmentation. In TSN, four data augmentation techniques, i.e. corner cropping, scale-jittering, random cropping and horizontal flipping are employed to prevent over-fitting. In this work, another technique is proposed, video clipping. In video clipping, we clip the long video more than 150 frames into diverse clips. For instance, a long video sample contains 200 frames. We segment it into 2 samples with the frame index 1-100, and 101-200. Using this technique, the scale of training data is augmented by about 30%.

3.3. Testing FightNet
Similar to the testing scheme of original two-stream ConvNets[19] and TSN [6], we extract 25 RGB frames and optical flow stacks or acceleration stacks from every video sample. Besides, 4 corners and 1 centre, and their horizontal flipping from sampled frames are employed to evaluate FightNet. We take a weighted average strategy to fuse the spatial and temporal networks. Larger weight is placed on the temporal stream as the motion information plays more important role on the fight than the appearance information. We set the weight as 0.25 for spatial network and 0.75 for temporal. When both optical flow fields and acceleration fields are applied, the fusion weights are 0.6 and 0.4 respectively.

4. Evaluation

4.1. Dataset
Nievas et al [7] introduced two datasets explicitly designed for assessing fight detection. As shown in figure 3-top, the first dataset ("Hockey") consists of 1000 clips, 500 fight ones and 500 non-fight ones

extracted from hockey games of National Hockey League. The second dataset ("Movies") contains 200 video samples in which fight clips were extracted from action movies and non-fight ones from public action recognition dataset, shown in figure 3-bottom. As the dataset about fight detection is too small to train deep ConvNets, we have collected a violent interaction dataset (VID), aggregating from four datasets, HMDB51, UCF101 and all the video samples of "Hockey" and "Movies". The proposed dataset contains 2314 videos, including 1077 fight ones and 1237 no-fight ones. In HMDB51 dataset, 206 violent interaction videos are picked out from "hit", "kick" and "punch", which show the fight between two persons. In UCF101 dataset, 271 fight videos from "Punch" and "SumoWrestling" are selected. The similar videos like "BoxingSpeedBag" and "BoxingPunchingBag" from UCF101 are considered as the non-fight samples. Examples of frames extracted from the proposed dataset [25] are shown in figure 4. It is obvious that this dataset is challenging involving illumination variations, view point changes and camera motions.

(a) (b) (c) (d)

Figure 3. Examples dataset of [7], where (a) and (b) are fight and non-fight images from "Hockey" dataset and (c) and (d) from "Movies".

Figure 4. Frame examples in violent interaction dataset (VID), with images in red boxes selected from violent interaction (positive) videos and those in green boxes from negative samples.

4.2. Implication Details

For the sake of the extraction of optical flow fields and acceleration fields, we employed the TVL1 optical flow algorithm [26] to finish this task implemented on OpenCV with CUDA. The same as TSN, FightNet is speeded up through a data-parallel strategy with a modified version of Caffe [27] and OpenMPI.

As mentioned in Section 3.2, we firstly construct the architecture of FightNet by cutting off the redundant layers of TSN network. Secondly, FightNet is pre-trained on UCF101 dataset and the pre-trained parameters are considered as the initialization values of FightNet. Thirdly, because we predict 2 classes in VID instead of 101 classes in UCF101, we change the "*InnerProduct*" layer in the model and change the name from *fc_action* to *fc_action_fight* in the prototxt file. Since there is no layer named that, this layer will begin training with random weights. In the solver prototxt, we decrease the *base_lr* to 0.0001 for spatial networks and 0.0005 for temporal networks, but boost the *lr_mult* on the *fc_action_fight* layer. This strategy is to have the rest layers change slowly with the new dataset, but let the *fc_action_fight* layer learn fast. Additionally, *test_iter* is adjusted in terms of the size of test data. In the dataset VID, around 20% samples are used as testing data and the others as training data. The training procedure stops at 2000 iterations for spatial networks with RGB images as input and 6000 iterations for temporal networks with optical flow and acceleration fields as input. With 4 TITANX GPUs, the whole training time on VID is around 15 minutes for spatial TSNs and about 1.5 hours for temporal TSNs.

4.3. Experimental Results and Evaluation

In this subsection, we focus on the study of the FightNet framework and the comparison with the state of the art. As the effect of fusion functions and ConvNet architectures have been studied in TSN [6], we quote the conclusion of TSN directly. The average pooling is chosen as the default fusion function and we consider BN-Inception [24] as the ConvNet architecture in accordance with its nice performance in the image classification task. In this work, we adopt the traditional definition of accuracy, as formed in equation (6).

$$Accuracy = \frac{TP + TN}{Total} \times 100\% \tag{6}$$

where *TP* and *TN* denote True Positives and True Negatives, *Total* being the total number of testing samples.

With all the design choice prepared for FightNet, different combinations of three input modalities are compared on the VID. As reported in table 1, the networks with three different input modalities achieve accuracy greater than 94% while the spatial network taking RGB images as input yields the highest accuracy to 96.81%. We guess that the RGB images provide not only appearance information but also motion features to better detect violent interaction. Then we observe that the combination of any two modalities boosts the performance compared to that of one modality. The fusion of RGB images and acceleration fields could improve the performance to 97.05% which is approximately equal to the combination of all three modalities (97.06%). We conjecture that the reason of high accuracy from the combination of RGB images and acceleration field is the comprehensive effect of the texture information from the former and motion information from the latter. It is also concluded that the proposed input modality, acceleration field is effective and suitable for violent interaction detection in ConvNet.

Table 1. Exploration of different input modalities for FightNet on the Violent Interaction Dataset.

Modality	RGB image	Optical flow	Acceleration	RGB & optical flow	RGB & acceleration	Optical flow & acceleration	All modalities
Accuracy	96.81%	94.23%	94.69%	96.85%	97.05%	95.33%	97.06%

After exploring the performance of different input modalities for FightNet, we are ready to compare FightNet with other fight detection algorithms. Specifically, we assemble three input modalities and test it on two widely used datasets, "Hockey" and "Movies". Since the number of videos in "Hockey" and "Movies" datasets is not enough to train FightNet, the model is trained on VID and fine-tuned on the corresponding dataset. In this work, 80% samples on the dataset are randomly selected as training data and the others as testing data. In additional, we want to classify that the testing data in "Hockey" and "Movies" is involved in the testing data of VID but there is no intersection with the training data in VID. Because there is still no deep learning approach on fight detection, we compare our method with some traditional representations, as shown in Table 2. In [7], by exploring STIP features and MoSIFT features on "Movies" and "Hockey" datasets, MoSIFT was considered as the best descriptor and the superiority of MoSIFT has also been proven in other applications of action recognition. In addition, violence detection using extreme acceleration features [17] and fight detection based on motion analysis [18] are employed to make comparison with the performance of FightNet.

Table 2. Comparison results on the "Hockey" and "Movies" datasets, the values in this table denoting the accuracy on the corresponding dataset.

Features	Classifier	Dataset	
		Movies	Hockey
BoW (STIP) [7]	SVM	82.5%	88.6%
	AdaBoost	74.3%	86.5%
BoW (MoSIFT) [7]	SVM	84.2%	91.2%
	AdaBoost	86.5%	89.5%
Extreme Acceleration [17]	SVM	85.4%	90.1%
	AdaBoost	98.9%	90.1%
Motion Analysis [18]	Random Forest	98.5%	84.5%
FightNet	Softmax	**100%**	**97.0%**

As shown in table 2, the previous performance on "Movies" dataset is about 98% in [17] and [18], but on "Hockey", motion analysis [18] yielded a lower accuracy by around 6% than extreme acceleration features [17]. The feature of MoSIFT achieved nice performance to more than 91% on the "Hockey" dataset, but it is not well on the "Movies" dataset. It is to say that these features are not robust enough to violence detection with different disturbance. However, the robustness is of vital importance for violent interaction detection. The performance of FightNet on "Movies" dataset reaches to 100% partly because the non-fight videos on this dataset are quite different from the fight ones. On the contrary, positive and negative data on "Hockey" are roughly similar which are all collected from the hockey sport, leading to the lower accuracy (97%) than that on "Movies". In terms of the performance of FightNet, it is concluded that deep ConvNets could capture more essential features and detect the violent interaction correctly with illumination variations, view point changes or camera motions.

5. Conclusion

In this work, we constructed a ConvNet named FightNet to model long-term temporal structure for violent interaction detection. By pre-training on UCF101 dataset of action recognition, the FightNet works well on the proposed violent interaction dataset (VID), which is collected from four public datasets, "Hockey", "Movies", HMDB51 and UCF101. A new modality, acceleration field is explored to capture the motion features and the contribution on accuracy is about 0.3%. Compared with other methods on fight detection, the FightNet achieves a higher accuracy with a reasonable computational cost. With the increasing scale of dataset, it is of vital importance to further explore efficient ConvNers to detect violent interaction in complex real-life scenarios in the future.

References

[1] Zhang T, Yang Z, Jia W, Yang B, Yang J and He X 2016 A new method for violence detection in surveillance scenes *Multimedia Tools and Applications* **75(12)** pp 7327-49

[2] Gracia I S, Suarez O D, Garcia G B and Kim T K 2015 Fast fight detection *PloS one* **10(4)** p e0120448

[3] Fu E Y, Va Leong H, Ngai G and Chan, S 2016 Automatic Fight Detection in Surveillance Videos *Proc. of the 14th Int. Conf. on Advances in Mobile Computing and Multi Media* ACM pp 225-34

[4] LeCun Y, Bottou L, Bengio Y and Haffner, P 1998 Gradient-based learning applied to document recognition *Proc. of the IEEE* **86(11)** pp 2278-324
[5] Zhang B, Wang L, Wang Z, Qiao Y and Wang H 2016 Real-time action recognition with enhanced motion vector CNNs *Proc. of the IEEE Conf. on Computer Vision and Pattern Recognition* IEEE pp 2718-26
[6] Wang L, Xiong Y, Wang Z, Qiao Y, Lin D, Tang X and Van Gool L 2016 Temporal segment networks: towards good practices for deep action recognition *In European Conf. on Computer Vision* Springer International Publishing pp 20-36
[7] Nievas E B, Suarez O D, García G B and Sukthankar R 2011 Violence detection in video using computer vision techniques *Int. Conf. on Computer Analysis of Images and Patterns* Springer Berlin Heidelberg pp 332-9
[8] Soomro K, Zamir A R and Shah M 2012 UCF101: A dataset of 101 human actions classes from videos in the wild *arXiv preprint arXiv* 1212.0402
[9] Kuehne H, Jhuang H, Garrote E, Poggio T and Serre T 2011 HMDB: a large video database for human motion recognition *2011 IEEE Int. Conf. on Computer Vision* IEEE pp 2556-63
[10] Lam V, Phan S, Le D D, Duong D A and Satoh S I 2016 Evaluation of multiple features for violent scenes detection *Multimedia Tools and Applications* pp 1-25
[11] Clarin C, Dionisio J, Echavez M and Naval PC 2005 Dove: Detection of movie violence using motion intensity analysis on skin and blood *ECCV* pp 150–6
[12] Liang-Hua C, Hsu HW, Wang LY and Su CW 2011 Violence detection in movies *Computer Graphics Imaging and Visualization (CGIV)* pp 119–24
[13] Derbas N and Quénot G 2014 Joint audio-visual words for violent scenes detection in movies *Proc. of Int. Conf. on Multimedia Retrieval* ACM p 483
[14] Goto S and Aoki T 2014 Violent scenes detection using mid-level violence clustering *Computer Science CSCP* pp 283-96
[15] Laptev I 2005 On space-time interest points *Int. journal of computer vision* **64(2-3)** pp 107-23
[16] Wang H, Kläser A, Schmid C and Liu C L 2013 Dense trajectories and motion boundary descriptors for action recognition *Int. journal of computer vision* **103(1)** pp 60-79
[17] Deniz O, Serrano I, Bueno G and Kim T K 2014 Fast violence detection in video 2014 *Int. Conf. on Computer Vision Theory and Applications (VISAPP)* IEEE **2** pp 478-85
[18] Fu E Y, Leong H V, Ngai G and Chan S 2015 Automatic Fight Detection Based on Motion Analysis *2015 IEEE Int. Symp. on Multimedia (ISM)* IEEE pp 57-60
[19] Simonyan K and Zisserman A 2014 Two-stream convolutional networks for action recognition in videos *Advances in neural information processing systems* pp 568-76
[20] Deng J, Dong W, Socher R, Li L J, Li K and Fei-Fei L 2009 Imagenet: A large-scale hierarchical image database *IEEE Conf. on Computer Vision and Pattern Recognition* IEEE pp 248-55
[21] Tran D, Bourdev L D, Fergus R, Torresani L and Paluri M 2015 Learning spatiotemporal features with 3d convolutional networks *ICCV* pp 4489-97
[22] Ji S, Xu W, Yang M and Yu K 2013 3D convolutional neural networks for human action recognition *IEEE Trans. on Pattern Anal. And Mach. Intell.* **35(1)** pp 221-31
[23] Varol G, Laptev I and Schmid C 2016 Long-term temporal convolutions for action recognition *arXiv preprint arXiv* p 1604.04494
[24] Ioffe S and Szegedy C 2015 Batch normalization: Accelerating deep network training by reducing internal covariate shift *arXiv preprint arXiv* p 1502.03167
[25] Link: *https://pan.baidu.com/s/1o8e5Wjo* Password: a66b
[26] Zach C, Pock T and Bischof H 2007 A duality based approach for realtime TV-L 1 optical flow *Joint Pattern Recognition Symp.* Springer Berlin Heidelberg pp 214-23.
[27] Jia Y, Shelhamer E, Donahue J, Karayev S, Long J, Girshick R, Guadarrama S and Darrell T 2014 Caffe: Convolutional architecture for fast feature embedding *ACM Multimedia* pp 675-8

IOP Conf. Series: Journal of Physics: Conf. Series **844** (2017) 012045 doi:10.1088/1742-6596/844/1/012045

Dehazed Image Quality Assessment by Haze-Line Theory

Yingchao Song[1,2,3,4], Haibo Luo[1,3,4], Rongrong Lu[1,2,3,4], Junkai Ma[1,2,3,4]

[1] Shenyang Institute of Automation, CAS, Shenyang, 110016, China
[2] University of Chinese Academy of Sciences, Beijing, 100049, China
[3] Key Laboratory of Opto-Electronic Information Processing, CAS, Shenyang, 110016, China
[4] The Key Lab of Image Understanding and Computer Vision, Liaoning Province, Shenyang, 110016, China

Abstract. Images captured in bad weather suffer from low contrast and faint color. Recently, plenty of dehazing algorithms have been proposed to enhance visibility and restore color. However, there is a lack of evaluation metrics to assess the performance of these algorithms or rate them. In this paper, an indicator of contrast enhancement is proposed basing on the newly proposed haze-line theory. The theory assumes that colors of a haze-free image are well approximated by a few hundred distinct colors, which form tight clusters in RGB space. The presence of haze makes each color cluster forms a line, which is named haze-line. By using these haze-lines, we assess performance of dehazing algorithms designed to enhance the contrast by measuring the inter-cluster deviations between different colors of dehazed image. Experimental results demonstrated that the proposed Color Contrast (CC) index correlates well with human judgments of image contrast taken in a subjective test on various scene of dehazed images and performs better than state-of-the-art metrics.

1. Introduction

Image dehazing has become a necessary pre-processing step in computer vision application. Recently, plenty of excellent dehazing algorithms have been proposed to restore degraded images, such as [1-18]. However, very few quality metrics were proposed to evaluate dehazing algorithms. Although image quality assessment has been extensively studied in the past few decades, traditional full-reference image quality indexes, such as the mean square error (MSE), the peak signal to niose ratio (PSNR), and the structural similarity index[20] (SSIM) are not appropriate for quality assessment of dehazed image as it is difficult to get a clear day reference image.

In this paper, an indicator of contrast enhancement is proposed basing on the newly proposed haze-line theory[1]. The theory assumes that pixels with similar colors in the haze-free image form a line in the corresponding hazy image. Finding out these haze-lines, we can cluster pixels of the dehazed image and assume pixels of each cluster have similar colors. Then we measure the inter-cluster deviation on each local image patch. A higher deviation implies a higher color contrast in the corresponding image patch.

To evaluate the performance of our proposed metric, a human subjective test was performed using a dataset of 50 hazy images and 150 corresponding dehazed images. As there is no available benchmark dataset to support our test, we assembled the dataset by ourselves. Fig.1 shows some examples of hazy image in the dataset. We collect these images from several professors' homepage[1-5,10-11,15] whom have proposed effective dehazing algorithms. Experimental results demonstrate that the proposed Color Contrast index (CC) correlates well with human judgments of image contrast on this images dataset.

Figure 1. Examples of hazy image in the dataset

2. Related works

2.1. Descriptor of visible edge

Among the few methods of dehazed image assessment, the most widely used is the indicator of visibility enhancement proposed by Hautiere et al[21]. The method includes two contrast descriptors namely the rate of new visible edges e and the gain of visibility level $\bar{r}$. The value of e evaluates the ability of dehazing method to restore edges which were not visible in hazy image but are in dehazed image. The value of $\bar{r}$ denotes the mean ratio of gradient norms before and after dehazing. They are defined as follow:

$$e = \frac{n_r - n_o}{n_o} \tag{1}$$

$$\bar{r} = \exp[\frac{1}{n_r}\sum_{P_i \in \wp_r} \log r_i] \tag{2}$$

Here, n_r and n_o denote respectively the numbers of visible edges in the dehazed image and in the original image. $\wp_r$ is the set of visible edge in the dehazed image. P_i denotes the pixels that belong to $\wp_r$. r_i is the ratio of the gradient at P_i in the dehazed image and in the original image.

The two descriptors have been applied by many dehazing methods[2,6,11] to evaluate their performance on visibility enhancement. However, the method obtains visible edge and solves gradient value by gray-scale images. It is not a good idea for color images, as the method of finding visible edges and solving gradient values is different from the one in gray-scale image. In addition, the two indicators often produce inconsistent evaluation results when assessing the same image.

2.2. Fog aware density evaluator

Fog Aware Density Evaluator[2] (FADE) is a newly proposed contrast descriptor which predicts the visibility of a foggy scene by measuring deviations from statistical regularities in natural scene foggy and fog-free images. A lower value of FADE implies better performance of visibility enhancement. The method extracts 12 fog-related features from the test image, and fits these features to a Multivariate Gaussian (MVG) model. Then, it computes the deviation using a Mahalanobis-like distance between the MVG fit of the test image and the MVG fit of a corpus of 500 fog-free images and another corpus of 500 foggy images, respectively. Each corresponding distance is defined as a foggy level D_f and a fog-free level D_{ff}. Finally, the perceptual fog density D of the test image is expressed as the ratio of the foggy level to the fog-free level.

$$D = \frac{D_f}{D_{ff} + 1} \tag{3}$$

The predicted fog density using FADE correlates well with human judgments of fog density taken in subjective experiments on a foggy image database[2]. However, sometimes a lower fog density is not exactly equivalent to a higher contrast.

2.3. CNC measurement system

The CNC measurement system (Contrast-Naturalness-Colorfulness), proposed by Guo et al[22], evaluates the dehazing effect from three aspects: contrast *e'*, Color Naturalness Index (CNI)[23] and Colorfulness Index (CCI)[23]. It is defined as:

$$CNC = e'^{\frac{1}{n_1}} CNI + CCI^{\frac{1}{n2}} CNI \tag{4}$$

Where, n_1 and n_2 are the adjustable parameters, $e'=n_r/n_o$ is a contrast descriptor, n_r and n_o denote respectively the numbers of visible edges[21] in the dehazed image and in the original image. The CCI index reflects the degree of color vividness, which is defined as the summation of average saturation and standard deviation of saturation of the test image. The CNI[23] index reflects the degree of color naturalness. It classify pixels of test image into three classes ('skin' 'grass' and 'sky') according to their hue, then measuring the differences between the average saturation and the prototypical saturation of each class through a Gaussian function.

The CNC index claims that it correlates well with human visual perception of the overall dehazing effect. However, colors of natural scene are rich and colorful, and will vary with time and light condition. It is impossible to use just three prototypical colors to represent the whole natural colors.

3. The proposed evaluation metric

3.1. Back ground

The widely used atmospheric scattering model can be expressed as[18]:

$$\mathbf{I}(x) = \mathbf{J}(x) \cdot t(x) + \mathbf{A} \cdot (1 - t(x)) \tag{5}$$

$$t(x) = e^{-\beta d(x)} \tag{6}$$

Where x is the pixel coordinates, $\mathbf{I}$ denotes the hazy image, $\mathbf{J}$ denotes the haze-free image, $\mathbf{A}$ denotes the airlight, and t is the medium transmission which denotes the proportion of scene radiance captured by camera and without being scattered. d is the scene depth, β is the scattering coefficient of atmospheric medium.

3.2. Haze-line theory

The haze-line theory[1] bases on the observation that a haze-free image can be expressed by a small number of distinct colors, which form tight clusters in RGB space. Each cluster has similar colors. The presence of haze make pixels that belong to the same color cluster end up with different acquired colors, since they have different scene depth and scattered by atmospheric medium in different degree. The color coordinates of these pixels are distributed along a line spanned by the original color $\mathbf{J}$ and the airlight $\mathbf{A}$. These are the haze-lines.

By transforming Eq.(5) to the linear form of Eq.(7), and express $\mathbf{I}_A(x)$ in spherical coordinates as Eq.(8), we can see from Eq.(9) that for given $\mathbf{J}$ and $\mathbf{A}$, changes in t affects only r without changing either θ or φ. Therefore, pixels have similar RGB values of $\mathbf{J}$ if their values of $[\theta,\varphi]$ are similar. Clustering $[\theta,\varphi]$ in the spherical coordinate system, we can get haze-lines.

$$\mathbf{I}_A(x) = \mathbf{I}(x) - \mathbf{A} = (\mathbf{J}(x) - \mathbf{A}) \cdot t(x) \tag{7}$$

$$\mathbf{I}_A(x) = [r(x), \theta(x), \varphi(x)] \tag{8}$$

$$r(x) = \|\mathbf{I}_A(x)\| = \|\mathbf{J}(x) - \mathbf{A}\| \cdot t(x) \tag{9}$$

3.3. The proposed Color Contrast index (CC)

The underlying principle of haze-line theory is that a haze-line in the hazy image respects a color cluster in the corresponding haze-free image, and pixels belong to the same color cluster have similar colors.

Our proposed method based on the observation that deviations of different colors in a local area can reflect the local contrast of color image. The larger the color deviation of different colors in the haze-free image, the higher the contrast. The proposed Color Contrast index (CC) is composed of four essential steps: 1) estimating the airlight, 2) clustering the pixels into haze-lines, 3) computing inter-cluster deviations in each local patch, and 4) computing the final CC of the whole image.

3.3.1. Estimating the airlight. Method [1] assumes that **A** is given by the method of airlight estimation illustrated in [19]. In fact, there are many approaches of airlight estimation used in field of single image haze removal[1-19]. Each has its own advantages and disadvantages, and maybe suitable for some particular scenes. In this section, for the sake of simplicity and generality, we introduce the widely used method proposed by He et al[10]. The method uses the dark channel to detect the most haze-opaque region, then picks up the top 0.1% brightest pixels of the most haze-opaque region and chooses the single pixel with the highest intensity in the hazy image as the estimation of **A**. He's method can always get a value of **A** with higher brightness and lower saturation, which will not produce large color distortion in the dehazed image. Fig.2 gives some examples of airlight estimation using He's[10] method. The red solid points in the top of images are the estimation points of **A**.

Figure 2. Airlight estimation using He's[10] method

3.3.2. Clustering the pixels into haze-lines We use method illustrated in literature [1] to cluster pixels into haze-lines. The method expresses physical model of hazy image in spherical coordinates as Eq.(8), then samples the unit sphere uniformly using 1000 points. Each point corresponds to a haze-line, and then groups the pixels of hazy image according to the closest sample point based on their values of $[\theta,\varphi]$.

3.3.3. Computing the inter-cluster deviations locally After obtain haze-lines, pixels of the haze-free image can be clustered. Then, we compute the inter-cluster deviations of haze-free image in each local patch. In literature [20], the standard deviation (the square root of variance) was used to measure image contrast. Here, we use the inter-cluster standard deviation in each local square window (patch p) to measure the local color contrast (CC_p) of haze-free image.

The CC_p is defined as:

$$\overline{\mathbf{J}_{pk}} = \frac{1}{N_{pk}} \sum_{x\in\wp_{pk}} \mathbf{J}(x) \tag{10}$$

$$\mathbf{J}_p = \frac{1}{N_k} \sum_{k=1}^{N_k} \overline{\mathbf{J}}_{pk} \tag{11}$$

$$CC_p = \sqrt{\frac{1}{N_k - 1} \sum_{k=1}^{N_k} (\overline{\mathbf{J}}_{pk} - \mathbf{J}_p)^2} \tag{12}$$

Where, k is the index of color cluster, p is the index of image patch, $\mathbf{J}(x)$ is the color value of haze-free image in coordinates x, $\wp_{pk}$ is a set of pixels which belong to the k-th color cluster in patch p, N_{pk} is the number of pixels in $\wp_{pk}$, $\overline{\mathbf{J}}_{pk}$ is the mean value of pixels in $\wp_{pk}$. N_k is the number of clusters in patch p,

IOP Conf. Series: Journal of Physics: Conf. Series **844** (2017) 012045 doi:10.1088/1742-6596/844/1/012045

$\mathbf{J}_p$ is the mean value of $\bar{\mathbf{J}}_{pk}$, CC_p is the unbiased estimate of inter-cluster standard deviation in patch p which denotes the Color Contrast of patch p.

3.3.4. Computing the overall Color Contrast (CC) Finally, we use the mean value of CC_p to measure the Color Contrast (CC) of the whole haze-free image:

$$CC = \frac{\frac{1}{N_p}\sum_{p=1}^{N_p} CC_p}{\sqrt{\frac{1}{H \cdot W - 1}\sum_{i=1}^{H}\sum_{j=1}^{W}(\mathbf{I}_{i,j} - \bar{\mathbf{I}})}} \tag{13}$$

Where, N_p is the number of patches in the whole haze-free image. $[H, W]$ denote the dimensions of the image. $\mathbf{I}_{i,j}$ is the color value of the hazy image in coordinates(i,j). $\bar{\mathbf{I}}$ is the mean color of the hazy image $\mathbf{I}$. The denominator of Eq.(13) is the standard deviation of the original hazy image, it used to normalize the value of CC.

4. Experiment and analysis

Since there is no feasible way to compare the performance of our proposed method with other contrast indexes, we evaluated the performance of our methods against the results of a human subjective test.

4.1. Human subjective test setup

4.1.1. Test images. As far as we know, there is no available benchmark dataset to support task of dehazed image assessment. In order to evaluate the effects of our method, we assembled a dataset by ourselves. The dataset contains 50 widely used hazy images and 150 corresponding dehazed results which produced by famous dehazing algorithms of [1-15]. We pick up these images from several professors' homepage[1-5,10-11,15], whom have proposed effective dehazing algorithms. Some examples of hazy image in the dataset are shown in Fig.1.

These test images form 50 groups, each with 1 hazy image and 3 corresponding dehazed results. If a hazy image has more than 3 dehazed results, we will split them into multiple groups. Similarly, if a hazy image has less than 3 dehazed results, we will use methods of [11] and [13] to produce dehazed images so that each group has 3 dehazed results.

4.1.2. Subjects. A total of 40 subjects attended the subjective experiments. They are students or staff in Shenyang Institute of Automation, Chinese Academy of Sciences. About half of them are major in image processing or related, and the others are naive. All subjects were between the ages of 20 and 40. No vision test was performed although a verbal confirmation of soundness of (corrected) vision was obtained from the subjects.

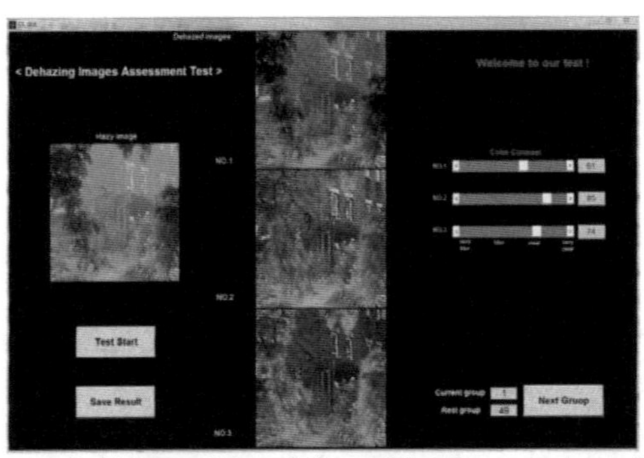

Figure 3. Screenshot of the subjective test interface

4.1.3. Test approach. We developed the user interface for the test on a Windows PC using the GUI tool of MATLAB. Fig.3 shows the subjective test interface. The scoring term, with 4 markings 'very blur', 'blur', 'clear', and 'very clear' to indicate the contrast of dehazed images, is the Color Contrast

(CC) index. Subjects were requested to score dehazed images of each group using the continuous slider bar. The subjective judgments were mapped to the integer interval [0,100] displaying on the right side of the slider bar. Operating instructions are interpreted before each subject starting the test.
4.1.4. Processing of subjective scores. All test data were used to form a Mean Opinion Scores[2] (MOS) for each image. The MOS is defined as:

$$MOS_j = \frac{1}{N_j}\sum_i S_{ij} \tag{14}$$

Here, S_{ij} denotes the score assigned by subject i to the test image j and N_j is the total number of scores received for test image j.

4.2. Performance in subjective test

We computed the Pearson's linear correlation coefficient (PLCC) and the Spearman's rank ordered correlation coefficient (SROCC) between evaluation results of our proposed CC index and values of MOS on each group of the test images. Then, we computed the mean value of PLCC and SROCC on the whole dataset of 50 groups test image. Table 1 tabulates the mean value of PLCC and SROCC over diverse patch sizes ranging from 16×16 to 128×128 pixels on the dataset. The results indicate that the best performing patch size for assessing contrast enhancement was 64×64.

Table 1. LCC and SROCC between CC scores and the MOS over different patch sizes.

Patch Size	16×16	32×32	48×48	64×64	96×96	128×128
PLCC	0.8039	0.8022	0.8351	**0.8462**	0.8188	0.8092
SROCC	0.8241	0.8241	0.8519	**0.8519**	0.8426	0.8056

4.3. Results of dehazed image assessment

In this section, we executed a quantitative evaluation of dehazing algorithms using the contrast enhancement indexes of Hautière et al[21], the perceptual fog density FADE[2], the CNC index and our proposed Color Contrast index(CC). Fig.4 and Fig.5 demonstrate several widely used hazy images and the corresponding dehazed results produced by dehazing algorithms of [1-15]. Objective evaluation results are shown in Tab2-Tab5, we use bold format to indicate the best result and underline format to indicate the worst result of each metric. Here, we compute CC value using a patch size of 64×64. Among these metrics, a lower value of FADE denotes a good job of contrast enhancement, while a higher value of other metrics denote good job of haze removal.
In hazy 1, values of CC denote that Choi, Gibson and Song do a better job on contrast enhancement, while Berman and He have a worse performance. These evaluation results correlate well with our subjective feeling. The values of e denote that Gibson produces the worst result on visibility enhancement, but values of $\bar{r}$ denote that Gibson produces the best result on visibility enhancement. The evaluation results of the two indicators (e and $\bar{r}$) are inconsistent. The value of FADE denotes that level of fog residual in He, Gibson, and Kratz are same, but they looks very different on foggy level. Values of CNC index denote that Song achieve the worst performance on the overall dehazing effect. However, result of Kratz looks even worse, as it tends to be over-saturated and color distort.
In hazy 2, Berman and Fattal14 perform better on contrast enhancement, while Nishino and Tarel produce worse results. Values of e denote that Tarel achieves the best performance on visible edge enhancement, but subjectively, it looks bad in visibility. Values of the CNC index denote that Nishino perform better to Berman, Fattal, Song, and He on the overall dehazing effect, but obviously this is not the case. Values of FADE denote that result of Nishino leave less fog than result of Fattal14, however, little fog residue does not represent better visibility and higher contrast.

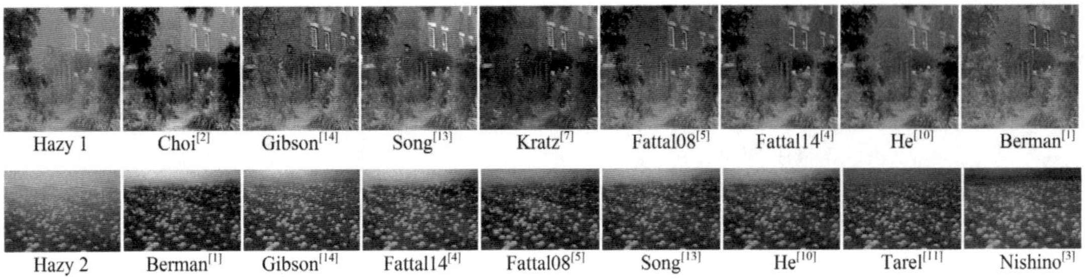

Figure 4. Dehazing results of several popular algorithms

Table 2. Objective assessment of Hazy 1

	Choi	Gibson	Song	Kratz	Fattal08	Fattal14	He	Berman
CC	**1.845**	1.758	1.671	1.483	1.314	1.236	1.127	1.088
e	0.06	0.04	0.09	0.12	0.09	**0.14**	0.11	0.11
$\overline{r}$	1.70	**2.21**	1.45	1.37	1.58	1.48	1.23	1.41
FADE	**0.13**	0.15	0.18	0.15	0.14	0.14	0.15	0.18
CNC	1.02	1.13	0.92	1.05	1.17	**1.54**	1.21	1.25

Table 3. Objective assessment of Hazy 2

	Berman	Gibson	Fattal14	Fattal08	Song	He	Tarel	Nishino
CC	**1.812**	1.685	1.650	1.596	1.561	1.512	1.471	1.353
e	0.24	0.37	0.18	0.14	0.26	0.29	**0.78**	0.38
$\overline{r}$	1.98	**3.43**	1.90	1.74	1.57	1.62	1.93	2.48
FADE	0.19	**0.16**	0.25	0.22	0.27	0.24	0.25	0.20
CNC	2.15	**2.25**	2.14	2.14	2.13	2.14	2.10	2.17

In hazy 3, values of CC, $\overline{r}$, and FADE denote that Tan get the best result on contrast enhancement, and result of e denotes that Tarel achieves the best result on visible edge enhancement. However, values of FADE denote that result of Fattal08 have little fog residue than He's result, but subjectively the reverse is true. Values of e denote that Fatta14 gets the worst result of visible edge enhancement, but subjectively, we think that result of Fattal14 is more clearly than results of Kopf and Fattal08.

In hazy 4, results of CC and FADE denote that Berman achieves the best performance and Tarel gets the worst result on contrast enhancement, which correlate well with our subjective feeling. However, values of e consider that Tarel achieves the best result on visible edge enhancement, but subjectively, we think Berman and Fattal14 perform better than Tarel.

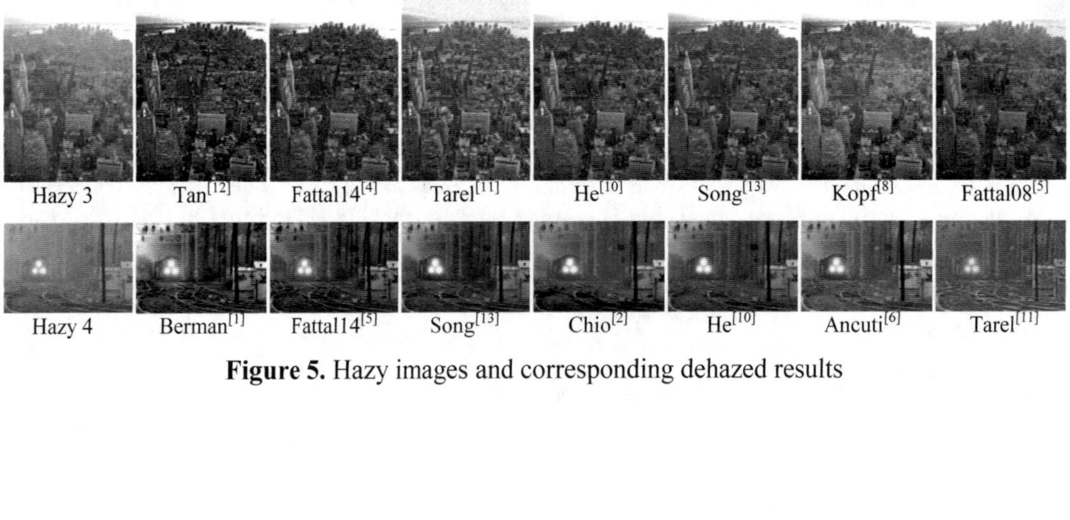

Figure 5. Hazy images and corresponding dehazed results

Table 4. Objective assessment of Hazy 3

	Tan	Fattal14	Tarel	He	Song	Kopf	Fattal08
CC	**1.834**	1.448	1.344	1.326	1.291	1.171	1.039
e	0.07	-0.16	**0.14**	0.04	0.07	0.03	-0.06
$\bar{r}$	**2.18**	1.83	1.76	1.40	1.26	1.41	1.29
FADE	**0.23**	0.24	0.32	0.47	0.47	0.53	0.32

Table 5. Objective assessment of Hazy 4

	Berman	Fattal14	Song	Choi	He	Ancuti	Tarel
CC	**1.566**	1.402	1.381	1.357	1.250	1.196	1.034
e	1.16	1.26	1.32	0.94	1.29	0.99	**1.58**
$\bar{r}$	**3.44**	3.13	2.05	1.84	1.93	2.10	2.50
FADE	**0.15**	**0.15**	0.31	0.27	0.35	0.45	0.45

To sum up, among metrics mentioned in this paper, the CNC index is sometimes inconsistent with the subjective judgment of overall dehazing effect as it based only on the dehazed image itself while does not take the original hazy image as the reference. The contrast descriptor e computes the visible edge in gray-scale image, which is not suitable for color image. The contrast descriptor $\bar{r}$ focuses on enhancement of gradient value which produces more reasonable evaluation results on contrast enhancement. As for FADE, it does good jobs on measuring the fog residue of dehazed image. However, a lower fog density not always equivalent to a higher contrast or good visibility. Our proposed Color Contrast index CC meanly focuses on contrast of different colors in local region, which could always produce relative reasonable evaluation results on contrast enhancement.

5. Conclusion

In this paper, we proposed an indicator of contrast enhancement of dehazed image basing on the haze-line theory. We demonstrated the performance of our proposed (CC) index by a subjective study on dehazed images of a variety of hazy scene. Experimental results demonstrate that the proposed CC index correlate well with human judgments of image contrast taken in the subjective test and performs better than state-of-the-art metrics.

However, there are also some deficiencies in our method. Firstly, the performance of our method relies on the clustering of haze-lines. If this step fails to produce good results, evaluation results achieved by our methods will be unreliable. Besides, purpose of image haze removal is not only to enhance the contrast, but also to restore the colorful color. Phenomenon of over-saturation or color distortion are can't be identified by existing contrast indicators. Therefore, color similarity evaluation is very necessary when assessing quality of dehazed image, and this will be the focus of our next research.

References

[1] Berman D, Treibitz T and Avidan S 2016 Non-Local Image Dehazing *IEEE Conference on Computer Vision and Pattern Recognition* pp 1674–1682
[2] Choi L K, You J and Bovik A C 2015 Referenceless Prediction of Perceptual Fog Density and Perceptual Image Defogging *IEEE Transactions on Image Processing* **24(11)** pp 3888-3901
[3] Nishino K, Kratz L and Lombardi S 2012 Bayesian Defogging *International Journal of Computer Vision* **98(3)** pp: 263-278
[4] Fattal R 2014 Dehazing using color-lines *ACM Transactions on Graphics* **34(1)** p 13
[5] Fattal R 2008 Single image dehazing *ACM Transactions on Graphics* **27 (3)** pp 72-80
[6] Ancuti O C and Ancuti C 2013 Single Image Dehazing by Multi-Scale Fusion *IEEE Transactions on Image Processing* **22(8)** pp 3271-3282
[7] Kratz L and Nishino K 2009 Factorizing scene albedo and depth from a single foggy image *IEEE 12th International Conference on Computer Vision* pp 1701–1708

[8] Kopf J, Neubert B, Chen B, Cohen M, Cohen D, Deussen O, Uyttendaele M and Lischinski D 2008 Deep photo: Model-based photograph enhancement and viewing *ACM Transactions on Graphics* **27(5)** p 116
[9] Song Y, Luo H, Hui B and Chang Z 2015 An improved image dehazing and enhancing method using dark channel prior. *The 27th Chinese Control and Decision Conference* pp 5840-5845
[10] He K, Sun J and Tang X 2011 Single image haze removal using dark channel prior *IEEE Transactions on Pattern Analysis and Machine Intelligence* **33 (12)** pp 2341-2353.
[11] Tarel J P and Hauti´ere N 2009 Fast visibility restoration from a single color or gray level image *The 12th IEEE International Conference on Computer Vision* pp 2201-2208
[12] Tan R T 2008 Visibility in bad weather from a single image *Proceedings of IEEE Conference on Computer Vision and Pattern Recognition* pp 1–8
[13] Song Y, Luo H, Hui B and Chang Z 2016 Haze removal using scale adaptive dark channel prior *Infrared and Laser Engineering* **45(9)** pp 286-297
[14] Gibson K B, Nguyen T Q 2011 Hazy image modeling using color ellipsoids, *IEEE International Conference on Image Processing* pp 1861–1864
[15] Gibson K B and Nguyen T Q 2013. Fast single image fog removal using the adaptive wiener filter *The International Conference on Image Processing* pp 714–718.
[16] Zhu Q, Mai J and Shao L 2015. A Fast Single Image Haze Removal Algorithm Using Color Attenuation Prior *IEEE Transactions on Image Processing* **24(11)** pp 3522-3533
[17] Zhang Z, Lu W and Lu J 2015 Dehazing Method Based on Fast Pseudosphere Filtering for Visual Navigation of Intelligent Tractor in Haze Weather Scene *ROBOT* **37(5)** pp 603-613.
[18] Narasimhan S.G, Nayar S.K 2003 Contrast restoration of weather degraded images. *IEEE Transactions on Pattern Analysis and Machine Intelligence* **25(6)** pp 713–724.
[19] Sulami M, Geltzer I, Fattal R and Werman M 2014 Automatic recovery of the atmospheric light in hazy images *IEEE International Conference on Computational Photography* pp 1-11
[20] Wang Z, Bovil A C, Sheikh H R and Simoncelli E P 2004 Image quality assessment: From error visibility to structural similarity *IEEE Transactions on Image Processing* **13(4)** pp 600–612
[21] Hauti`ere N, Tarel J P and Aubert D 2008 Blind contrast enhancement assessment by gradient ratioing at visible edges *Image Analysis and Stereology Journal* **27(2)** pp 87–95
[22] Guo F and Cai Z 2012. Objective Assessment Method for the Clearness Effect of Image Defogging Algorithm[J]. Acta Automatica Sinica. 38(9) pp: 1410-1419.
[23] Yendrikhovskij SN, Blommaert F JJ, Ridder HD 1998. Perceptually optimal color reproduction, *SPIE Human Vision and Electronic Imaging*. **3299** pp: 274–281.

IOP Conf. Series: Journal of Physics: Conf. Series **844** (2017) 012046 doi:10.1088/1742-6596/844/1/012046

Discriminative feature selection for visual tracking

Junkai Ma[1,2,3], Haibo Luo[1,3], Wei Zhou[4], Yingchao Song[1,2,3], Bin Hui[1,3] and Zheng Chang[1,3]

[1] Shenyang Institute of Automation, Chinese Academy of Sciences, Shenyang, China
[2] University of Chinese Academy of Sciences, Beijing, China
[3] Key Laboratory of Opto-Electronic Information Processing CAS, Shenyang, China
[4] AVIC Jiangxi HONGDU Aviation Industry Group LTD, Nanchang, China

E-mail: junkai.ma@hotmail.com

Abstract. Visual tracking is an important role in computer vision tasks. The robustness of tracking algorithm is a challenge. Especially in complex scenarios such as clutter background, illumination variation and appearance changes etc. As an important component in tracking algorithm, the appropriateness of feature is closed related to the tracking precision. In this paper, an online discriminative feature selection is proposed to provide the tracker the most discriminative feature. Firstly, a feature pool which contains different information of the image such as gradient, gray value and edge is built. And when every frame is processed during tracking, all of these features will be extracted. Secondly, these features are ranked depend on their discrimination between target and background and the highest scored feature is chosen to represent the candidate image patch. Then, after obtaining the tracking result, the target model will be update to adapt the appearance variation. The experiment show that our method is robust when compared with other state-of-the-art algorithms.

1. Introduction

Visual tracking plays an important role in various computer vision tasks such as video surveillance, action recognition, traffic monition etc. Although a variety of tracking algorithms are proposed during recent decades, the robustness of the tracking is still a challenge. This is mainly caused by the illumination variation, clutter background, pose changes, scale variation et al in the tracking processing. To solve these problems, it is necessary to design robust and efficient tracking algorithms.

The existing tracking algorithms can be categorized into two groups in term of their appearance model. One is called generative model [1, 2], which construct an appearance model based on the target appearance information, and track the target by finding the most similar candidate in the every frame of the video. Mei et al. proposed a tracking method with sparse model and represented the target by a template dictionary [2]. Ross et al. present a method that learns an incrementally a low-dimensional subspace as the representation of the target, this method can copy with the appearance changes during tracking [1]. The other is called discriminative model [3, 4], which consider the tracking as classification problem to distinguish the target from the background. Henriques et al. exploit the property of circulant matrices and proposed an algorithm called kernelized correlation filter (KCF) [3]. KCF learned a regressive function in frequency domain and incorporate multi-channel feature what makes the algorithm efficient and robust. Hare et al. employed a structured support vector machine as the classifier to distinguish the target and background [4]. This structure output classifier can output the accurate position of the target rather than the positive and negative label of the image patch, so it attained a preferable tracking result.

Most algorithms proposed recently are focus on the observation model, which determines whether a candidate is the target or the background. As an important component in tracking algorithm, the feature used to represent the target is also closely related to the performance of the tracking algorithm. The proper feature can enhance the performance of the tracker. On the contrary, improper feature will degrade the performance of the tracker. Numerous works studied the feature selection in tracking [5, 6, 7, 8]. In this paper we employed an on-line feature selection mechanism to choose the most discriminative feature during the tracking processing to make the tracker more accurate.

2. Proposed method

2.1. Algorithm overview

An overview of the propose method is shown in Fig. 1. As the common implement in tracking processing, the target is selected in the 1-st frame and the tracking result is given by algorithm iteratively frame by frame. When the t-th frame is coming, the search region is assigned around l_{t-1}, the target location in the previous frame. Firstly, some candidate is sampled in the search region and the most discriminative features, which is provided by the previous frame, are extracted form each candidate. Then, the confidence score of each candidate is calculate by the classifier according to the target appearance model. The image patch which has the highest confidence score is considered as the target. Then some positive and negative samples are extracted. At last by using the negative and positive examples, the most discriminative feature is chosen according to the feature selection mechanism to represent the target in the next frame and the target appearance model can also be update depended on these samples.

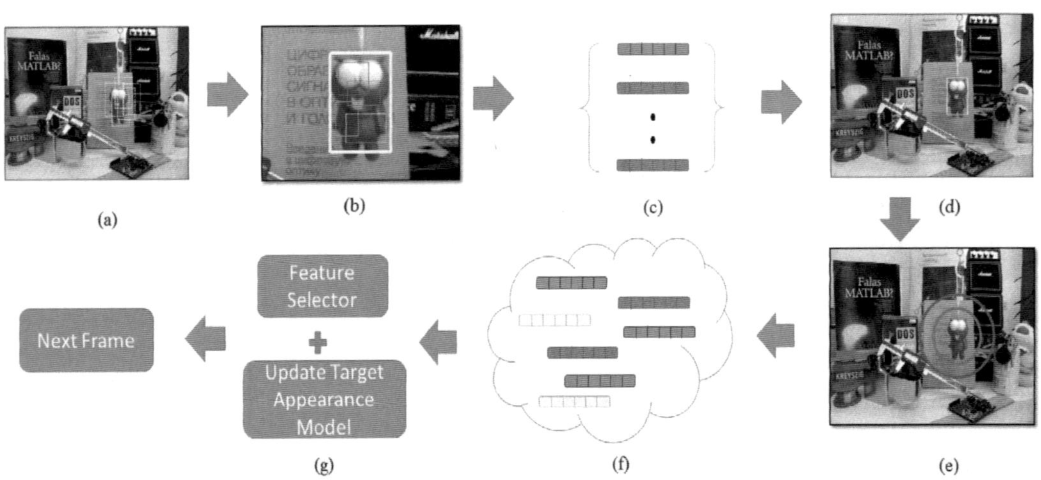

Figure 1. Overview of proposed algorithm. (a) Extract candidate patches according to the search method. (b)-(c) Extract the top k most discriminative features following the previous frame. (d) Calculate the confidence score of each candidate image patch and determine the target. (e) Extracted the positive and negative samples near the target. The feature selector choose the . (f)-(g) Establish the feature pool for each candidate and update the target appearance model.

The MIL tracker[9] is the baseline of the proposed method. Following by the MIL, a strong

IOP Conf. Series: Journal of Physics: Conf. Series **844** (2017) 012046 doi:10.1088/1742-6596/844/1/012046

classifier is build to discriminate target from the background. The strong classifier consist of several weak classifier

$$H(x) = \sum_{i=1}^{K} h_i(x), \tag{1}$$

where x is the sample. Each weak classifier is related to a feature instance in the feature pool. The detail of the proposed method is shown in the Algorithm 1.

Algorithm 1 Proposed Tracking Algorithm

```
Point the location target manually in the 1-st frame l_0.
Sample the image patch sets S^+ = {x| ||l_x - l_0|| < α}, S^- = {x| β < ||l_x - l_0|| < γ} as the
  positive and negative sample set respectively.
Generate the feature rectangles randomly and extract the feature pool F = {f_1, f_2, ..., f_n}
  in every sample according to the random rectangles.
Establish the n weaker classifier h_k, i = 1, 2, ...n for each f_i in the feature pool F.
Top k most discriminative features are be selected based on fisher criterion.
for i = 2 to end of the sequence do
    Sample some candidate image patches in the search region near the l_i.
    Extract the top k features of every candidate.
    Use the weak classifiers which are related to the k feature(selected by the last step) to
      constitute the strange classifier H^i = Σ_{i=1}^{k} h_k
    Calculate confidence score of each candidate image patch using H^i.
    Set l_i = arg max H^i(x(l)) to the location of the target in current frame.
    Sample the image patch sets S^+ = {x| ||l_x - l_i|| < α}, S^- = {x| β < ||l_x - l_i|| < γ} as the
      positive and negative sample set respectively.
    Create the feature pool on the sample set. Rank the features in the feature pool and select
      the top k most discriminative feature for next tracking.
    Update the all of the n weak classifiers.
end for
```

2.2. Feature pool

Different kind of feature has variant performance to represent the image according to different scenario. So more the diverse the feature is, more robust the feature can represent the target. In our algorithm, the feature pool consist of the gradient information(HoG) [10], local contrast information (Haar) [11], which shows good performance in studies [4, 12] in recent years.

In order to make the HoG and the Haar features fit to the tracking algorithm, we use the modified HoG and Haar feature proposed by the studies [12, 13, 8]. In traditional HoG, the histogram of gradient is summed up in a small rectangle, which is called cell, in the image patch. The feature of the holistic image patch is composed by the histogram in a set of overlapping cell. Fig. 2(a) show the detail. In our method we replace this regular cell selection with a randomly selection method, which is shown in the Fig. 2(b). By this method, the histogram of each random cell is considered as feature instance of the image patch. For Haar feature, the base implement is shown in Fig. 3(a), each rectangle indicates a type of Haar feature, whose value is the weighted sum of the pixel value in the rectangle. The pixels in the white rectangle have the weight of 1 and those in the black rectangle have the weight of -1. For a image cell, p types of Haar feature can be obtained. As the same in the HoG feature, the Haar feature of whole region is represented by a set of overlapping cells. In this case the types of feature is fixed, We also using the randomly selected cells to replace them. As shown in Fig. 3(b), a Haar feature instance is composed of several sub-rectangles, which are randomly selected in the target image patch. The weight of each sub-rectangle is also a random value between -1 and 1. Every feature element is the weighted sum of the pixel value in those sub-rectangle.

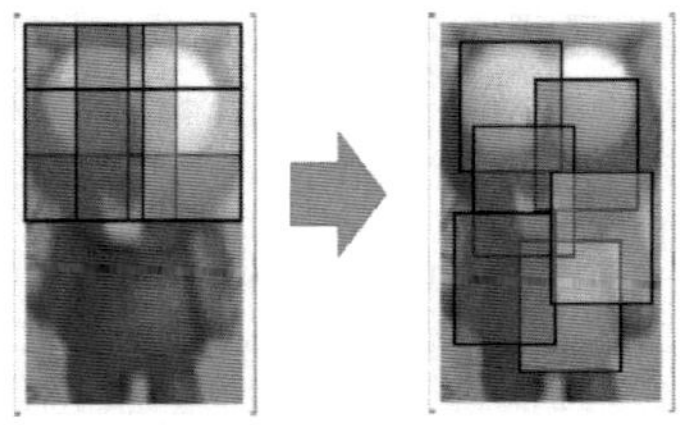

Figure 2. (a)The locations of cells in the original HoG are distributed in a regular pattern. (b)The locations of cells in our method are distributed in a random pattern.

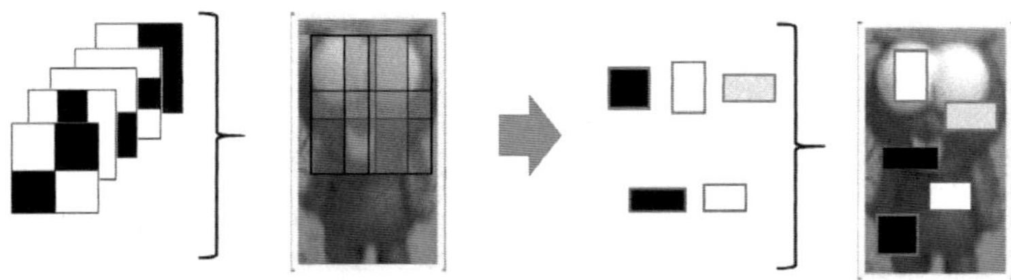

Figure 3. (a) An example of the original Haar feature. Five typical types of the Haar feature are list in the left. Applied these type of feature in every small region in the target bounding box and obtain a feature vector. (b) The type of Haar feature,the location and the weight of every small patch, is designed in a random way. A group of rectangles have same color belong to a type of Haar feature.

Calculating these features is a time consuming processing in tracking algorithm. The computing time is increasing along with the growing of the feature regions number, while a large number of the feature region is necessary to guarantee the diverse of the feature. So, a fast feature extraction algorithm is needed. In [14] a general framework, which is called integral channel, is proposed to fast calculate the region-based feature. By using the integral channel, both Haar feature and HoG feature can be computed efficiently.

Given an input image I, a channel of I is the output of a channel generation function Ω, which is a linear or non-linear transformation of the input. So the channel of I is defined as $C = \Omega(I)$. The widely used channels are R-G-B channels, Gabor filter channel, and DoG (Difference of Gaussian) filter channel etc. An integral image of these channels can be used to speed up the computation of the region based feature. The HoG feature is a weighted histogram where bin index is determined by the gradient angle and weight by gradient magnitude, so the channels are given by $Q_\theta(x,y) = G(x,y) \cdot \mathbf{1}[\Theta(x,y) = \theta]$, where $G(x,y)$ and $\Theta(x,y)$ are gradient magnitude and quantized gradient angle, respectively. $\mathbf{1}(\cdot)$ is the indicate function. Based on this way, integral channels can be used to simplify the computation of the feature. The histogram of a given region can be obtain by plus or minus the value of four corners in every integral channel. The Haar feature can be obtained in the same way.

2.3. Discriminative feature ranking
By giving the feature pool of the samples set. The discrimination of each feature is measured by fisher criterion in a supervised way. We labeled the target as the positive and background as negative. The i-th feature extracted from target and background is defined as f_{i+} and f_{i-} respectively. The fisher discrimination score is given by:

$$p_i = \frac{|\mu_{i,+} - \mu_{i,-}|^2}{\sigma_{i,+}^2 + \sigma_{i,-}^2} \tag{2}$$

where μ_i and σ_i are the mean and standard deviation of i-th feature. For the total number of n features, the fisher discrimination score is calculate by Eq. 2. Then those features are ranked based on their fisher discrimination score. In the end, the top k most discriminative feature is selection to represent the target.

3. Experiment
We experiment our method on several sequences in the OTB-50 dataset[15]. There are some details of our experiment. We set the total number of the feature to $n = 100$ and using the top $k = 20$ most discriminative feature to represent the target. We test our algorithm on five typical sequences,*david3*, *subway*, *suv*, *coke* and *trellis* in the dataset.

In order to shown performance of the proposed method, we compare the result with other 4 state-of-art tracker ASLA[16], L1APG[17], IVT[1] and MIL[9]. The tracking results on the 5 test sequences are shown in Fig. 4 to Fig. 8. From the result, we can find that our method is more robust than the compared methods. In the *trellis* sequence, when the illumination changes our tracking can catch the target correctly. In the *coke* and *david3* sequence the target undergoes deformation and in-plane rotation, our method can still tracking the target. In the *suv* and *subway* sequences, the target have some slightly occusion, the result shown that our method can handle this kind of challenge.

Figure 4. Tracking results comparison on *Trellis* sequence

The ground truth of the target location is given in the OTB-50 dataset. For evaluate the propose method more accurate, we plot the center location error(CLE) of each tracker on every sequence in Fig. 9. The CLE is widely used to evaluate the performance of tracker. If the ground truth location l_g and the tracking result l_t are given, the CLE is defined as:

$$CLE = d(c_g, c_t), \tag{3}$$

in which the $d(\cdot, \cdot)$ is the Euclidean distance and c_g, c_t are the center of the ground truth and tracking result, respectively.

Figure 5. Tracking results comparison on *coke* sequence

Figure 6. Tracking results comparison on *suv* sequence

Figure 7. Tracking results comparison on *david3* sequence

From the Fig. 9, we can conclude that our method has the best result among the compared trackers. Even in the end of each sequence, other trackers undergo heavily drift while our method has low center location error. Table .1 shows the average center location error of all the 5 trackers in every sequence.

4. Conclusion

In this paper, we propose an robust tracking algorithm which can select the discriminative features dynamically during the tracking. A rich feature pool which contains different information of the image is built. A features ranked method is used to choose the most discriminative ones between target and background. The experiment show that our method outperform some state-of-the-art algorithms in some challenge tracking scenarios.

Figure 8. Tracking results comparison on *subway* sequence

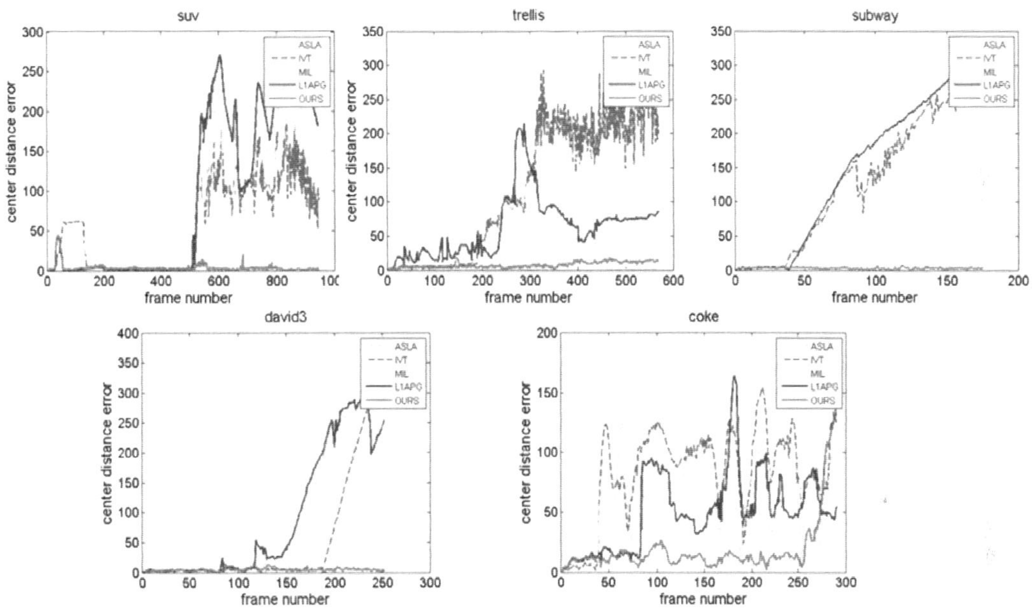

Figure 9. The plot of center distance error on the 5 sequences

Table 1. Average center location error of 5 trackers on 5 sequences

	ASLA	IVT	MIL	L1APG	Ours
Trellis	7.6	119	71	62	7.7
coke	60	82	46	50	18
david3	87	51	29	90	4.5
subway	138	130	7.5	147	2.9
suv	74	57	82	93	3

References

[1] Ross D A, Lim J, Lin R and Yang M 2008 Incremental Learning for Robust Visual Tracking *International Journal of Computer Vision* **77** 125–141

[2] Mei X and Ling H 2011 Robust Visual Tracking and Vehicle Classification via Sparse Representation *IEEE Transactions on Pattern Analysis and Machine Intelligence* **33** 2259–2272
[3] Henriques J F, Caseiro R, Martins P and Batista J 2015 High-Speed Tracking with Kernelized Correlation Filters *IEEE Transactions on Pattern Analysis and Machine Intelligence* **37** 583–596
[4] Hare S, Golodetz S, Saffari A, Vineet V, Cheng M, Hicks S L and Torr P H S 2015 Struck: Structured Output Tracking with Kernels *IEEE Transactions on Pattern Analysis and Machine Intelligence* **38** 2096–2109
[5] Collins R T, Liu Y and Leordeanu M 2005 Online selection of discriminative tracking features *IEEE Transactions on Pattern Analysis and Machine Intelligence* **27** 1631–1643
[6] Danelljan M, Khan F S, Felsberg M and De Weijer J V 2014 Adaptive Color Attributes for Real-Time Visual Tracking *In Proceedings of the IEEE Conference on Computer Vision and Pattern Recognition* 1090–1097
[7] Roffo G, Melzi S and Cristani M 2015 Infinite Feature Selection *IEEE International Conference on Computer Vision* 4202–4210
[8] Yang J, Zhang K and Liu Q 2016 Robust object tracking by online Fisher discrimination boosting feature selection *Computer Vision and Image Understanding* **153** 100–108
[9] Babenko B, Yang M and Belongie S 2009 Visual tracking with online Multiple Instance Learning *In Proceedings of the IEEE Conference on Computer Vision and Pattern Recognition* 983–990
[10] Dalal N and Triggs B 2005 Histograms of oriented gradients for human detection *In Proceedings of the IEEE Conference on Computer Vision and Pattern Recognition* **1** 886–893
[11] Viola P and Jones M 2001 Rapid object detection using a boosted cascade of simple features **1** 511–518
[12] Benenson R, Mathias M, Tuytelaars T and Van Gool L 2013 Seeking the Strongest Rigid Detector *In Proceedings of the IEEE Conference on Computer Vision and Pattern Recognition* 3666–3673
[13] Zhang K, Zhang L and Yang M H 2013 Real-Time Object Tracking Via Online Discriminative Feature Selection *IEEE Transactions on Image Processing* **22** 4664–4677 ISSN 10577149
[14] Dollar P, Tu Z, Perona P and Belongie S 2009 Integral Channel Features *British Machine Vision Conference*
[15] Wu Y, Lim J and Yang M H 2013 Online Object Tracking: A Benchmark *IEEE Conference on Computer Vision and Pattern Recognition*
[16] Jia X, Lu H and Yang M 2012 Visual tracking via adaptive structural local sparse appearance model 1822–1829
[17] Bao C, Wu Y, Ling H and Ji H 2012 Real time robust L1 tracker using accelerated proximal gradient approach 1830–1837

Measuring the diffusion coefficient of ganglioside on cell membrane by fluorescence correlation spectroscopy

Shiqing Dong, Minghai You, Jianling Chen, Jie Zhou, Shusen Xie, Hongqin Yang*

Institute of Laser and Optoelectronics Technology, Fujian Provincial Key Laboratory for Photonics Technology, Key Laboratory of OptoElectronic Science and Technology for Medicine of Ministry of Education, Fujian Normal University, Fuzhou 350007

*Corresponding author, e-mail: hqyang@fjnu.edu.cn

Abstract The fluidity of proteins and lipids on cell membrane plays an important role in cell's physiological functions. Fluorescence correlation spectroscopy (FCS) is an effective technique to detect the rapid dynamic behaviors of proteins and/or lipids in living cells. In this study, we used the rhodamine6G solution to optimize the FCS system. And, cholera toxin B subunit (CT-B) was used to label ganglioside on living Hela cell membranes. The diffusion time and coefficients of ganglioside can be obtained through fitting the autocorrelation curve based on the model of two-dimensional cell membrane. The results showed that the diffusion coefficients of ganglioside distributed within a wide range. It revealed the lateral diffusion of lipids on cell membrane was inhomogeneous, which was due to different microstructures of cytoplasmic membrane. The study provides a helpful method for further studying the dynamic characteristics of proteins and lipids molecules on living cell membrane.

Keywords: diffusion coefficient, gangliosides, Hela cells, fluorescence correlation spectroscopy

1. Introduction

Lipid bilayers are vital self-assembled structure of cell to segregate different constituents and environments from the surrounding. It is essential to ensure life activities can be carried on in the cells. One of the essential properties is the fluidity. The fluidity of the lipids has great effects on the social life of cells, which include intracellular or extracellular signaling, transmembrane signal transduction, cell–cell or cell–extracellular matrix adhesion etc. [1]. Therefore, how does the fluidity of membrane regulate the physiological function of the cell has become the focus of the biology. However, the membrane lipids have various dynamic properties even in the same cell. The main reasons may come as follow: Due to the difference structure of lipid such as the head group, the length of the carbon chain, the degree of saturation of fatty acids, etc., there are thousands of known lipids in the membrane [2]. Because of entropy effects, the interaction between lipid tail and head group will lead the same lipids are more likely to act as similar dynamic characteristics. Another reason is the protein in the membrane can segregate specific groups of lipids. Hence, there is a difference in dynamic properties even the same lipid [3]. Finally, the diffusion of lipids will also be changed by the cytoskeleton and other structures in the process of cell migration, Endocytosis and exocytosis [4]. In summary, the dynamics of lipids are closely related to the physiological functions of membrane. Exploring the dynamic characteristics of lipids have gradually become an effective hotspot to understand the fluidity of lipid bilayers in the molecular level and reveal the physiological processes and regulatory mechanisms of the membrane.

The velocity of lipid molecules is too fast to capture in diffusion process. It requires that technique's temporal resolution is above millisecond. Fluorescence Recovery After Photobleaching (FRAR) is an important tool in previous studies to detect the fluidity of plasma membrane [5]. Fluorescence correlation spectroscopy (FCS) is also a versatile molecule technique [6]. We can obtain the diffusion coefficient of molecules, concentration, chemical reaction rates, and other parameters though autocorrelation analysis of the fluorescence intensity in the spot. Confocal-based FCS has become an effective mean to detect cell dynamics at the molecular level because of high temporal and spatial resolution and weak invasiveness.

Ganglioside is a kind of sphingolipid containing sialic acid. It exists in the mammalian cell membrane to protect cell membrane and promote the activity of enzyme. Ganglioside closely relate to membrane function which has attracted the attention of biologists. The diffusion coefficient of ganglioside was measured in this study. The parameters of facula were optimized by the rhodamine6G solution at the beginning of the experiment. And, cholera toxin B subunit (CT-B) was used to label the ganglioside on living HeLa cell. The living cell was measured by FCS and the diffusion coefficient of ganglioside was obtained by autocorrelation calculation to achieve the dynamic characteristic of specific lipid.

2. Materials and methods

2.1 Materials

The system was TCS SP8 (Germany, Leica) equipped with a 63× water immersion objective (NA=1.2). It was composed by a confocal microscope, a couple of avalanche photo-detectors and correlators. The fluorophore was excited with a 488nm laser. The size of the FCS focus was limited by pinholes, which reduced the intensity of the background and improve the signal to noise ratio.

Two kinds of dyes were used for systematic calibration and lipid measurements on the membrane in this article. The Rhodamin6G was used to measure the volume of the spot in order to optimize the system. The aqueous solution is scarlet and glow a yellow-green fluorescence excited with a 488nm laser. The dyes for the measurement of lipid in the membrane are divided into two species, which are lipophilic dyes and specific marker dyes, respectively. The specific marker dye can attach to specific lipids in the membrane. However, the lipophilic dyes cannot reflect the diffusion coefficient of specific lipids and diffuse into certain areas. Cholera toxin from vibrio cholera is comprised of two subunits, A and B. The nontoxic B subunit is important to protein complex as it allows the protein to bind to cellular surfaces via the pentasaccharide chain of ganglioside G_{M1} to get the diffusion coefficient of ganglioside in living cell [7].

2.2 Methods

There are two ways to obtain the parameters of the confocal volume. It included the effective volume, lateral and longitudinal waist radius, which are related to the excitation wavelength, laser power, objective and imaging settings. One of the methods is calibrate the system with the known diffusion coefficient of solution. According the obtained diffusion time we can acquire the effective volume by FCS. The other is calibrated with a known concentration solution to obtain the number of molecules. We can calculate the size of the detected volume based on the concentration of sample. The former method was used in this study. Where the diffusion coefficient of Rhodamin6G solution was D = 2.8 $*10^{-6}cm^2/s$ at 22 ℃ [8].

We use auto-correlate curve to analyze the fluctuation of fluorescence intensity, so we need to select the appropriate fitting model for different samples. The normalized autocorrelation function (ACF) can be written as

$$G(t) = \frac{\langle \delta F(t)\delta F(t+\tau)\rangle}{\langle F(t)\rangle^2} \tag{1}$$

where the G(t) represent the correlation of fluctuations. <F(t)> is the average of fluorescence intensity. The simplest example of a diffusion process with a Gaussian propagator is simple Fickian diffusion, with a mean squared displacement linear in time,

$$MSD(t) = 6Dt \tag{2}$$

where MSD(t) is mean squared displacement, the diffusion coefficient is given by the Stokes–Einstein relationship,

$$D = \frac{KT}{6\pi\eta R} \tag{3}$$

where R is the hydrodynamic radius of the particle, K is the fluid viscosity, η is Boltzmann's constant and T is the absolute temperature. In this case, the normalized autocorrelation function obtained as

$$G(t) = \frac{1}{N}(1+\frac{t}{\tau_D})^{-1}(1+\frac{t}{S^2\tau_D})^{-1/2} \tag{4}$$

the S is the ratio of the longitudinal waist to the lateral waist. The characteristic decay time associated with the diffusion of the fluorophore

$$\tau_D = \omega_0^2/(4D) \tag{5}$$

Since the triplet stage and the diffusion are not at the same time scale, we measure the data with a delay time about 0.008ms. We can obtain the arithmetic of volume though Eq (5) according to the time of lateral diffusion and the known diffusion coefficient of Rhodamin6G solution.

$$S = \frac{z_0}{w_0}; V_{eff} = \pi^{3/2} w_0^2 z_0; D_i = \frac{w_0^2}{4\tau_D}; \tag{6}$$

The two-dimensional detection volume can be calculated as

$$V_{2D} = \pi w_0^2 / 2 \tag{7}$$

The diffusion of lipid molecules in the membrane will deviate from the simplest Fickian diffusion under the influence of a variety of components and complex structures. A frequently discussed anomalous diffusion process with a Gaussian propagator is fractional Brownian motion [9]. A fractional Brownian motion process has a mean-squared displacement

$$MSD(t) = 6Dt^{\alpha} \tag{8}$$

The normalized autocorrelation function can be simplified as

$$G(t) = \frac{1}{\overline{N}}\left[1+\left(\frac{t}{\tau_D}\right)^{\alpha}\right]^{-1} \tag{9}$$

We have quantified the anomaly α and the residence time[10]

$$\tau_D = \left[\omega_0^2 / (4D)\right]^{1/\alpha} \tag{10}$$

The diffusion coefficient also can be acquired. It has been shown to be a good approximation for a large class of anomalous diffusion processes when studied with single-scale FCS [10].

3. Results and Discussion

Before the experiment, the parameters of spot were measured by the rhodamine6G solution. We selected the different measurement points in the concentration of 2nM rhodamine6G solution and repeated the measurement several times. The autocorrelation curve was shown in Fig. 1. When the autocorrelation amplitude was attenuated to half, the diffusion time was obtained. The small residuals indicated that the model selection was appropriate. It is reliable to obtain the relevant parameters such as diffusion coefficient and diffusion time from the autocorrelation curve. The parameters were showed as table 1 according the Eq. (2) and Eq. (6). The results indicated the size of the focus was limited by pinholes. The system was good to measure the diffusion on the membrane.

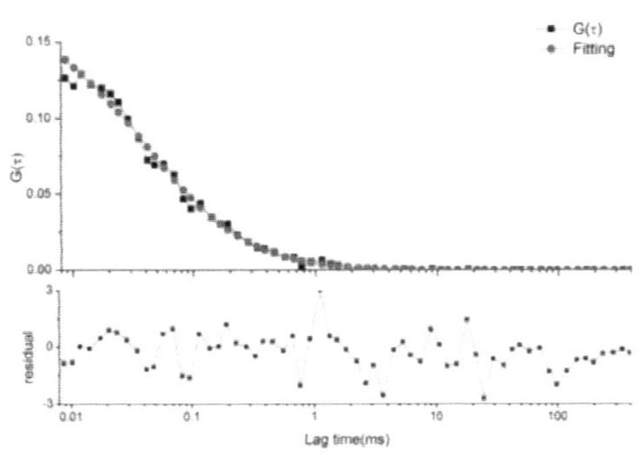

Fig.1. Autocorrelation curve of Rhodamin6G in solution.

Tab. 1. The data of referred volume V_{eff}

samples	τ_d (ms)	ω_0 (um)	S	Z_0 (um)	V_{eff} (fl)
1	0.023	0.225	4.051	0.911	0.091
2	0.018	0.201	3.591	0.719	0.057
3	0.035	0.28	3.832	1.073	0.165
4	0.036	0.285	4.546	1.296	0.207
5	0.052	0.341	3.248	1.109	0.254
6	0.041	0.303	3.775	1.145	0.207
Average	0.034 ±0.012	0.027 ±0.051	3.841 ±0.438	1.042 ±0.201	0.163 ±0.076

The Hela cells were stained for 30mins by cholera toxin subunit B conjugates. We measured the diffusion coefficient of lipid on the membrane in an optimized FCS system. The different autocorrelation curve showed different diffusion time in Fig. 2. The curves were fitted with extend model as Eq (9). Therefore we obtained the distribution of diffusion coefficients according hundreds of data (show in Fig.3). The specific lipid has a characteristic diffusion coefficient in model membrane. On the contrary, the results showed that the diffusion coefficients of ganglioside were inhomogeneous on the cell membrane. The diffusion coefficient obtained by lipid measurement in living cell was distributed in two orders of magnitude, indicating that there were two significant different kinds of diffusion in the membrane. There was a small amount (about 20%) of the data stay at a faster level, and the number of data greater than $0.6um^2s^{-1}$ was significantly reduced. It indicated not all of ganglioside diffused at slow level. On the contrary, the diffusion coefficients of ganglioside were aggregated in the range of 0.1-$0.3um^2s^{-1}$, which was similar to the protein in the membrane. The reason is the diffusion properties of specific marker dyes are usually associated with the type and location of labeled lipid. The B subunit of cholera toxin has proven to be a powerful tool for mark the lipid domains. The domain can associate with protein which usually segregate specific groups of lipids and form the platform to lead the cellular signaling. Hence, the slow diffusion kinetics may be associated with protein signaling in the membrane.

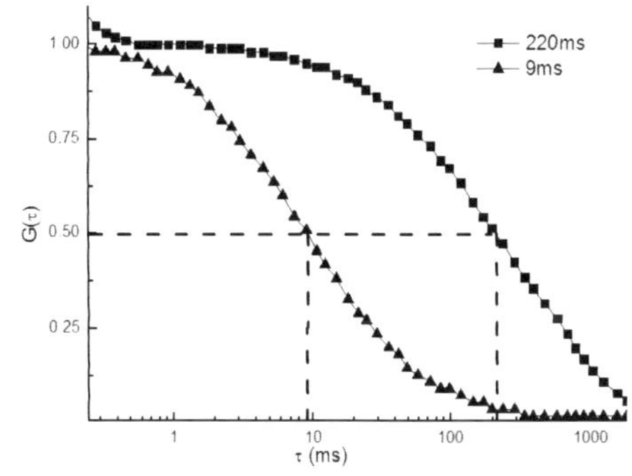

Fig.2. Autocorrelation curve of ganglioside on the membrane.

The results showed that fluorescence correlation spectroscopy can be used to quantitatively measure the diffusion coefficient of the cell membrane. Although membrane behavior is determined not only by the behavior of individual lipids but also the emergent properties of lipid collectives, a large number of measurements can also obtain the dynamic characteristic for the specific lipid. This study provided an effective technical support for further study of the dynamic characteristics of proteins and lipid molecules in living cell membranes.

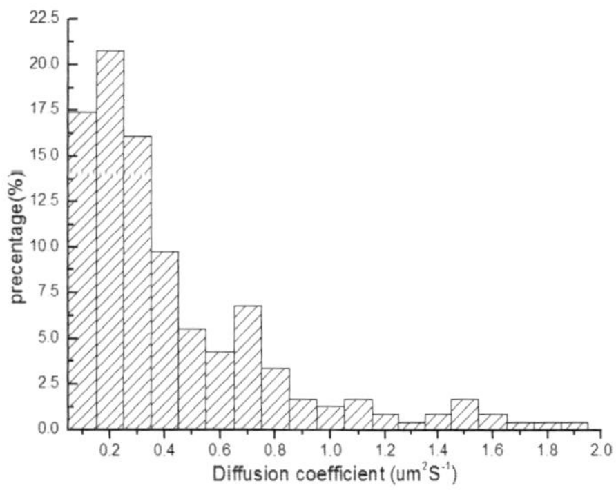

Fig.3. The distribution of diffusion coefficient.

4. Conclusions

We built up fluorescence correlation spectroscopy based on confocal microscope and discussed the feasibility of FCS to explore the molecular diffusion in the membrane. The results showed that the heterogeneity of the diffusion coefficient of ganglioside was obtained in a wide range. Then it was verified that there were various diffusion coefficients in the cell membrane even the same lipid. Different diffusion rates may correspond to the different physiological state of cell membranes. It also reflects the complexity and diversity of cell membrane structure. Above all, FCS is a promising technique to investigate the dynamic characteristic of membrane in living cell.

Acknowledgments

This work was partly supported by the National Key Basic Research Program of China (No. 2015CB352006), the National Natural Science Foundation of China (No.61335011) and the Program for Changjiang Scholars, Innovative Research Team in University (IRT_15R10).

References

[1] Markus Andersson Trojer, Gerald Brezesinski. "Self-assembly of lipid domains in the extracellular leaflet of the plasma membrane and models thereof." Current Opinion in Colloid & Interface Science. 22, 65–72, (2016).

[2] Mélanie Carquin, Ludovic D'Auria, Hélène Pollet, Ernesto R. Bongarzone, Donatienne Tyteca, "Recent progress on lipid lateral heterogeneity in plasma membranes:From rafts to submicrometric domains." Progress in Lipid Research, 62,1–24, (2016).

[3] Lawrence Rejendren, Kai Simons, "Lipid rafts and membrane dynamics." Journal of Cell Science. 118(6), 1099-1102, (2005).

[4] Pierre-Francois Lenne, Laure Wawrezinieck, Fabien Conchonaud, Olivier Wurtz,Annie Boned, Xiao-Jun Guo, Herve Rigneault, Hai-Tao He, Didier Marguet, "Dynamic molecular confinement in the plasma membrane by microdomains and the cytoskeleton meshwork." The EMBO Journal, 25, 3245–3256, (2006).

[5] Ilya Levental, Sarah L. Veatch, "The Continuing Mystery of Lipid Rafts." Journal of Molecular Biology, 428, 4749–4764, (2016).

[6] Oleg Krichevsky, Grégoire Bonnet. "Fluorescence correlation spectroscopy: the technique and its applications." Reports on Progress in Physics. 65(2), 3-35, (2002).

[7] Simon Van Heyningen, "Cholera toxin." Bioscience Reports, 2(3), 135-181, (1982).

[8] Shiqing Dong, Jie Zhou, Xuemei Ding, Yuhua Wang, Shusen Xie, Hongqin Yang, "Parameter Estimation and Analysis Model Selections in Fluorescence Correlation Spectroscopy", Proc. of SPIE, 10024(33), (2016).

[9] Jean Pierre Boon, James F. Lutsko, "Nonlinear Theory of Anomalous Diffusion and Application to Fluorescence Correlation Spectroscopy." Journal of Statistical Physics, 161, 1366–1378, (2015).

[10] Daniel S. Banks, Charmaine Tressler, Robert D. Peters, Felix Hofling, "Characterizing anomalous diffusion in crowded polymer solutions and gels over five decades in time with variable-lengthscale fluorescence correlation spectroscopy." The Royal Society of Chemistry, 12, 4190-4203, (2016).

Low-loss nanowire and nanotube plasmonic waveguide with deep subwavelength light confinement and enhanced optical trapping forces

Xiaogang Chen[1], Qijing Lu[1], Xiang Wu[2], Hongqin Yang[1], Shusen Xie[1]

[1]Key Laboratory of Optoelectronic Science and Technology for Medicine of Ministry of Education, Provincial Key Laboratory for Photonics Technology, Institute of Laser and Optoelectronics Technology, Fujian Normal University, Fuzhou, 350007, China. xgchen01@139.com, qjlu@fjnu.edu.cn, hqyang@fjnu.edu.cn, ssxie@fjnu.edu.cn.

[2]Key Laboratory for Micro and Nano Photonic Structures (Ministry of Education), Department of Optical Science and Engineering, Fudan University, Shanghai, 200433, China. wuxiang@fudan.edu.cn.

Abstract With the rapid development of the micro/nano fabrication technology, the semiconductor nanowires and nanotubes with size and dimensions controllable realize wide applications in nanophotonics. In this talk, we propose two kinds of hybrid plasmonics waveguides, one is consisting of nanowires, another is consisting of nanotubes. By employing the simulating with different geometric parameters, the basic waveguiding properties, including the effective mode area, the propagation length, the mode character and the optical trapping forces can be achieved. Compared with previous plasmonic waveguide with plane metal substrate, current plasmonics waveguides with ease of fabrication have the advantage of long propagation length and effectively optical trapping of nanoparticles with deep subwavelength light confinement, which may be very useful for nanophotonic integrated circuits, nanolasers and biosensing.

1. Introduction

With the nanometer fabrication technology rising, the semiconductor nanowires/nanotube with size and dimensions controllable realize wide applications in nanophotonics, such as waveguides, sensors, photodetectors, and lasers [1, 2]. However, confined to the diffraction limited, the nanowire/nanotube waveguide is difficult to achieve deep subwavelength optical scale. Breaking the shackles of the diffraction limited and manipulating photons under deep subwavelength are very important for the next-generation integrated photonic devices. On one hand, the subwavelength optical confinement can enhance the interaction between light and matter. On the other hand, the strong optical confinement can enhance optical field strength and gradient of light field, which will highly enhance the optical force in the nanoscale region [3-5].

Surface plasmon polaritons (SPPs) are a coherent oscillation of free charges in the surface of metals in resonance with the incident light wave, which have the distinguishing capabilities of enhancing the local electric field intensity as well as of confining the optical energy within the nanoscale domain [6-11]. As one of the most promising candidates for nanophotonic circuits, SPP-based nanoscale optical devices have attracted enormous interest in recent years. However, SPP based waveguides suffer huge propagation loss because of the intrinsic ohmic loss. It is difficult to balance between these two opposites to strong optical confinement and propagation loss. In recent report, a novel hybrid plasmonic waveguide consisting of a dielectric waveguide on one side of a metal surface has been proposed by R.F.Oulton et al [12]. It is able to support long range propagation length, while maintaining an ultra-small mode size due to the coupling between the dielectric cylindrical waveguide mode and the SPP mode at the metal-dielectric interface [13-19].

In this article, we propose the nanowire/nanotube hybrid waveguide by replacing the metal film with a metal nanowire. Through theory simulation and analysis, we can draw a conclusion that the hybrid mode in this structure can achieve deep-subwavelength mode, while maintaining propagation distance as well as that in the SPP. Moreover, the combination of ultra-low loss and strong coupling strength between the dielectric waveguide mode and SPP mode would lead to a much larger optical force [5]. On the whole, the proposed structure provides an excellent platform for nanoscale optical devices, such as nanolasers and nanotweezers.

2.Modal properties of the nanowire plasmonic waveguide

The nanowires hybrid waveguide geometry shown in Fig. 1, which consists of two cylindrical nanowires with a small gap distance h. A high-permittivity semiconductor nanowire is Si with the relative permittivity $\varepsilon_s = 12.25$. Another one is silver with diameter of D = 200 nm and permittivity of ε_m = -129+3.3i at λ = 1550 nm. The surrounding dielectric layer is a low-permittivity dielectric of SiO_2 (ε_d = 2.25). In the follow study of nanowires hybrid waveguide, we vary the semiconductor nanowire diameter, d, and the gap distance, h, to adjust the mode field distribution, the propagation distance, L_m, and the mode area, A_m/A_0. The modal properties are investigated by the finite-element method using COMSOL 4.2.

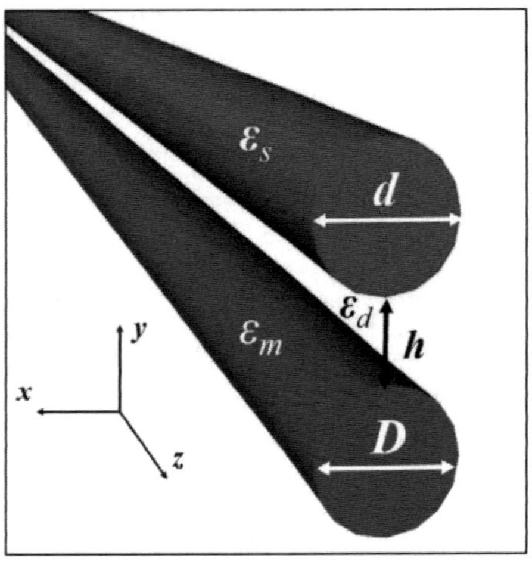

Figure 1. The geometry of the nanowire hybrid surface plasmonic waveguide. Two cylindrical nanowires with a small gap distance h.

Figure 2 shows the dependence of the normalized modal area, A_m/A_0, the propagation length, L_m, and the distributions of electromagnetic energy density on different dielectric diameter, d, and gap distance, h. In this paper, we defined the propagation length and modal area as same as that in [20]. The propagation length is given by

$$L_m = \lambda / [4\pi \operatorname{Im}(n_{eff})] \quad (1)$$

where n_{eff} is the effective index of the hybrid mode. λ is the wavelength of light transmission.

The effective mode area, A_m, is given as follows

$$A_m = \iint W(r)dA / \{\max(W(r)\} \quad (2)$$

Here, the $W(r)$ is the energy density for the hybrid plasmonic waveguide (per unit length along the direction of propagation). Due to the dispersion and material loss, the $W(r)$ can be calculated by

$$W(r) = \frac{1}{2}\left(\frac{d(\varepsilon(r)\omega)}{d\omega}|E(r)|^2 + \mu_0 |H(r)|^2\right) \quad (3)$$

$|E(r)|$ and $|H(r)|$ are the electric and magnetic fields, respectively. $\varepsilon(r)$ is the permittivity, and μ_0 is the vacuum magnetic permeability.

The normalized effective mode area is defined as the ratio of the effective mode area and the diffraction-limited area of vacuum,

$$A = \frac{A_m}{A_0} = \frac{A_m}{\lambda^2/4} \quad (4)$$

The normalized effective mode area was used for consistently quantifying the mode confinement, and $A < 1$, indicates the confinement of the hybrid plasmonic mode is subwavelength.

As shown in Fig. 2(a), with a certain value of diameter of dielectric nanowire, the strong confinement in hybrid plasmonic waveguide results in an ultra-small mode area down to $0.0019(\lambda/2)^2$. As shown in Fig. 2(b), the hybrid plasmonic mode can travel through a long distance, which is more than 24 μm. Figs. 2(c-f) show the electromagnetic field distributions of energy density $W(r)$. For a small gap distance, the hybrid mode confine light energy in the gap and the hybrid mode displays both the characteristics of the dielectric mode and the SPP mode [Figs. 2(c, d)]. For a large diameter of dielectric nanowire and gap distance, the hybrid waveguide supports a dielectric-like mode that the electromagnetic energy is confined in the dielectric nanowire [Fig. 2(f)]. In this case the propagation loss of the electromagnetic wave in the waveguide is low, but the localization of the dielectric-like mode is weak. What is interesting is that, for the hybrid mode, the normalized modal area and the propagation length are improved remarkably.

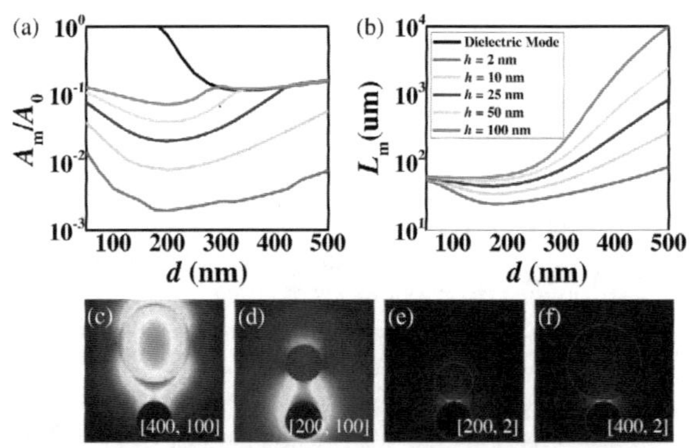

Figure 2. Propagation distance, mode area and field distributions of the hybrid mode. (a) Mode area, A_m/A_0 versus diameter of dielectric nanowire, d, for different gap width, h. (b)

The hybrid mode's propagation distance. (c–f) electromagnetic energy density distribution of the hybrid mode at 1550 nm. (c) $[d, h] = [400, 100]$, (d) $[d, h] = [200, 100]$, (e) $[d, h] = [200, 2]$, (f) $[d, h] = [400, 2]$.

In order to gain a deeper understanding, the dependence of the effective index of the hybrid mode n_{hyb} on the diameter of dielectric nanowire, d, and gap distance, h, is shown in Fig. 3(a). The hybrid mode (coloured lines) is beyond that of the pure dielectric mode (black solid line). When the diameter of dielectric nanowire, d, is fixed, the effective index will increase with reducing the gap distance, h. For a fixed gap distance, h, the mode's effective index increases as the diameter of dielectric nanowire, d, increases. This is because that the waveguide mode has a relationship between the diameter of dielectric nanowire and gap distance.

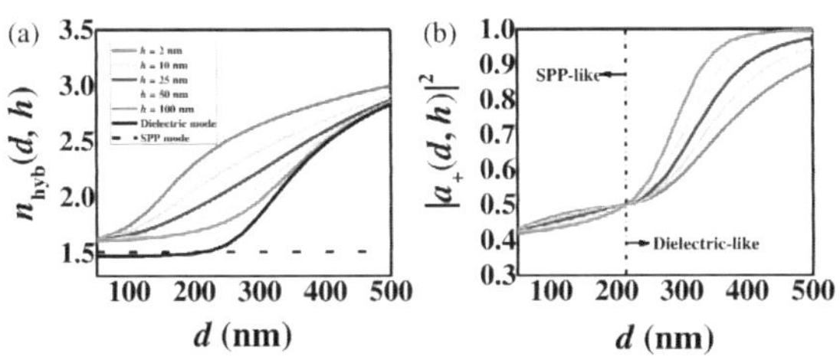

Figure 3. The hybridization of the dielectric mode and SPP mode is modelled by coupled mode theory. (a) The effective index of the hybrid waveguide for a range of gap width, h, and diameter of dielectric nanowire, d, (coloured lines) compared with the analytical effective index of the pure dielectric mode, $n_{die}(d)$ (black solid line), and pure SPP mode, n_{spp} (black broken line). (b) The mode character derived from Eq. (5).

A mode character, $|a_+(d, h)|^2$, is employed to describe the superposition of the dielectric waveguide mode and the SPP mode [12]. Then it can be used to evaluate the degree to which the guided mode is dielectric-like or SPP-like.

$$|a_+(d,h)|^2 = \frac{n_{hyb}(d,h) - n_{spp}}{(n_{hyb}(d,h) - n_{die}(d)) + (n_{hyb}(d,h) - n_{spp})}. \quad (5)$$

Where $n_{die}(d)$, n_{spp} are the pure dielectric-mode and SPP-mode, respectively. n_{spp} can be calculated by the following formula

$$n_{spp} = \sqrt{\frac{\varepsilon_m \varepsilon_d}{\varepsilon_m + \varepsilon_d}}. \quad (6)$$

Figure 3(b) shows the dependence of the mode character derived from Eq. (5) on the diameter of dielectric nanowire, d, for different gap distance, h. Here, we define the mode is dielectric-like for $|a_+(d, h)|^2 > 0.5$ and SPP-like otherwise. If $n_{die}(d)$ is larger than n_{spp}, $|a_+(d, h)|^2 > 0.5$ that the hybrid mode more bias the dielectric waveguide modes. In this case, the light energy is confined in the dielectric nanowires. On the contrary, when the $|a_+(d, h)|^2 < 0.5$, the hybrid mode is more like a SPP mode. When h is fixed, the mode character increases with the increase of the diameter of dielectric nanowire (Fig. 3(b)). A larger d and h lead to a larger mode character. The hybrid mode is a dielectric-like waveguide mode. On the other hand, with the decrease of the diameter of dielectric nanowire, a smaller mode character can be achieved, then the hybrid mode is a more SPP-like mode. For a moderate side length, d_m, the hybrid mode displays both the characteristics of the dielectric and the SPP mode ($|a_+(d, h)|^2 = 0.5$). This case is corresponded to the condition $n_{die}(d) = n_{spp}$, indicating that polarization charge and plasma oscillations move in phase and maximize the effective optical capacitance of the waveguide [12].

3. The influences of structure dimension on the performance parameters of nanotube hybrid plasmonic waveguide

Figure 4 schematically shows the geometry of the nanotube hybrid surface plasmonic waveguide, where a high index semiconductor nanotube was embedded in a low index dielectric and placed on a metal nanowire with a small gap distance, h. In this section, the relative permittivity of material and mode character are defined as the same as that in nanowire hybrid surface plasmonic waveguide.

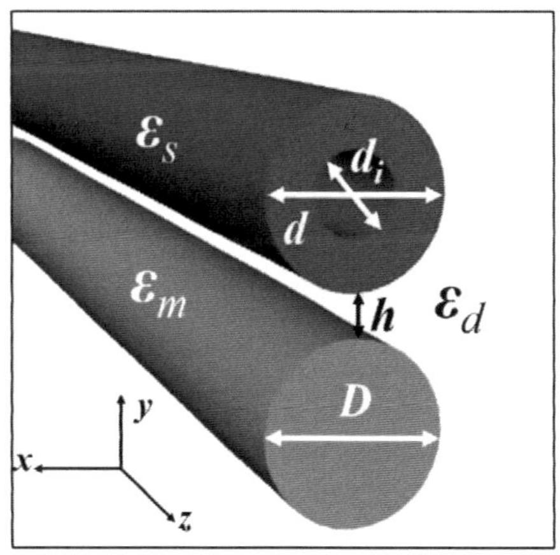

Figure 4. The geometry of the nanotube hybrid surface plasmonic waveguide. A semiconductor nanotube place on a metal nanowire substrate with a small gap distance.

In the following, we defined the diameter of the nanotube outer surface and inner surface as d and d_i, respectively. Figs. 5(a, b) show the propagation length and modal area of the nanotube hybrid mode on the diameter of the Si nanotube outer surface, d, for the different gap distance h. For a fixed diameter of the nanotube inner surface, d_i = 50 nm, the modal area and the propagation length are decline at first and then gradually increased. We can see that there exists a point with d = 150 nm, that both of the modal area and propagation length is minimum.

IOP Conf. Series: Journal of Physics: Conf. Series **844** (2017) 012048 doi:10.1088/1742-6596/844/1/012048

According to the coupled mode theory, in this point, the coupling strength gets its maximum value and the dielectric nanotube mode and SPP mode satisfy the phase-matched condition $n_{die} = n_{spp}$. Then, we will further explore the changed the inner surface diameter of nanotube to confirm the influence of the air hole on field enhancement. As shown in the Figs. 5(c, d), the outer surface diameter of nanotube is fixed of 200 nm, and the diameter of air hole is from 0 nm to 180 nm. Certain monotonous variation of the modal area and the propagation length curve appear with the different size of air hole, d_i. The modal area and propagation length can be increase by enhancing the diameter of air hole, d_i, for a fixed gap distance, h. This can be explained that the optical confinement in hybrid mode couples is weaken gradually as d_i increase.

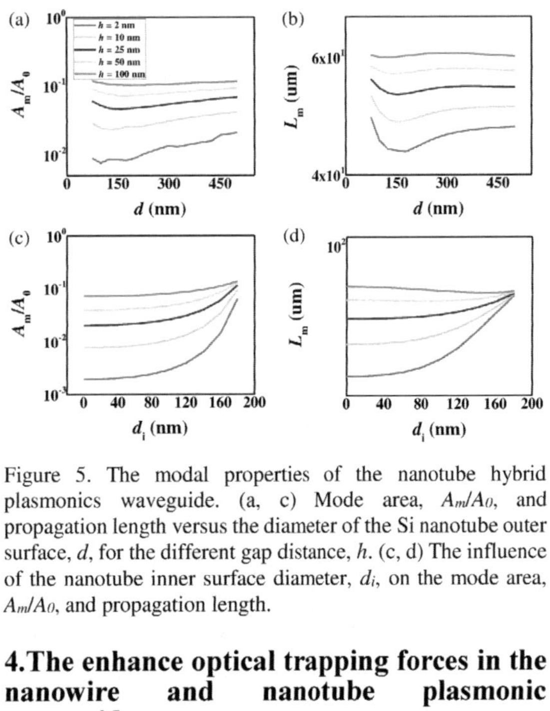

Figure 5. The modal properties of the nanotube hybrid plasmonics waveguide. (a, c) Mode area, A_m/A_0, and propagation length versus the diameter of the Si nanotube outer surface, d, for the different gap distance, h. (c, d) The influence of the nanotube inner surface diameter, d_i, on the mode area, A_m/A_0, and propagation length.

4.The enhance optical trapping forces in the nanowire and nanotube plasmonic waveguide

Next, we analyzed the optical trapping forces in the proposed hybrid plasmonic waveguide. The optical trapping forces are proportional to the gradient of the electrical field distribution, which can be described as [21]

$$F = -\nabla U = -\frac{n_s}{2}\alpha \nabla E^2 , \qquad (7)$$

where U is the optical trapping potential, n_s is the refractive index of environmental medium ($\varepsilon_s = n_s^2$), and α is the polarizability of the nanoparticle. For a dielectric sphere of radius r and dielectric permittivity of ε_n, the polarizability can be written as

$$\alpha = 4\pi r^3 \varepsilon_s \varepsilon_0 \frac{\varepsilon_n - \varepsilon_s}{\varepsilon_n + 2\varepsilon_s} . \qquad (8)$$

The strong optical confinement of the hybrid mode in the gap region, will enhance the optical trapping forces and enable to trap single nanoparticle. Figure 6 shows the system of the optical trapping with a dielectric nanoparticle. In this paper, we used a polystyrene nanoparticle with diameter of 5 nm and refractive index n_p = 1.59 for studying the optical trapping forces. The environmental medium surrounding the waveguide is water with refractive index of n_s = 1.33 [22].

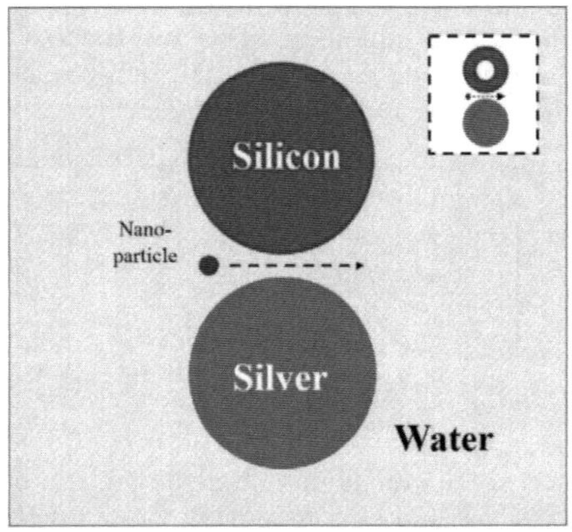

Figure 6. The schematically of optical trapping. The nanoparticle in figure was a polystyrene nanoparticle with refractive index n_p = 1.59 with diameter of 5 nm. The medium around the waveguide is water with refractive index of n_s = 1.33. The illustration shows the nanotube trapping system.

According to Eq. (7), the optical trapping force, at the position where the optical field gradient become largest, was maximized. Figure 7(a) shows the nanowire optical trapping force dependence on D = 200 nm, d = 200 nm, for h = 10, 20, and 30 nm, respectively. As the gap distance reduces, the optical trapping force increases. At h = 10 nm, the maximum of optical trapping force per unit input optical power is 1187 fN/W for the proposed hybrid plasmonic mode. The optical trapping potential profiles along x direction with trapping force integration (Fig. 7(a)) was shown in Fig. 7(b). We define the zero-point potential as the particle is 100 nm away from the waveguide center (x = -100 nm). The optical trapping potential U_x is around 10.1 K_BT/W right underneath the waveguide at h = 10 nm, which means that the kinetic energy of the optical trapping force was 10.1 times larger than the kinetic energy of Brownian motion ($K_BT = 4.1 \times 10^{-21}$ J at room temperature, where K_B is the Boltzmann constant and T is absolute temperature [22]). In this case, when the particle moves across the x = 0 point, the particle can be trapped effectively in the gap region.

Figures 7(c, d) shows the influence of the air hole on the optical trapping force and corresponding trapping potential along x-direction. In a simulation, we set up the D = 200 nm, d = 200 nm and h =10 nm. As the inner surface diameter of nanotube increases, the optical trapping force decreases. The optical trapping force gets its maximum value at the inner surface diameter of nanotube d_i = 0 nm. As mentioned above, the effective index of the nanotube decreases monotonously along with the inner surface diameter increasing. As a result, the effective index mismatch between the dielectric mode and the SPP mode, the coupling strength become weaken gradually. What is interesting is that, for the nanowire/nanotube hybrid mode, the optical force can be significantly larger than that of previous hybrid plasmonic structures [5]. In summary, the combination of low loss and strong coupling strength between the dielectric waveguide mode and SPP mode would lead to a large averaged optical force per unit propagation length along the waveguide.

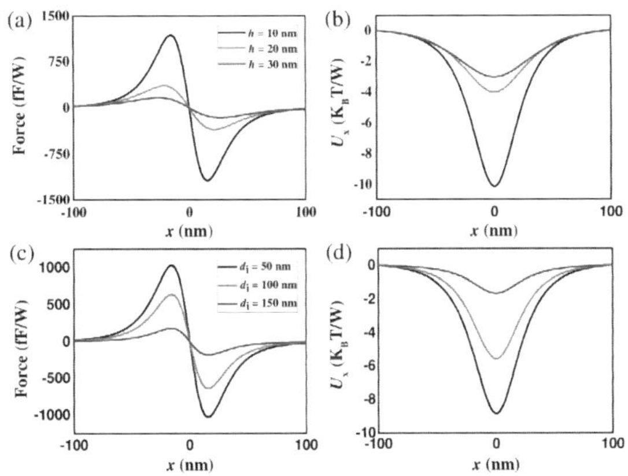

Figure 7. The optical trapping forces along x direction F_x applied on a single polystyrene nanoparticle of 5 nm diameter for the hybrid plasmonic mode. (a) The dependence of the optical trapping forces on the nanowire system with d = 200 nm, D = 200 nm, for h = 10, 20, and 30 nm, respectively. (c) The dependence of the optical trapping forces on the size of the air hole, d_i. (b, d) The corresponding trapping potential along x-direction U_x for (a, c).

5. Conclusion

We have proposed a hybrid plasmonics waveguide system consisting of nanowire/nanotube. Simulation results based on the FEM method, the basic waveguide properties, including the effective mode area, the propagation length, the mode character and the optical trapping forces, can be achieved. Compared with previous plasmonic waveguide with plane metal substrate, current plasmonics waveguides with ease of fabrication have the advantage of long propagation length and effectively optical trapping of nanoparticles with deep subwavelength light confinement. Additionally, we discussed the effect of the size of nanowire and nanotube air hole on the modal properties of the hybrid plasmonics waveguide. In future experimental, the nanowire/nanotube hybrid waveguide can be fabricated easily by current nanofabrication technology and be useful for nanophotonic integrated circuits, nanolasers and biosensing.

Acknowledgment

This work was supported by the National Key Basic Research Program of China (973 project) under Grant No. 2015CB352006 and the Key Natural Foundation by Fujian Province under Grant No. JZ160423. National Natural Science Foundation of China under Grants Nos. 61335011 and 61405035, and Program for Changjiang Scholars and Innovative Research Team in University under Grant No. IRT_15R10. And National High Technology Research and Development Program of China (863 Program-No. 2015AA020508).

References

1. X. Duan, Y. Huang, R. Agarwal, and C. M. Lieber, "Single-nanowire electrically driven lasers," Nature 421(6920), 241-245 (2003).
2. R. Yan, D. Gargas, and P. Yang, "Nanowire photonics," Nat. Photonics 3(10), 569-576 (2009).
3. H. Li, J. W. Noh, Y. Chen, and M. Li, "Enhanced optical forces in integrated hybrid plasmonic waveguides," Opt. Express 21(10), 11839-11851 (2013).
4. A. H. J. Yang, S. D. Moore, B. S. Schmidt, K. Matthew, L. Michal, and E. David, "Optical manipulation of nanoparticles and biomolecules in sub-wavelength slot waveguides," Nature 457(7225), 71-75 (2009).
5. X. Yang, Y. Liu, R. F. Oulton, X. Yin, and X. Zhang, "Optical Forces in Hybrid Plasmonic Waveguides," Nano Lett. 11(2), 321-328 (2011).
6. W. L. Barnes, A. Dereux, and T. W. Ebbesen, "Surface plasmon subwavelength optics," Nature 424(6950), 824-830 (2003).
7. E. Ozbay, "Plasmonics: merging photonics and electronics at nanoscale dimensions," Science 311(5758), 189-193 (2006).
8. J. A. Schuller, E. S. Barnard, W. Cai, Y. C. Jun, J. S. White, and M. L. Brongersma, "Plasmonics for extreme light concentration and manipulation," Nat. Mater. 9(3), 193-204 (2010).
9. H. Lu, J. Zhao, and M. Gu, "Nanowires-assisted excitation and propagation of mid-infrared surface plasmon polaritons in graphene," J. Appl. Phys. 120(16), 83-91 (2016).

10. D. F. P. Pile, and D. K. Gramotnev, "Channel plasmon–polariton in a triangular groove on a metal surface," Opt. Lett. 29(10), 1069-1071 (2004).

11. S. A. Maier, P. G. Kik, H. A. Atwater, S. Meltzer, E. Harel, B. E. Koel, and A. A. G. Requicha, "Local detection of electromagnetic energy transport below the diffraction limit in metal nanoparticle plasmon waveguides," Nat. Mater. 2(4), 229 232 (2003).

12. R. F. Oulton, V. J. Sorger, G. Bartal, and X. Zhang, "A Hybrid Plasmonic Waveguide for Subwavelength Confinement and Long Range Propagation," Nat. Photonics 2(8), 496-500 (2008).

13. R. Salvador, A. Martinez, C. Garcia-Meca, and R. Ortuno, "Analysis of Hybrid Dielectric Plasmonic Waveguides," IEEE J. Sel. Top. Quant. 14(6), 1496-1501 (2008).

14. D. Dai, and S. He, "A silicon-based hybrid plasmonic waveguide with a metal cap for a nano-scale light confinement," Opt. Express 17(19), 16646-16653 (2009).

15. Y. Zhao, and L. Zhu, "Coaxial hybrid plasmonic nanowire waveguides," J. Opt. Soc. Am. B 27(6), 1260-1265 (2010).

16. D. Chen, "Cylindrical hybrid plasmonic waveguide for subwavelength confinement of light," Appl. Optics 49(36), 6868-6871 (2010).

17. S. Dahiya, S. Kumar, and B. K. Kaushik, "Hybrid plasmonic waveguide with centimeter-scale propagation length for nanoscale optical confinement," Appl. Optics 55(36), 10341-10346 (2016).

18. L. Chen, T. Zhang, X. Li, and W. Huang, "Novel hybrid plasmonic waveguide consisting of two identical dielectric nanowires symmetrically placed on each side of a thin metal film," Opt. Express 20(18), 20535-20544 (2012).

19. J. Tian, and M. Sun, "Modal properties of novel hybrid plasmonic waveguide consisting of two identical dielectric nanotubes symmetrically placed on both sides of a thin metal film," Eur. Phys. J. D 70(1), 1-7 (2016).

20. R. F. Oulton, G. Bartal, D. F. P. Pile, and X. Zhang, "Confinement and propagation characteristics of subwavelength plasmonic modes," New J. Phys. 10(10), 105018-105032 (2008).

21. H. Xu, and M. Käll, "Surface-plasmon-enhanced optical forces in silver nanoaggregates," Phys. Rev. Lett. 89(24), 246802-246805 (2002).

22. L. Chen, T. Zhang, and X. Li, "Enhanced Optical Forces by Hybrid Long-Range Plasmonic Waveguides," J. Lightwave Technol. 31(21), 3432-3438 (2013).

Photophysical properties of a novel axially substituted tetra-α-(pentyloxy) Titanium(IV) Phthalocyanine - Hematoxylin

Yufeng Jiang [1], Huafei Lv [1], Xinxin Yu [1], Sujuan Pan [1], Tiantian Zhang [1],

Yide Huang[2], Yuhua Wang[2], Shusen Xie[2] , Hongqin Yang[2*] , Yiru Peng [1*]

[1]College of Chemistry and Chemical Engineering, Fujian Normal University, Fuzhou 350007, China
[2]Key Laboratory of Optoelectronic Science and Technology for Medicine of Ministry of Education, Fujian Provincial Key Laboratory for Photonics Technology, Fujian Normal University, Fuzhou 350007, China
*Corresponding author, hqyang@fjnu.edu.cn, yirupeng@fjnu.edu.cn
Yufeng Jiangand [1] Huafei Lv [1] equally contributed to the paper.

Abstract Metal phthalocyanines (MPcs) are a very important class of widespread application compounds. They are not only used as dyes but also used as catalysts, data storage, electron charge carriers in photocopiers, photoconductors in chemical sensors, and photo-antenna devices in photosysthesis, photovoltaic cells. A particular application of Pcs is that it was used as a photosensitizers (PS) for treatment of certain cancer by photodynamic therapy of cancer and non-cancer diseases. However, the molecular aggregation of phthalocyanines, which is an intrinsic property of these large π-π conjugated systems, provides an efficient non-radioactive energy relaxation pathway, thereby shortening the excited state lifetimes, and reducing the photosensitizing and target efficiency. To overcome these problems, the introduction of axial ligands to phthalocyanine can prevent the formation of aggregation to some extend. In this paper, hematoxylin axially substituted tetra-α-(pentoxy) titanium (Ⅳ) phthalocyanine ($TiPc(OC_5H_{11})_4$-Hematoxylin) were characterized by elemental analysis, IR, ^{1}H NMR, UV-Vis, fluorescence spectra. No obviously Q band change was observed after the hematoxylin was substituted at the peripheral position of substituted phthalocyanine ring. Because of the rigidity structure of the hematoxylin, the fluorescence intensity of hematoxylin peripheral substituted phthalocyanine decreased compared with free substituted phthalocyanines. The fluorescence lifetimes of axially substituted phthalocyanine was fitted to be 3.613 ns. This compound may be considered as a promising photosensitizer for PDT.

1. Introduction

Photodynamic therapy (PDT) is now a well-recognized modality for the treatment of a variety of malignant tumors and wet age-related macular degeneration [1-3]. It utilizes the combined action of three individually nontoxic components, namely, a photosensitizer, light, and molecular oxygen, to induce oxidative cellular and tissue damage. Phthalocyanines are known to have the potential for use as second-generation photosensitizers for the photodynamic therapy of cancer. Phthalocyanines used for the photodynamic therapy of cancer are required to have a long life-time triplet sate and show strong absorption of far-red light between 600 and 800 nm, which has a greater penetration of tissue, and satisfactory photosensitization of singlet oxygen [4-9]. A problem for phthalocyanine is that it has limited accumulation in the tumor tissues, which is an intrinsic property of this large π-π conjugated systems, which reduce the photosensitizing and target efficiency of phthalocyanine. The aggregation of phthalocyanine can be reduced efficiently through intrudcing hinderance structure to peripheral position of phthalocyanine ring . Catechol (1,2-dihydroxybenzene) units are often present in enzymes, antioxidants, and hormones, which play an important role in organisms [10]. Hematoxylin were introduced in the Ps as the portion of axial ligands would gain higher PDT efficacy and bioavailability.

2. Materials and methods

2.1.Materials

NMR spectra were recorded on a Bruker 400 MHz FT-NMR spectrometer. MALDI-TOF MS was recorded on a Micromass Q-TOF micro (model). The IR spectra were acquired on a Thermo Nicolet 5700 spectrometer. The steady-state and time-resolved fluorescence spectra were measured on the FL900/FS920 steady-state fluorescence spectrometers. UV/Vis spectra was recorded on the Perkin-Elmer Lambda 9 UV/Vis spectrometers.

The fluorescence decay curves were analyzed by a re-convolution of the instrumental response function. Upon excitation at 350 nm, the decay curves of the emission were obtained and the lifetime could be fitted by exponential equation: Fit=A+B1e-t/τ1+ B2e-t/τ2+ B3e-t/τ3+ B1e-t/τ4.

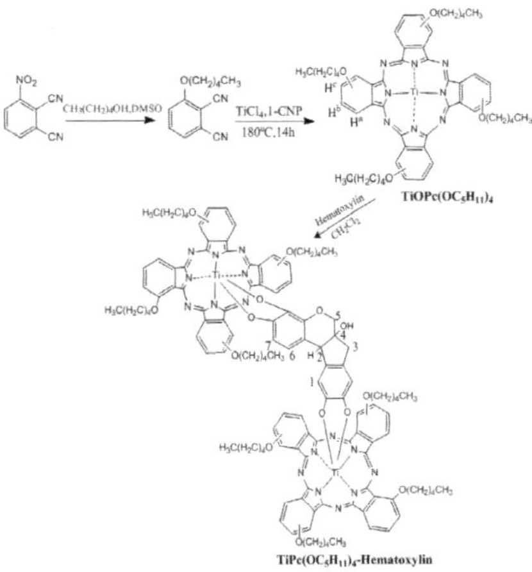

Fig .1 Scheme of the synthesis of $TiPc(OC_5H_{11})_4$-Hematoxylin

2.2 Synthesis and characterization

2.2.1 Synthesis of 3-pentoxy phthalonitrile

3-pentoxy phthalonitrile were prepared by dissolving 3-nitrophthalonitrile (5.0 g, 29 mmol) in the mixture of dimethyl sulfoxide (DMSO) (100 mL), dry pentanol (15 mL), and K_2CO_3 (10 g). The mixture was refluxed under nitrogen for 5h at 80℃. The reactions was monitored by thin layer chromatography (TLC).The crude products was washed with water (500mL) and the resulting solids were collected by filtration and purified by recrystal in methanol. Yield: 50%. m.p.112℃~114℃; FT-IR (KBr, cm^{-1}): 3087, 2989, 2866, 2233, 1245~1052(C-O-C); ^1H NMR(400Hz, $CDCl_3$, ppm): δ=7.216~7.235 (d, J=9.5Hz, 1H, H^2), 7.325~7.339 (s, 1H, H^1), 7.612~7.627 (s, 1H, H^3), 4.12 (d, J=6Hz, 2H, H^4), 1.887(s, 2H, H^5), 1.47 (d, J=6Hz, 2H, H^6), 1.39 (d, J=6.5Hz, 3H, H^8), 0.93 (d, J=6.5Hz, 2H, H^7). Anal. Calcd. for $C_{13}H_{14}ON_2$; C 72.89, H 6.54, N 13.08; Found: C 72.56; H 6.23; N 13.65.

2.2.2 Synthesis of tetra-α-(pentyloxy) titanium(IV) phthalocyanine ($TiOPc(OC_5H_{11})_4$)

A mixture of 3-pentoxy phthalonitrile (1.5622 g, 7.3 mmol), 1-chloronaphthalene · (8 mL), titanium tetrachloride (0.26 mL, 2.4 mmol) were refluxed under N_2 atmosphere for 14 h at 180℃. After being cooled to room temperature, n-hexane was added in order to precipitate the product. After using soxhlet extracted for 24 h by n-hexane and dichloromethane, the resulting s crude products were purified by flash chromatography on silica (v/v= 10:1 trichloromethane/methanol). Yield: 48%; m.p.>300℃; FT-IR (KBr, cm^{-1}): 2855~2950, 2230; 1590, 1487, 1459 (C＝C, C＝N), 1230~1055, 961 (Ti＝O), 738 (-$(CH_2)_4$-); ^1H NMR(400Hz, $CDCl_3$, ppm): δ=9.062 (s, 4H, H^a), 7.712~7.664 (d, 4H, H^b), 8.156 (s, 4H, H^c), 4.942~4.743 (d, 8H, H^1), 2.465~2.352 (d, 8H, H^2), 2.120 (s, 8H, H^3), 1.764~1.519 (t, 8H, H^4), 1.616~0.933 (t, 12H, H^5). UV–vis ($CHCl_3$) λmax(logε) = 327 (4.72), 660 (4.56), 736 (5.29); Anal. Calcd. for $C_{52}H_{56}N_8O_5Ti$; C 67.83; H 6.09; N 12.17; Found: C 67.90; H 6.07; N 12.28; MS: m/z: Calcd: 920; Found: [M＋H]＋921.

2.2.3 Synthesis of hematoxylin axially substituted tetra-α-(pentyloxy) Titanium(IV) Phthalocyanine ($TiPc(OC_5H_{11})_4$-Hematoxylin)

Tetra-α-(pentyloxy) titanium(IV) phthalocyanine ($TiOPc(OC_5H_{11})_4$) (46 mg, 0.05 mmol), hematoxylin (9 mg, 0.374 mmol) were dissolved in dichloromethane (4 mL). The mixture was reacted under nitrogen for 3 days at room temperature. Trichloromethane and water was added in order to precipitate the product. The resulting solids were collected by filtration and purified by recrystalliation (3~4 times) in methylene chloride/n-hexane. Yield: 50%; m.p.> 300℃; FT-IR (KBr, cm^{-1}): 2855~2950, 2230; 1590, 1487, 1459 (C＝C, C＝N), 1230~1055, 954 (Ti-O), 738 (-$(CH_2)_4$-); ^1H NMR (400Hz, $CDCl_3$, ppm): δ= 9.136 (s, 8H, H^a), 8.272~8.030 (m, 8H, H_c), 7.718~7.577 (m, 8H, H^b), 7.428～7.410 (t, 2H, H^1), 5.300(s, 1H, H^2), 4.202~4.168 (t, 2H, H^5), 2.511~2.406 (d, 1H, H^4); UV–vis ($CHCl_3$) λmax(logε) = 328 (4.87), 660 (4.75), 738 (5.28); Anal. Calcd. for C120H122N16O14Ti2; C 68.38; H 5.79; N 10.64; Found: C 68.40; H 5.77; N 10.62; MS: m/z: Calcd: 2107; Found: [M＋H]＋2108.

3. Results and Discussion

3.1 Photophysical properties of a novel Titanium(IV) Phthalocyanine

The UV/Vis spectra of $TiOPc(OC_5H_{11})_4$ and $TiPc(OC_5H_{11})_4$-Hematoxylin in $CHCl_3$ were showed

IOP Conf. Series: Journal of Physics: Conf. Series **844** (2017) 012049 doi:10.1088/1742-6596/844/1/012049

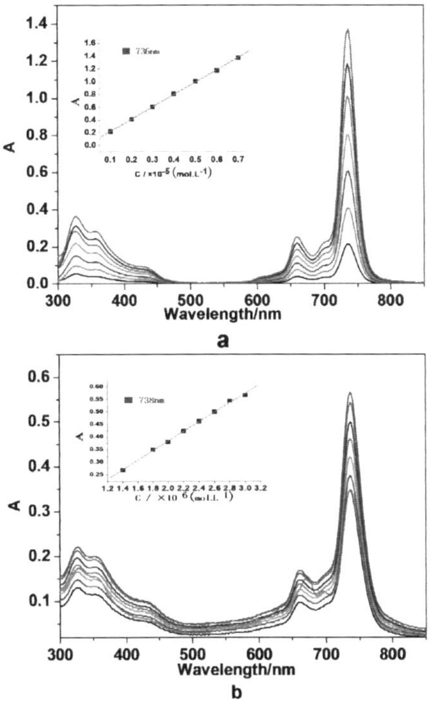

Fig.2 UV/Vis spectra of $TiOPc(OC_5H_{11})_4$ and $TiPc(OC_5H_{11})_4$-Hematoxylinin $CHCl_3$

$TiOPc(OC_5H_{11})_4$ (a) and $TiPc(OC_5H_{11})_4$-Hematoxylin (b)

(Inset: relationship between the absorbance and the concentration of $TiOPc(OC_5H_{11})_4$ (a) and $TiPc(OC_5H_{11})_4$-Hematoxylin (b))

Compound	$CHCl_3$(Qmax/nm)	CH_2Cl_2(Qmax/nm)	DMF(Qmax/nm)	THF(Qmax/nm)
$TiOPc(OC_5H_{11})_4$	736	734	739	727
$TiPc(OC_5H_{11})_4$-Hematoxylin	738	737	–	–

Table.1 Absorption spectra of $TiOPc(OC_5H_{11})_4$、$TiPc(OC_5H_{11})_4$-Hematoxylin in different solvents

in Fig.2. It exhibited B-bands is about 327 nm and a strong Q-band at 736 nm and 738 nm, respectively.The maximum absorption wavelength of Q band of $TiOPc(OC_5H_{11})_4$ and $TiPc(OC_5H_{11})_4$-Hematoxylin increasing with the increase of the electron donating ability of solvents (Table.1). It is probably due to the electronic donating ability of solvents binding to the metal ions. Because of electron donating ability slightly weakened the coordination ability of the phthalocyaninering, thus the Q band of phthalocyanine was blue-shift [11].

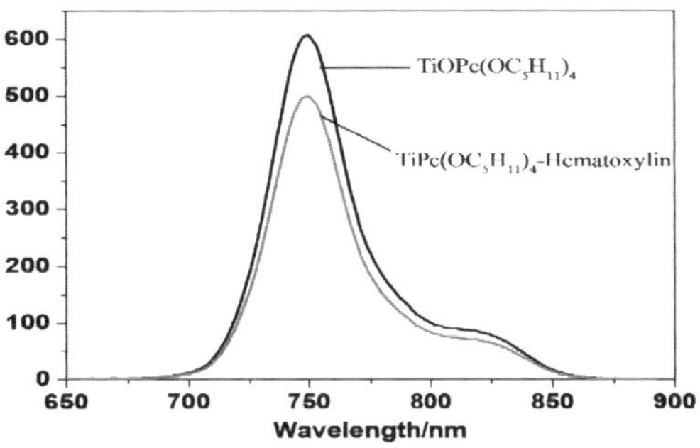

Fig.3 Fluorescence spectra of $TiOPc(OC_5H_{11})_4$、$TiPc(OC_5H_{11})_4$-Hematoxylin in $CHCl_3$ λex= 610nm

The fluorescence spectra of $TiOPc(OC_5H_{11})_4$ and $TiPc(OC_5H_{11})_4$-Hematoxylin in $CHCl_3$ was showed in Fig.3. Fluorescence emission peak was observed at 750 nm for $TiOPc(OC_5H_{11})_4$ and $TiPc(OC_5H_{11})_4$-Hematoxylin upon excitation at 610 nm. $TiOPc(OC_5H_{11})_4$ has a relatively higher fluorescence intensity than $TiPc(OC_5H_{11})_4$-Hematoxylin. It is probably due to the rigidity of hematoxylin, fluorescence intensity of axially substituted phthalocyanine were decreased compared with free phthalocyanines [12].

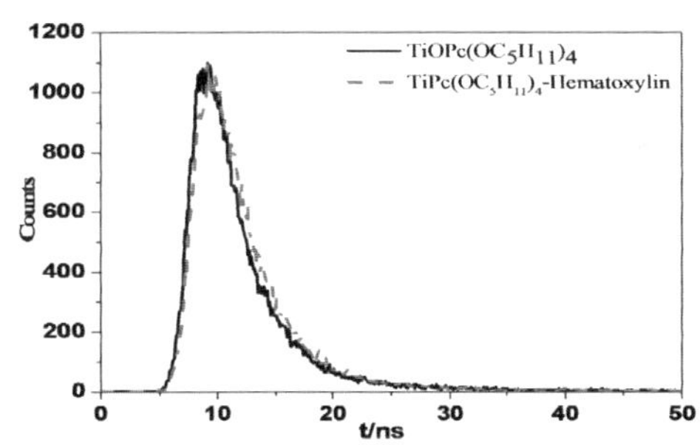

Fig.4 Fluorescence decay curve of $TiOPc(OC_5H_{11})_4$ and $TiPc(OC_5H_{11})_4$-Hematoxylin in $CHCl_3$

(λex=350nm, $C=1.0\times10^{-5}$ mol/L)

The fluorescence decay curves of $TiOPc(OC_5H_{11})_4$ and $TiPc(OC_5H_{11})_4$-Hematoxylin in $CHCl_3$ were showed in Fig.4. The fluorescence lifetimes of axially substituted phthalocyanine was fitted to be 3.613 ns was higher compared to unsubstituted phthalocyanine (3.417 ns).

4. Conclusions

The synthesis and characterization of tetra-α-(pentyloxy)

titanium(IV) phthalocyanine ($TiOPc(OC_5H_{11})_4$) and hematoxylin axially substituted tetra-α-(pentoxy) titanium(IV) phthalocyanine ($TiPc(OC_5H_{11})_4$-Hematoxylin) have been successfully accomplished. The UV-Vis absorption and fluorescence spectra of the compounds in $CHCl_3$ were examined. Their photophysical properties and steady-state fluorescence in $CHCl_3$ were also studied. It was found that no obviously Q band change was observed between the axially substituted phthalocyanine and unsubstituted phthalocyanine. Because of the introducing of hematoxylin on the axially position, fluorescence intensity of $TiPc(OC_5H_{11})_4$-Hematoxylin were decreased compared with free phthalocyanines. However, the fluorescence lifetimes of axially substituted phthalocyanine was higher compared to unsubstituted phthalocyanine.

Acknowledgment

This work is supported by the National Key Basic Research Program of China (973 project) (2015CB352006), National Natural Science Foundation of China (61335011, 21274021), Natural Science Foundation of Fujian (2015J01040) and National Undergraduate Training Programs for Innovation and Entrepreneurship under Grant No. 201610394006.

Reference

[1] Dolmans D E J G J, Dai F, Jain R K, "TIMELINE: Photodynamic therapy for cancer," Nature Reviews Cancer. 3(5), 380-387 (2003).

[2] Brown S B, Brown E A, Walker I, "The present and future role of photodynamic therapy in cancer treatment," Lancet Oncol. 5(8), 497-508 (2004).

[3] Celli J P, Spring B Q, Rizvi I, et al, "Imaging and photodynamic therapy: mechanisms, monitoring, and optimization," Chemical Reviews. 110(5), 2795-838 (2010).

[4] Kolarova H, Nevrelova P, Bajgar R, et al, "In vitro photodynamic therapy on melanoma cell lines with phthalocyanine," Toxicology in Vitro. 21(2), 249-253 (2007).

[5] Haywood-Small S L, Vernon D I, Griffiths J, et al, "Phthalocyanine-mediated photodynamic therapy induces cell death and a G0/G1 cell cycle arrest in cervical cancer cells," Biochemical & Biophysical Research Communications. 339(2), 569-576 (2006).

[6] Kolarova H, Nevrelova P, Bajgar R, et al, "In vitro photodynamic therapy on melanoma cell lines with phthalocyanine," Toxicology in Vitro. 21(2), 249-253 (2007).

[7] Banfi S, Caruso E, Buccafurni L, et al, "Zinc phthalocyanines-mediated photodynamic therapy induces cell death in adenocarcinoma cells," Journal of Organometallic Chemistry. 692(6), 1269-1276 (2007).

[8] Huang H F, Chen Y Z, Wu Y, et al, "Efficacy of ZnPcS2P2 photodynamic therapy solely or with tumor vaccines on mouse tumor models," Photodiagnosis & Photodynamic Therapy. 4(2), 100-105 (2007).

[9] Kolarova H, Nevrelova P, Bajgar R, et al, "In vitro photodynamic therapy on melanoma cell lines with phthalocyanine," Toxicology in Vitro. 21(2), 249-253 (2007).

[10] Atsuya Muranaka, Masako Okuda, Nagao Kobayashi, et al, "Recognition of Chiral Catechols Using Oxo−Titanium Phthalocyanine," Journal of the American Chemical Society. 126(14), 4596 (2004).

[11] Margaron P, Langlois R, Lier J E V, et al, "Photodynamic properties of naphthosulfobenzoporphyrazines, novel asymmetric, amphiphilic phthalocyanine derivatives," J Photochem Photobiol B. 14(3), 187-99 (1992).

[12] Rihter B D, Kenney M E, Ford W E, et al, "Synthesis and photoproperties of diamagnetic octabutoxyphthalocyanines with deep red optical absorbance," Cheminform. 22(5), 8064-8070 (1991).

IOP Conf. Series: Journal of Physics: Conf. Series **844** (2017) 012050 doi:10.1088/1742-6596/844/1/012050

Algorithm for Evaluation of Temperature 3D-Distribution of a Vapor Cell in a Diode End-pumped Alkali Laser System

J H Han, Y Wang*, H Cai, G F An, K P Rong, H Yu, S Y Wang, H Y Wang, W Zhang, L P Xue and J Zhou

Southwest Institute of Technical Physics, Chengdu, Sichuan 610041, China

*Corresponding Author. E-Mail: youwang_2007@aliyun.com

Abstract. We develop a new 3D-model to evaluate the light characteristics and the thermal features of a cesium-vapor laser end-pumped by a laser diode. The theoretical model is based on the principles of both heat transfer and laser kinetics. The 3-dimensional population density distribution and temperature distribution are both systematically obtained and analyzed. The methodology is thought to be useful for realization of a high-powered diode-pumped alkali laser (DPAL) in the future.

1. Introduction

With narrow linewidth, high Stokes-efficiency, compact size, good thermal performance, non-toxic system, etc., a diode-pumped alkali laser (DPAL) becomes one of the most hopeful high-powered laser sources of the next generation and has been rapidly developed since the beginning of the 21th century [1-5]. However, the thermal effects will bring about some serious problems in nonuniformity of the temperature distribution for a high-powered DPAL system because the thermal conductivity of a gas-state medium is so small that the generated heat cannot be transferred outside efficiently [6-9]. Unlike a conventional electrically-excited gas-state laser, the number densities of alkali vapors (gain gas) and buffer gases (generally being helium and small hydrocarbons such as methane and ethane) in a DPAL often exhibit inhomogeneous distributions resulting from the temperature gradient inside a vapor cell [10, 11]. Actually, the inhomogeneous of the alkali vapor directly influences the output performances of a DPAL. Furthermore, the nonuniformity of buffer gases affects the line-center collisionally broadened cross sections of both the D_1 line ($n^2P_{1/2} \rightarrow n^2S_{1/2}$) and the D_2 line ($n^2S_{1/2} \rightarrow n^2P_{3/2}$) of an alkali atom as well as the finestructure mixing rate ($n^2P_{3/2} \rightarrow n^2P_{1/2}$) [12-15].

2. Experimental Setup

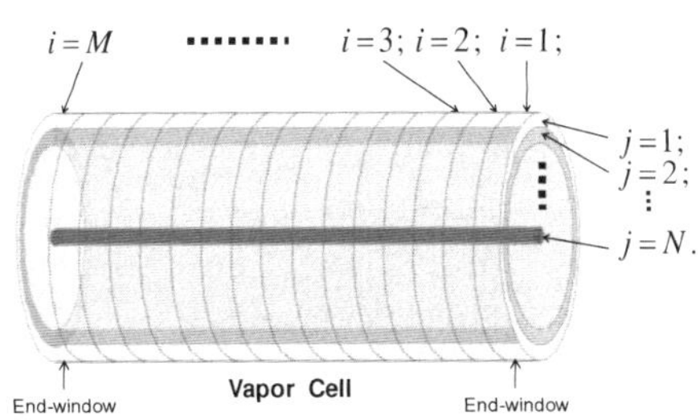

Figure 1. Schematic illustration for segmentation of a gain media cell.

For an end-pumped DPAL configuration, the optical axis of a pump laser diode coincides with that of the output laser. As shown in Figure 1, we divide the cylindrical gas vapor cell into many cylindrical annuli along the optical axis direction with $i = 1, 2, 3\ldots M$ and the transversal surface direction with $j = 1, 2, 3 \ldots N$. Every small cylindrical annulus is thought as a heat source and a gain source. Both the gain and the heat generated in the cylindrical annulus are different with each other.

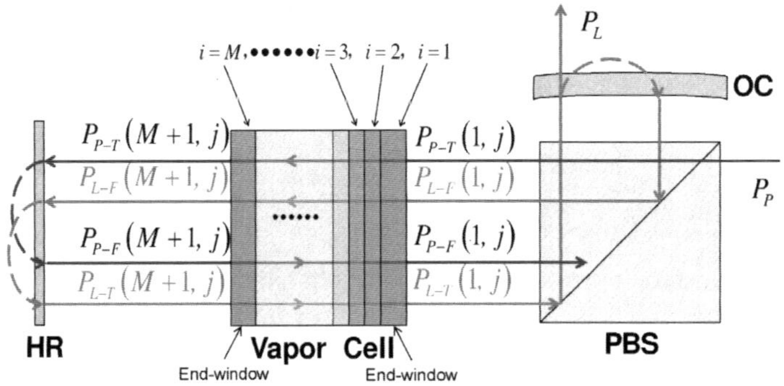

Figure 2. Schematic illustration showing pump and laser propagation directions at various points as the pump and laser fluence traverses the gain media in a double-pass geometry.

The pump and laser energies at the outside surface of the end-windows as the pump and laser fluence goes through the gain medium in double-pass geometry can be expressed by

$$\begin{cases} P_{P-T}(1,\ j) = \eta_{del} P_P^j \\ P_{P-F}(M+1,\ j) = R_{\text{HR}} \cdot P_{P-T}(M+1,\ j) \end{cases}, \tag{1}$$

and

$$\begin{cases} P_{L-F}(1,\ j) = R_{\text{OC}} \cdot P_{L-T}(1,\ j) \\ P_{L-T}(M+1,\ j) = R_{\text{HR}} \cdot P_{L-F}(M+1,\ j) \\ P_L^j = (1 - R_{\text{OC}}) \cdot P_{L-T}(1,\ j) \end{cases}. \tag{2}$$

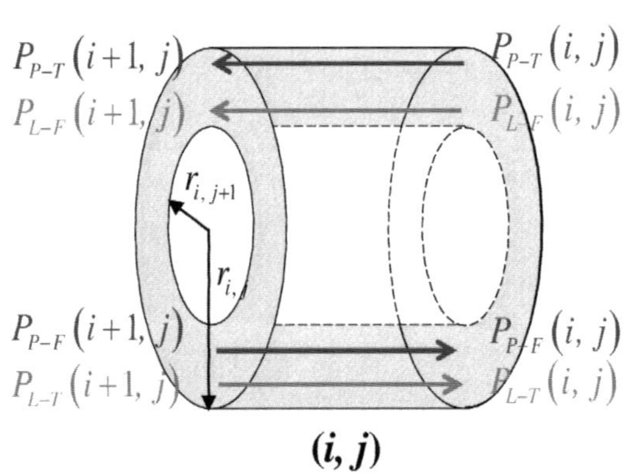

Figure 3. Diagram describing the power changes when the pump and laser fluence traverses the (*i*, *j*) cylindrical annulus.

As shown in Figure 3, we select an arbitrarily cylindrical annulus (*i*, *j*) among the segments to analyze the laser and heat features. $P_{P\text{-}T}(i, j)$ and $P_{P\text{-}F}(i, j)$ are the pump power corresponding to the (*i*, *j*) cylindrical annulus transferred towards the left and right (reflected back from the high reflector), respectively. $P_{L\text{-}F}$ (i, j) and $P_{L\text{-}T}$ (i, j) are the laser power corresponding to the (i, j) cylindrical annulus transferring towards the left (reflected back from the output coupler), respectively.

The relationships between $P_{P\text{-}T}$ (i, j) and $P_{P\text{-}T}$ $(i+1, j)$, $P_{P\text{-}F}(i, j)$ and $P_{P\text{-}F}$ $(i+1, j)$, $P_{L\text{-}T}$ (i, j) and $P_{L\text{-}T}$ $(i+1, j)$, $P_{L\text{-}F}(i, j)$ and $P_{L\text{-}F}$ $(i+1, j)$, can be respectively given by

$$\begin{cases} P_{P-T}(\mathrm{i},j)\exp\left[-\left(n_1^j-\frac{1}{2}n_3^j\right)\sigma_{D_2}(T_j,\lambda)\frac{\mathrm{L}}{\mathrm{M}}\right]=P_{P-T}(\mathrm{i}+1,j) \\ P_{P-F}(\mathrm{i}+1,j)\exp\left[-\left(n_1^j-\frac{1}{2}n_3^j\right)\sigma_{D_2}(T_j,\lambda)\frac{\mathrm{L}}{\mathrm{M}}\right]=P_{P-F}(\mathrm{i},j) \end{cases} , \tag{3}$$

and

$$\begin{cases} P_{L-T}(i+1,j)\exp\left[\left(n_2^{(i,j)}-n_1^{(i,j)}\right)\sigma_{D_1}^{He-broadened}(i,j)\frac{\mathrm{L}}{\mathrm{M}}\right]TT^{1/N}=P_{L-T}(i,j) \\ P_{L-F}(i,j)\exp\left[\left(n_2^{(i,j)}-n_1^{(i,j)}\right)\sigma_{D_1}^{He-broadened}(i,j)\frac{\mathrm{L}}{\mathrm{M}}\right]TT^{1/N}=P_{L-F}(i+1,j) \end{cases} . \tag{4}$$

The equated power of the pump and the laser power can be simply expressed by

$$P_P(i,j)=P_{P-T}(i,j)+P_{P-F}(i+1,j) \quad , \tag{5}$$

and

$$P_L(i,j)=P_{L-T}(i,j)+P_{L-F}(i+1,j) \quad . \tag{6}$$

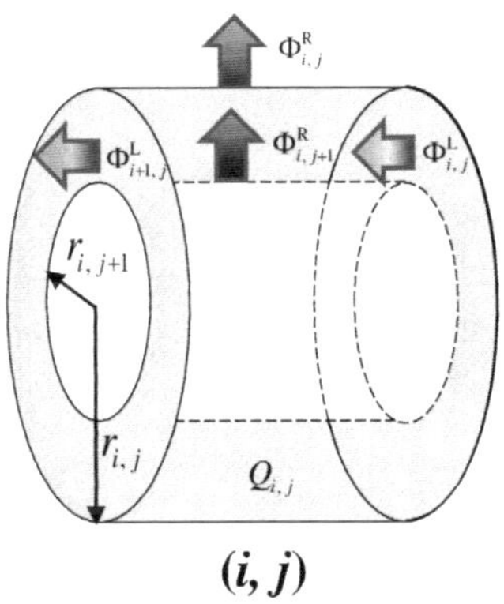

Figure 4. Drawing of illustrating heat generated and transferred for the (i, j) cylindrical annulus of a vapor cell.

Next, the heat characteristics can be shown as Figure 4. The relationship between the *input heat* and the *output heat* as well as the *generated heat* the can be described as

$$\Phi^{R}_{i,j}+Q_{i,j}+\Phi^{R}_{i,j+1}=\Phi^{L}_{i+1,j}+\Phi^{L}_{i,j} \quad . \tag{7}$$

Then, we calculate the population distribution in a three energy-level system of the (i, j) cylindrical annulus (see Figure 3) by using the well-known rate equations as follows [5, 11, 12]:

$$\begin{cases} \dfrac{dn_1(i,j)}{dt}=-\Gamma_P(i,j)+\Gamma_L(i,j)+\dfrac{n_2(i,j)}{\tau_{D_1}}+\dfrac{n_3(i,j)}{\tau_{D_2}} \\ \dfrac{dn_2(i,j)}{dt}=-\Gamma_L(i,j)+\gamma_{32}(T_{i,j})\left[n_3(i,j)-2n_2(i,j)\exp\left(-\dfrac{\Delta E}{k_BT_{i,j}}\right)\right]-\dfrac{n_2(i,j)}{\tau_{D_1}} \\ \dfrac{dn_3(i,j)}{dt}=\Gamma_P(i,j)-\gamma_{32}(T_{i,j})\left[n_3(i,j)-2n_2(i,j)\exp\left(-\dfrac{\Delta E}{k_BT_{i,j}}\right)\right]-\dfrac{n_3(i,j)}{\tau_{D_2}} \end{cases} . \tag{8}$$

IOP Conf. Series: Journal of Physics: Conf. Series **844** (2017) 012050 doi:10.1088/1742-6596/844/1/012050

We calculate the volume density of generated heat of the *(i, j)* cylindrical annulus by using the following formula:

$$Q_{i,j} = V_{i,j} \cdot \gamma_{{}^2P_{3/2} \to {}^2P_{1/2}} \left(T_{i,j}\right) \left[n_3(i,j) - 2n_2(i,j)\exp(-\frac{\Delta E}{k_B T_{i,j}}) \right] \Delta E \quad . \tag{9}$$

By utilizing the theoretical model, the thermal features and the physical characteristics of a DPAL can be obtained.

3. Results and Discussions

By using the theoretical model introduced in Section 2, we calculate the physical characteristics of a DPAL and the thermal features of a cesium cell. The relevant parameters are listed in Table 1.

Table 1. Parameters for evaluating temperature distribution of a Cs vapor cell

Parameter Description	Values
Partial pressure of He	500 Torr
Partial pressure of C_2H_6	100 Torr
η_{del}	90%
L	25 *mm*
R	5 *mm*
T_W	100 ℃
Pump power	100 W
End-windows thickness	1 *mm*
R_{OC}	30%
Room temperature	20 ℃
Pump FWHM	30 GHz
Center wavelength of the pump	852.3 *nm*
D_1 transition wavelength	894.6 *nm*
Laser beam waists	0.5 *mm*
Pump beam waists	1.1 *mm*

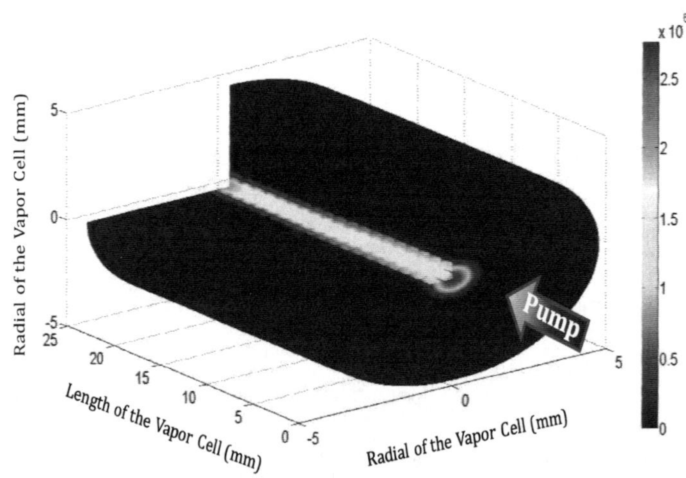

Figure 5. 3-D distribution of absorbed pump density inside a cesium cell.

We first analyze the distribution of absorbed pump density inside a vapor cell. As shown in Figure 5, the absorption of gas medium nearby the two end-windows is much stronger than those around the central area. The reason is that the pump power is high enough to make the gas medium around the central core be absorbed adequately, and the gas medium density nearby the end-windows is higher than that around the central core since the temperatures nearby the end-windows are lower than those in the other places. The laser density of the pump input side is higher than another side (see Figure 6) as the gain of the input side is higher than another one.

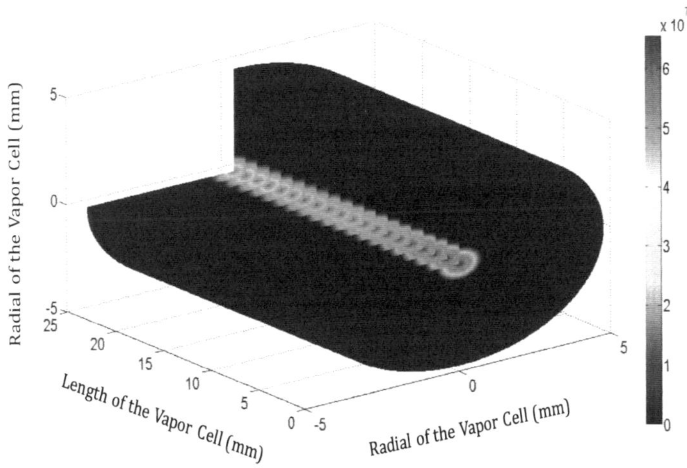

Figure 6. 3-D density distribution of laser intensity inside a cesium cell.

As shown in Figure 7, the temperature distribution exhibits a distinct gradient and achieves the maximum values at the central axis of the cell. Such tendencies can be explained by a fact that the thermal conductivity of a gas-state medium is so small that the generated heat cannot be transferred outside efficiently. It can be found that the absorbed pump power leads to a more obvious temperature rise for a static medium. The maximum temperature rise even reaches about 550 K at the cell axis. However, such a temperature gradient can be effectively decreased with a flowing approach. The published literatures have proven that the flowing-gas procedure is essential and effective to construct an even-higher DPAL system.

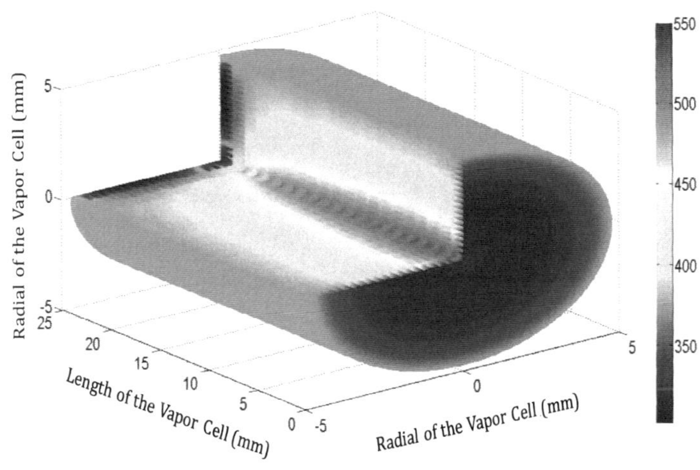

Figure 7. 3-D temperature distribution of a cesium cell.

IOP Conf. Series: Journal of Physics: Conf. Series **844** (2017) 012050 doi:10.1088/1742-6596/844/1/012050

4. Conclusions

In this report, we present the results of theoretical studies on the thermal features of a cesium vapor cell pumped by a laser diode. A theoretical model is developed by combining the procedures of laser kinetics and heat transfer together.

The temperature around the cell axis is somewhat higher than the other positions inside the cell. The highest temperature is located at the central optical axis near the pump input side. The absorption efficiency, gain intensity and generated thermal power near the pump input side are much stronger than those near the opposite side.

References

[1] Krupke W F, Beach R J, Kanz V K and Payne S A 2003 *Opt. Lett.* **28** 2336–2338
[2] Page R H, Beach R J, and Kanz V K5 2006 *Opt. Lett.* **31** 353–355
[3] Zhdanov B V and Knize R J 2007 *Opt. Lett.* **32** 2167–2169
[4] Wang Y, Niigaki M, Fukuoka H, Zheng Y, Miyajima H, Matsuoka S, Kubomura H, Hiruma T, and Kan H 2007 *Phys. Lett. A* **360** 659–663
[5] Cai H, Wang Y, Xue L P, Zhang W, Han J H, Wang H Y, and An G F 2014 *Appl. Phys.* B **117** 1201–1210
[6] Zhu Q, Pan B L, Chen L, Wang Y J, and Zhang X Y 2010 *Opt. Commun.* **283** 2406–2410
[7] Barmashenko B D and Rosenwaks S, 2013 *Opt. Soc.* Am. B **30** 1118–1126
[8] Barmashenko B D and Rosenwaks S 2013 *Appl. Phys. Lett.* **102** 141108–141111
[9] Rosenwaks S, Barmashenko B D, and Waichman K 2014 *Proc. SPIE* ***8962*** 311–316
[10] Steck D A 2010 Available online at http://steckus/alkalidata
[11] Han J H, Wang Y, Cai H, Zhang W, Xue L P, and Wang H Y 2014 *Opt. Express.* **22** 13988–14003
[12] Beach R J, Krupke W F, Kanz V K, and Payne S A 2004 *J. Opt. Soc. Am. B* **21** 2151–2163
[13] Yang Z, Wang H, Lu Q, Hua W, and Xu X 2011 *Optics. Express* **19** 23118
[14] Bogachev A V, Garanin S G, Dudov A M, Yeroshenko V A, Kulikov S M, Mikaelian G T, Panarin V A, Pautov V O, Rus A V, and Sukharev S A 2012 *Quantum Electronics* **42** 95–98
[15] Rice C A, Lott G E, and Perram G P 2012 *Appl. Opt.* **51** 8102–8110

IOP Conf. Series: Journal of Physics: Conf. Series **844** (2017) 012051 doi:10.1088/1742-6596/844/1/012051

Photoacoustic imaging of early gastric cancer diagnosis based on long focal area ultrasound transducer

Huaqin Wu, Zuoran Li, Lantian Liu, Zhifang Li, Shulian Wu, Hui Li*
Key Laboratory of Optoelectronic Science and Technology for Medicine, Ministry of Education, Fujian Normal University, Fuzhou, China

E-mail: hli@fjnu.edu.cn

Abstract We illustrated a novel imaging method to diagnose gastric neoplasms via photoacoustic tomography (PAT). Depending on the structural characteristics of gastric cavity, we used column diffusion fiber to irradiate the stomach tissue through the esophagus, and the externally placed telecentric focus ultrasonic transducer detected photoacoustic signals from the gastric tissue. We reconstructed the distribution of light energy deposition of the simulated gastric tumor, and obtained the location and size information of gastric tumor.

1 Introduction

Over the past two decades, the mortality of gastric cancer showed a continuous downward trend. This decline was the result of advances in detection and treatment of early gastric cancer. Fighting against gastric cancer needs continuous clinical and basic research—enhancing detection technology and methods of treatment. In addition, creative detection technology is indispensable[1].

An important sign of gastric cancer is oxygen overrun—high metabolism. Photoacoustic imaging technology can detect tissue oxygen metabolism. Photoacoustic imaging is an effective detection method for early gastric cancer because early gastric cancer has not spread that the diagnosis does not need contrast agent. Owing to the morphology and physiology pathological features of tumor, Photoacoustic imaging is used in early gastric cancer functional detection shows significant advantage than other advanced imaging methods which are used in various clinical applications. Therefore the photoacoustic imaging applys in early gastric cancer diagnosis, staging and nondestructive positioning shows that it is irreplaceable.

Photoacoustic imaging is based on the photoacoustic effects. Photoacoustic effects, discovered from solid by Bell in 1880[2]. Photoacoustic effects were applied in spectroscopic studies to form photoacoustic spectroscopy in 1970s [3][4][5][6]. In the 1980s, photoacoustic effects were applied in biological tissue imaging [7][8][9[. PAT[10] is a nondestructive and lossless method in biological photon imaging. PAT is based on the difference optical absorption of biological tissue and the ultrasound-mediated which combines high contrast ratio from pure optical imaging and high penetration depth from pure Ultrasonic imaging. In principle, the use of ultrasonic detectors instead of optical photon detection avoid the effect of optical scattering, so it can provide high contrast and high resolution in biological tissue imaging. PAT provides an important means of detection in biomedical diseases[11][12], such as in physiological characteristics [13][14], pathological features [15][16], metabolic function [17][18]. Photoacoustic imaging has wide application prospect in biomedical clinical diagnosis not only in vivo tissue structure but also in functional imaging.

2 PAT Principle and Setup

2.1 Principle

Figure.1 showed the schematic diagram of the PAT. A pulsed laser irradiates the tissue, light absorption tissue absorbs the light energy and expands thanks to photothermal effect that the ultrasound has generated. The external ultrasonic transducer receives the photoacoustic signal and stores the photoacoustic signal to the computer then analyzes the photoacoustic signal which carries the absorber information.

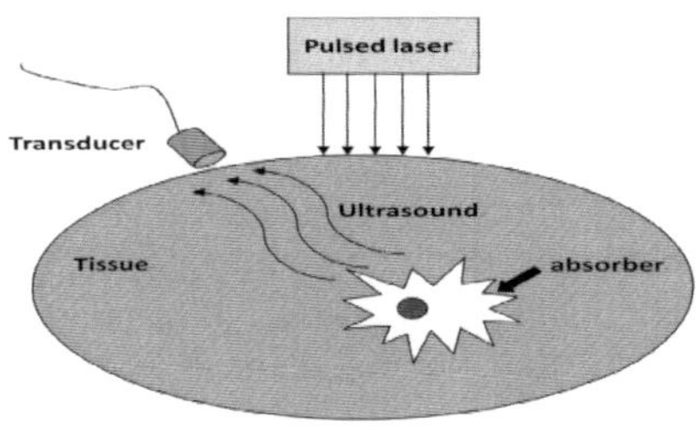

Figure.1 Schematic diagram of PAT.

Due to the different distance between absorber locations and transducer, the photoacoustic signals are received after different delay time. The position of absorber is determined by the product of time-resolved signals which detected by transducer and ultrasonic velocity. The size of the absorber can be calculated by using the distance between the anode and cathode of photoacoustic bipolar signal [18]. The amplitude of photoacoustic signal reflects the light energy deposition of the absorber. The relationship between in situ photoacoustic pressure and photon energy deposition as follow:

$$P_0(x) = \Gamma \mu_a F(x)$$

Where Γ represents the Grignon coefficient (biological tissue is 0.25), μ_a means the absorption coefficient of the absorber, F(x) is the light energy distribution at position x. The distribution of optical energy flow rate can be determined by the photoacoustic signals when the optical parameters are known. And then the absorption distribution in the tissue can be detected.

2.2 Experimental Setup

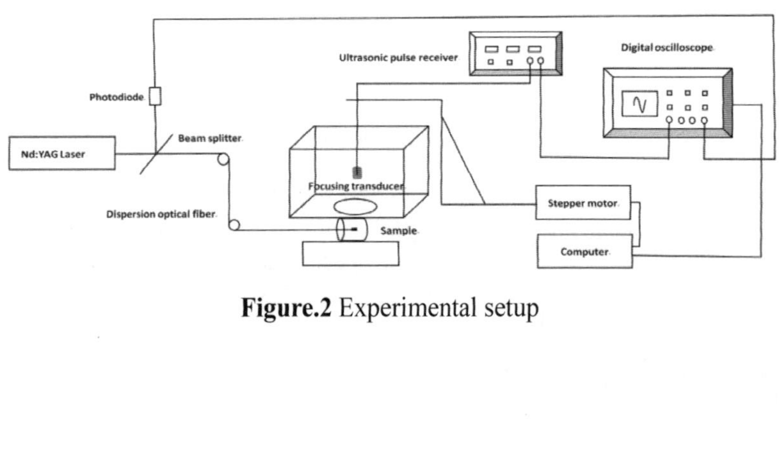

Figure.2 Experimental setup

The experimental setup was shown in Figure 2. The Q-switched Nd: YAG pulsed laser (Sutelite I-10, Continnum) was used to generate 532nm laser with 6ns pulse width and 10Hz repetition frequency. The short pulse laser was coupled to the customized multimode columnar dispersion fiber by optical coupler. According to the anatomical characteristics and the positional characteristics of stomach in the human body, the pulsed light was imported into the filling gastric cavity through the esophagus by fiber, and then irradiated the gastric tumor to produce photoacoustic signal. The signal was collected in the form of spatial scanning by the telephoto focus ultrasonic transducer which was placed in the left abdomen of the human body to achieve the detection of gastric cancer tissue.

3 Data Analysis

Figure 3 (b) and Figure 3 (c) showed the typically longitudinal photoacoustic pressure curve and the two-dimensional photoacoustic image which was obtained by scanning 5,000 times along the Y direction for figure 3 (a). It could be directly observed the location of simulated tumor in figure3(c), and the result showed that photoacoustic imaging has great potential value in non-destructive detecting for gastric tumor.

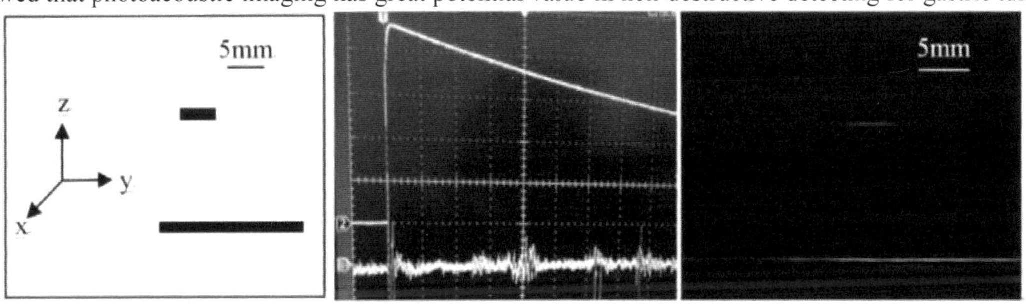

(a) (b) (c)

Figure.3 Simulated photoacoustic imaging of gastric tumor.

4 Summary and Outlook

The application range of emerging non-destructive biological tissue imaging technology photoacoustic imaging has been expanding very fast, however, there are few researches about the monitoring and diagnosis of gastric cancer. Owing to rich of blood vessels, the cancerous tissue is an excellent source of fresh blood. The blood has a strong absorption on the 532nm wavelength green light, there is a significant difference between normal tissue. Accordingly, photoacoustic imaging is very suitable for the detection of gastric cancer with the highly sensitive of difference in tissue optical absorption.

This paper took advantage of the ability of PAT, which combines the advantage of optical imaging and acoustic imaging, to reconstruct the light energy deposition distribution of gastric cancer, and obtained the size and location information. However, because of the acoustic signal reflection and mechanical vibration, the system will appear artifacts and errors. Thus more effort is needed to diminish the system error and improve the quality of photoacoustic image further.

Acknowledgment

This work was funded by the National Natural Science Foundation of China (No. 61675043).

Reference

[1] Siegel R L, Miller K D, Jemal A 2017 *CA-Cancer J Clin* **67** 7–30
[2] Bell A G 1880 *Am J Sci* **118** 305-324

[3] Rosencwaig A 1973 *Sci* **181** 657-658
[4] Rosencwaig A 1973 *Opt Commun* **7** 305-308
[5] Rosencwaig A, Gersho A 1976 *J Appl Phys* **47** 64-69
[6] Rosencwaig A 1980 *Wiley*
[7] Oraevsky A A, Jacques S L, Esenaliev R O 1998 *U.S. Pat* 5,840,023
[8] Hoelen C G A, De Mul F F M, Pongers R 1998 *Opt letters* **23** 648-650
[9] Hoelen C G A, de Mul F F M 2000 *Appl Opt* **39** 5872-5883
[10] Nagata Y, Koda T 1991 *U.S. Patent* 5,070,733
[11] Oladipupo S, Hu S, Kovalski J 2011 *Proc Nat Acad Sci* **108** 13264-13269
[12] Oladipupo S S, Hu S, Santeford A C 2011 *Blood* **117** 4142-4153
[13] Zhang E Z, Povazay B, Laufer J 2010 *Biomed opt express* **2** 2202-2215
[14] Hu S, Rao B, Maslov K 2010 *Opt letter* **35** 1-3
[15] Erpelding T N, Kim C, Pramanik M 2010 *Radiology* **256** 102-110
[16] Kim C, Cho E C, Chen J 2010 *ACS nano* **4** 4559-4564
[17] Hu S, Yan P, Maslov K 2009 *Opt letters* **34** 3899-3901
[18] Jansen K, Van Der Steen A F W, van Beusekom H M M 2011 *Opt letters* **36** 597-599

IOP Conf. Series: Journal of Physics: Conf. Series **844** (2017) 012052 doi:10.1088/1742-6596/844/1/012052

Continuous-Wave Alkali Vapor Laser Pumped by a Ti-sapphire Laser with Narrow Linewidth

H Cai[1], G F An[1], K Dai[2], Y Wang[1]*, W Zhang[1], J H Han[1], K P Rong[1], H Yu[1], S Y Wang[1], H Y Wang[1], L P Xue[1] and J Zhou[1]

[1]Southwest Institute of Technical Physics, Chengdu, Sichuan 610041, China
[2]School of Physics Science and Technology, Xinjiang University, Urumqi 830046, China

*Corresponding Author. E-Mail: youwang_2007@aliyun.com

Abstract. We have experimentally demonstrated the continuous wave rubidium and cesium lasers pumped by a Ti-sapphire laser with the linewidth of about 5 MHz. The pump and laser beams were orthogonally polarized and they can be separated by a polarized beam splitter (PBS). Two 4-cm-long cells were respectively filled with metallic rubidium and cesium as well as 300 Torr ethane as a buffer gas. A series of output couplers at different cell temperatures have been used and the optimal parameters have been found for earning the highest output. As a result, we have achieved a maximum output power of 111 mW with the optical to optical efficiency of 18.4% for a rubidium laser and a maximum output power of 136 mW with the optical to optical efficiency of 30% for a cesium laser, respectively. By considering there are no anti-reflection coatings on the surfaces of two cell end-windows, the output should be improved if the transmission attenuation is effectively decreased in the future.

1. Introduction

The concept of a diode pumped alkali laser (DPAL) was first proposed by William F. Krupke [1]. In the recent years, DPALs have drawn a lot of attentions because they combine the positive features of both gas lasers (GLs) and diode pumped solid state lasers (DPSSLs) [2-4]. As a gas-state laser, the DPAL contains several kinds of buffer gases and the vapor medium of three kinds of neutral alkali such as potassium (K), rubidium (Rb), and cesium (Cs) [5]. In a DPAL, an alkali atom along with the two electric-dipole-allowed transitions participates in the laser action and the typical three electronic-level is formed as shown in figure 1 [6]. Until now, several reports on the experimental studies of CW alkali lasers pumped with narrowband Titanium surrogate lasers have been published for the gain media of cesium and potassium vapors [7-10]. But the experimental report about the rubidium vapor laser pumped by a light source with the linewidth less than 100 MHz is rare [11-12]. In this study, we experimentally analyze the output features of rubidium and cesium alkali lasers with an end-pumped configuration. The output power of our optically pumped resonance transition rubidium laser achieves 111 mW, which should be the first experimental demonstration of the rubidium laser with the linewidth of a pump source narrower than 100 MHz worldwide.

IOP Conf. Series: Journal of Physics: Conf. Series **844** (2017) 012052 doi:10.1088/1742-6596/844/1/012052

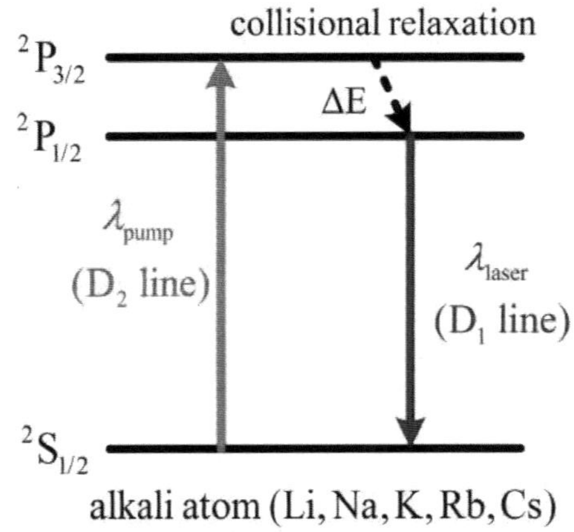

Figure 1. Energy level diagram and kinetic processes for an alkali laser.

2. Experimental Setup

The schematic diagram of the experimental setup of an optically pumped alkali laser is presented in figure 2. A tunable Ti:sapphire laser with a linewdith of about 5 MHz is used to longitudinally pump the gain media inside a Rb cell and a Cs cell in the experiment. Adopting a narrow linewidth pump source is helpful to fit the absorption band of alkali gain media. In addition, the shift of the central wavelength of such a pump source with the linewidth smaller than 10 MHz is much more insensitive to the output performances of a DPAL than a normal narrow-linewidth LD (usually about ~100 GHz). The pump and laser beams were orthogonally polarized so that they can be separated by a polarized beam splitter (PBS). Both Rb and Cs cells are filled with 300 Torr ethane as the buffer gas to broaden the absorption line (the D_2 line) and hasten the fine-structure mixing rate between the $^2P_{3/2}$ and $^2P_{1/2}$ levels [13]. By using such configurations, we have experimentally realized the narrow-linewidth output of both Rb and Cs vapor lasers. As both end-windows of the cell are not coated with anti-reflection films, the cavity transmission losses would bring about some negative effects on the laser output performance.

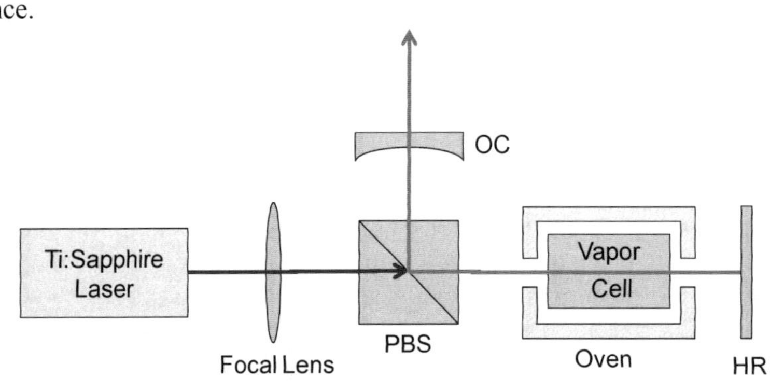

Figure 2. Schematic diagram of the experiment.

3. Results and Discussions

3.1. Output Features of a Rb laser

As the number density of alkali vapor closely relates to the cell temperature, the output power of a Rb laser will be strongly governed by the vapor temperature. Figure 3 shows the output power versus the cell temperature for different reflectances R_{oc} of an output coupler (OC) in a Rb laser system. The data have been obtained when the pump power was kept as 569 mW. We can see that there are the optimal temperatures in all curves which are 120, 113, 113, and 109 °C for the OC reflectances of 4, 50, 70, and 90%, respectively. The peak position corresponding to the maximum output shifts to the low temperature when the reflectance of an OC becomes higher.

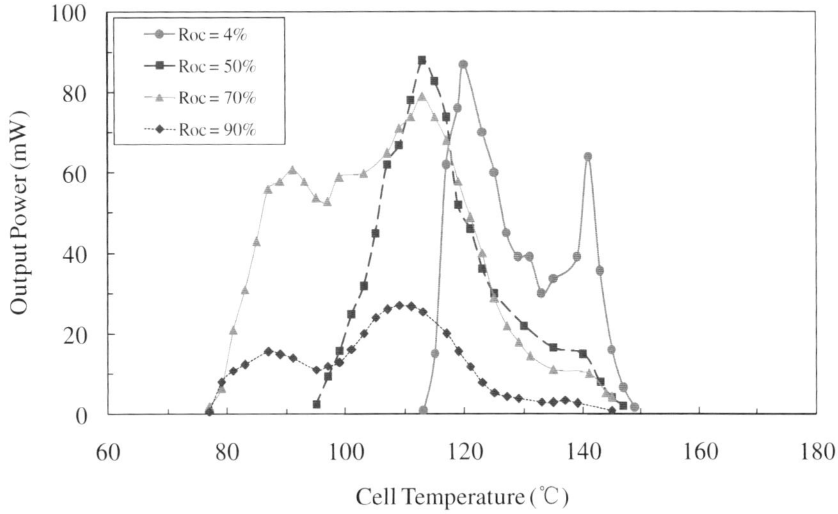

Figure 3. Output power versus the cell temperature by different reflectances of the output coupler in a Rb laser system.

In the study, we also investigate the output performances changing with the pump power. The results are depictured in figure 4. In the plot, the output power increases with the pump intensity without an evident saturation tendency for all curves. It can be seen that the higher the slope efficiency, the smaller the reflectance of an OC. One can deduce that the gain of a DPAL should be much higher than those of most conventional DPSSLs. We have achieved a maximum power of 111 mW with a reflectance of 4% of an OC. The corresponding optical to optical efficiency is 18.4% and the laser intensity profile is presented in figure 5. Such efficiency is not very high since the power of the pump source is relative small. We deduce the maximum optical to optical efficiency should become higher if the pump power is further increased. To the best of our knowledge, this is the first Rb laser pumped by a narrowband (<100 MHz) laser source in the world.

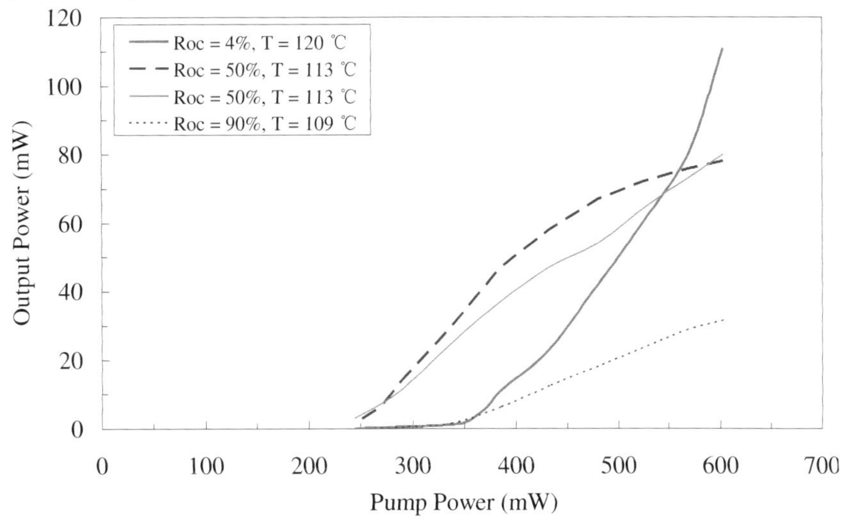

Figure 4. Output power of a rubidium laser as a function of the pump power at different experimental conditions.

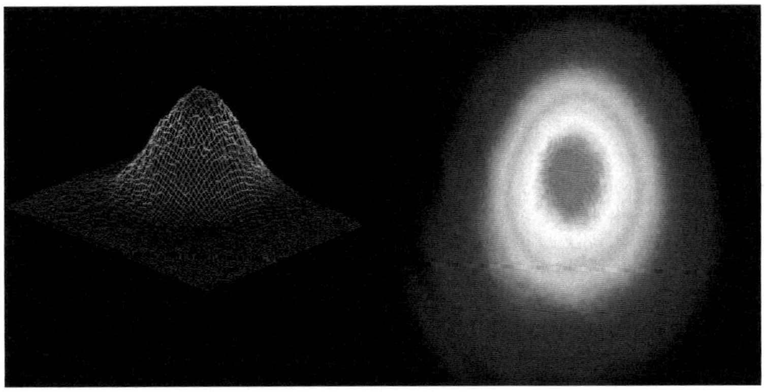

Figure 5. Beam profiles of a Rb laser in both three dimensions (left) and two dimensions (right).

According to the kinetic model we established before [14-15], we have also undertaken the theoretical simulation. The calculated optical to optical efficiency for the rubidium laser is as high as 26.2%, which is higher than the measured one (18.4%) under the same conditions. We think the difference between the theoretical results and the experimental ones are mainly aroused by four uncoated surfaces of two side-windows of the cell.

3.2. Output Features of a Cs laser

Similar to subsection 3.1, we also experimentally tested the output performance of a Cs laser. In figure 6, the output power is expressed as a function of the cell temperature with different R_{oc} while the pump power is 409 mW. We can see that the output power first increases and then falls as the cell temperature becomes higher. The optimal cell temperatures are 91, 93, 113, 100, and 103 °C with the corresponding reflectance of an OC of 4, 20, 50, 70, and 90%, respectively. As shown in figure 6, the maximum power is obtained when R_{oc} is set to 4%. Some jaggednesses of the curves in both figure 3 and figure 6 can be seen and are thought to be mainly caused by the short waiting time in the measurement. We are planning to increase the waiting time during the measurement in the upcoming experiments.

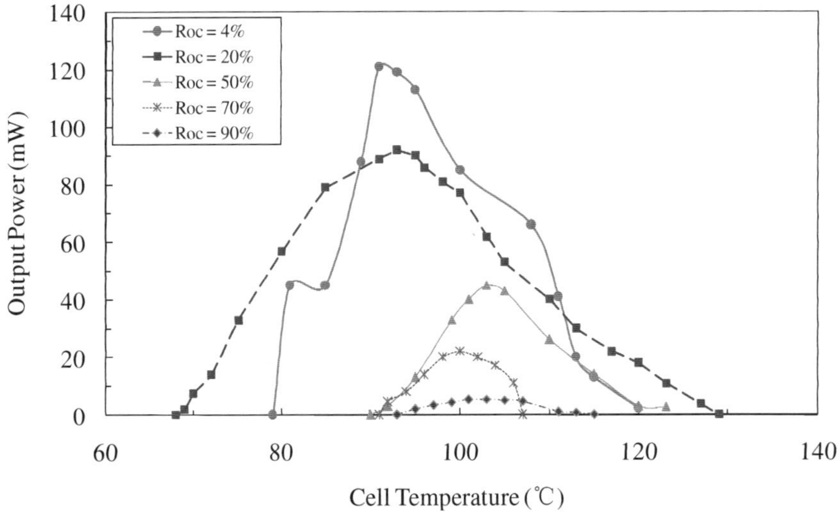

Figure 6. Output power versus the cell temperature by different reflectances of the output coupler in a Cs laser system.

The output power of a Cs laser at different pump conditions is diagrammed in figure 7. We can see that the slope efficiency decreases with increase of Roc. In addition, the lasing threshold is the lowest for Roc = 4% among five curves. Thus, an output coupler with the reflectance as low as possible should be probable for a Cs laser system. In this study, the maximum output power of a cesium laser has been achieved to 136 mW with the optical to optical efficiency of 30%. The intensity distribution of a Cs laser beam is recorded as shown in figure 8. The distribution profile of the cesium laser implies that the beam propagation factor is near to diffraction-limited.

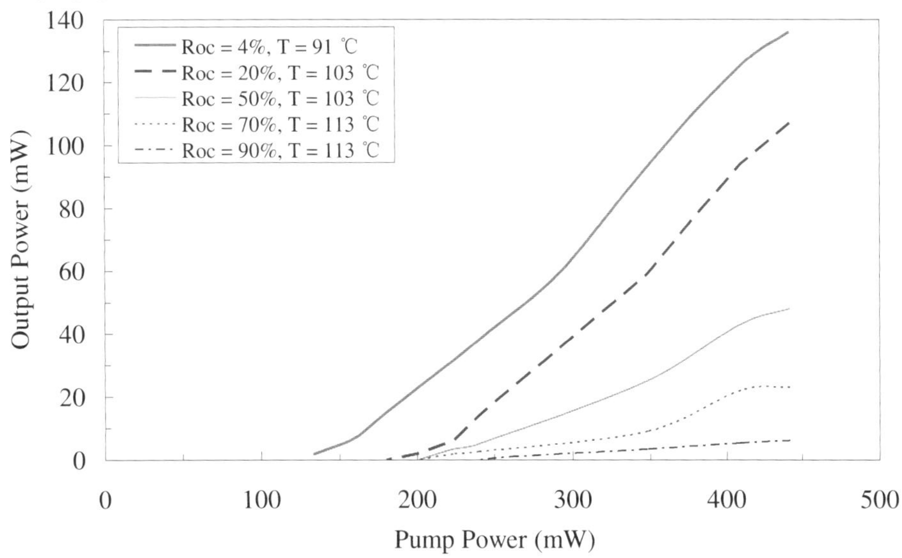

Figure 7. Output power of a cesium laser as a function of the pump power at different experimental conditions.

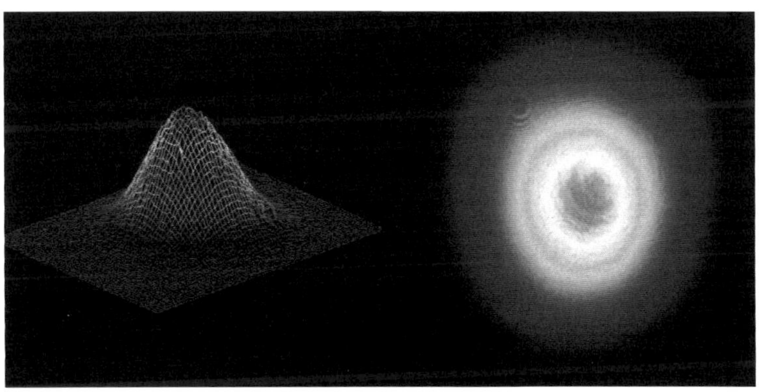

Figure 8. Beam profiles of a Cs laser in both three dimensions (left) and two dimensions (right).

4. Conclusions

In this study, we have realized both cesium and rubidium vapor lasers pumped by a CW Ti:Sapphire surrogate laser. According to the experimental results, we discuss the influence of both the cell temperature and the OC reflectance on output features of an optically-pumped alkali laser system. It can be seen that the optimum cell temperature changes with the reflectance of an output coupler, which is similar to the conclusion from several prior researchers. We also find that the smaller output coupler reflectance is, the higher slope efficiency becomes. The research is thought to be helpful for design of a high-powered DPAL system with the narrow linewidth.

References

[1] Krupke W F, Beach R J, Payne S A, Kanz V K and Early J T 2004 *AIP Conference Proceedings* **702** 367–77

[2] Wang Y, Niigaki M, Fukuoka H, Zheng Y, Miyajima H, Matsuoka S, Kubomura H, Hiruma T and Kan H 2007 *Phys. Lett.* A **360** 659–63

[3] Zhang W, Wang Y, Cai H, Xue L P, Han J H, Wang H Y and Liao Z Y 2014 *App. Opt.* **53** 4180–86

[4] Han J H, Wang Y, Cai H, Zhang W, Xue L P and Wang H Y 2014 *Opt. Express* **22** 13988–14003

[5] Yang Z N, Wang H Y, Lu Q S, Liu L, Li Y D, Hua W H, Xu X J and Chen J B 2011 *J. Phys.* B: *At. Mol. Opt. Phys.* **44** 085401

[6] Miller W S, Sulham C V, Holtgrave J C and G. P. Perram 2011 *Appl Phys.* B **103** 819–24

[7] Zhdanov B V, Ehrenreich T and Knize R J 2006 *Opt. Commun.* **260** 696–98

[8] Zhdanov B V, Maes C, Ehrenreich T, Haveko A, Koval N, Meeker T, Worker B, Flusche B and Knize R J 2007 *Opt. Commun.* **27** 353–55

[9] Zhdanov B V, Kontur F, Phipps S, Hallada F, Elsbernd P, Miller W, Peay A and Knize R J 2007 *Opt. Commun.* **280** 161–64

[10] Zhdanov B V, Stooke A, Boyadjian G, Voci A and Knize R J 2008 *Electron. lett.* **44** 735–36

[11] Krupke W F 2012 *Progress in Quantum Electron.* **36** 4–28

[12] Zhdanov B V and Knize R J 2013 *Opt. Eng.* **52** 021010

[13] Beach R J, Krupke W F, Kanz V K and Payne S A 2004 *J. Opt. Soc. Am.* B **21**, 2151–63

[14] Wang Y, Cai H, Zhang W, Xue L P, Wang H Y and Han J H 2014 *Proc. of SPIE* **9142** 914227

[15] Han J H, Wang Y, Cai H, An G F, Zhang W, Xue L P, Wang H Y, Zhou J, Jiang Z G and Gao M 2015 *Opt. Express* **23** 9508–15

IOP Conf. Series: Journal of Physics: Conf. Series **844** (2017) 012053 doi:10.1088/1742-6596/844/1/012053

Determination of fluorescent whitening agent in emulsion paint based on fluorescence spectrophotometry

Lin Zhang*[1], Youlin Zhao[2]

[1]College of Science, Hohai University, Nanjing, P. R. China, 211100, hhuzhangl@126.com

[2]Business school , Hohai University, Nanjing, P. R. China, 211100, sobzyl@hhu.edu.cn

Abstract: Fluorescence spectrophotometry was applied to the determination of fluorescent whitening agent in emulsion paint. At temperature of 30℃, linear relationship was found between fluorescence intensity and concentration of fluorescent whitening agent in the range of 0.025-0.25mg/ml when the emission wavelength was 437nm and excitation wavelength was 347nm, correlation coneficient was 0.998, detection limit was 0.0007mg/ml, relative standard deviation ranged from 0.49% to 0.72%, recovery of standard addition ranged from 98.3% to 104.5%. The influence of storage time and exposure time were also studied here. Measurement results indicated the feasibility and precision of using this method for quantitative analysis of fluorescent whitening agent.

Key words: fluorescent whitening agent; emulsion paint; fluorescence intensity

1. Introduction

Fluorescent whitening agent (FWA) is an important functional auxiliary. The whitening mechanism of FWA can be described as follows: The FWA absorb ultraviolet light (340nm-380nm) which has the same characteristic frequency then emit visible light in the region of blue purple light (410nm-460nm). The white light can be obtained based on the complementary relationship between the emission ray and faint yellow light in the base material. So the whitening effect can be achieved[1]-[2]. So far, there are about 2500 kinds of FWA which belongs to 15 basic types. FWA is widely used in the textile, paper, plastics, paint and other industries[3].

Emulsion paint (EP) is a new type of decoration materials. Many transparent or white paint has an inherent yellow hue. Therefore, FWA has been widely used in the white and undertint EP. Based on the mixed light of blue purple light and white light, FWA can seem to be lily and brightsome[4]. Household interior-wall EP must be water-soluble and have a good compatibility with the other components. At the same time, the component ratio between FWA and EP must be suitable in order to obtain good whitening effect[5]. If the content of FWA in EP is overproof, the whitening effect can be destroyed. When FWA gets on the skin, there will be an impact on the health[6]. Therefore, it's of great significance to provide impartial, accurate and fast measurement of FWA in EP.

Currently, the quantitative analysis methods for FWA consist of whiteness method[7], thin layer chromatography method[8]-[9], capillary electrophoresis method[10], high efficiency liquid chromatography method[11]-[12], ultraviolet spectrophotometry[13]-[14], fluorescence spectrophotometry[15]-[16]. The range of applications includes paper, water, detergent, textile, food and so on. But there is no reports about the measurement method of FWA in EP.

Based on fluorescence spectrophotometry, FWA content in EP is measured in this paper. According to the optical properties of the FWA, the best measuring wavelengh and temperature is selected. The standard curve can be obtained through the fluorescence spectrum measurement of the FWA-EP mixture. The percent recovery experiment is carried on here to verify the measurement accuracy of this method. Meanwhile, the influences of storage time and exposure time on the measurement results are also studied in this paper. Measurement results show that the method of fluorescence spectrophotometry can effectively realize the quantitative analysis of FWA in EP with high accuracy and credibility.

2. Experiment

2.1 Instruments and samples

All the fluorescence spectra were performed on a RF-5301PC spectrofluorophotometer (Shimadzu, Japan) equipped with a 150-W Xenon lamp, using a quartz cell of 1.0 cm path length. All the absorption spectra were performed on a UV-3600 UV-VIS spectrophotometer (Shimadzu, Japan).

The Chinese name of the FWA studied in this paper is 4BK and the foreign name is Tinopal 5BM which belongs to the kind of acid double triazine FWA (anionic). The appearance is light yellow powder. The acid resistance is PH=2. The basic structure of 4BK is shown in figure 1.

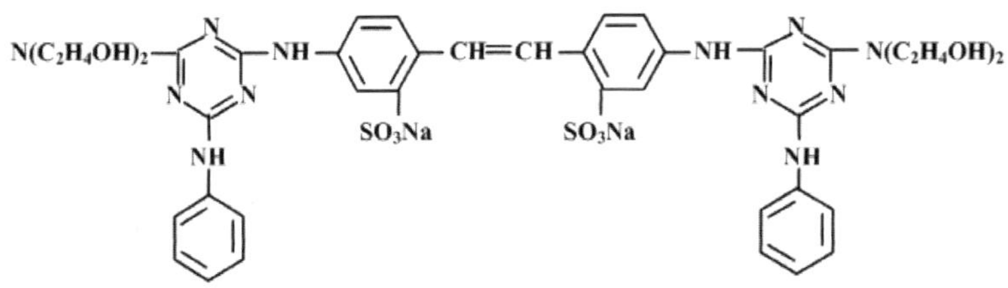

Figure 1. Chemical formula of FWA 4BK

The EP used in the experiment is household water dilution interior-wall coating. The ingredients consist of Palm oil, white latex, 107 glue powder, talcum powder, waterproof, ash calcium powder, light calcium carbonate, benzoic acid, sodium carbonate and ethylene glycol, tributyl phosphate and deodorant.

The water used in the experiment is sub boiling water. Analytical balance precision is 0.0001g, constant temperature water bath precision is 0.1℃, two kinds of brown conical flasks capacity are 250ml and 50ml, respectively, pipette capacity is 1ml.

2.2 Experiment method

Accurately weigh 0.1 g FWA 4BK and set the sample in the brown conical flask (250 ml). Then add sub boiling water to keep the capacity 200ml. This solution is the standard liquid of FWA with concentration of 0.5 mg/ml.

Accurately weigh 1ml, 2ml, 3ml.......10ml standard liquid in turn and set these sampes in ten brown conical flask (25ml). Then add EP to keep the capacity 20ml. The FWA concentrations of these calibration samples are 0.025mg/ml, 0.05mg/ml, 0.075mg/ml, 0.1mg/ml, 0.125mg/ml, 0.15mg/ml, 0.175mg/ml, 0.2mg/ml, 0.225mg/ml, 0.25mg/ml.

Before measurement, the temperature is set at 30℃ through the water bath. Add 3ml FWA standard sample into the cuvette with pipette. Absorption spectrum can be obtained with the UV3600 spectrophotometer with scanning range from 220nm to 400 nm. Fluorescence spectrum can be obtained with RF5301 fluorescence spectrophotometer with scanning range from 370nm to 560nm. The slit width is 1.5nm. Record the fluorescence intensity of the sampleS. The whole operation process is carried on in a dark room.

3. Results and discussions

3.1 Selection of the measuring wavelength

The UV3600 spectrophotometer is used to scan the absorption spectrum of the FWA standard liquid. The RF5301 fluorescence spectrophotometer is used to scan the fluorescence spectrum and excitation spectrum. Measuring results are shown in figure 2.

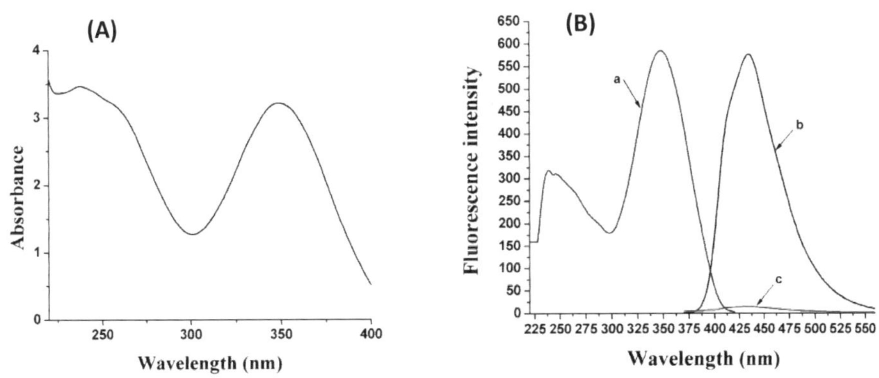

Figure 2. Absorption spectrum and fluorescence spectrum of the FWA standard liquid ((A)-absorption spectrum, (B): a-excitation spectrum of the FWA standard liquid, b- fluorescence spectrum of the FWA standard liquid, c-fluorescence spectrum of the EP)

It can be seen from the figure (A) that the strongest absorption peak of FWA standard liquid is at 347nm. It can be seen from the figure (B) that when the excitation wavelength is 347nm, there is an obvious fluorescence peak of FWA at 437nm while there is no fluorescence peak of EP here. According to the figure B(b), when the emission wavelength is at 437nm, the strongest excitation wavelength is 347nm. This result is in accordance with the results of absorption spectrum. Based on this results, the selected excitation wavelength is $\lambda_{ex} = 347nm$, the emission wavelength is $\lambda_{em} = 437nm$.

3.2 Selection of the measuring temperature

The fluorescence intensity of the FWA standard liquid is measured at the temperature of 20℃, 25℃, 30℃, 35℃, 40℃, 45℃, 50℃ and 55℃ when the excitation wavelength is 347nm. The whole operation process is carried on in a dark room. The results is shown in figure 3.

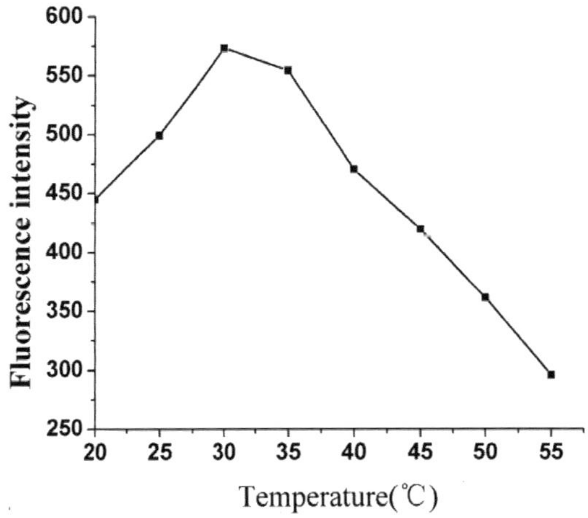

Figure 3. The fluorescence intensity of the FWA standard liquid at different temperatures

It can be seen from the figure that along with the increase of temperature, the fluorescence intensity of the FWA standard liquid at 437 nm increases first and then decreases, and reached the maximum at 30℃. Therefore, 30℃ is selected as the optimum measuring temperature.

3.3 Establishment of standard curve

The fluorescence spectrum of 10 samples is measured at 30℃ when the excitation wavelength is 347nm. The whole operation process is carried on in a dark room. The results is shown in figure 4.

It can be seen from the figure that along with the increase of the concentration of FWA, the fluorescence intensity will also increases. The fluorescence intensity of the mixture of FWA-EP at 437nm (take off the fluorescence intensity of EP at 437nm) and the concentration of FWA are used to obtain the linear standard equation. The fitting equation is Y=1000X+12, where X is the concentration of FWA (mg/ml) and Y is the fluorescence intensity of the sample (take off the fluorescence intensity of EP). The correlation coefficent is 0.998.

According to the research results in reference [17], when the ratio of the FWA and EP is 0.02%, the whitening effect is optimal. Based on the this optimal ratio, when the concentration of FWA is not more than 0.2mg/ml, we can get the best whitening effect. So, according to the measurement results, when the concentration of FWA ranges from 0.025mg/ml to 0.25mg/ml, there is a good linear relationship between the concentration and fluorescence intensity.

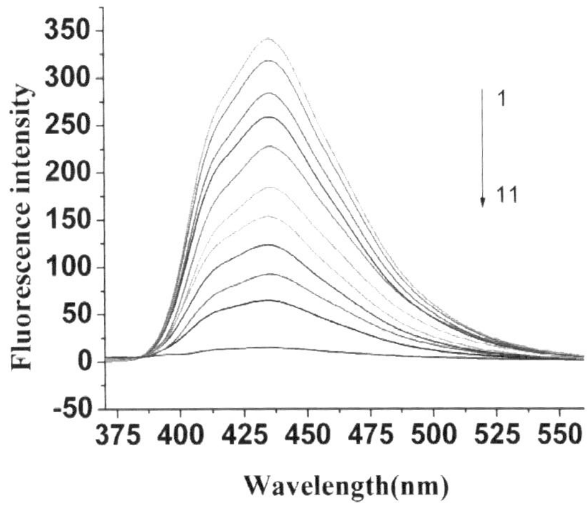

Figure 4. Fluorescence spectrum of FWA-EP (Top to bottom, 1-10: mixture of FWA-EP, the concentration of FWA is 0.25mg/ml, 0.225mg/ml, 0.2mg/ml, 0.175mg/ml, 0.15mg/ml, 0.125mg/ml, 0.1mg/ml, 0.075mg/ml, 0.05mg/ml, 0.025mg/ml, respectively, 11: EP).

3.4 Percent recovery experiment

Accurately weigh 1.5ml, 2.5ml, 3.5ml, 4.5ml, 5.5ml, 6.5ml, 7.5ml, 8.5ml, 9.5ml standard liquid in turn and set these sampes in ten brown conical flask (25ml). Then add EP to keep the capacity 20ml. The fluorescence intensity of the samples at 437nm are measured at the temperature of 30℃. According to the standard equation, the concentration of FWA can be calculated. The percent recovery (R%) and relative standard deviation (RSD) can also be obtained at the number of measurements n=4. The whole operation process is carried on in a dark room. The results is shown in table 1.

Table 1 Results of percent recovery experiment

Standard value (mg/ml)	Average measurement value (mg/ml)	RSD (n=5)	R%
0.0375	0.0392	0.49%	104.5%
0.0625	0.0643	0.52%	102.9%
0.0875	0.0889	0.64%	101.6%
0.1125	0.1106	0.69%	98.3%
0.1375	0.1362	0.72%	99.1%
0.1625	0.1609	0.59%	99%
0.1875	0.1902	0.63%	101.4%
0.2125	0.2141	0.62%	100.7%
0.2375	0.2365	0.71%	99.6%

It can be seen from the table that RSD ranges from 0.49% to 0.72% and the percent recovery ranges from 98.3% to 104.5%. This shows that the method is good at the precision and repeatability. The measurement error at high concentration is bigger than that at low concentration.

3.5 Measurement of detection limit

According to the fluorescence intensity of blank reagent (EP without FWA) at 437nm (Measurement of 24 times), the average measurement value is X_B=11.865 and standard deviation

S_B=0.226. Based on the ratio of three times of standard deviation and slope of the standard equation, we can calculate the detection limit of this method is 0.0007mg/ml.

3.6 Stability of the standard sample

The standard liquid of FWA is set in the refrigerator lucifugally. The fluorescence intensity at 437nm is measured for ten consecutive days. Befoure measurement, the temperatures of the samples are kept at 30℃ with water-bath. The results are shown in figure 5.

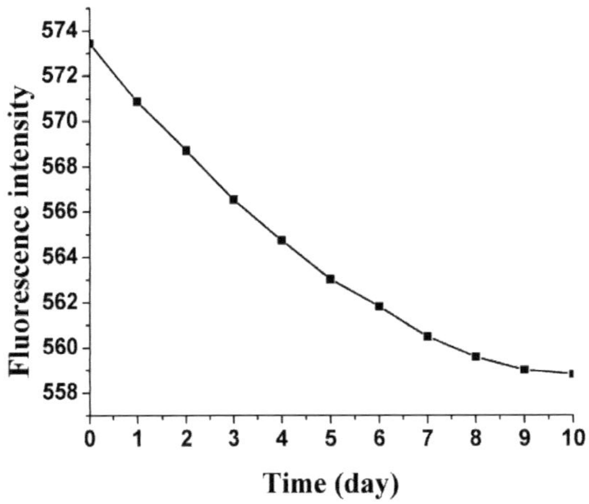

Figure 5. Stability of the standard liquid of FWA

It can be seen from the figure that fluorescence intensity reduces by 14.634 on the tenth day. According to the standard equation, the measurement results can be influenced by 0.003mg/ml. So, when the sample is configured, the measurement should carry on the same day to avoid influence on the result of measurement.

3.7 Influence of exposure time

Standard liquid of FWA is studied in the following three exposure environment: (1) Outdoor sunlight exposure; (2) UV lamp with the center wavelength of 365nm and 0.1mw power; (3) Common daylight lamp. After exposure, the temperature of the sample is kept at 30℃. The fluorescence intensity at 437nm is measured. The results are shown in figure 6.

It can be seen from the figure that when the FWA solution is set under the outdoor sunlight, the fluorescence intensity decreases with exposure time rapidly. When the exposure time is more than 80s, the fluorescent intensity is basically stable; When the FWA solution is set under UV lamp, the fluorescence intensity decreases with the exposure time. But the variation rate is lower than that in the sunlight. When the exposure time is more than 100s, the fluorescent intensity is basically stable; When the FWA solution is set under common daylight lamp, the fluorescence intensity retains stable basically.

The research results show that when FWA is used in the EP, it is only suitable for the application in interior wall coating and not suitable for exposure outdoor.

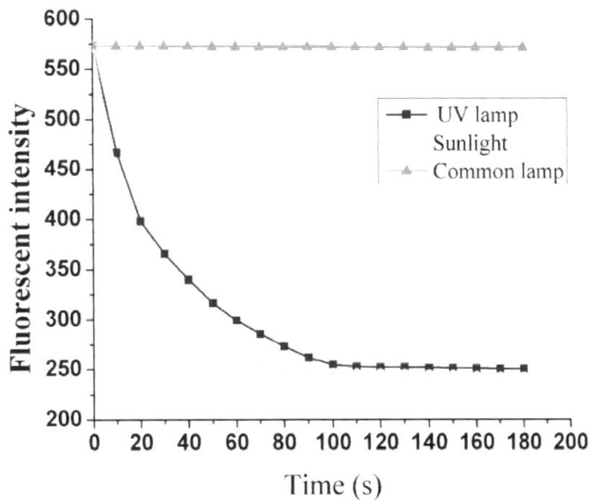

Figure 6. Influence of exposure time on fluorescence intensity

4. **Conclusions**

In this paper, fluorescence spectrophotometer was used to measure the FWA 4BK in household inside wall EP. Excitation wavelength is selected at 347 nm, emission wavelength is selected at 437nm and the measuring temperature is 30℃. Measurement results show that there is a good linearity between the fluorescence intensity and the FWA concentration in the range of 0.025 mg/ml to 0.25 mg/ml. The correlation coefficent is 0.998 and the detection limit is 0.0007 mg/ml. RSD ranges from 0.49% to 0.72% and the percent recovery ranges from 98.3% to 104.5%.

The influence of exposure time and storage time on measurement results are also studied in this paper. The results show that the fluorescence intensity of FWA solution changes obviously under UV lamp and sunlight while keeps stable under the common lamp. When the FWA solution stores for 10 days, the fluorescence intensity reduces by 14.634 which caused the measurement error 0.003mg/ml.

The research results in this paper show that the proposed method implements quantitative analysis of the FWA in EP. During the measurement, the sample should keep away from the sunlight. The measurement should carried on after the sample is configuration completed. We should also pay attention to maintain a constant temperature of the sample. The research results in this paper provides a reliable basis for the research and application of FWA in EP.

ACKNOWLEDGMENTS

This work was financially supported by the Nature Science Foundation of Jiangsu province [No. BK20150807], Fundamental Research Funds for the Central Universities [No. 2014B04714], Fundamental Research Funds in Key Research Areas for the Central Universities [No. 2015B09614].

References

[1] Li ZH, Wei Y, Fan Y 2013 *Henan Science* **1** 46-48.
[2] Liu QJ, Gao L, Wang L 2009 *Spectroscopy and Spectral Analysis* **10** 2810-2814.
[3] Sun BG, Xu LX 2003 *Papermaking in Heilongjiang* **4** 32-34.

[4] Dong ZS 2006 *Practical technology of fluorescent whitening agent* (China Textile and Apparel Press) p 444-445

[5] Park MY, Lee YK, Lim BS 2007 *Dental Materials* **6** 731-735.

[6] Yu K, Shi HZ, Zhang CM 2013 *Chinese Journal of Spectroscopy Laboratory* **1** 112-116.

[7] Guo RH, Zhou YH 2007 *China Pulp&Paper* **10** 33-35.

[8]Anders G 1975 *Environmental quality and safety* **4** 104-110.

[9] Fuzhou commodity inspection bureau of tea room 1980 *Fujian Tea* **Z1** 50-52.

[10] Damant AP, Castle L 1999 *Journal of microcolumn separations* **4** 259-262.

[11] Kim JS, Kim DH, Kim K 2012 *Bulletin of the korean chemical society* **12** 3971-3976.

[12] Wang Y, Yao XY, Han YH 2007 *Tobacco Science & Technology* **11** 40-44.

[13] Dong ZS 2009 *Industry standard and quality of petroleum and chemical industry in China* **11** 15-19.

[14] Wang SG, Yang QB, Du JS 2007 *Journal of Applied Polymer Science* **4** 2382-2386

[15] Lee JY, Youn HJ, Lee HL 2012 *Bioresources* **1** 315-326.

[16] Pan KL, Li SW, Yang L 2011 *Journal of analytical science* **4** 542-544.

[17] Jiang T, Wu FZ, Wang YH 1989 *Wuhan Chemical Engineering* **3** 46-49.

IOP Conf. Series: Journal of Physics: Conf. Series **844** (2017) 012054 doi:10.1088/1742-6596/844/1/012054

Laser-induced self-assembly gold nanoparticles on the silanized surface of a tapered fiber and its application as a SERS probe

Huajian Liu, Zhenyi Chen, Na Chen*, Shupeng Liu, and Tingyun Wang

Key Laboratory of Specialty Fiber Optics and Optical Access Networks, Shanghai University, 149 Yanchang Road, Shanghai 200072, China

*E-mail: na.chen@shu.edu.cn

Abstract. Fiber surface-enhanced Raman scattering (SERS) probes have some advantages including flexibility, remote sensing capability, etc. In this paper, we proposed a new way of making fiber SERS probes. Desirable sizes of gold nanoparticles were absorbed on coated 3-aminopropyltrimethoxysilane (APTMS) surface of tapered fiber by laser. We studied the relationship between the light exposure time and the Raman intensity, and compared the Raman enhancement effects of probes made by different processes. The formed SERS probes have high repeatability and Raman activity. Its prepared time was much shorter, and the size of absorbed Au nanoparticles could be well arranged.

1. Introduction

Fiber SERS probes have a wide range of applications of chemistry [1], biology [2] and environmental science [3] due to its high sensitivity and great flexibility. To obtain high enhancement factor in a fiber SERS probe, the fiber is shaped into a variety of geometries, including flat fiber [4], tapered fiber [5] and D-shaped fiber [6]. Among them, the tapered fiber probe has the advantages of high light transmission efficiency and large interaction areas for excitation light and SERS materials [7]. At present, chemical method [8] and laser-induced chemical deposition method (LICDM) [9] are two main ways of fabricating fiber SERS probe. Chemical method could fix desirable size of nanoparticles to fiber surface, but have to consume longer time. LICDM possess simple and cost-efficient characteristic, however, this method could not control the size of particles on the taper fiber surface that affect the Raman enhancement effect. Combining the advantages of these two methods, we can quickly prepare the fiber SERS probe with the desirable size of the particles.

This paper proposed using laser absorption desirable sizes of gold nanoparticles on the silanized surface to make the fiber SERS probes. We studied the relationship between the light exposure time and the Raman intensity. Also, we compared the Raman intensity of fiber SERS probes made by different processes, such as without the silanized support or without laser application.

2. Fiber probe preparation

2.1. *Experimental reagents*

HF acid (40%), chlorauric acid, trisodium citrate, rhodamine 6G (R6G) and anhydrous ethanol were purchased from Sinopharm Chemical Reagent. Potassium hydroxide (95%) and isooctane were

IOP Conf. Series: Journal of Physics: Conf. Series **844** (2017) 012054 doi:10.1088/1742-6596/844/1/012054

purchased from Aladdin Reagent online. Milli-Q deionized water (resistivity = 18.2 MΩ cm^{-1}) was used throughout the experiment.

2.2. Fiber probe preparation and SERS detection

Fig. 1 shows the fiber probe preparation process. First, surface of the taper fiber (50/125μm) after a series of chemical treatments were silanized by 3-Aminopropyltrimethoxysilane (APTMS). Then the silanized tapered fibers were immersed in the prepared colloidal gold nanoparticles with the diameter about 50 nm. 13 mW-power of 532 nm laser were induced in the fiber. As shown in Fig 1, in the experiments, we change different immersion time, i.e. laser application time, in the process.

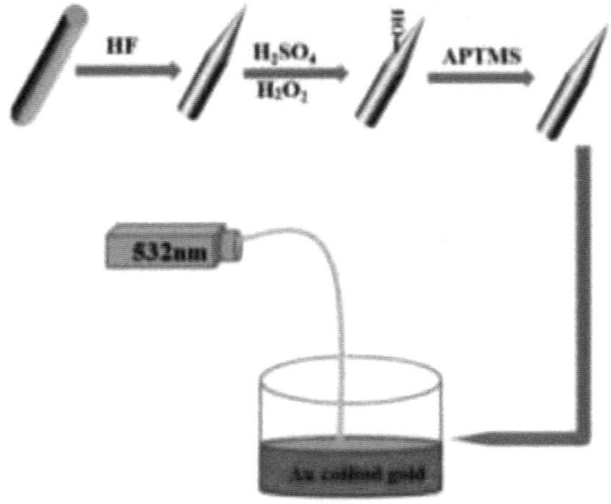

Fig. 1. Scheme illustration of the fiber pretreatment and the preparation of SERS-active layer on the silanized fiber taper by the laser-induced reaction.

A confocal micro-Raman spectrometer (Renishaw inVia plus) was used to detect the Raman spectra of R6G. Excitation wavelength was 633 nm and the excitation power was 10% of the maximum laser power. Integration time was 10 s and the accumulation number was 1.

3. Result and discussion

In order to eliminate the influence of cone angle on Raman intensity, we prepared the same cone angle tapers fiber with the hydrofluoric acid etching fabrication technique [10]. As shown in Fig. 2(a), flat-end fiber was shaped into a tapered fiber after 42 min etching in HF. During the process of fabricating the SERS substrate on the probes, we controlled different laser application time: 1 min, 2 min, 2.5 min, and 3 min, respectively. Fig. 3 shows their detected Raman spectra of 10^{-5} M R6G aqueous solution. It can been seen that Raman intensity increased with laser irradiation time less than 2 min, and the strongest intensity achieved at 2 min. However, Raman intensity continued to decrease if the laser irradiation time is beyond 2 min. Fig. 2(b) shows the SEM picture of the fiber SERS probe with the laser irradiation time of 2 min.

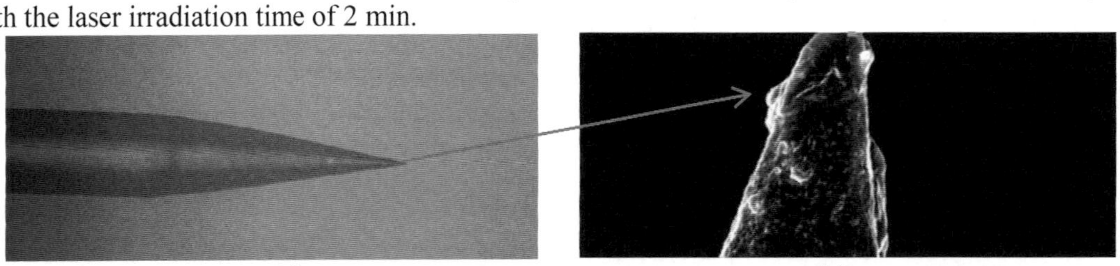

(a) (b)

Fig. 2. (a) A tapered optical fiber was obtained after 42 minutes of etching. (b) SEM picture of the taped fiber SERS probe with the laser irradiation time of 2 min.

With the time increase, more and more nanoparticles were adsorbed on the tapered fiber, Raman intensity rises from low to high, and then drops from high to low [11-12]. We think that the rate of incorporation of nanoparticles into the optical fiber became much faster under the influence of APTMS and the laser.

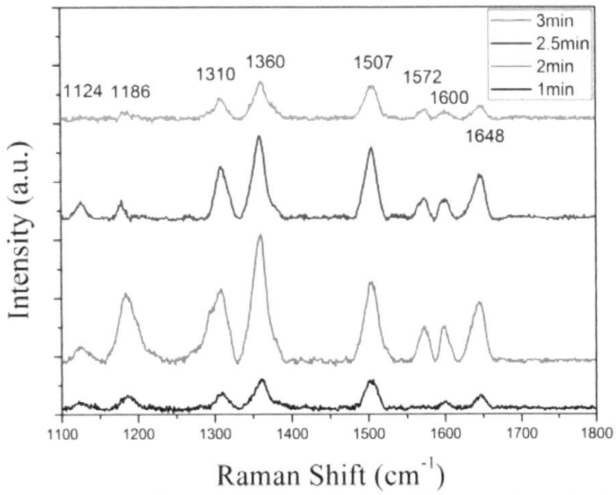

Fig. 3. Dynamic SERS spectra of 10^{-5} M R6G recorded at different time during the fabrication of the fiber probes with a laser.

We did a comparative study to show the effects of the APTMS support and the laser application on the probe performance. The time for preparing the fiber probes was set 2 min. The results of three different processes were shown in Fig. 4. The red line corresponds to the conventional method of chemical self-assembly without laser application. And the blue line corresponds to directly laser adsorption nanoparticles without silanized support. We found out that in both cases, the detection limit of the fiber SERS probes was weaker than the combined effects of silanized surface and laser application.

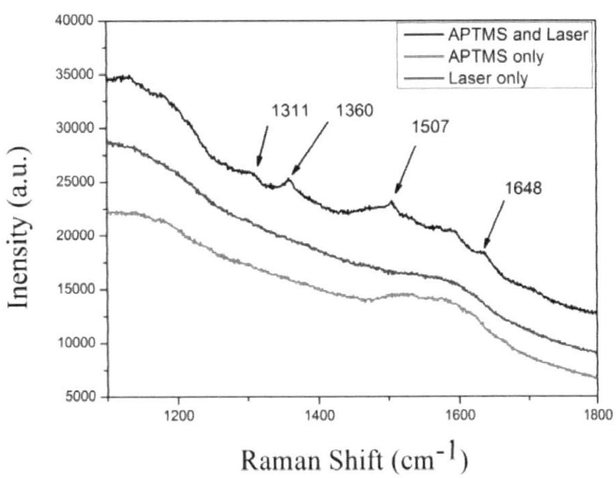

Fig. 4. Probes prepared only by APTMS and onlyby laser were compared with the same measurement conditions. The concentration of the R6G solution was 10^{-7} M.

At last, to show the repeatability of this method, five fiber SERS probes were made under the same condition. Their SERS detection performances were compared by measuring a 10^{-6} M R6G solution as shown Fig. 5.

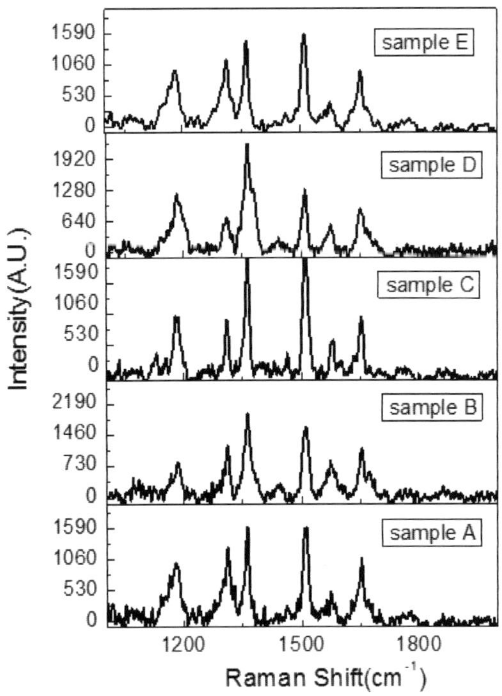

Fig. 5. Raman intensity of R6G collected on six as-prepared fiber SERS probes

4. Conclusion

The fiber SERS probe produced by this method has high repeatability, Raman activity and short fabrication cycle. Through the optimization of the laser application time, we found that the best time for the SERS substrate with 50nm-diameter gold nanoparticles is about 2 min.

Acknowledgment

This work is supported in part by National Natural Science Foundation of China (Project Nos: 61575120, 61027015, 61475095, and 61635006).

References

[1] Fang Y, Seong N H, Dlott D D 2008 Measurement of the distribution of site enhancements in surface-enhanced Raman scattering *Science* **321**:388.

[2] Kang D, Kwon J Y, Cho H, et al. 2012 Oxidation resistance of iron and copper foils coated with reduced graphene oxide multilayers *Acs Nano* **6**:7763-9.

[3] Cao J, Zhao D, Lei X, et al. 2014 One-pot hydrothermal synthesis of silver nanoplates on optical fiber tip for surface-enhanced Raman scattering *Appli. Phys. Lett.* **104**:388.

[4] Lan X, Han Y, Wei T, et al. 2009 Surface-enhanced Raman-scattering fiber probe fabricated by femtosecond laser *Optics Letters* **34**:2285-7.

[5] Xie G, Cheng J, Li Y, et al. 2012 Fluorescent graphene oxide composites synthesis and its biocompatibility study *Journal of Materials Chemistry* **22**:9308-14.

[6] Y. Zhang, C. Gu 2005 Surface-enhanced Raman scattering sensor based on D-shaped fiber *Appl. Phys. Lett.* **87**: 123105.

[7] Viets C, Hill W. 2001 Fibre-optic SERS sensors with conically etched tips *Journal of Molecular Structure* **563**:163-166.

[8] Zhang J, Chen S, Gong T, et al. 2016 Tapered Fiber Probe Modified by Ag Nanoparticles for SERS Detection *Plasmonics* **11**:743-751.

[9] Fan Q. 2013 Investigations of the fabrication and the surface-enhanced Raman scattering

detection applications for tapered fiber probes prepared with the laser-induced chemical deposition method *Applied Optics* **52**:6163.

[10] Viets C, Hill W. 2001 Fibre-optic SERS sensors with conically etched tips *Journal of Molecular Structure* **s563–564**:163-166.

[11] Cao J, Zhao D, Mao Q. 2015 Laser-induced synthesis of Ag nanoparticles on the silanized surface of a fiber taper and applications as a SERS probe *Rsc Advances* **5**:99491-97.

[12] Liu Y, Huang Z, Zhou F, et al. 2016Highly sensitive fibre surface-enhanced Raman scattering probes fabricated using laser-induced self-assembly in a meniscus *Nanoscale* **8**:10607-614.

IOP Conf. Series: Journal of Physics: Conf. Series **844** (2017) 012055 doi:10.1088/1742-6596/844/1/012055

Gold nanoparticles modified double-tapered fiber for SERS detection

Chunqin Zhao, Na Chen*, Shupeng liu, Zhenyi Chen and Tingyun Wang

Key Laboratory of Specialty Fiber Optics and Optical Access Networks, Shanghai University, 149 Yanchang Road, Shanghai 200072, China

*E-mail: na.chen@shu.edu.cn

Abstract. Double-tapered fiber probes modified with gold nanoparticles were fabricated for the surface-enhanced Raman scattering (SERS) sensing. Its performance was compared with the fiber SERS probes of flat end and of single-tapered structure. The remote detection performances of three fiber probe structures were compared by detecting rhodamine 6G (R6G) aqueous solutions with the different concentrations. The results of remote detection of R6G aqueous solution indicate that double-tapered fiber probe with the detection limit of 10^{-9} M is the most sensitive.

1. Introduction

As a SERS sensor, its ability to detect solutions with low concentration is an issue parameter to judge its performance. The use of flat-end fiber (FEF) probe realized the fiber SERS sensors from double fiber detection (one fiber propagates excited laser and the other collects Raman scattering light) to single fiber detection [1]. However, the performance of FEF probe in remote detection cannot reach the optimum effect. Single-tapered fiber (STF) probe with an optimal cone angle can achieve superior performance for optical fiber SERS detection, due to it increasing the SERS active surface [2], transmitting the evanescent field to the surroundings with an increased magnitude [3], and good collection efficiency [4]. Also, the STF probe with the needle structure would be used to biomedical techniques for living cells, and obtain the lives period information of cells. STF tips with silver nanoplates have been fabricated to detect the 10^{-7} M 4-ATP [5], and STF tips with gold nanoparticles could detect the 10^{-8} M solution of R6G [6]. Single molecule detection was demonstrated by the SERS technique in different environments [7], even in single living cell [8]. It is worth noting that the silver nanoparticles is deleteriousness for cells or biological tissue. Therefore, the gold nanoparticles were used as our SERS enhancement substrates, though the gold nanoparticles perform worse than silver nanoparticles in enhancement substrates. In order to increase the SERS active surface and detect lower concentration solution, the double-tapered fiber (DTF) was proposed. It was confirmed that with the optimized geometry, fiber-optic sensor could detect the SERS spectra from a 10^{-9} M solution of crystal violet in distilled water both with measurements performed in solution and with a “dip and dry” technique [2]. That the large area of column increase the interaction opportunity between the detection molecules and evanescent wave, would be the reason.

In this paper, we prepared three types of fiber SERS probes with FEF, STF and DTF. Their performance for SERS sensing were compared. All fiber probes were modified with gold nanoparticles by the electrostatic self-assembly technology. Raman spectra of R6G aqueous solutions with different concentrations were detected by these fiber probes in remote mode.

2. Materials and methods

2.1. Experimental reagents

HF acid (40%), chlorauric acid, trisodium citrate, rhodamine 6G (R6G) and anhydrous ethanol were purchased from Sinopharm Chemical Reagent. Potassium hydroxide (95%) and isooctane were purchased from Aladdin Reagent online. Milli-Q deionized water (resistivity = 18.2 MΩ cm^{-1}) was used throughout the experiment.

2.2. Probe preparation and SERS detection

A multimode fiber with a cladding of 125 μm and a core of 50 μm were used in our experiment. There were two steps to fabricate our three types of SERS fiber probe. The first step was to fabricate the fiber probes. Three multimode fibers were removed the coating layers and were cleaned with anhydrous ethanol. The FEF probe was obtained by directly cutting the fiber end with a flat surface. For the DTF probe, we put the FEF into a beaker with 40 % hydrofluoric acid aqueous solution covered with isooctane over-layer for etching 55 min, lift up 1 mm for etching 8 min, and finally obtain the double-tapered structure. After etching, the DTF was rinsed with deionized water. The method of prepared STF was similar with that of reference [6]. The second step was to modify these fiber probes with gold nanoparticles. The gold nanoparticles with around 55 nm in diameter were prepared by citratereduction of chloroauric acid. And the gold nanoparticles were solidified on these tapered fiber probes by electrostatic self-assembly technology [9].

Take the DTF probe as an example to introduce the solidified progress. Firstly, the DTF tip should be totally cleaned by dipping the DTF tip into a piranha solution (3:1 mixture of 96 % concentrated sulfuric acid and 30 % hydrogen peroxide) for 30 min, and then rinsing twice in deionized water and ethanol respectively. Secondly, the cleaned DTF tip was immersed in the mixture solution with 5 % v/v deionized water, 5 % v/v (3-Aminopropyl) trimethoxysilane (APTMS 97 %) and 90 % ethanol for 30 min. Again, it was rinsed twice in deionized water and ethanol respectively to remove the residual APTMS. Thirdly, it was kept in the incubator at 90 °C for 30 min. Finally, the DTF was cleaned in the ethanol and then immersed into the gold colloids for 48 h. Finally, the gold nanoparticles modified double-tapered fiber probes were obtained.

The final microscope photograph of double-tapered structure with SEM images inset is shown in figure 1.

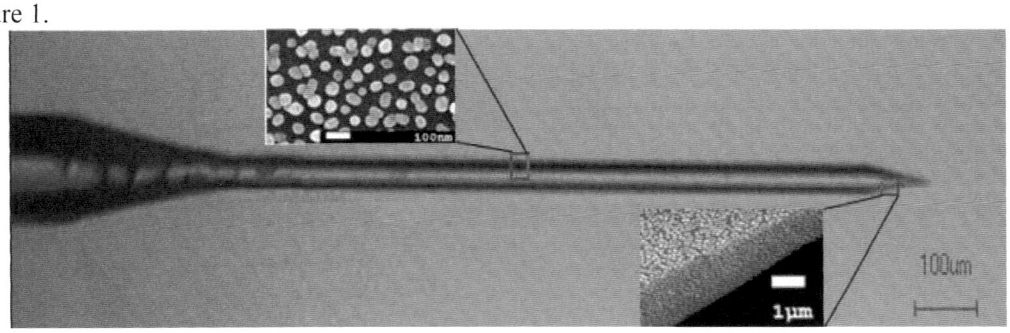

Figure 1. Optical and SEM images of the DTF SERS probe.

The schematic of remotely detecting R6G was shown in figure 2. A confocal micro-Raman spectrometer (Renishaw inVia plus) was used to detect the Raman spectra by the DTF SERS probe with a segment around 28 cm long. Through the 10× microscope objective, 633 nm He-Ne laser as excitation was injected into the fiber from the other flat end. The backward SERS signals of the R6G molecules on the probe surface were collected by the same probe. The exposure time of Raman spectrometer was set to 10 s for each measurement. Raman signals were collected in the spectral range from 1000 to 1800 cm^{-1}. The detection of the other two types of fiber SERS probes was the same as above method.

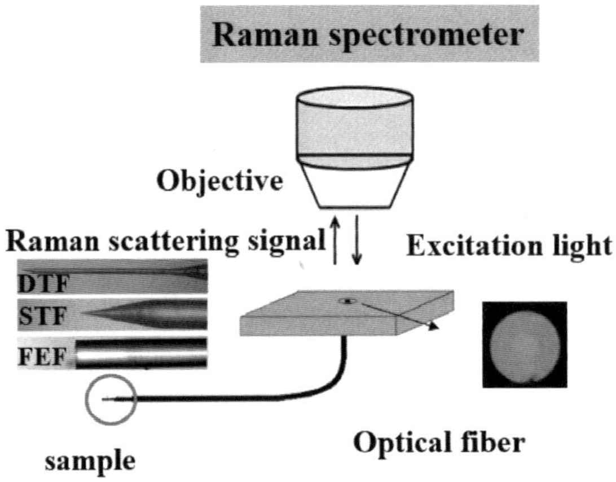

Figure 2. Schematic diagram of remotely detecting R6G solution.

3. Results and discussions

The probes of DTF, STF and FEF were used to detect R6G aqueous solution with several concentrations in remote mode, their results shown in figure 3.

For the detection of DTF probe from figure 3 (a), the main Raman characteristic peaks of R6G were clearly recognized, and the intensity of each peak lowers with the concentrations decreasing. The detection limit reached 10^{-9} M. From figure 3 (c) and figure 3 (d), it can be concluded that for the detection of STF probe and FTF probe, these detection limits reached 10^{-8} M and 10^{-6} M, respectively. Comparing the detection limit of three fiber probes from figure 3 (b), we could find that the detection intensity of FEF probe (detection limit corresponding to 10^{-6} M) is the highest, and the DTF probe is the lowest. The detection limit of FEF probe is, after all, the highest, although its detection sensitivity is the worst. Therefore, the detection intensity of FTF probe is the strongest.

It is obviously that the tapered fibers have lager effect surface area than that of flat end fibers, and the tapered structure increases the contact opportunity between detection molecules and evanescent wave. Furthermore, the tapered structure can be collected more Raman signals from the detection molecules. Therefore, tapered structure is superior to flat end structure. Apparently, the effect surface area of double-tapered structure is the largest, so as for the most sensitivity in the remotely Raman detection.

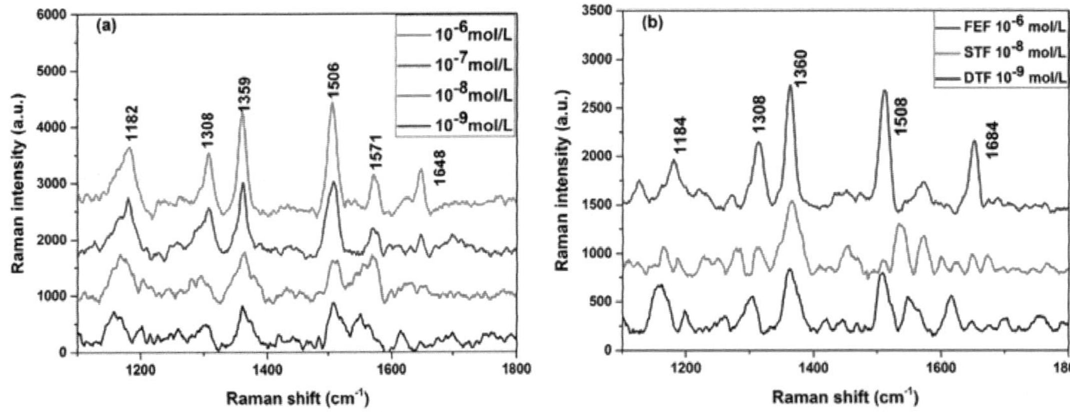

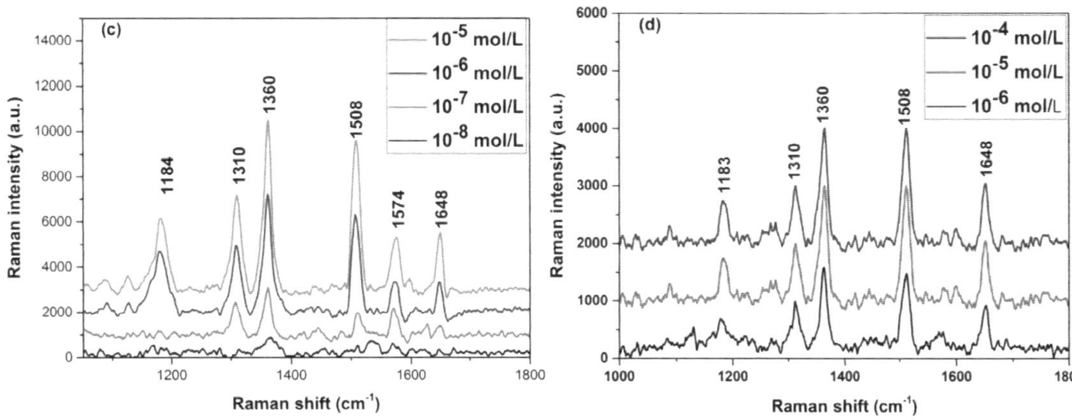

Figure 3. (a) Raman spectra of R6G aqueous solution with different concentrations remotely measured by DTF probe; (b) Comparison of the detection limit to R6G with DTF, STF and FEF; (c) Raman spectra of R6G aqueous solution with different concentrations remotely measured by STF probe; (d) Raman spectra of R6G aqueous solution with different concentrations remotely measured by FTF probe.

4. Conclusion

The probes of DTF, STF and FEF, modified with gold nanoparticles by electrostatic self-assembly technology, were successfully fabricated. When they were used to detect the standard sample of R6G aqueous solution, the detection limits of DTF, STF and FEF SERS sensor were 10^{-9} M, 10^{-8} M and 10^{-6} M, respectively. The DTF probe has the largest sensitivity, due to the double-tapered structure increasing the contact opportunity between the detection molecules and the evanescent wave. Furthermore, the tapered structure can also increase the collecting efficiency. In the same conditions, the flat end fiber probe is two-order lower than tapered probe because of its little interaction area and lower collecting efficiency. The high sensitive remote SERS detection of DTF will provide a new tool for detecting Raman spectra of kinds of biomolecules.

Acknowledgment

This work is supported by the National Natural Science Foundation of China (Project Nos: 61575120, 61027015, 61475095, and 61635006).

References

[1] D. L. Stokes, T. Vo-Dinh 2000 Development of an integrated single-fiber SERS sensor *Sens. and Actuators B* **69** 28-36

[2] Lucotti, A., and G. Zerbi 2007 Fiber-optic SERS sensor with optimized geometry *Sens. Actuators B Chem.* **121** 356-364

[3] Sharma, Anuj K., Rajan Jha, and B. D. Gupta 2007 Fiber-optic sensors based on surface plasmon resonance: a comprehensive review *IEEE Sens.* J. **7** 1118-1129

[4] Jayawardhana, S., A. P. Mazzolini, and P. R. Stoddart 2012 Collection efficiency of scattered light in single-ended optical fiber sensors *Opt. let.* **37** 2142-2144

[5] Cao J., Zhao D., Lei X., Liu Y. and Mao Q 2014 One-pot hydrothermal synthesis of silver nanoplates on optical fiber tip for surface-enhanced Raman scattering *Appl. Phys. Lett.* **104** 201906

[6] Zhenyi Chen, Zhangmin Dai, Na Chen, Shupeng Liu, Fufei Pang, Bo Lu, and Tingyun Wang 2014 Gold nanoparticles-modified tapered fiber nanoprobe for remote SERS detection *IEEE Photonics Technol. Lett.* **26** 777-780

[7] Pieczonka, N.P.W. Aroca, R.F 2008 Single molecule analysis by surfaced-enhanced Raman scattering *Chem. Soc. Rev.* **37** 946–954

[8] Scaffidi, J.P. Gregas, M.K. Seewaldt, V. Vo-Dinh, T 2009 SERS-based plasmonic nanobiosensing in single living cells *Anal. Bioanal. Chem.* **393** 1135-1141

[9] E. Polwart, R. L. Keir, C. M. Davidson, W. E. Smith, and D. A. Sadler 2000 Novel SERS-active optical fibers prepared by the immobilization of silver colloidal particles *Appl. Spectroscopy* **54** 522-527

IOP Conf. Series: Journal of Physics: Conf. Series **844** (2017) 012056 doi:10.1088/1742-6596/844/1/012056

Amplified spontaneous emission of an end-pumped cesium vapor laser

Guofei An, You Wang*, He Cai, Juhong Han, Shunyan Wang, Kepeng Rong, Hang Yu, Liangping Xue, Wei Zhang, Hongyuan Wang, and Jie Zhou

Southwest Institute of Technical Physics, Chengdu, Sichuan 610041, China

youwang_2007@aliyun.com

Abstract. Diode pumped alkali lasers (DPALs) provide a significant potential for construction of high-powered lasers. A series of models have been established to analyze the DPAL's kinetic process and most of them are based on the algorithms in which the amplified spontaneous emission (ASE) effect has not been considered. However, ASE is harmful in realization of a high-powered DPAL since the gain is very high. Usually, ASE becomes serious when the volume of the gain medium is large and the pump power is high. Basically, the conclusions we obtained in this study can be extended to other kinds of laser configurations.

1. Introduction

In the recent years, diode-pumped alkali laser (DPAL) has been extensively studied due to the potential for their excellent physical features [1–3]. The oscillator in a DPAL usually contains vapors of neutral alkali (K, Rb, Cs) and several kinds of buffer gases, such as helium and hydrocarbon with small molecular weight. Since the D2 ($n^2S_{1/2} \rightarrow n^2P_{3/2}$, where n = 4, 5, 6 for K, Rb, and Cs, respectively) and D1 ($n^2P_{1/2} \rightarrow n^2S_{1/2}$) transitions are spectrally close, the quantum defect in a DPAL is extremely small. In addition, DPAL has narrow linewidth, compact size, non-toxic system, etc [4,5]. With these advantages, DPAL has become one of the most hopeful candidates for realizing a high-powered laser system.

It is well–known that amplified spontaneous emission (ASE) can be a major source of upper laser level loss in high gain lasers [6]. The ASE will surely decrease the gain, and become more serious when the gain is higher. Thus the ASE must be considered in realization of a high-powered DPAL as the gain is very high. There have been few research groups referring the physical characteristics of the ASE in different kinds of laser systems including a DPAL system [7–9]. For the paper that studied ASE in an alkali laser, it mainly discussed a side-pumped alkali vapor amplifier [9].

In this paper, the ASE of an end-pumped DPAL is systemically studied by using a theoretical model. The influences of cell radius, cell length, and cell wall temperature on the optical-to-optical efficiency, fluorescence, and ASE have been evaluated.

2. Theoretical analyses

In this paper, we consider the configuration of a typical end-pumped continuous-wave DPAL with static media as schematically shown in Figure 1. After passing a polarized beam splitter (PBS), a pump beam with power of Pp enters into a cylindrical vapor cell with the length of *L* and the radius of *R* in which metallic cesium and buffer gases of ethane and helium have been beforehand sealed. The cavity consists of a high reflector with the reflectivity of 100% and an output coupler with the transmittance of 70%.

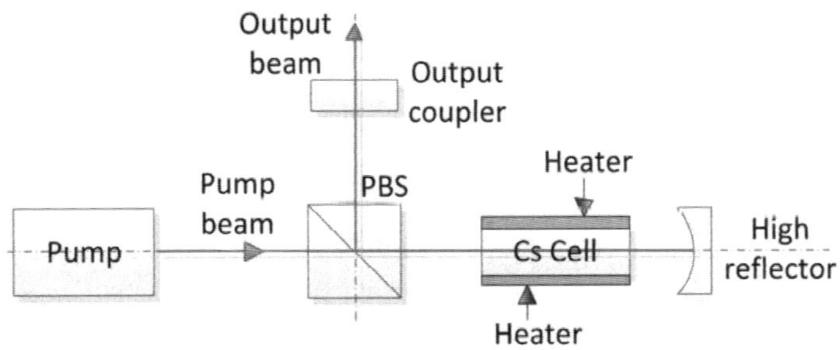

Figure 1. Schematic diagram of a static DPAL.

During the simulation, we made the following assumptions with purpose of the algorithm simplification:

(1) The transverse pump distribution holds out a Gaussian intensity profile and keeps unchanged along the optical axis;

(2) The temperature of every cylindrical annulus is a constant along the optical axis;

(3) The temperature distribution at the transverse section of a vapor cell is symmetrical.

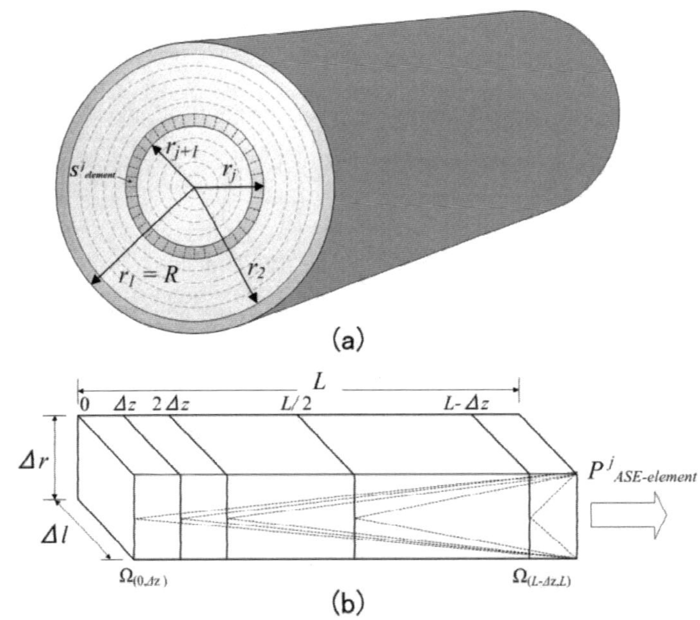

Figure 2. (a) Segmented configuration of a vapor cell and (b) schematic diagram of longitudinal ASE effect in a volume element in the *j*th cylindrical annulus.

In this model, we divide a cylindrical vapor cell into many coaxial cylindrical annuli, in which the temperature is treated as a constant along the optical axis as shown in Figure 2(a). Every cylindrical annulus is thought as both a heat source and a lasing source in the meantime. With respect to the heat transfer between the coaxial cylindrical annuli, we use the same theory of our previous paper [10–13]. To examine the effect of ASE and fluorescence of a DPAL, we make the second division of every cylindrical annulus. Each divided element has dimensions of $\Delta r \times \Delta l \times L$, where l is the perimeter of the jth annulus.

Figure 2(b) shows the schematic diagram of longitudinal ASE effect in an element. The element is further divided into many sub-segment with length of Δz. Each sub-segment has a solid angle relative to the end face of Ω. The ASE rate in an element is then expressed as [9]

$$\Gamma^j_{ASE-element} = \frac{2P^j_{ASE-element}}{h\nu_{21}V^j_{element}} = \frac{2n_2^j}{\tau_{D_1}\left(n_2^j - n_1^j\right)}\frac{V^j_{element}}{\pi L^4}\int l(\lambda)\frac{\exp\left[\left(n_2^j - n_1^j\right)\sigma_{21}L\right]-1}{\sigma_{21}}d\lambda, \quad (1)$$

where $P^j_{ASE-element}$ is the ASE power that emitted from an end of an element, $V^j_{element}$ is the volume of an element, n_i^j (i=1,2,3) is the number densities of the $6^2S_{1/2}$, $6^2P_{1/2}$, $6^2P_{3/2}$ energy level, τ_{D_1} is the D_1 ($6^2P_{1/2}\rightarrow 6^2S_{1/2}$) radiative lifetime, σ_{21} is the atomic emission transverse section, $l(\lambda)$ is the normalized Lorentzian function that describes the spectral intensity distribution of spontaneous emission [14]

$$l(\lambda) = \frac{E}{\Delta\lambda}\cdot\frac{1}{1+\left(\frac{\lambda-\lambda_0}{\Delta\lambda/2}\right)^2}, \quad (2)$$

where E is the normalization constant, $\Delta\lambda$ is the full width at half-maximum of the emitted light, respectively. To simplify the algorithm of this model, the transverse ASE between the cylindrical annuli are not considered, and the transverse ASE in an element that induced by the other element in the same cylindrical annulus are assumed to have the same effect as a single element.

The appropriate laser rate equations of the *jth* cylindrical annulus (see Figure 2(a)) in a DPAL are

$$\begin{aligned}
\frac{dn_1^j}{dt} &= -\Gamma_P^j + \Gamma_L^j + \Gamma_{ASE}^j + \frac{n_2^j}{\tau_{D_1}} + \frac{n_3^j}{\tau_{D_2}}, \\
\frac{dn_2^j}{dt} &= -\Gamma_L^j + \gamma_{32}\left(T_j\right)\left\{\left[n_3^j - n_2^j\right] - \left[2\exp\left(-\frac{\Delta E}{k_B T_j}\right)-1\right]n_2^j\right\} - \Gamma_{ASE}^j - \frac{n_2^j}{\tau_{D_1}}, \\
\frac{dn_3^j}{dt} &= \Gamma_P^j - \gamma_{32}\left(T_j\right)\left\{\left[n_3^j - n_2^j\right] - \left[2\exp\left(-\frac{\Delta E}{k_B T_j}\right)-1\right]n_2^j\right\} - \frac{n_3^j}{\tau_{D_2}},
\end{aligned} \quad (3)$$

where Γ_P^j is the stimulated absorption transition rate caused by pump photons, Γ_L^j is the transition rate of laser emission, $\gamma_{32}\left(T_j\right)$ is the fine-structure relaxation rate, ΔE is the energy gap between the $^2P_{3/2}$ and $^2P_{1/2}$ levels with the value of 554 cm^{-1}, k_B is the Boltzmann constant, τ_{D2} is the D_2 ($6^2S_{1/2}\rightarrow 6^2P_{3/2}$) radiative lifetime, respectively.

3. Results and discussions

We first analyze the influence of the cell length on the optical-to-optical efficiency, fluorescence efficiency, and ASE efficiency. The cell radius, cell wall temperature, and pump power are set as 5 mm, 383 K, and 100 W, respectively. The results are shown in Figure 3. It can be seen that there is an optimal cell length corresponds to the maximal value of the optical-to-optical efficiency. For the curve of fluorescence, the fluorescence efficiency increases monotonously with the cell length. This is because that the longer cell length corresponds to the much more cesium atoms. On the contrary, the ASE efficiency decreases as the cell length increases. When the cell length exceeds 30 mm, the decrease of the ASE is not obvious.

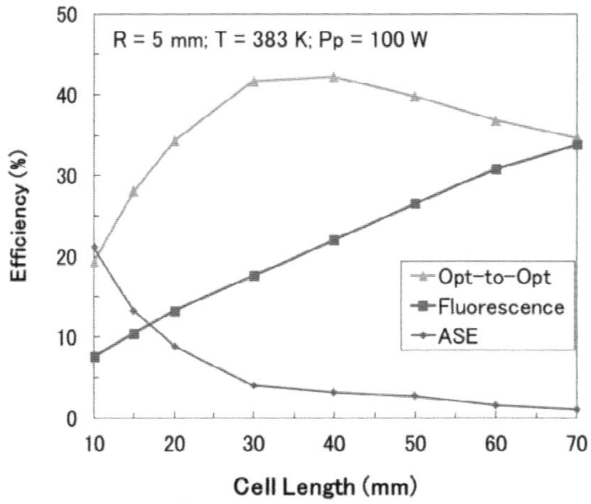

Figure 3. The influence of cell length on the optical-to-optical efficiency, fluorescence efficiency, and ASE efficiency with cell radius of 5 mm at cell wall temperature 383 K under 100 W-pumping.

Figure 4 shows the optical-to-optical efficiency, fluorescence efficiency, and ASE efficiency as a function of the cell radius, while the cell length, cell wall temperature, and pump power are set to be 40 mm, 383 K, and 100 W, respectively. It can be seen from Figure 4 that the cell radius has little influence on both the optical-to-optical efficiency and the fluorescence efficiency. The ASE increases obviously with the cell radius. Based on the results of Figure 3 and 4, the ASE effect might be suppressed to some extent by adjusting the cell dimensions.

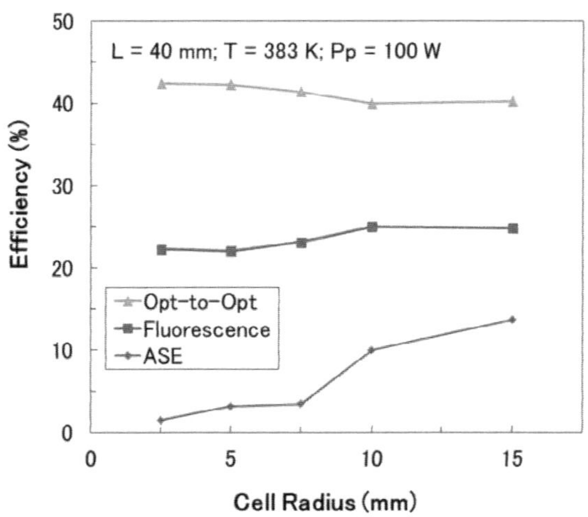

Figure 4. The influence of cell radius on the optical-to-optical efficiency, fluorescence efficiency, and ASE efficiency with cell length of 40 mm at cell wall temperature 383 K under 100 W-pumping.

Next, we analyze the influence of the cell wall temperature on the optical-to-optical efficiency, fluorescence efficiency, and ASE efficiency, respectively. The cell length, cell radius, and pump power are set to be 40 mm, 5 mm, and 100 W, respectively. Given a certain cell dimension, the total number of the cesium atom will change with the cell wall temperature. When the cell wall temperature is lower than the optimal value, the output power cannot achieve the peak value because of the shortage of the total number of cesium atoms. On the other hand, when the cell wall temperature becomes higher than the optimal value, the excessive cesium vapor will also lead to the reabsorption of the stimulated radiation in the cell. Since the fluorescence is in direct proportion to the cesium atom number, the fluorescence efficiency increases obviously with the cell wall temperature. For the case of ASE, its change is slightly.

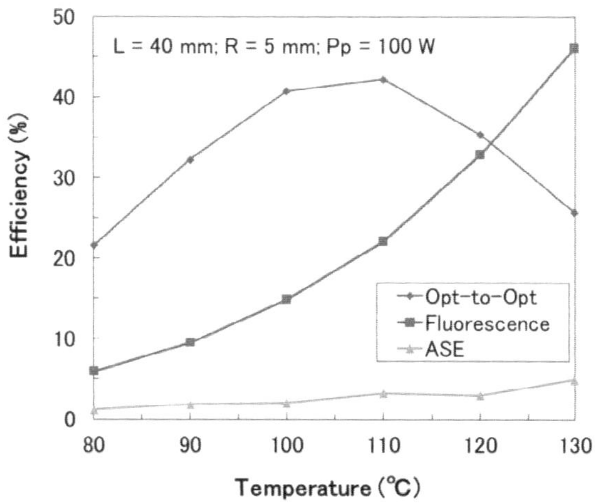

Figure 5. The influence of cell wall temperature on the optical-to-optical efficiency, fluorescence efficiency, and ASE efficiency with cell length and cell radius of 40 and 5 mm under 100 W-pumping, respectively.

In Figure 6, the output laser power, fluorescence power, and ASE power are given as a function of the pump power, respectively. The cell length, cell radius, and cell wall temperature are set to be 40 mm, 5 mm, and 383 K, respectively. It can be observed that the ASE and fluorescence power changed little with the pump power increases. This might could be ascribed to the cell dimension setting, suggesting that ASE can be suppressed to some extent.

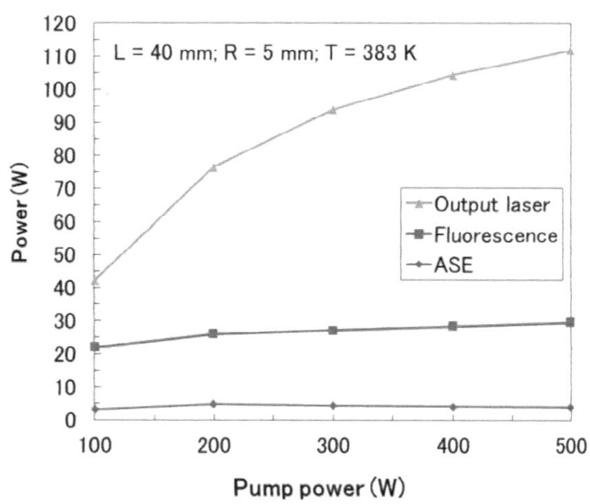

Figure 6. The influence of pump power on the power of output laser, fluorescence and ASE with cell length and cell radius of 40 and 5 mm at 383 K, respectively.

4. Summary

In this study, we develop a theoretical model to study a static-state DPAL by taking the process of amplified spontaneous emission (ASE) into account. The influences of the cell radius, cell length, and cell wall temperature on output features of a DPAL including ASE, fluorescence, and output power are systemically investigated. The results show that ASE can be effectively decreased by adjusting the cell dimension. And the output optical-to-optical efficiency can be optimized at the same time.

Acknowledgments

This work was supported by the National Natural Science Foundation of China under Grant No.61605178.

References

[1] Beach R J, Krupke W F, Kanz V K, and Payne S A 2004 *J. Opt. Soc. Am. B* **21** 2151
[2] Wang Y, Kasamatsu T, Zheng Y, Miyajima H, Fukuoka H, Matsuoka S, Niigaki M, Kubomura H, Hiruma T, an Kan H 2006 *Appl. Phys. Lett.* **88** 141112
[3] Krupke W F, 2012 *Prog. Quantum Electron.* **36** 4
[4] Zhdanov B V, Ehrenreich T, and Knize R J 2006 *Opt. Commun.* **260** 696
[5] Waichman K, Barmashenko B D and Rosenwaks S 2017 *J. Opt. Soc. Am. B* **34** 279
[6] Su H and Wang X 2012 *Proc. of SPIE* **8235** 82351Q
[7] Goren C, Tzuk Y, Marcus G and Pearl S 2006 *IEEE J. Quantum Electron.* **42** 1239
[8] Speiser J 2009 *J. Opt. Soc. Am. B* **26** 26
[9] Yang Z N, Wang H Y, Lu Q S, Hua W H, and Xu X J 2011 *Opt. Express* **19** 23118
[10] Han J H, Wang Y, Cai H, Zhang W, Xue L P and Wang H Y 2014 *Opt. Express* **22** 13988
[11] Han J H, Wang Y, Cai H, An G F, Zhang W, Xue L P, Wang H Y, Zhou J, Jiang Z G and Gao M 2015 *Opt. Express* **23** 9508
[12] An G F, Wang Y, Han J H, Cai H, Zhou J, Zhang W, Xue L P, Wang H Y, Gao M and Jiang Z G 2015 *Opt. Express* **23** 26414
[13] An G F, Wang Y, Han J H, Cai H, Wang S Y, Yu H, Rong K P, Zhang W, Xue L P, Wang H Y and Zhou J 2017 *Opt. Express* **25** 4335
[14] Copeland D A 2013 *Proc. of SPIE* **8599** 85991P

IOP Conf. Series: Journal of Physics: Conf. Series **844** (2017) 012057 doi:10.1088/1742-6596/844/1/012057

Global Algorithm Applied on Single Photon Detection

Hua LIU [1,2,3], Quanxin DING [1,2], Helong Wang[1,2], Hongliang Chen[1,2], Chunjie GUO[1,2] and Liwei ZHOU [3]

[1] Science and Technology on Electro-optic Control Laboratory, China
[2] Luoyang Institute of Electro-optical Equipment of AVIC, China
[3]Beijing Institute of Technology, Beijing, China

Corresponding author's e-mail: Melody.h.liu@vip.163.com

Abstract. There are three major contributions. Firstly, applied study on the theory and experiment of single photon detection , including the project and experiment of quantum key distribution. Secondly, based on methods of the selection of detector, main photo electronic system configuration, design and creation, along with the relationship between these to system characteristics have been studied. Thirdly, based on the considerations on the research of image sensor systems on single photon detection, and the total system characteristics are evaluated and discussed in quantity. The results of simulation experiments and theory analyzing demonstrate that proposed method could advance the system validity effectively, and theory analysis and experiment shows the method is reasonable and efficient.

Keywords: Single photon detection, Imaging sensors, System engineering

1. Introduction

The research on the theory and experiment of single photon detection, including the project and experiment of quantum key distribution, what has become the research direction in these years. Based on the selection of detection methods, the configuration, the reality and the creation of the sensor system, has become an important application direction. The study of the relationship among the components in these systems, which based on the quantum theory and the basic mechanism of single photon detection, have increased as the implementation methodology .

Application and modern imaging sensors need to pay special attentions to the special technology in the development process of photoelectric system, research on single photon detection imaging sensor system is a very active direction based on. In particularly, in many fields, focused mainly on to find the objects description, to increase the detection probability, and to reduce the processing speed. Find by hard and thorough search, detection can be traced to the development of the quantum theory of long rich history.

Quantum theory is one of the two most important cornerstones of modern physics, quantum theory provides the representation methods and ways of thinking about the nature of the novel, reveals the basic rules in the material world, and laid the theoretical foundation for the atomic physics, solid state physics, and physics and plasma physics. It can well explain the atomic structure, the regularity of the atomic spectrum, the nature of the chemical element, light attraction and radiation and so on.

In classical physics theory of energy is continuously changing, which can take any value.These phenomena should arouse our attentions on that the background and mechanism of quantum cryptography, and the recent research and development. Through analyzing the development of single photon detectors and their respective advantages and disadvantages, especially focus on single photon detection technology and development, and the special principle of core device APD and for the modes of single photon detection, we can recognize the value direction, such as to realize the principle of single photon pulse, and fiber optic coupler to achieve a single photon pulse output. Thus, single photon detection sensor system configuration, comprehensive consideration and system, lead us into the future promising and challenging research.

When the design of the sensor system is beginning, what requires systematic considerations, because the overall system features need to be evaluated and discussed in terms of quantity. The results of simulation and theoretical analysis are need to evaluated and determined by. Therefore, we propose that the global optimization method which establish correlative algorithms can effectively improve the effectiveness of the system. Theoretical analysis and experimental results show that the method is reasonable and effective in these two aspects.

2. Methods

2.1. Detection analysis

Quantum state has its particularity. Based on the special state of the quantum state of the information carrier, we give a global optimization algorithm.

From the point of view of quantum mechanics， that quantum state as the information carrier, through the quantum channel transport between the users. The security of quantum cryptography is guaranteed by the principle of quantum mechanics. The so-called absolute safety is that even if the eavesdropper may be smart enough that the most sophisticated eavesdropping measures, may use the most advanced means of measurement, the transfer is still the key security.

Usually, the eavesdropper intercepted by key method has two kinds, which means that one kind of method is measured by the quantum state to carry information by which the measurement results will to extract key information. However, the basic principle of quantum mechanics tells us that the measurement of the quantum state will cause the wave function collapse, so that the nature will change quantum state, and the sender and the receiver can be checked by their communications evaluating, because this kind of eavesdropping is bound to stay with obvious features of quantum measurement traces.

The second method is to avoid the direct quantum measurement, which the device with copy function, so that the information transformed by can be intercepted and copied. Then, the evaluator and the quantum state transfer of the original can be use. If a copy of the quantum state for evaluators can be in measurement and analysis, and in order to steal information. So, the evaluator should do not leave any traces. But determined by quantum coherence, quantum no-cloning theorem shows that quantum copying device may allow any physically impossible to clone and the input quantum state exactly.

2.2. Detection options

Base on the Quantum key distribution model and Quantum communication channel model, global algorithm can be applied on single photon detection. Correlation methods applied and analyzed can be demonstrated in Figure1 to Figure 2.

IOP Conf. Series: Journal of Physics: Conf. Series **844** (2017) 012057 doi:10.1088/1742-6596/844/1/012057

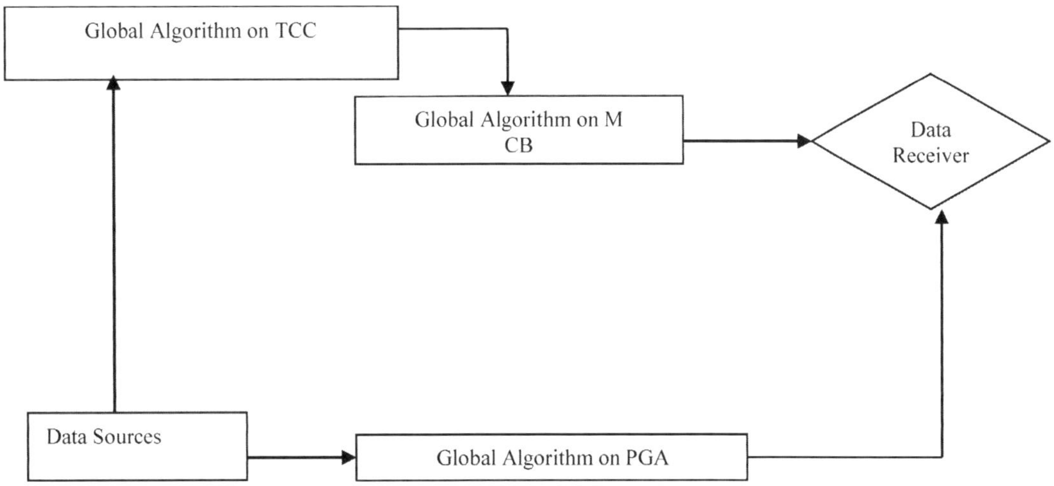

Figure 1. Global algorithm applied on single photon detection A

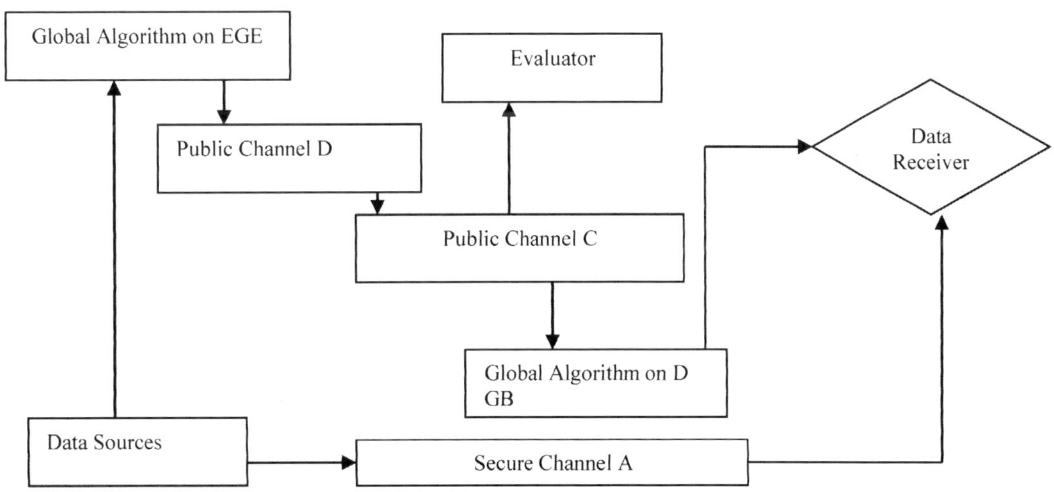

Figure 2. Global algorithm applied on single photon detection B

3. Results

Base on the judge by reason that particles in the micro world are strictly abide by the above two kinds of quantum mechanics, so in the quantum secure communication, we use the above characteristics to test the theory and experimentation which we design. According to that quantum communication protocol is based on quantum communication channel, which divided into different approach in communication.

According to the methods discussed above, single quantum state, quantum communication and quantum key distribution simulation analysis have been studied by some experimentation we design, which mainly using the properties of light quantum in the micro world, high-speed transmission through a single photon or entangled photon pairs as the carrier of information to achieve information, can be said that the transmitter large capacity information theoretically, and is capable of generating cannot decipher key, and to encrypted data information communication using the key.

Based on theoretical analysis, principle and process of this protocol, the simulation software simulation model of quantum circuits on agreement and intercept and resend method based on structural reliability and quantitative agreement, effective mutual information analysis model of transmission line, efficacy and safety of different interference model and noise model, quantitative. The experimental results show that. See Figure 3 and 4. And the Results on criterion MTF of sensor system can be evaluated by Figure 5, 6 .

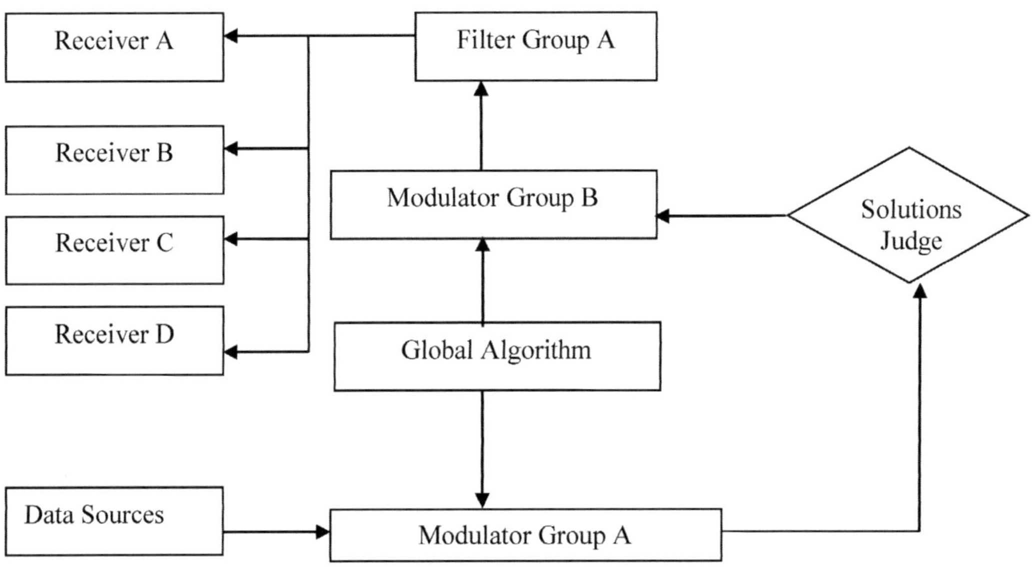

Figure 3. Global algorithm applied on sensor systems A

IOP Conf. Series: Journal of Physics: Conf. Series **844** (2017) 012057 doi:10.1088/1742-6596/844/1/012057

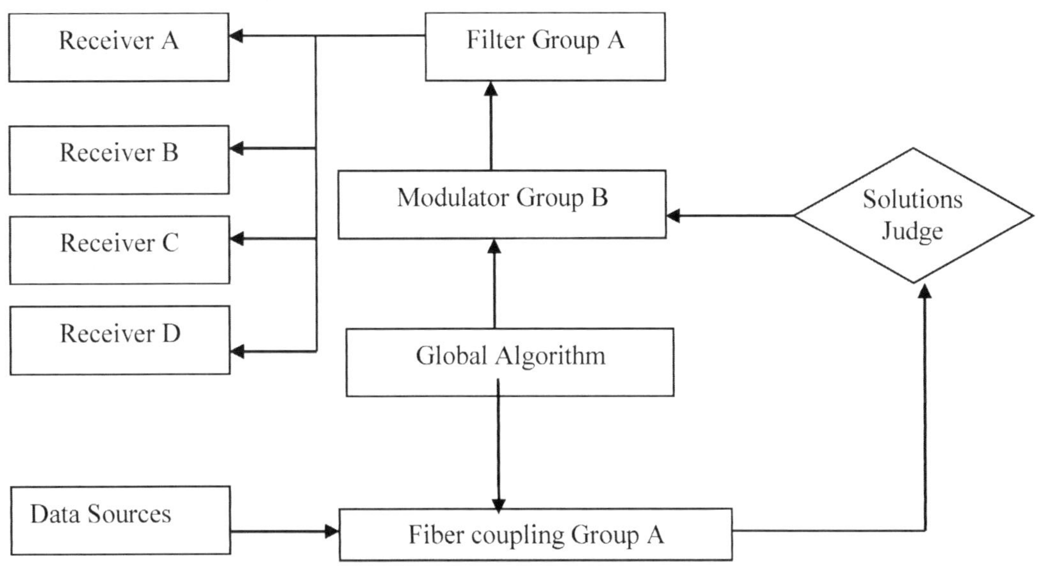

Figure 4. Global algorithm applied on sensor systems B

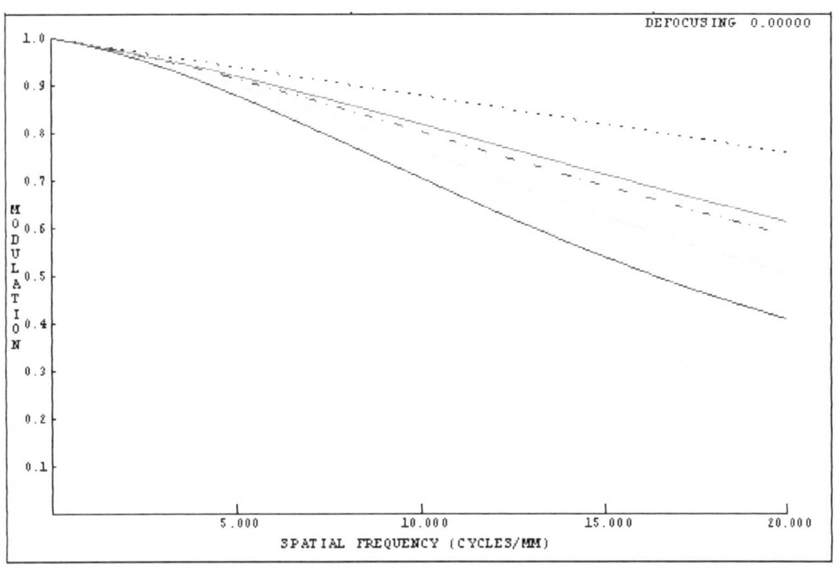

Figure 5. Results on criterion MTF of sensor system A

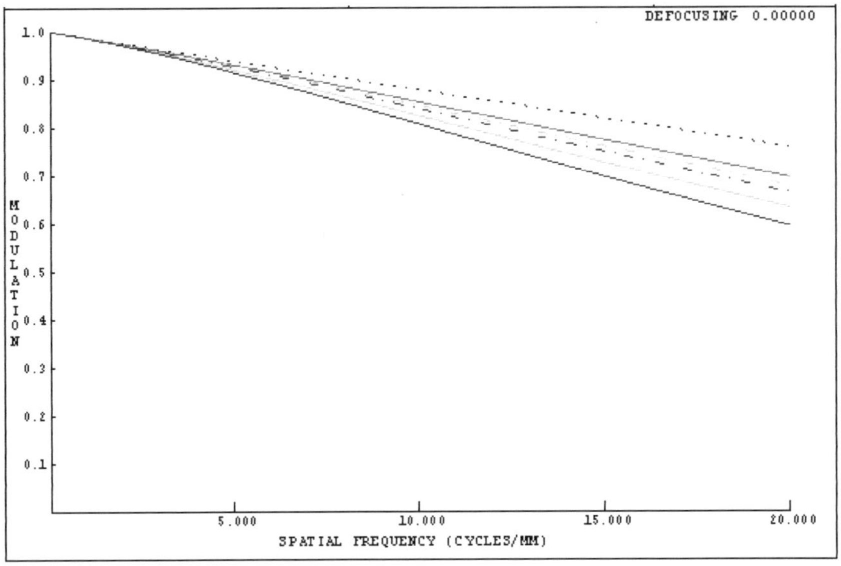

Figure 6. Results on criterion MTF of sensor system B

4. Conclusions

The results of simulation experiments and theory analyzing demonstrate that proposed method could advance the system validity effectively, and theory analysis and experiment shows the method is reasonable and efficient. Firstly, applied study on the theory and experiment of single photon detection , including the project and experiment of quantum key distribution. Secondly, based on methods of the selection of detector, main photo electronic system configuration, design and creation, along with the relationship between these to system characteristics have been studied. Thirdly, based on the considerations on the research of image sensor systems on single photon detection, and the total system characteristics are evaluated and discussed in quantity.

References

[1] Born M. and Wolf E., Principles of Optics, Pergamon Press, Oxford, 2005.
[2] Leslie E. Ballentine, Quantum Mechnanics, World Scientific Pub Co., 1998.
[3] Landau L., Quantum Mechanics Non relativistic Theory, Butterworth-Heinemaann, 1981.
[4] Lim Y., Problems and solutions on quantum mechanics, World Scientific Pub Co Inc, 1998.
[5] Joseph W. Goodman, Introduction to Fourier Optics, McGraw-Hill, New York, 1996.
[6] Zheng J. Y., Introduction to quantum mechanics, Science Press, 2003.

IOP Conf. Series: Journal of Physics: Conf. Series **844** (2017) 012058 doi:10.1088/1742-6596/844/1/012058

The dynamic process of laser drawing germanium core optical fiber

Wei Zhang, Na Chen*, Zhenyi Chen, Ziwen Zhao, Qiang Guo, Fufei Pang and Tingyun Wang

Key Laboratory of Specialty Fiber Optics and Optical Access Networks, Shanghai University, 149 Yanchang Road, Shanghai 200072, China

*E-mail: na.chen@shu.edu.cn

Abstract. The Laser drawing process of germanium core optical fiber is studied numerically and experimentally in this paper. We analysed the changes in the maximum temperature, maximum velocity and surface profile over time. The final profile of dynamic simulation is in good agreement with the experimental result.

1. Introduction

As a member of semiconductor core fibers, germanium core optical fiber is attracting more and more attention due to its mid-wave infrared transparency and high third-order nonlinear coefficients. It has been reported that the germanium core fiber preform can be fabricated by the molten-core method [1], high-pressure chemical vapor deposition [2], and so on. After that, the prepared preform was drawn into a fiber with a graphite furnace. However, because of the long heating zone of the graphite furnace, the semiconductor core is easily oxidized during the drawing process, and the molten core material is easily deposited under the action of gravity at the bottom of the softened area before the drawing process.

Laser drawing has many advantages such as short heating time and small action area, and the neck-down region of the preform during the drawing process is totally observable in real time. It appears to be a better solution. Here we present a numerical model to study the dynamic changes in the laser drawing process, and then we studied the change in temperature and velocity, at the end we compared the final profile obtained by the experiment and the simulation.

2. Models and experimental conditions

2.1. Simulation conditions

Optical fiber drawing process involves heat transfer, mass transfer, and laminar flow phenomena. This problem is generally considered to be axisymmetric, so a 2-D numerical model was used to reduce the scale of the model. The intersection of the preform bottom line and the axis is the origin of the coordinate, the axis of symmetry is set to z axis, and the germanium core is at the top of the center of the preform.

Softened silica and molten germanium are treated as Newton's incompressible fluids, which are governed by Navier-Stokes equations:

IOP Conf. Series: Journal of Physics: Conf. Series **844** (2017) 012058 doi:10.1088/1742-6596/844/1/012058

$$\frac{\partial \rho}{\partial t} + \nabla \cdot (\rho \boldsymbol{u}) = 0, \tag{1}$$

$$\rho \frac{\partial \boldsymbol{u}}{\partial t} + \rho(\boldsymbol{u} \cdot \nabla)\boldsymbol{u} = \nabla \cdot [-p\boldsymbol{I} + \boldsymbol{\tau}] + \boldsymbol{F}, \tag{2}$$

$$\rho C_p\left(\frac{\partial T}{\partial t} + (\boldsymbol{u} \cdot \nabla)T\right) = -(\nabla \cdot \boldsymbol{q}) + \boldsymbol{\tau}{:}\boldsymbol{S} - \frac{T}{\rho}\frac{\partial \rho}{\partial T}\Big|_p \left(\frac{\partial p}{\partial t} + (u \cdot \nabla)p\right) + Q, \tag{3}$$

where $\boldsymbol{u}$ is the velocity vector, p is pressure, ρ is density, $\boldsymbol{\tau}$ is the viscous stress tensor, $\boldsymbol{F}$ is the volume force vector, C_p is the specific heat capacity at constant pressure, T is the absolute temperature, $\boldsymbol{q}$ is the heat flux vector, $\boldsymbol{S}$ is the strain-rate tensor, $\boldsymbol{I}$ is unit matrix, Q is heat sources. This is an intuitive formulation that facilitates boundary condition specifications. Equ. (1) is the continuity equation and represents conservation of mass. Equ. (2) is a vector equation which represents conservation of momentum. Equ. (3) describes the conservation of energy, formulated in terms of temperature. Besides that, $\mathrm{a{:}b}$ stands for $\sum_{\mathrm{n}}\sum_{\mathrm{m}} \mathrm{a}_{\mathrm{nm}}\mathrm{b}_{\mathrm{mn}}$. In this paper, we set the feed speed as 1 mm/min, the same as experimental condition.

The heat transfer process can be expressed by the following equation:

$$\rho C_p \frac{\partial T}{\partial t} + \rho C_p \boldsymbol{u} \cdot \nabla T + \nabla \cdot (\boldsymbol{q} + \boldsymbol{q}_r) = Q, \tag{4}$$

where $\boldsymbol{q}$ is the heat flux by conduction, $\boldsymbol{q}_r$ is the heat flux by radiation, we use a homemade CO_2 laser (10.6 μm) drawing instrument to heat the preform. The schematic of the laser heating path is shown in Fig. 1. The laser will be absorbed in a very short distance on the surface of the preform, so we can treat it as a boundary heat source.

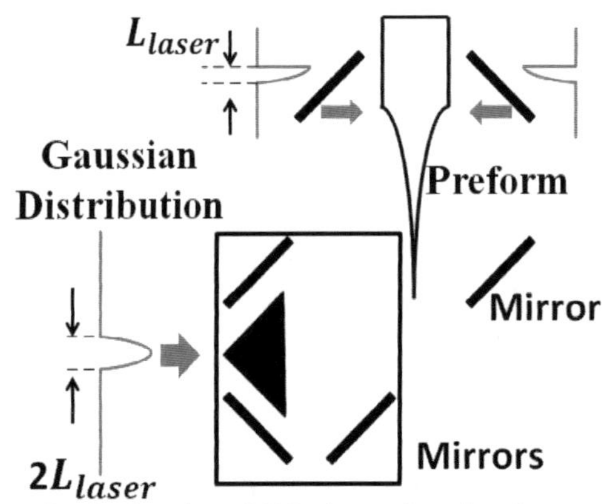

Fig. 1. Optical paths of CO_2 laser drawing instrument.

The power of laser is initially a Gaussian distribution. After the conversion of the optical paths, it reaches the preform all around with a near half-Gaussian distribution. When the total power is 320 W, the power distribution obtained by Zemax simulation is shown in the Fig. 2.

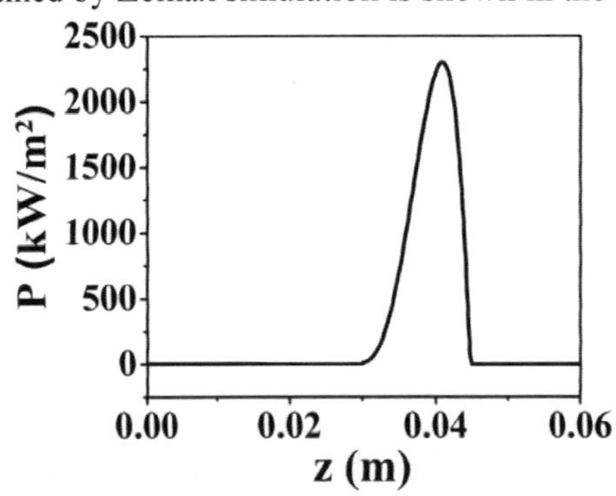

Fig. 2. Power distribution on the surface of preform.

In reality, the power density increase as the radius of the preform decrease, so the laser power function must multiply a scale factor r/r_0, where the r_0 equals to the initial radius of the preform. The preform is cooled by radiation and air convection of the surface. The air convection can be simplified as a boundary convective heat flux. It can describe as follows:

$$\boldsymbol{q} = h(T_{ext} - T), \quad (5)$$

where h is heat transfer coefficient, and T_{ext} is external temperature. The radiation heat flux can be described as:

$$-\mathbf{n} \cdot \boldsymbol{q}_r = \varepsilon\sigma(T_{ext}^4 - T^4), \quad (6)$$

where ε is the surface emissivity, σ is the Stefan-Boltzmann constant, and **n** is the normal vector on the boundary. In this model, the surface and the bottom of the preform have suffered both radiation heat flux and air convection, the external temperature is set as 20 centigrade. The material parameters involved here can be found in [3-6]. The equations are solved numerically with Comsol Multiphysics using a finite-element method, and the free surface is solved with an Arbitrary Lagrangian-Eulerian formulation.

2.2. Experiment conditions

In addition to theoretical simulation, we also experimentally validated this process. The fiber preform was prepared by putting a single crystal Ge-rod with OD around 220 μm into a silica tube with OD around 8 mm. The whole drawing process could be divided into three states: heating state, unstable drawing state, and stable drawing state. In the first state, the preform has been heated by the CO_2 laser. After being well heated, the preform starts to deform under the effect of gravity. The preform is tapering and eventually forming a neck-down shape. In the final state, the shape of preform stop changing, and the drawing process becomes steady under a fixed preform feed speed and a fixed fiber drawing speed.

During the drawing process, we set the initial feed speed to 1 mm/min and initial power to approximately 100 W, gradually increasing the power until the preform starts to deform. By tuning the laser power, the preform feeding and fiber drawing speeds, the steady drawing condition can be finally reached. After the part containing the germanium core is completely drawn, the laser is turned off and the entire drawing process ends.

3. Result and discussion

For comparison, the preform parameters and the drawing conditions in the simulation are set as the same with the experiment. When laser power reaches 320 W, the preform is tapering and eventually forms a neck-down shape. The maximum temperature and maximum velocity vary with time, as shown in Fig. 3.

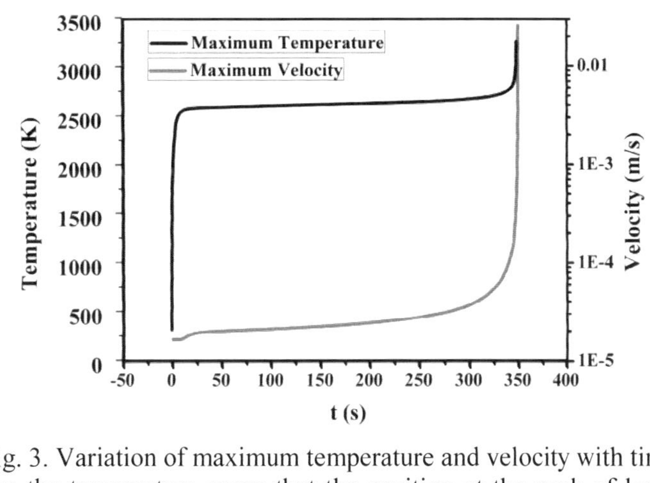

Fig. 3. Variation of maximum temperature and velocity with time.

It can be seen from the temperature curve that the position at the peak of laser power is rapidly heated to higher than 2500 K just in a few seconds because of the high absorption of the silica material at the wavelength of 10.6 μm. As time goes on, heat conduction and accumulation let more and more silica material soften, and the preform begins to deform with the decreasing of silica's viscosity. The maximum velocity increases greatly after 300 s, which denotes the increasing deformed rate. This will lead to a rapid temperature increase, and finally causes an increasing velocity. At the last 25 s, the deformation rate is so large that the preform is quickly falling down just like free falling.

Fig. 4 shows the preform profile changes with the laser heating time. In the first 50 s, the deformation is not obvious. Starting from 100 s, the deformation speed increases as time passed by. From 345 s to 349.13 s, the rapidly falling happens and a necked-down profile forms at 349.13s.

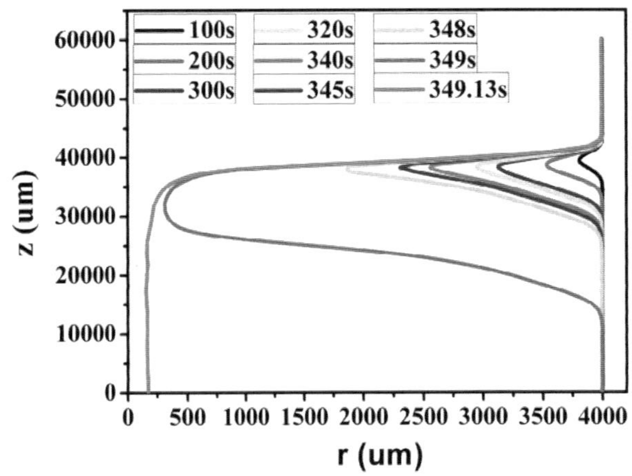

Fig. 4. Variation of preform surface profile with time.

In the experiment, the feed speed of preform and the power of laser were adjusted to make sure that the preform is fully softened but not vaporized. In this way, we successfully reached the steady drawing state, and the part containing the germanium core was fully drawn into Ge-core optical fiber in the end. Both the picture of the drawn preform and the neck-down profile from above simulation are shown in Fig. 5. We can see from the figure that the simulation result has a good agreement with experiment results.

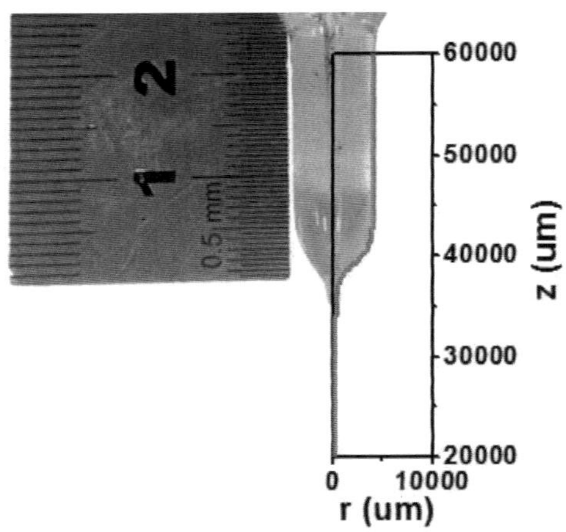

Fig. 5. The final neck-down profile of the fiber preform obtained by the experiment and the simulation.

4. Conclusion

A model to simulate the dynamic process of laser drawing pure Ge-core fiber is presented. The variation curves of the temperature, the velocity, and the surface profile of the preform with time were calculated and discussed. The final profile of the preform has a good agreement with the profile obtained by the experiment.

Acknowledgment
This work is supported by the National Natural Science Foundation of China (Project Nos: 61227012, 61575120, 61422507, 61475095, and 61635006).

References

[1] Ballato J, Hawkins T, Foy P, et al. 2011 Silica-clad crystalline germanium core optical fibers. *Optics Lett.* **36**: 687-688.

[2] Mehta P, Krishnamurthi M, Healy N, et al. 2010 Mid-infrared transmission properties of amorphous germanium optical fibers. *Appl. Phys. Lett.* **97**: 071117.

[3] Doremus R H. 2002 Viscosity of silica. *Journal of Applied Physics* 92: 7619-7629.

[4] Rhim W K, Ishikawa T. 2000 Thermophysical Properties of Molten Germanium Measured by a High-Temperature Electrostatic Levitator1. *International Journal of Thermophysics* **21**: 429-443.

[5] Brückner R. 1970 Properties and structure of vitreous silica. I. *Journal of non-crystalline solids* **5**: 123-175.

[6] Sato Y, Nishizuka T, Tachikawa T, et al. 2000 Viscosity and density of molten germanium. *High Temperatures High Pressures(UK)* **32**: 253-260.

IOP Conf. Series: Journal of Physics: Conf. Series **844** (2017) 012059 doi:10.1088/1742-6596/844/1/012059

Strain Distribution in Silica-Clad Crystalline-Germanium-Core Fiber

Dongyang Wang, Na Chen*, Ziwen Zhao, Zhenyi Chen and Tingyun Wang

Key Laboratory of Specialty Fiber Optics and Optical Access Networks, Shanghai University, 149 Yanchang Road, Shanghai 200072, China

*E-mail: na.chen@shu.edu.cn

Abstract: In this paper, we fabricated the crystalline germanium core silica cladding fiber with rod-in-tube method. Its strain distribution inside the fiber was analysed using Raman spectroscopy. The results reveal that the core of the prepared fiber has relatively high degree of crystallization and a relatively uniform strain distribution inside the germanium core.

1. Introduction

The merger of germanium with its high nonlinearity, high Raman gain, infrared transparency and rich optoelectronic functionality into optical fiber opens the door to a nascent and cutting-edge filed.

These good features attracted many researchers starting to study the manufacture of the germanium core optical fiber. Some excellent method, such as molten-core method [1], high-pressure micro-fluidic chemical vapor deposition (HPMFCVD) [2] and powder-in-tube method [3] have been used in the semiconductor core fiber fabrication. In these methods, HPMFCVD is easy to introduce reaction intermediates. Powder-in-tube method is preferred to form an oxide. Due to the big difference of the core and cladding materials, large strain is usually induced during the fiber drawing process, and this will affect the fiber properties a lot. In this paper, we present a highly crystalline germanium core fiber with a uniform strain distribution fabricated by the rod-in-tube method.

Although many ways were presented to produce the fiber, more problems still exist to be solved. Some challenging question, such as how to reduce the transmission loss or analysis the crystal characteristics of the optical fiber, inspired the researchers to devote themselves into these research. After years of unremitting efforts, a lot of methods have been put forward to solve these problems. By annealing [4] or tapering [5], the degree of crystallization of the core can be improved. Through changing the core size [6] or improving the purity of the fiber core, the transmission loss was reduced. But few people studied the strain distribution of the fiber core. One of the difficulties of fabrication is that the thermal expansion coefficient of the cladding material and core material does not match. In this paper, we present a study on the strain distribution inside the highly crystalline germanium core fiber fabricated by the rod-in-tube method.

2. Fiber fabrication

Firstly, we fabricated Ge-core silica-clad fiber with rod-in-tube method. The high-purity single crystal Ge rod with an outer diameter 2.5 mm was put into a silica tube, of which an inner diameter slightly larger than 2.5 mm. The outer diameter of the silica tube is 20 mm. In order to melt the Ge rod evenly during the fiber draw, the silica tube had been pre-draw and shaped. It is also for this reason that we can get a uniform distribution of strain in the core. Using graphite furnace, we can obtain the uniform hot zone with the temperature higher than 2200°C that the pre-fabricated fiber preform needed. Throughout the drawing process, the temperature and the drawing speed are monitored and controlled by the drawing tower. Stable external environment and constant stretching speed are more conducive to create a uniform strain distribution and high crystalline core.

As is shown in microscope (Fig. 1), we found that in the fiber center emerged a perfect round core and the interface between core and cladding is smooth without micro-cracks or gaps, which ensure the performance of fiber. The diameter of Ge core is 19 μm.

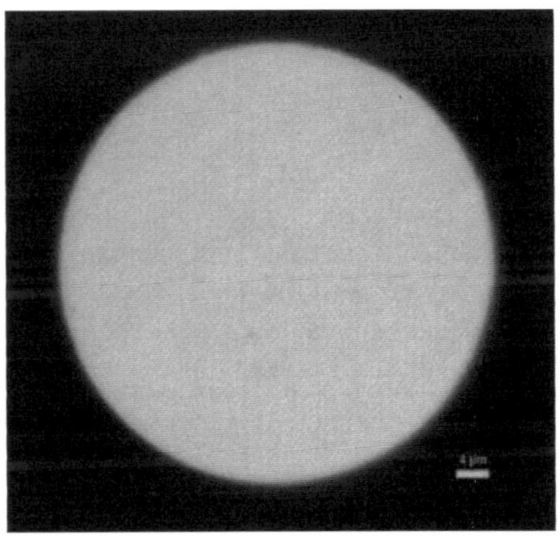

Fig. 1. Optical micrograph of the germanium core cross-section

3. Strain analysis

Fig. 2 shows the Raman spectra of the single crystal Ge rod that we used as the core material and the drawn fiber core. Comparing both spectra, we find that the peak of Ge rod (red line) is 300.994 cm^{-1}, and the peak of fiber (black line) is 300.321 cm^{-1}, while the FWHM of Ge rod and the fiber are both 2.609 cm^{-1}. This indicates that the fiber has a high crystalline Ge core, but it may be due to the stretching strain that the Raman peak of Ge-Ge slightly shifts toward lower wave number.

IOP Conf. Series: Journal of Physics: Conf. Series **844** (2017) 012059 doi:10.1088/1742-6596/844/1/012059

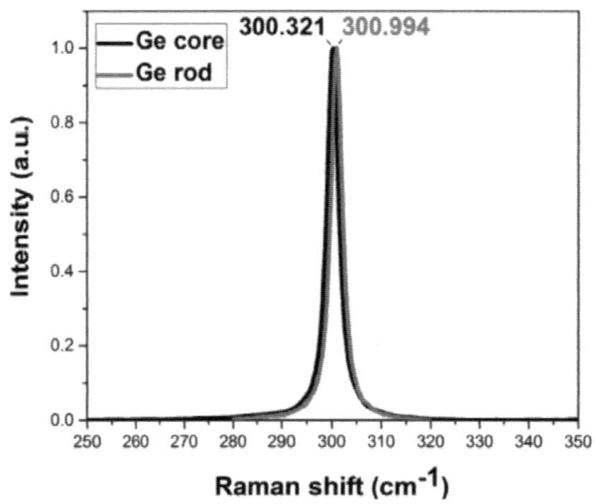

Fig. 2. Raman spectra of Ge rod (red) and Ge core fiber (black)

For detecting the strain distribution, we obtained the Raman spectra at the points A, B and C, on the flat end face of the fiber. In Fig. 3 the Raman shift of Ge gradually shift to a higher wavenumber from A of 300.363 cm^{-1} to C of 301.346 cm^{-1}, contrastively the silica shift to a lower wavenumber from B of 488.995 cm^{-1} to C of 489.134 cm^{-1}. It shows that stretching strain has been formed in the core compared with the pressure strain in the cladding. The intensity of Ge Raman shift in the core is obviously stronger than that of the cladding, which is in line with the actual situation. From the Raman spectra analysis, we also found that the line width of the Raman peak is narrower in the core than in the cladding. This reflects that from interface to center the crystallinity of Ge is tend to be single-crystalline, while the core presents stretching strain.

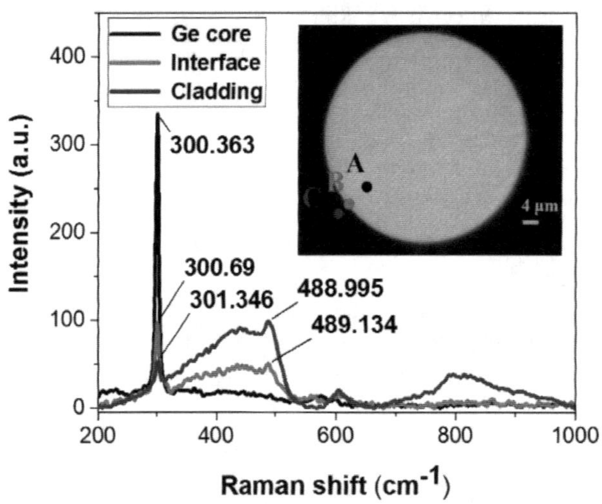

Fig. 3. Raman spectra of the Ge core, core/cladding interface and the cladding.

Using Raman spectroscopy, we performed a mapping on the end of the fiber. We scanned 50 μm × 49 μm area with the step size 1 μm. The results are shown in Fig. 4.

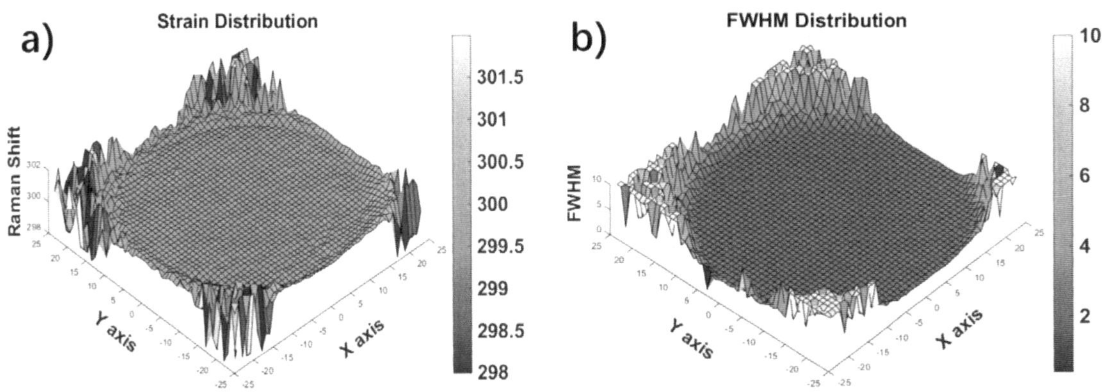

Fig. 4. (a). Strain distribution; (b). FWHM distribution.

From the peak position distribution of measurement points, we can obtain the shifts of the peak position. Thus the strain change of the fiber can be conveyed. From Fig. 3 (a), we notice that there is no obvious peak drift inside the core. And near the interface, core has a clear red shift, which is the performance of the pressure strain. In the core, the peak concentrates upon 301 cm^{-1}, which exactly is the peak of Ge rod. It is because the silica tube and Ge rods have a mismatch thermal expansion coefficient.

Likewise, we analysed the full width at half maximum (FWHM) of each point, there was no significant saltation in the core as shown in Fig. 3 (b). The FWHM just goes wide when close to the interface. In the core, the FWHM is 2.609 cm^{-1}. This indicates that Ge in the core tends to be single-crystalline. The results of the above analysis are consistent to the result of peak analysis.

4. Conclusion

In conclusion, we fabricated the Germanium core optical fiber using the rod-in-tube method. A smooth junction formed between the core and the cladding. The Raman spectrum reveals uniform strain distribution of Ge core that is overall highly crystalline. This study may give a good insight for the future manufacture of the Ge core fiber.

Acknowledgment

This work is supported by the National Natural Science Foundation of China (Project Nos: 61575120, 61227012, 61422507, 61475095, and 61635006).

Reference

[1] Ballato J, Hawkins T, Foy P, et al. 2009 Glass-clad single-crystal germanium optical fiber *Optics Express* **17**: 8029-35.
[2] Mehta P, Krishnamurthi M, Healy N, et al. 2010 Mid-infrared transmission properties of amorphous germanium optical fibers *Appl.Phys. Lett.* **97**: 071117.
[3] Scott B, Wang K, Caluori V, et al. 2009 Fabrication of silicon optical fiber. *Optical Engineering* **48**: 100501-3.

[4] Lagonigro L, Healy N, Sparks J R, et al. 2010 Low loss silicon fibers for photonics applications. *Appl. Phys Lett.* **96**: 041105.
[5] McMillen C, Brambilla G, Morris S, et al. 2012 On crystallographic orientation in crystal core optical fibers II: Effects of tapering *Optical Materials* **35**: 93-96.
[6] Morris S, McMillen C, Hawkins T, et al. 2012 The influence of core geometry on the crystallography of silicon optical fiber *Journal of Crystal Growth* **352**: 53-58.

IOP Conf. Series: Journal of Physics: Conf. Series **844** (2017) 012060 doi:10.1088/1742-6596/844/1/012060

PbS Quantum Dots Filled Photonic Crystal Fiber for All-fiber Amplifier

Zhanbing Wang, Yana Shang*,Fufei Pang, Huanhuan Liu, Na Chen, Yan Wu, Yanan Kang

[1]The key Lab of Specially Fiber Optics and Optical Access Neyworks, Shanghai University, Shanghai,20072

ynshang@shu.edu.cn

Abstract. In this paper, we have proposed a novel type of fiber amplifier by filling the PbS semiconductor quantum dots into the holes of photonic crystal fibers (PCFs) for the first time. Based on simulation results, we have found that the loss of PCF filled with PbS is slightly increased compared with the one without PbS at wavelength of 1310 nm. Furthermore, we have successfully fabricated the PbS-filled PCF with selective air-hole cladding by a new perfusion technique that can optimize the overall loss.

1. Introduction

In recent years, semiconducting quantum dots (QDs) based optical fibers for optical amplifier are As an important support platform for information technology, optical fiber communication plays an important role in the future information society. In optical communication, it is vital to enhance the utilization of optical band. With the development of low loss of anhydrous fiber, people reconsidered using 1310 nm wavelength to communicate[1], currently, mainly propose three types of fiber amplifiers, doped rare earth fiber[2], raman fiber [3]and quantum dot fiber[4,5] ,but an important problem is the development of optical amplifier used in 1310nm is not mature so that a stable and high gain optical amplifier supporting optical communication in 1310nm is needed.

On the other hand, quantum dots(QDs) show great potential in the application of optical amplifier[6-9]. Due to the size effect as well as quantum confinement effect, the semiconductor quantum dots based optical amplifier possesses the advantages of wide spectrum of amplification, low pumping conditions and high gain[10]. In particular, the Lead sulfide(PbS) QDs are attractive because the emission spectrum covers the telecommunication band ranging from 1200 nm to 1600 nm [11,12].In addition, PbS quantum dots have high quantum yield (QYs) and good photo thermal stability in practical applications [13,14]. At present, the processing technique of quantum dots becomes mature, but the research of quantum dots as the gain medium to compose quantum dot device for quantum dot fiber, especially for , still needs to go deeper.

In this work, we have proposed and demonstrated a novel method to fabricate optical amplifier with high gain by combining semiconductor quantum dot. Based on the capillary pressure, PbS quantum dots are filled into the air hole of photonic crystal fiber (PCF). Such PbS filled PCF enables to build optical amplifier system that isolates quantum dot from oxygen. We have established the model of the end face of the optical fiber in COMSOL and the mode field distribution of photonic crystal fiber and quantum dot photonic crystal fiber. We improved the filling method by dropping the overall loss of 3.5 dB.

2. Simulation of PbS Quantum Photonic Crystal Fiber

Through the model analysis, we can obtain the mode field distribution of the quantum dot photonic crystal fiber (QDPCF) model, according to the mode field distribution of the fundamental mode of PCF which is evenly distributed in the fiber core. Considering the weak fundamental mode penetrating in the cladding, the energy is mainly concentrated in the core in optical transmission with a low transmission loss. The quantum dots poured into the holes of cladding, and then dried into quantum dot films, which is equivalent to plate with a layer of PbS quantum dot films in the air holes. We build a QDPCF face model with Comsol, and simulated the mode field with adding 5 layers of air holes. The mode field distribution is as shown in Fig.1.

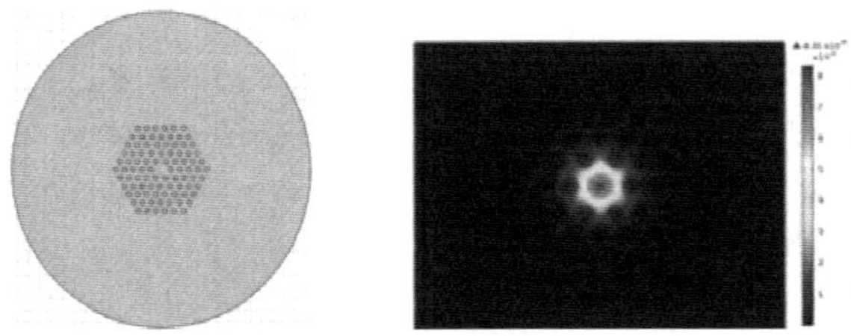

Fig. 1. (a) The schematic diagram of solid-core photonic crystal fiber (PCF); (b)The mode field distribution of the quantum dot filled in PCF.

When the quantum dot coated film leaning to the left cladding, the mode field distribution is as in Fig. 2(a). Oppositely, when the quantum dot coated film leaning to the right cladding, the mode field distribution is as in Fig. 2(b).

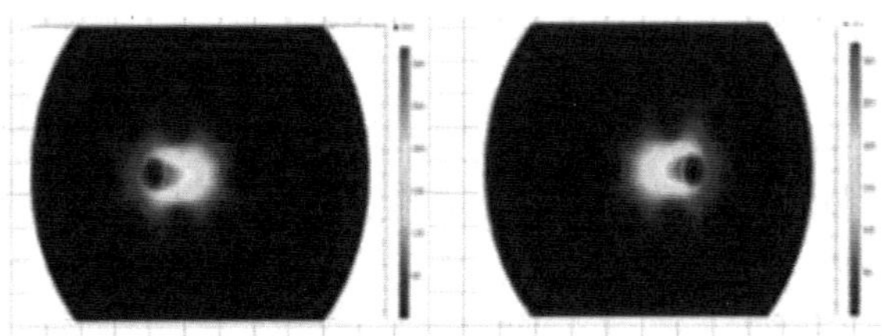

(a) (b)

Fig.2. The mode field distribution (a).The left side of the refractive index is high; (b).The right side of the refractive index is high

It is can be found that the fundamental mode meets the condition of total reflection in the transmission. Because the refractive index of cladding is increased after adding quantum dot films into cladding, which results in weakened bound of fundamental mode. So the energy leakage ratio of fundamental mode will be increased, and the energy fundamental mode is leaning to the quantum dot coated film. We calculate the limiting loss of the optical fiber by the imaginary part of the effective refractive index of the fundamental mode based on method reported by previous report . We find that the confinement loss of the PbS-filled PCF is slightly increased by 1.3 dB/km compared with the PCF without PbS at operation wavelength of 1310 nm.

3. Preparation of PbS Quantum Photonic Crystal Fiber

The PbS QDs were synthesized according to the Cademartiri method[8]. They were then modified by In the experiment, we have fabricated two kinds of PbS-filled PCF: one is PCF with all of air-hole cladding filled with PbS; the other is PCF with selective air-hole cladding filled with PbS. Both PCF devices are made by capillary action. The quantum dots are prepared by ultrasonic vibration at first, and then poured into air hole of photonic crystal fiber. The PCF should be put into a vacuum oven at once and dried vacuum for 12 hours at 25 ℃.

The fabricated PCF with all the air-hole cladding filled by PbS is measured by Scanning Electron Microscope (SEM) and the results are shown in Fig. 3. silicon (Si) and Oxygen (O) refer to photonic crystal fiber itself. and S (shown in Fig. 3(c))and Pb (shown in Fig. 3(d)) appear on end face of the fiber. This indicates that PbS quantum dots have been filled into air holes of photonic crystal fiber.

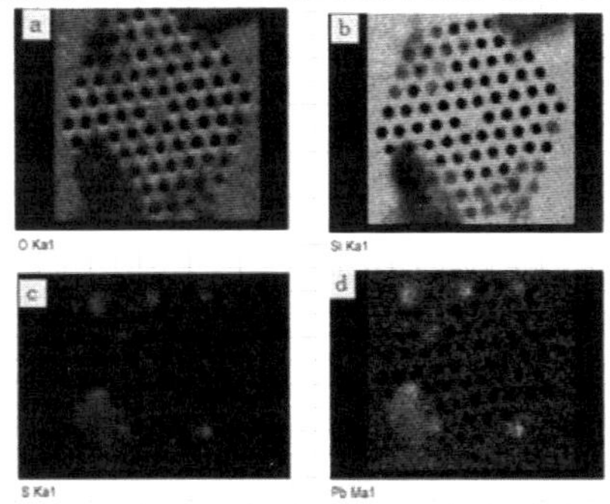

Fig.3. The elemental analysis by scanning the end face: (a)O of elements (b) Si of element (c) S of element(d) Pb of element

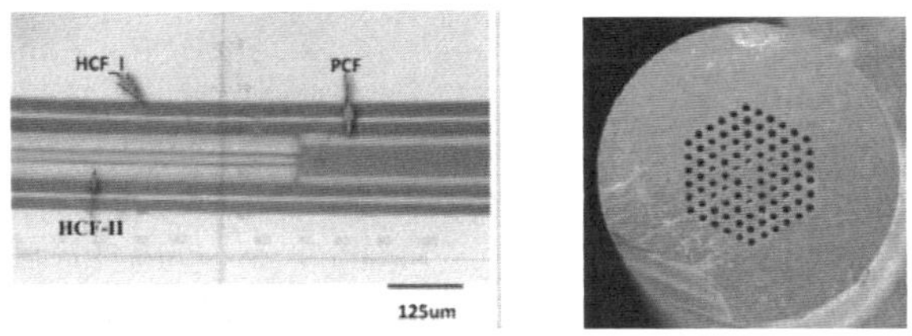

Fig. 4. (a) Microscopic image of PCF for filling PbS in selective air hole cladding: the hollow core fiber (HCF-I) is used to fix the fibers; the HCF-II is used to deliver the PbS to the targeted air-hole cladding. (b) Scanning Electron Microscope image of end-facet of the fabricated PCF with selective air-hole cladding filled with PbS.

Moreover, we can selectively fill the PbS into air holes of PCF. The microscopic image is shown in Fig. 4(a): the hollow core fiber (HCF-I) is used to fix the fibers; the HCF-II is used to deliver the PbS to the targeted air-hole cladding. The end-facet of fabricated PbS-filled PCF is shown in Fig. 4(b). In this way, we can complete the selective filling of three layer of air holes. Then we take out the fiber and fuse the ends with single-mode fiber pigtail to measure the overall loss.

In the absence of pumping light injection, optic transmission spectrum of 1310 nm in two cases as shown in Fig.5(Without filling shown in Solid line,Selectively filling shown in second line,Ordinary filling shown in lower line) , the overall loss is approximately dropped 3.5dB in same of fiber length.

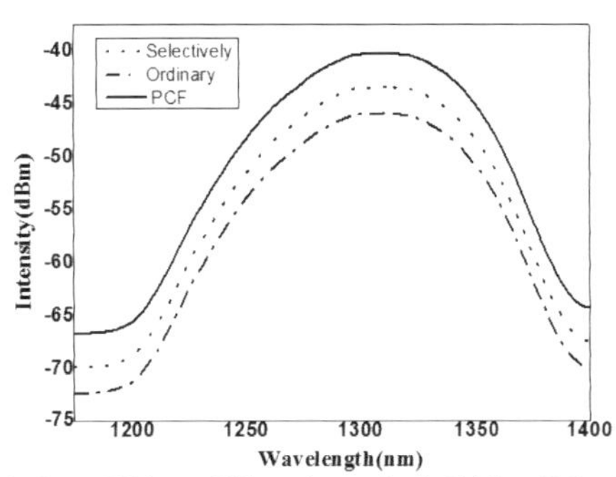

Fig. 5. The transmission spectrum at 1310nm(Without filling shown in Solid line,Selectively filling shown in second line, Ordinary filling shown in lower line)

4. Conclusion

In this work, the quantum dot is injected into the holes of the 10 cm photonic crystal fiber , and it is dried under the normal temperature vacuum condition, then the photonic crystal fiber is pigtailed by single-mode fiber, using 980nm laser whose power is 2mW as a sign light source. Moreover, this method is more simple and controllable compared with growth method where the quantum dot in solution is unstable so that it is worth studying to improve the growth method. We find that the confinement loss of the PbS-filled PCF is slightly increased by 1.3 dB/km compared with the PCF without PbS at operation wavelength of 1310 nm. Selectively filling compared with ordinary filling that the overall loss is approximately dropped 3.5dB in same of fiber length.

References

[1] Turkiewicz, Jarosław Piotr. "Cost-effective n × 25 Gbit/s DWDM transmission in the 1310 nm wavelength domain." Optical Fiber Technology 17.3(2011):179-184.\
[2] Seo Y S, Lim C H, Fujimoto Y, et al. 9.6 dB Gain at a 1310 nm Wavelength for a Bismuth-doped Fiber Amplifier[J]. Journal of the Optical Society of Korea, 2007, 11(2):63-66.
[3] Pang, F., Sun, X., Guo, H., Yan, J., Wang, J., & Zeng, X., et al. (2010). A pbs quantum dots fiber amplifier excited by evanescent wave. Optics Express, 18(13), 14024-14030.
[4] Cheng C, Hu N. Broadband Pb Se Quantum-Dot-Doped Fiber Amplifiers from 1250 nm to 1370 nm[J]. Acta Optica Sinica, 2016, 36(4):0406002.
[5] Cheng C, Hu N, Cheng X. Experimental realization of a PbSe quantum dot doped fiber amplifier with ultra-bandwidth characteristic[J]. Optics Communications, 2017, 382:470-476.
[6] Sapra, S, et al. "Unraveling internal structures of highly luminescent PbSe nanocrystallites using variable-energy synchrotron radiation photoelectron spectroscopy. " Journal of Physical Chemistry B 110.31(2006):15244-50.
[7] Lobo, A, et al. "Photoelectron spectroscopic investigations of chemical bonding in organically stabilized PbS nanocrystals. " Journal of Physical Chemistry B 109.37(2005):17422-8.
[8] Peng, Xiaogang, et al. "Epitaxial Growth of Highly Luminescent CdSe/CdS Core/Shell Nanocrystals with Photostability and Electronic Accessibility." Journal of the American Chemical Society 119.30(1997).
[9] Akiyama T, Kuwatsuka H, Simoyama T, et al. Nonlinear gain dynamics in quantum-dot optical amplifiers and its application to optical communication devices[J]. IEEE Journal of Quantum Electronics, 2001, 37(8):1059-1065.
[10] Cheng Cheng, Lin Yanguo, and Yan Jinhua. "Red Shift of Photoluminescence Spectrum of CdSe/ZnS-Quantum-Dot Doped Fiber." Acta Optica Sinica 31.4(2011):0406002.
[11] Bakueva L, Musikhin S, Hines M A, et al. Size-tunable infrared (1000–1600 nm) electroluminescence from PbS quantum-dot nanocrystals in a semiconducting polymer[J]. Applied Physics Letters, 2003, 82(17):2895-2897.
[12] Klem E J D, Levina L, Sargent E H. PbS quantum dot electroabsorption modulation across the extended communications band 1200-1700 nm[J]. Applied Physics Letters, 2005, 87(5):053101-053101-3.
[13] Hatami S, Würth C, Kaiser M, et al. Absolute photoluminescence quantum yields of IR26 and IR-emissive Cd(1-x)Hg(x)Te and PbS quantum dots--method- and material-inherent challenges.[J]. Nanoscale, 2015, 7(1):133-43.
[14] Lee K R, Homan S B, Kodaimati M S, et al. Near-Quantitative Yield for Transfer of Near-Infrared Excitons within Solution-Phase Assemblies of PbS Quantum Dots[J]. 2016, 120(39).

IOP Conf. Series: Journal of Physics: Conf. Series **844** (2017) 012061 doi:10.1088/1742-6596/844/1/012061

Study of focal ratio degradation for fibers

M Zhang, H Zou, J Liu and L Wang

School of Science, Hohai University, Nanjing, 211100, China

E-mail: zouhua96@163.com

Abstract. Along with the development of spectroscopy, the focal ratio degradation of fibers is the most important for instrumental applications in astronomy. Fibers are very susceptible to the effects of focal ratio degradation, especially the shape and stress of fibers. The focal ratio degradation of fibers caused through bending (including C-shape, S-shape and circle of the fiber) and stress is analyzed and studied in this paper. Finally, we get a conclusion that stress and bending of fibers have serious influences on the focal ratio degradation. The analyses of focal ratio degradation have great significance for the construction and application of fibers in astronomy.

1. Introduction

Fibers have been applied in astronomical systems generally since Angel et al used fibers in astronomical telescopes in 1977[1]. The applications of fibers are for Multi-object spectroscopy (MOS) [2] and Integral field spectroscopy (IFS) [3].The focal ratio degradation (FRD) and transmission of fibers are the main factors that affect the measurement of multi-object spectroscopy technology[4], and the FRD has the greatest effect on the star light's transmission quality in astronomy. The reasons for FRD include the following aspects, wavelength[5-6], incident focal ratio[7], diameter of fibers[1], length of fibers[8-9], bending and stress[10-11], smoothness of fibers[12], temperature[4] and so on. In order to improve the observation quality of astronomical telescopes, we need to do more in-depth studies of the FRD phenomenon.

The Large Sky Area Multi-Object Fiber Spectroscopic Telescope (LAMOST) adopts an innovative active optics technique that changes mirror surface continuously to achieve a series different reflecting Schmidt systems at different moments[13-14]. The properties of the circular fiber, the straight fiber, the stressed fiber, the C-shaped fiber and the S-shaped fiber are studied according to the actual conditions of fibers (diameter is 320um and numerical aperture is 0.2) used in the LAMOST in this paper. The analyses of FRD have great significance for application of fibers in astronomy.

2. FRD measurement setup

IOP Conf. Series: Journal of Physics: Conf. Series **844** (2017) 012061 doi:10.1088/1742-6596/844/1/012061

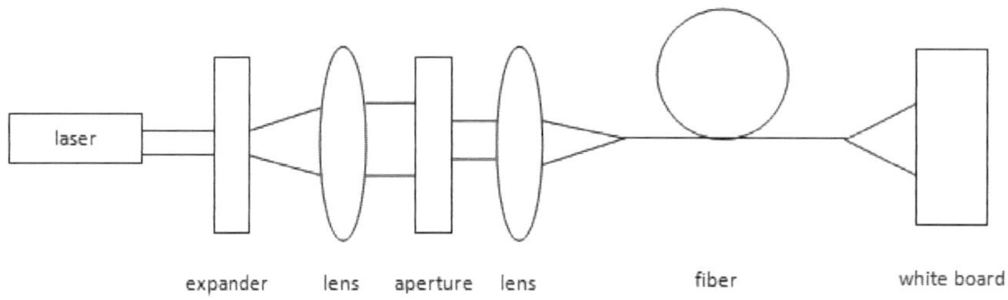

Figure 1. FRD measurement setup.

FRD measurement setup is shown in Figure 1. The expander expands the light emitted through the laser and then the beam is collimated through the first converging lens. The aperture is used to limit the effective spot diameter and the collimated beam is then coupled into the fiber through the second lens. Finally, a clear and diffused spot is re-imaged on to the white board.

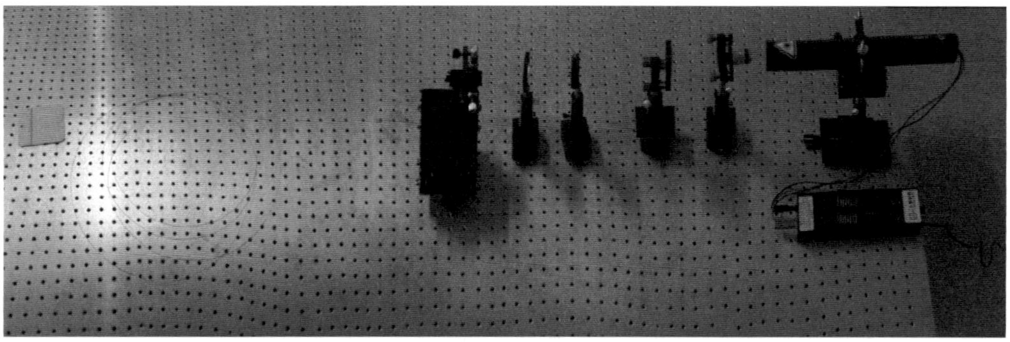

Figure 2. Top view of the experimental setup.

The experimental setup is shown in Figure 2. The setup is built in the experimental platform and the center of the input spot is adjusted so that the spot is transmitted into the fiber through the center of lens, aperture and other devices.

3. Experiments and data analysis

To measure the FRD properties of a fiber it is necessary to illuminate the test fiber with a known incident focal ratio. The incident focal ratio of the test fiber is f/5 in this paper. We measure the FRD degree of the straight fiber and then analyze the effects of stress through imposing different stress on the same fiber. The effects of bending are studied, which is compared with output focal ratios of the straight fiber, the C-shaped fiber, the S-shaped fiber and the circular fiber in this paper.

3.1. The FRD of the straight fiber

To obtain the corresponding output focal ratio, the distance d (5cm, 10cm, 15cm, 20cm, 25cm, 30cm, 35cm, 40cm) is adjusted between the end of this fiber and the white board according to Figure 1. The images of the output spot are shown in Figure 3 in different distance.

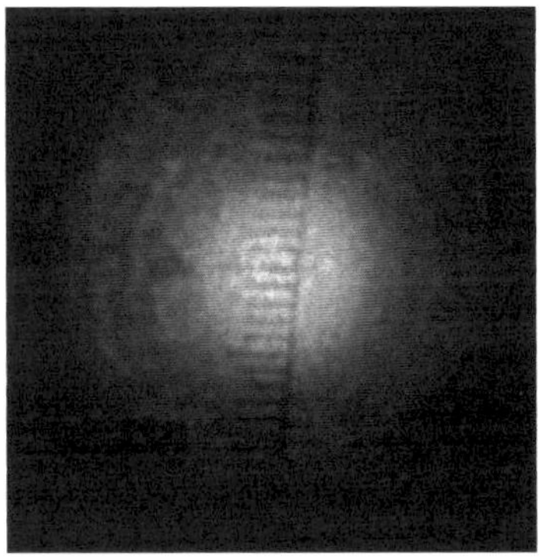
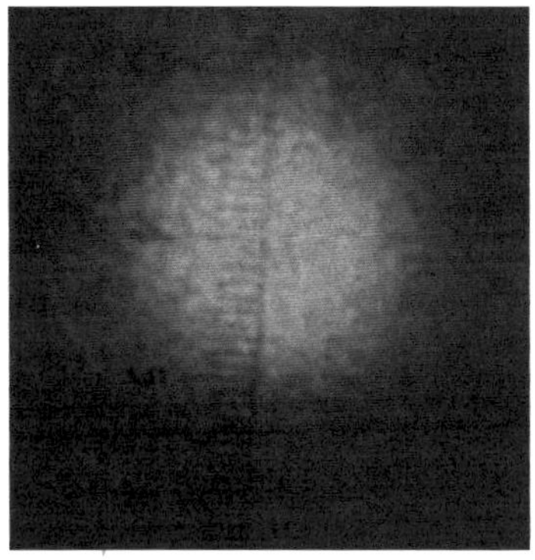

5cm 10cm

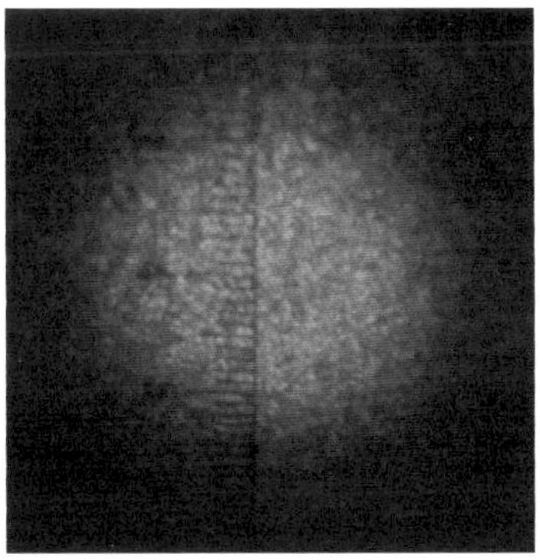
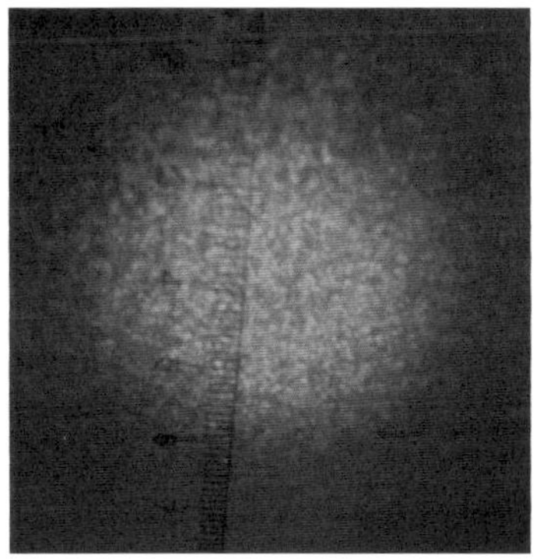

15cm 20cm

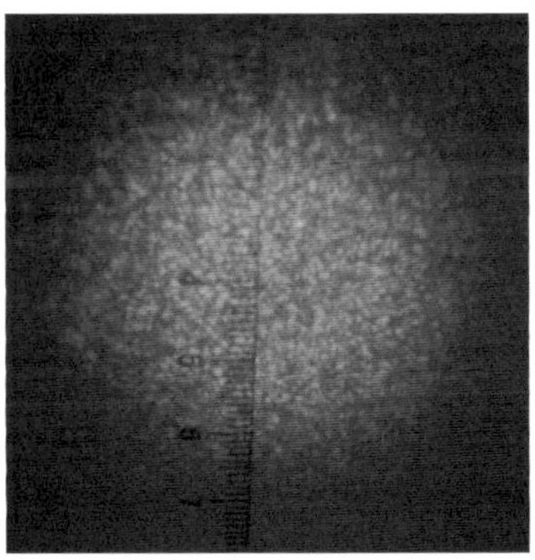
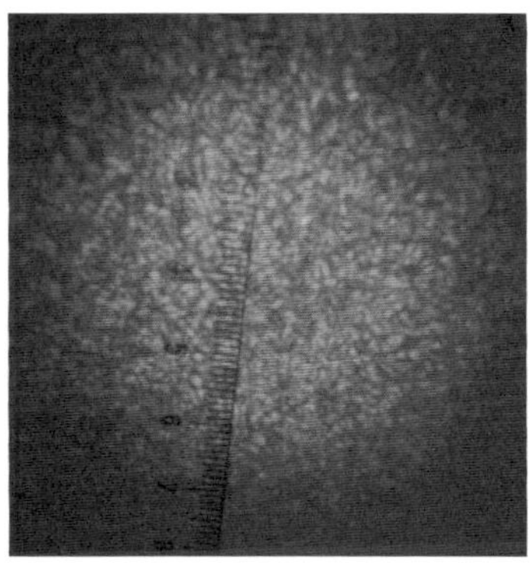

25cm 30cm

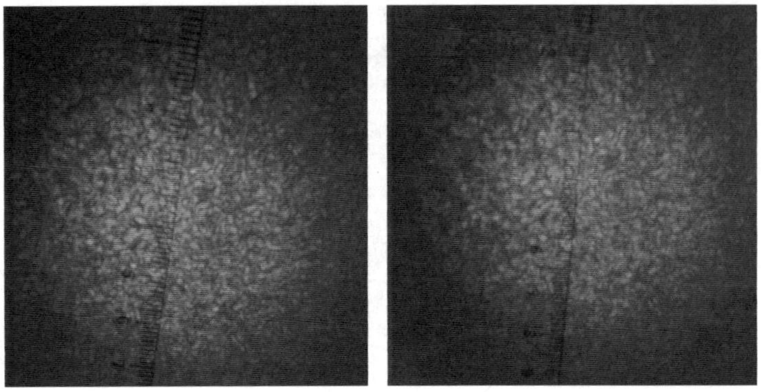

35cm 40cm

Figure 3. The images of the output spot in different distance.

According to Figure 3, the diameter D_i' of output spot and output focal ratio are shown in Table 1 in different distance.

Table 1. The data of the straight fiber.

Distance d(cm)	Output diameter D_i' (cm)	Output focal ratio ($d(D_i')^{-1}$)
5	1.10	4.55
10	2.15	4.65
15	3.30	4.55
20	4.35	4.60
25	5.45	4.59
30	6.20	4.84
35	7.50	4.67
40	8.45	4.73
average		4.65

From the Table 1, we can get that the FRD degree of the straight fiber is 7%.

3.2. *Stress effects on FRD*

The stress varies with changing the number of objects which is a square object with a weight of 9.7 N and pressed against the straight fiber. The distance d between the end of the fiber and the white board is 8cm. The images of the output spot are shown in Figure 4 under different stress.

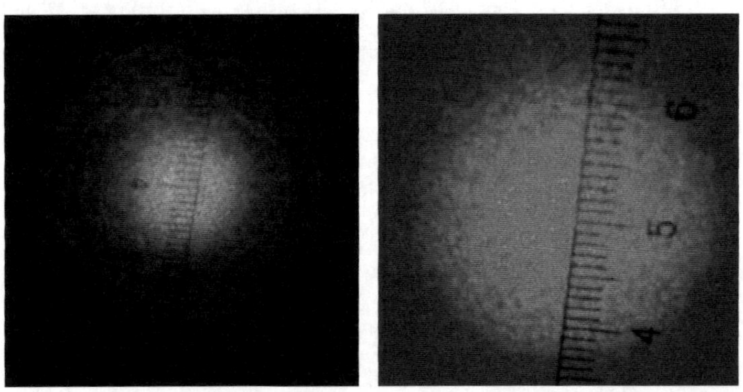

0N 9.7N

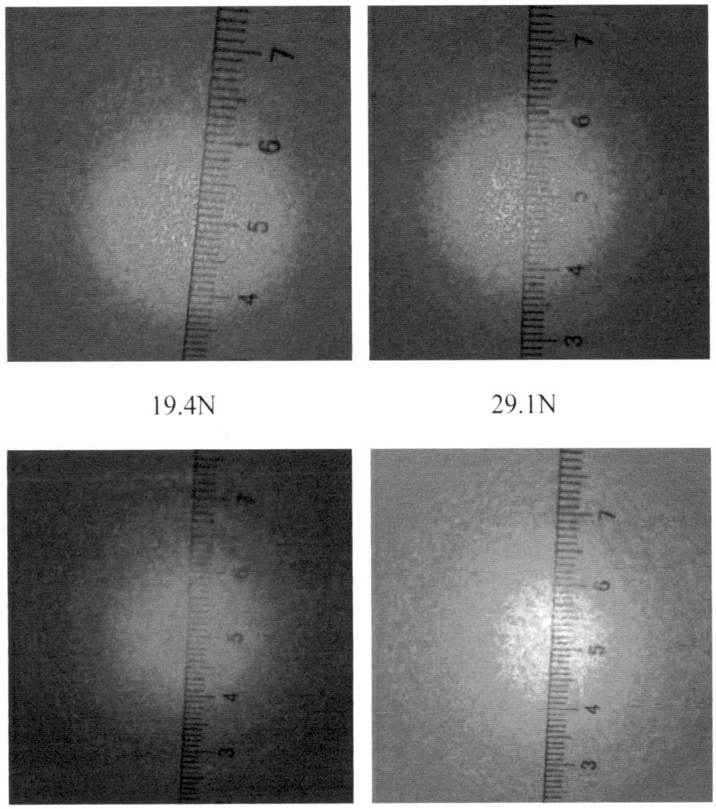

19.4N 29.1N

38.8N 48.5N

Figure 4. The images of the output spot under different stress.

According to Figure 4, the diameter D_i' of output spot and output focal ratio are shown in Table 2 under different stress.

Table 2. The data of the fiber under different stress.

Stress (N)	Output diameter D_i' (cm)	Output focal ratio ($d(D_i')^{-1}$)
0	1.65	4.85
9.7	2.60	3.08
19.4	2.80	2.86
29.1	3.08	2.60
38.8	3.30	2.42
48.5	3.50	2.29

As shown in Table 2, when the stress is 0 N, the output focal ratio of this fiber is 4.85; when the stress is 9.7N, the output focal ratio is 3.08. Above the data, we can conclude that FRD rises with the stress increased so the stress has a serious influence on the FRD in optical fiber.

3.3. Bending effects on FRD

The effects of bending on the FRD of fibers are studied through comparing with output focal ratios of the straight fiber, the C-shaped fiber, the S-shaped fiber, the circular fiber. The distance d between the end of the fiber and the white board is 8cm. The images of the output spot are shown in Figure 5 in different bending.

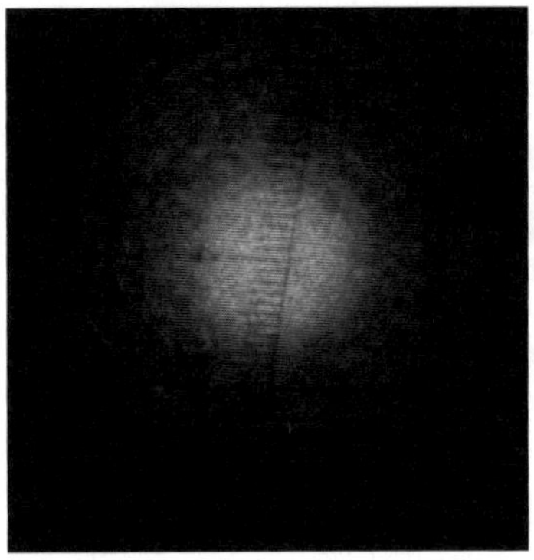

the straight fiber

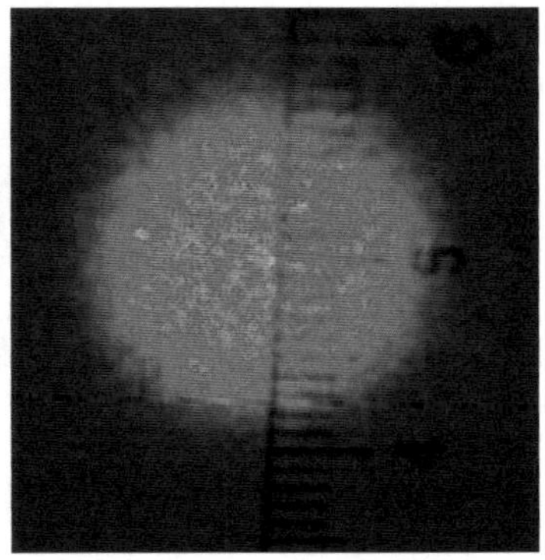

the C-shaped fiber

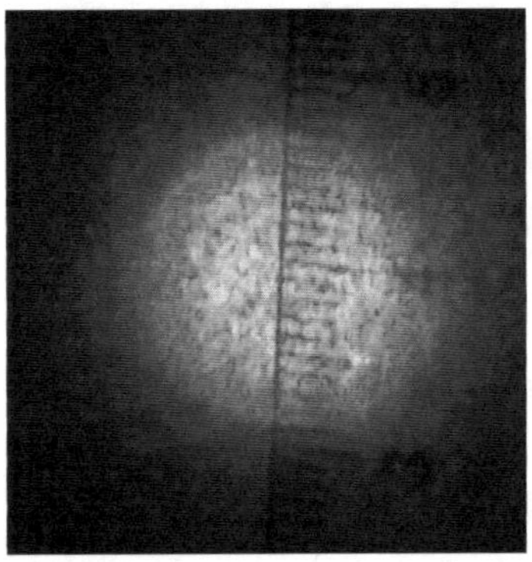

the S-shaped fiber

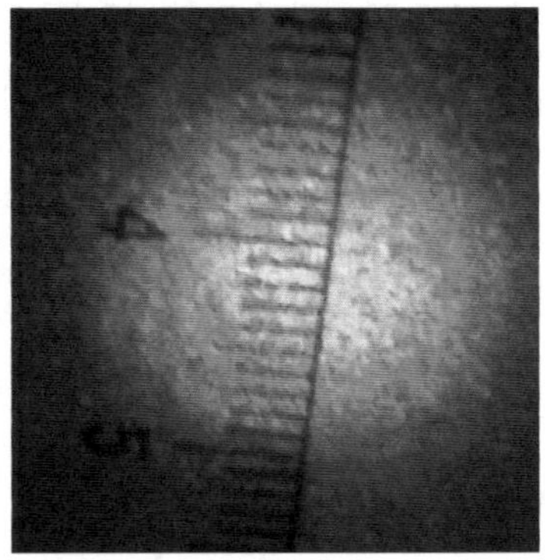

the circular fiber

Figure 5. The images of the output spot in different bending.

According to Figure 5, the diameter D_i' of output spot and output focal ratio are shown in Table 3 in different bending.

Table 3. The data of the fiber with different bending.

Shape of the fiber	Output diameter D_i' (cm)	Output focal ratio ($d(D_i')^{-1}$)
straight	1.65	4.85
C-shape	1.75	4.57
S-shape	1.85	4.32
circle	2.00	4.00

It can be seen from Table 3 that when incident focal ratios of different shapes of fibers are the same, the output spot diameter of the straight fiber is smaller and the degree of FRD is smaller. The FRD degree of the C-shaped and S-shaped fibers is bigger than that of the straight fiber, but it is smaller than FRD of the circular fiber, and FRD of the S-shaped fiber is more obvious than FRD of the C-shaped fiber. Finally we can conclude that the degree of bending of fibers is greater, the FRD phenomenon is sharper.

In a word, the experimental results show that FRD is evident when fibers are bent. With FRD degree of the circular fiber, the C-shaped fiber, the S-shaped fiber and the straight fiber compared, it indicates that FRD of fibers is more serious when the bending degree of fibers is larger (the circular fiber> the S-shaped fiber > the C-shaped fiber > the straight fiber).

4. Conclusion

The FRD of fibers caused through bending and stress is analyzed and studied in this paper. We conclude that FRD is unavoidable and when the stress that imposed on fibers is increased, the diameter of output spots is much bigger than the diameter without stress and FRD rises with the stress increased. Bending of fibers has a serious influence on the FRD, that is to say, the FRD of fibers with greater bending degree (smaller bending radius) is more obvious. Therefore, it is necessary to control the stress to reduce the degree of FRD and improve the efficiency of fibers except controlling the shape (bending degree) of fibers in installing fibers, at the same time the analyses of FRD have great significance for the construction and application of fibers in astronomy.

Acknowledgement

This work is supported by the National Natural Science Foundation of China (NSF) (1014/515029111); The Fundamental Research Funds for the Central Universities (2013/B15020271).

References

[1] Angel J R P, Adams M T, Boroson T A and Moore R L 1977 *ApJ* **218** 776
[2] Hill J M, Angel J R P, Scott J S, Lindley D and Hintzen P 1980 *ApJ* **242** 69
[3] Parry I R 2006 *New Astron. Rev.* **50** 301
[4] Grupp F 2006 *New Astron. Rev.* **50** 323
[5] Poppett C L and Allington-Smith J R 2007 *Mon. Not. R. Astron. Soc.* **379** 143
[6] Bershady M A, Andersen D R, Verheijen Marc A W, Westfall K B, Crawford S M and Swaters R A 2005 *The Astrophysical Journal Supplement Series* **156** 311
[7] Barden S C 1998 *Fiber Optics in Astronomy III. ASP Conf. Ser.* Vol 152 p 14
[8] Avila G, Singh P and Albertsen M 2006 *Proc. SPIE* Vol 6269 p 1
[9] Murphy J D, MacQueen P J, Hill G J, Grupp F, Kelz A, Palunas P, Roth M and Fry A 2008 *Proc. SPIE* Vol 7018 p 1
[10] Dong Y and Zhu B 2007 *Journal of University of Science and Technology of China* **37** 606
[11] Wan X, Ge J, Guo P, Zhao B, Hariharan L and McDavitt D 2006 *Proc. SPIE* Vol 6269 p 1
[12] Speziali R, Liberi D, Cortecchia F, D'Alessio F, Causi G L, Lorenzetti D, Pedichini F and Vitali F 2004 *Proc. SPIE* Vol 5492 p 1520
[13] Shi J 2016 *Chin. Sci. Bull.* **61** 1330
[14] Cui X et al 2012 *Research in Astronomy and Astrophysics* **12** 1197

IOP Conf. Series: Journal of Physics: Conf. Series **844** (2017) 012062 doi:10.1088/1742-6596/844/1/012062

Lumped Parameter experiments for Single Mode Fiber Laser Cutting of Thin Stainless Steel Plate

Shengying Lai [1,2], Ye Jia [2], Bing Han [3], Jun Wang [2], Zongkai Liu [2], Xiaowu Ni [1,2], Zhonghua Shen [1], Jian Lu [1*]

1. School of Science, Nanjing University of Science and Technology, Nanjing Jiangsu 210094, China
2. MIIT Key Laboratory of Advanced Solid Laser, Nanjing University of Science and Technology, Nanjing Jiangsu 210094, China
3. Laser-Material Interaction Lab,2011Co-innovation Center, Nanjing University of Science and Technology, Nanjing Jiangsu 210094,China

Corresponding author's e-mail address: lujian@njust.edu.cn (Jian Lu)

Abstract. The present work reports the parameters on laser cutting stainless steel including workpiece thickness, cutting speed, defocus length and assisting gas pressure. The cutting kerf width, dross attachment and cut edge squareness deviation are examined to provide information on cutting quality. The results show that with the increasing thickness, the cutting speed decrease rate is about 27%. The optimal ranges of cutting speed, defocus length and gas pressure are obtained with maximum quality. The first section in your paper

1. Introduction

Stainless steel is being increasingly used in automotive and aerospace structures for its good corrosion resistance and high strength. Due to the advantages of the high brightness, high directivity, laser has been applied to various fields of social productions since it was born in 1960s especially in laser material processing such as laser drilling[1], laser welding[2] and laser cutting[3~10].Laser cutting is an important partition method in such applications. There are several kinds of industrial lasers available at present, including the conventional CO_2 and Nd:YAG lasers as well as recently available high power fiber lasers. In recent decades, fiber lasers are playing an increasing role in laser processing for high photoelectric conversion efficiency, better beam absorption, good beam quality, good cutting reliability, strong flexible light effect and low cost compared to traditional CO_2 laser and YAG laser cutting. For example, Paul A. Hilton et al. presented a diffractive optical element design for thick C-Mn steel plate cutting by fiber laser and concluded that using the DOE and lens combination can reduce the power density in the residual defocused laser beam compared to that from the 500mm lens alone[11]. S. O. Al-Mashikhi et al. studied the effect of cutting speed and sheet thickness on surface oxidation and heat affected zones for laser-oxygen cutting of mild steel with a fiber laser and found that the HAZ is wider at the bottom of the cut than that at the top and there is no simple direct relationship between HAZ width and surface oxidation[12]. Catherine Wandera et al. experimentally investigated high power fiber laser cutting thick-section stainless steel and aluminum with inert gas assisted and obtained the parameter window[13].

In the present study, an investigation into the process characteristics and mechanisms involved in single mode CW fiber laser cutting of 316L stainless steel is carried out with the consideration of

IOP Conf. Series: Journal of Physics: Conf. Series **844** (2017) 012062 doi:10.1088/1742-6596/844/1/012062

cutting speed, defocus length and assist gas pressure. Further, the performance characteristics kerf width, dross attachment and cut edge squareness deviation are optimized for the evaluation of cut quality. The aim of this work is to perform a detailed analysis on the characteristics of single mode fiber laser cutting.

2. Experimental set up

In this study, a single mode CW fiber laser was used with maximum power of P=2000W at wavelength λ=1080nm. As illustrated in Fig.1 there is schematic of the setup used in experiments. The cutting head is controlled by the ABB robot arm using the assembly language to realize the controlling of laser output power, cutting speed, cutting distance and the cutting direction et.al. The lens including collimating lens and focus lens and protection lens are assembled in the cutting head. The protection lens is located below the focus lens to avoid the metal spattering. The focal length is 124mm and the focus is located at the nozzle outlet. Oxygen with a purity of 99.9% is used as a cutting assist gas and it can be adjusted up to 50bar with the cutting nozzle of 1.5mm diameter. The experimental material is 316L stainless steel plate of 1.5mm, 2.0mm, 3.0mm in thickness. The thermodynamic parameters of the metal plate are showed in Table 1. In the experiments, the metal plates are fixed into aluminum frame while the moving program is realized by cutting head.

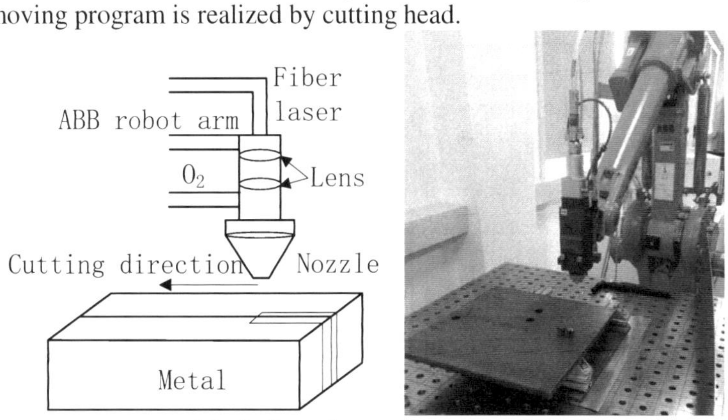

Fig.1. Experimental setup for fiber laser cutting

During the laser cutting process, the nozzle keeps a constant distance about 1.0mm to the sample surface. Assist gas pressure are varied from 0.4 to 2.2 bar in the experiments. Two replications for each experimental run have been performed to make sure the results to be repeatable and valid. After the cutting, the cut quality is initially evaluated by visual observation and microscopic observation and measurement subsequently using optical microscopy. Kerf width and dross attachment and cut edge squareness deviation are measured from these images taken from the cuts. The top and bottom kerf widths are measured using the optical measuring microscope (AFTVISION) at 4* magnifications. To make this more clear schematics illustration of the top view and bottom view and the cross sectional view of laser cut surface are shown in Fig. 2. Here, *d1* is the input kerf width from the top view, *d2* is the output kerf width from the bottom view, *h* is the thickness of the workpiece and θ is the cut edge squareness deviation which can be described as[14]:

$$\theta = \frac{(d1-d2)*180}{2*\pi*h} \qquad (1)$$

Table1. Thermodynamic parameters of stainless steel

Density (kg/m^3)	7870
Thermal conductivity (W/m/K)	15.1
Specific heat (J/kg/K)	502
Absorptivity	0.307
Melting point (K)	1370
Tensile strength (MPa)	480

According to the dross adhering slag at the lower cutting surface in the cutting morphology, the dross attachment is divided into five level and each level corresponds to a numeric value: 1 equals to dross free, 2 equals to little dross, 3 equals to medium dross, 4 equals to heavy dross and 5 equals to not cut through.

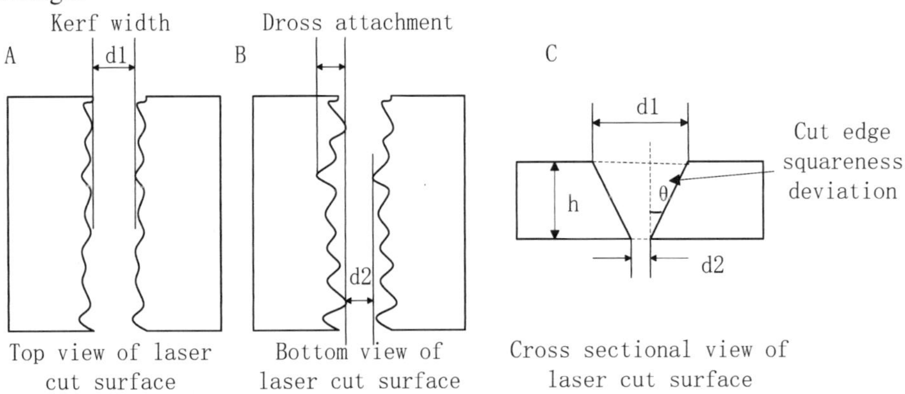

Fig.2 Schematic view of kerf width, dross attachment and cut edge squareness deviation

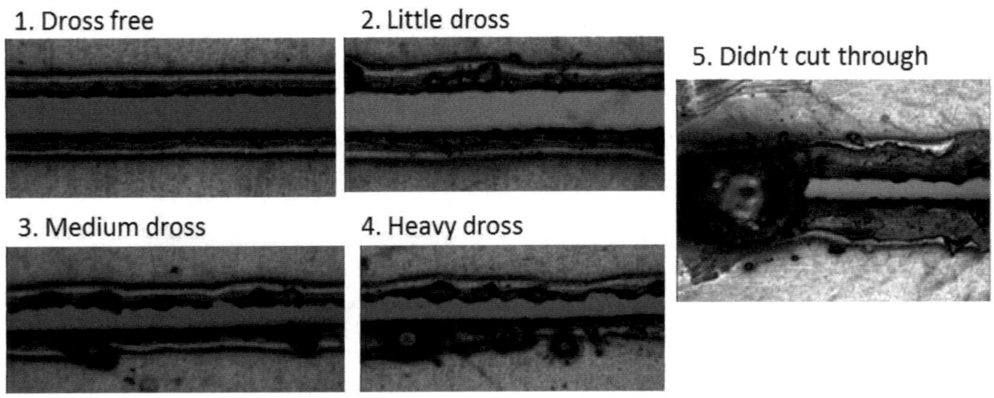

Fig.3 Cutting results for dross attachment level

3. Results and discussion

There are many factors that may affect the cutting quality, for example cutting speed, defocus length, assist gas pressure, laser output power, nozzle diameter et.al. Experiments taken in this paper include three parameters of cutting speed, defocus length and assist gas pressure. In order to secure comparable results, only one parameter is varied in each cut, the other parameters being kept constant.

3.1 Effects of the cutting speed

The effects of cutting speed on cut performance (kerf width, dross attachment, cut edge squareness deviation) are shown in Fig.4. In these experiments, the output power is 2000W, the assist gas pressure is 0.7bar, the defocus length is 5.5mm.

Fig.4A shows the kerf width with the effects of cutting speed for the given three workpiece thicknesses. The primary trend can be obtained that the kerf width is decreasing with the increasing cutting speed and workpiece thickness. Fig.4B describes the kerf width relative to the thickness (w/h) which shows a more clear view of the difference between kerf width and workpiece thickness. The lower cutting speed may increase the interaction time between laser and workpiece which can consequently increase the molten metal removal induced by laser melting and splashing resulting in the wider kerf. Due to the oxidation properties of stainless steel, during the O_2 assisted laser cutting process, lower cutting speed will increase the heat accumulation induced by workpiece oxidation reaction consequently increase the kerf width and cause heavy dross attachment at the bottom kerf as shown in Fig.4C. On the contrary, when the cutting speed is too high, due to less time for the interaction between laser and workpiece, the heat accumulation can't afford to melting or splashing the workpiece which will result in a heavy dross at the bottom kerf and may also cause the un-cut through cases as shown in Fig.5. Therefore, the reasonable control of the cutting speed is one of the keys to achieve high quality and high precision laser cutting.

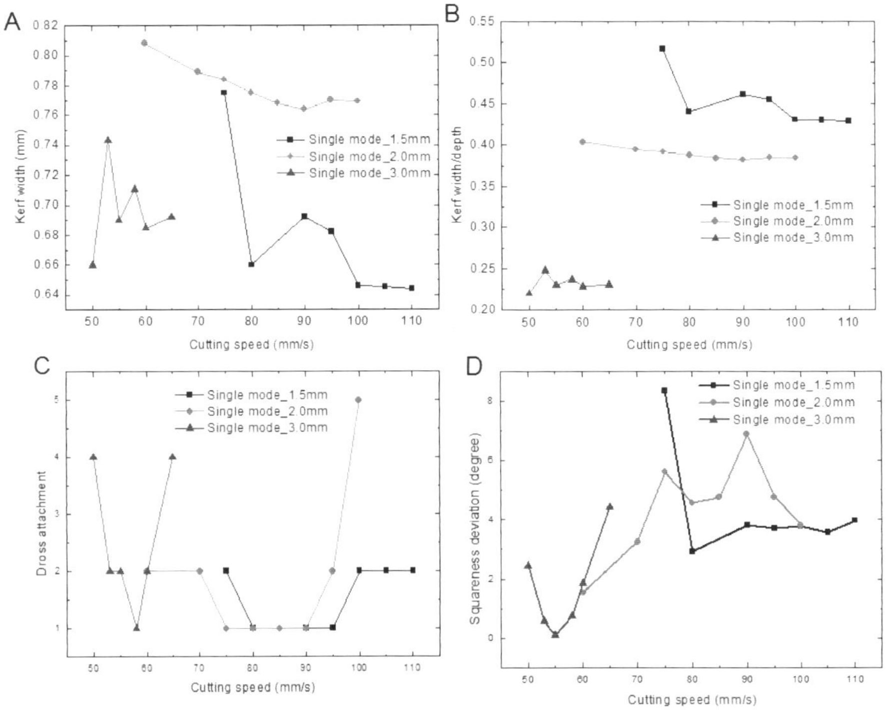

Fig.4 Variation of (A)kerf width(B)kerf width/depth(C)dross attachment(D)cut edge squareness deviation with cutting speed

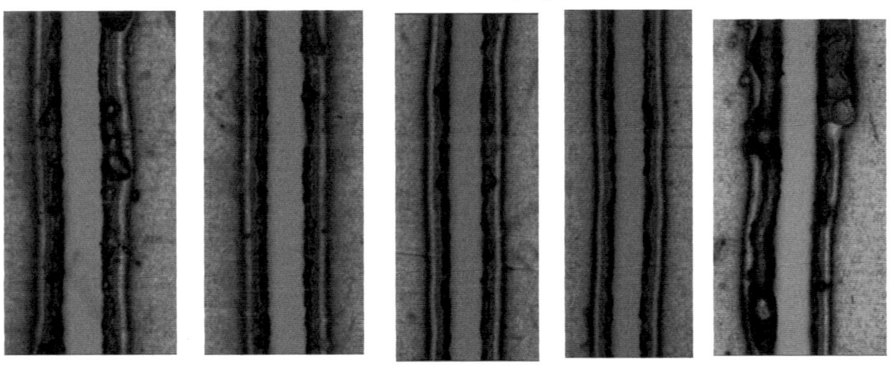

(a) u=60mm/s (b) u=70mm/s (c) u=80mm/s (d) u=90mm/s (e) u=100mm/s

Fig.5 Bottom view of 2.0mm stainless steel cutting results with different cutting speeds (a~e)

Taken together, the optimal ranges of cutting speed with small kerf width, no dross slag and small cut edge squareness deviation are as follows: 1.5mm for 90~100mm/s, 2.0mm for 80~90mm/s and 3.0mm for 55~60mm/s. The results show that with the increasing thickness, the cutting speed decrease rate is about 27%.

3.2 Effects of the defocus length

During the cutting performance, the defocus length plays an important role for the cutting quality. The defocus length is related to workpiece thickness and the assisted gas species. In these experiments, the output power is 2000W and the assist gas pressure is 0.7bar for 2.0mm and 3.0mm plate and 1.0bar for 1.5mm plate. The cutting speed is varied with the workpiece thickness as discussed in 3.1 section. The cutting speeds are 1.5mm at 85mm/s, 2.0mm at 60mm/s and 3.0mm at 58mm/s. The results are shown in Fig.6. It can be found that the kerf width is increasing with the increased defocus length and the workpiece thickness as shown in Fig.6A. However, the kerf width relative to the thickness (w/h) indicates that the width-depth ratio is decreasing with the increased defocus length and thickness as shown in Fig.6B.

For a certain focusing lens used in this equipment, the increasing defocusing length will change the laser beam spot size and then reduce the energy distribution at the workpiece surface thus resulting in a wider kerf width or even uncut-through. On the contrary, the decreasing defocusing length will increase the laser beam spot size and increase the energy distribution which will result in a narrower cut kerf width and deeper cut depth. For another aspect, small laser beam spot will decrease the oxidation heat and affects the cut quality. The exothermic competition of oxidation and laser heating determines the cutting quality including kerf width, cut depth, dross attachment and the cut edge squareness deviation. The results show that the optimal ranges of defocus length are as follows: 1.5mm for 5.5~7.0mm, 2.0mm for 6.5~7.5mm and 3.0mm for 5.5~7.0mm.

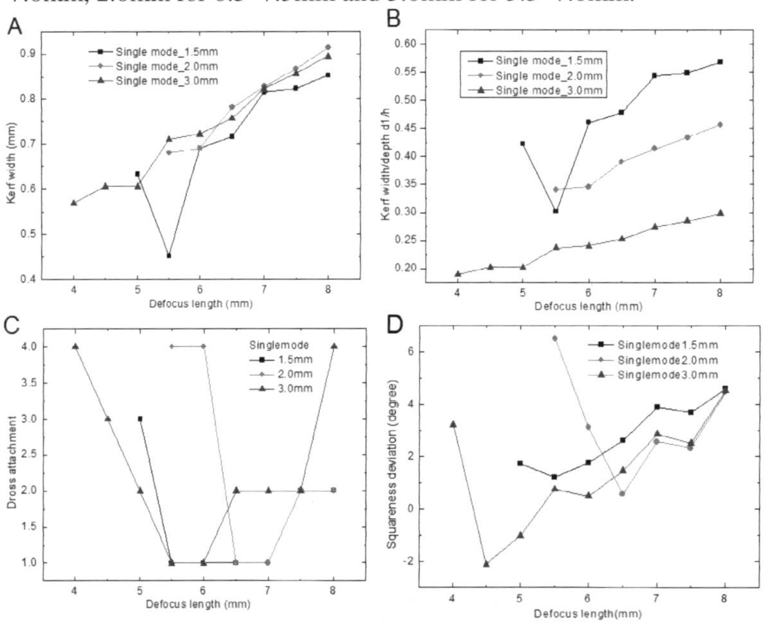

Fig.6 Variation of (A)kerf width(B)kerf width/depth(C)dross attachment(D)cut edge squareness deviation with defocus length

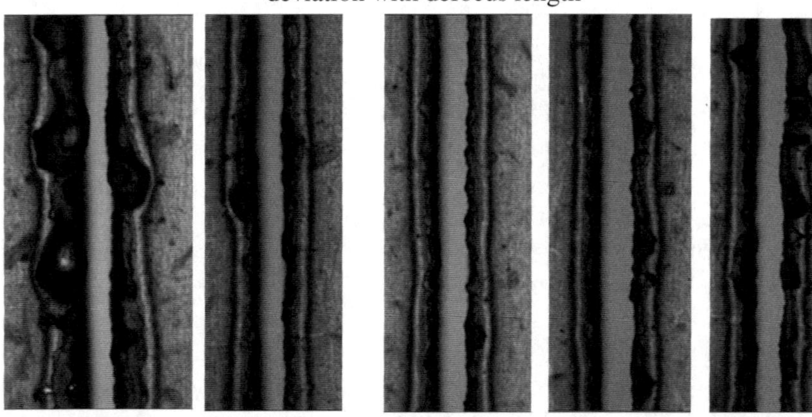

(a) f=4mm (b) f=5mm (c) f=6mm (d) f=7mm (e) f=8mm

Fig.7 Bottom view of 3.0mm stainless steel cutting results with different defocus length (a~e)

3.3 Effects of the assist gas pressure

The assist gas pressure is varied from 0.4bar to 2.2bar for the three thickness workpieces by single mode laser cutting as shown in Fig.8. From Fig. 8A and B it can be found that gas pressure has less influence on kerf width or the ratio of kerf width and workpiece thickness compared to cutting speed and defocusing length. Fig. 8C shows the variation of dross attachment with gas pressure. Dross free cuts are seldom found with low pressure cutting (p<0.7bar) in this study, which may be due to the low drag force on the cut front under low pressure. It can also be found that dross free cuts are not occurred with high assist gas pressure (p>1.6bar) and it also differs with workpiece thickness. The maximum pressure is decreasing with the thickness for 1.5mm of 1.9bar, 2.0mm of 1.6bar and 3.0mm of 1.3bar, respectively. The dross attachment is worse when the pressure is higher over 1.6bar or below 0.4bar. When the pressure is high, the gas flow will form vortex at the workpiece surface which may decrease the drag force to remove the melted material and resulting in a rough cut surface, wider kerf width and heavier dross attachment. The results are shown in Fig.9.

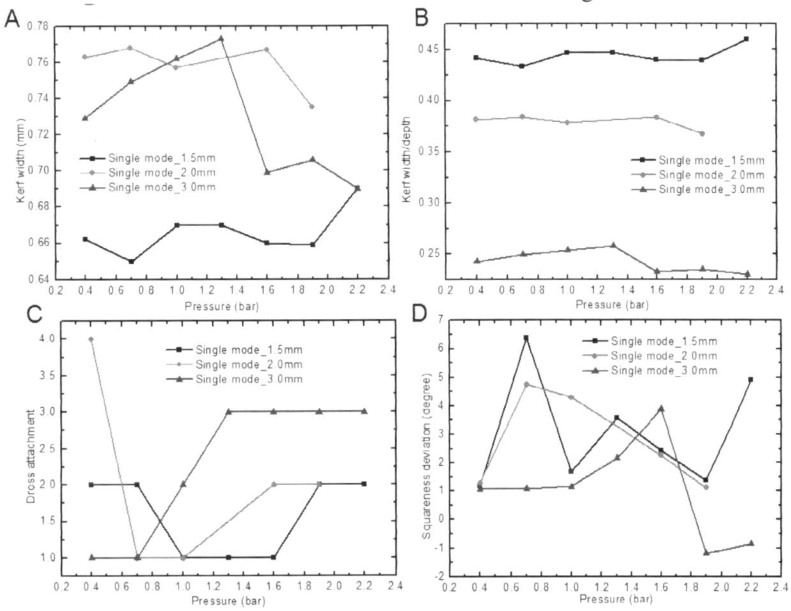

Fig.8 Variation of (A)kerf width(B)kerf width/depth(C)dross attachment(D)cut edge squareness deviation with gas pressure

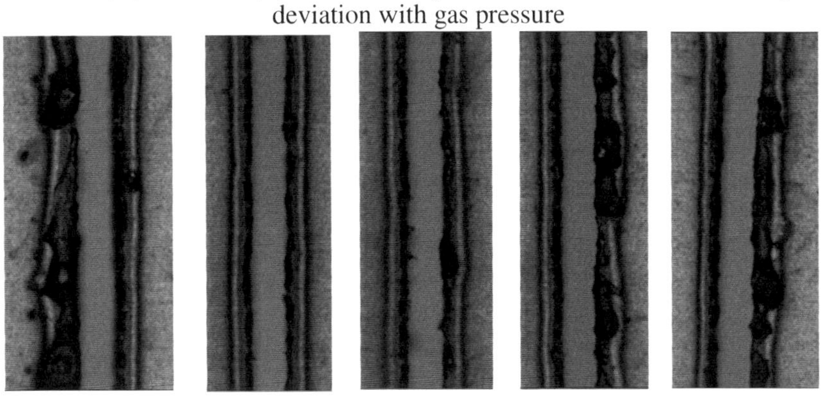

(a) 0.4bar (b) 0.7bar (c) 1.0bar (d) 1.6bar (e) 1.9bar

Fig.9 Bottom view of 2.0mm stainless steel cutting results with different assist gas pressure (a~e)

On the other hand, since oxygen is used as the assisted gas, the oxidation heats make great contributions to laser cutting progresses. When the pressure is low, the oxidation heats make lower contributions to the material melting and vaporizing which may to some extent decrease the cut depth. Reasonably, the higher pressure can obtain more oxidation heat. However, with the gas pressure increasing, oxidation heat will reach a saturated value which means that the pressure above the saturated value will not be used to supply the oxidation progress. Therefore, suitable gas pressure has an important influence on the cut quality. When the other parameters are kept constant, the optimum pressure is obtained about 1.5mm for 0.7~1.3bar, 2.0mm for 0.7~1.0bar and 3.0mm for 0.4~0.7bar.

4. Conclusion

In present study, four parameters including cutting speed, defocus length, assisted gas pressure and workpiece thickness have been taken into consideration in single CW fiber laser cutting of stainless steel workpieces. Kerf width, dross attachment and cut edge squareness deviation have been estimated as the references of cut quality. The results indicate that faster cutting speed will lead to less time for melting and result in uncut-through while slower cutting speed will cause heavier dross slag for oxidation heat supplied by assist gas. The optimal cutting speed ranges are 1.5mm for 90~100mm/s, 2.0mm for 80~90mm/s and 3.0mm for 55~60mm/s. The higher defocus length will cause larger laser beam spot size and result in larger kerf width while lower ones will cause smaller spot and higher energy focus which lead to more dross slag by the assist gas support. The optimal defocus length ranges are 1.5mm for 5.5~7.0mm, 2.0mm for 6.5~7.5mm and 3.0mm for 5.5~7.0mm. The assist gas pressure has a saturated value to keep a high cutting quality and the optimal ranges are wider with the decreasing workpiece thickness which are shown as follows: 1.5mm for 0.7~1.3bar, 2.0mm for 0.7~1.0bar and 3.0mm for 0.4~0.7bar.

Acknowledgment
This work is supported by the "National Natural Science Foundation of China for Young Scholars (No.11402120)", the "Jiangsu Natural Science Foundation for Young Scholars (No. BK20140796)", the "Fundamental Research Funds for the Central Universities (No.30915015104)", the Fundamental Research Founds for the Central Universities (NO.30916014112-008) and Jiangsu province Doctor Innovation Fund (NO.KYLX16_0429).

References

[1] Low D K Y, Li L, Corfe A G. Effects of assist gas on the physical characteristics of spatter during laser percussion drilling of NIMONIC 263 alloy [J]. Applied surface science, 2000, 154: 689-695.
[2] Zhu J, Li L, Liu Z. CO 2 and diode laser welding of AZ31 magnesium alloy[J]. Applied Surface Science, 2005, 247(1): 300-306.
[3] Yilbas, B.S., Turbulent boundary layer approach allowing chemical reactions for CO, laser oxygen-assisted cutting process. Proc Instn Mech Engrs, 1994. 208.
[4] Yilbas, B.S., Oxygen assisted laser cutting mechanism -a layer approach including the combustion process[J]. Optics and Laser Technology, 1995. 27(3).
[5] Yilba, B.S., Experimental investigation into CO 2 laser cutting parameters[J]. Journal of Materials Processing Technology, 1996. 58(2): p. 323 - 330.
[6] Chen, S., The effects of gas composition on the CO2 mild steel[J]. Journal of Materials Processing Technology, 1998. 73.
[7] CHEN, S., The effects of high-pressure assistant-gas flow on high-power CO2 laser cutting[J]. Journal of Materials Processing Technology, 1999. 88(1-3): p. 57 - 66.
[8] Ghany, K.A. and M. Newishy, Cutting of 1.2mm thick austenitic stainless steel sheet using pulsed and CW Nd:YAG laser[J]. Journal of Materials Processing Technology, 2005. 168(3): p. 438-447.
[9] Golnabi, H. and M. Bahar, Investigation of optimum condition in oxygen gas-assisted laser cutting[J]. Optics & Laser Technology, 2009. 41(4): p. 454-460.
[10] Yilbas B S. Laser heating process and experimental validation[J]. International journal of heat and mass transfer, 1997, 40(5): 1131-1143.
[11] Hilton, P.A., D. Lloyd and J.R. Tyrer, Use of a diffractive optic for high power laser cutting[J]. Journal of Laser Applications, 2016. 28(1): p. 012014.

IOP Conf. Series: Journal of Physics: Conf. Series **844** (2017) 012062 doi:10.1088/1742-6596/844/1/012062

[12] Al-Mashikhi, S.O., et al., Heat affected zones and oxidation marks in fiber laser–oxygen cutting of mild steel[J]. Journal of Laser Applications, 2011. 23(4): p. 042003.
[13] Wandera, C., A. Salminen and V. Kujanpaa, Inert gas cutting of thick-section stainless steel and medium-section aluminum using a high power fiber laser[J]. Journal of Laser Applications, 2009.
[14] Sharma A, Yadava V. Modelling and optimization of cut quality during pulsed Nd: YAG laser cutting of thin Al-alloy sheet for straight profile [J]. Optics & Laser Technology, 2012, 44(1): 159-168

Gain stability in fluorine polymer modified PbS Quantum dot fiber amplifier

Qi Liang[1], Nana Li[1], Bin Zhou[1], Hanning Zhang[1], Xiaolan Sun[1] and Alan R. Kost[2]

[1]The key Lab of Specially Fiber Optics and Optical Access Neyworks, Shanghai University, Shanghai,20072
[2]College of Optical Sciences, The University of Arizona, Tucson, AZ85721-0094

xiaolansun@shu.edu.cn

Abstract. The fluorescence intensity drop rate of PbS quantum dot with fluorine-containing polymer is 9.76% and the gain drop rate of quantum dot fiber amplifier (QDFA) with fluorine-containing polymer is 57.66% separately with temperature, which is more stable than no fluorine polymer QDFA.

1. Introduction

A fiber amplifier is a key component for optical fiber communication systems featuring direct in-line amplification. The erbium-doped fiber amplifier (EDFA) is one of the most dominating commercial products in the fiber amplifier market[1] and has been applied in various fiber communication systems. However, the gain bandwidth of EDFA only covers the C band (1525nm-1560nm) due to the limited band width of erbium[2]. Therefore, there is a need to find alternative new gain media to improve the performance of fiber amplifiers. In recent years, the quantum dot fiber amplifier (QDFA) has drawn great attention on account of its broad band, tunable bandwidth, compact size, simple fabrication process and compatibility with standard optical fiber communication systems[3-7].

In this paper, a novel optical amplifier having a fused bi-conical single mode fiber (SMF) coupler is demonstrated, as shown in Fig. 1[5], on which PbS QDs capped with fluorine-containing polymers (FCPs) were coated. The multifunctional copolymers were designed to realize simultaneous QDs ligand exchange, encapsulation, dispersion, and protection. A QDFA modified by such FCPs exhibits exceptional gain stability.

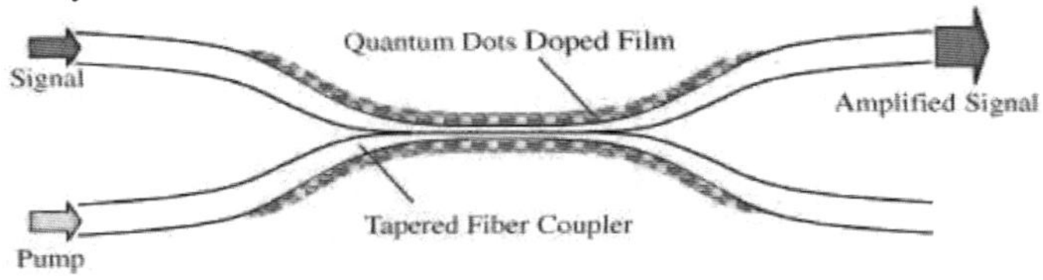

Fig. 1. Schematic device structure of the semiconductor quantum dots fiber amplifier (SQDFA)[5].

2. Fluorescence thermal stability of PbS quantum dot

Adopting 980 nm laser as pump source, we research fluorescence thermal stability of PbS QDs with temperature increasing. Fluorescent variety of PbS QDs modified by three fluorine content polymer takes on different downtrend. As Fig. 2 shown, fixed pump power and other experimental factors, the fluorescence intensity at 1550 nm of PbS QDs modified by 0%, 15% and 30% fluorine-containing polymer drop by 26.27%, 11.13% and 9.76% respectively from 25℃ to 65℃. It is obvious that the growing of fluorine content contributes to improving surrounding invulnerability. The reason why consistent trend of QDs with 15% and 30% fluorine-containing polymer coincides by and large is weakening binding affinity of QDs and polymer.

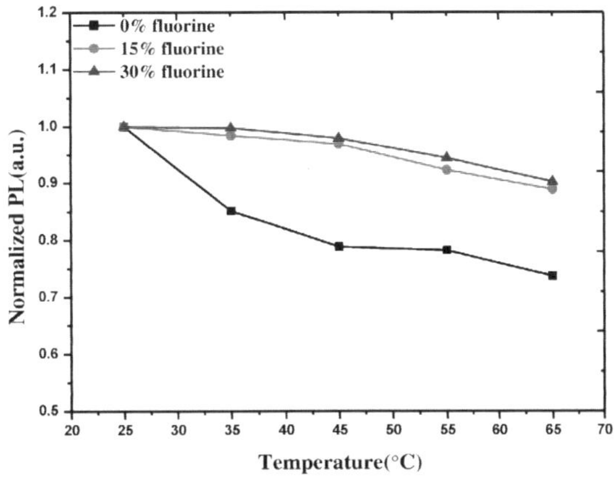

Fig. 2. The curve of fluorescence at 1550 nm of PbS QDs modified by different fluorine content polymers with temperature

3. Gain stability of quantum dot fiber amplifier with temperature

The PbS QDs were synthesized according to the Cademartiri method[8]. They were then modified by FCPs to improve surface characteristics and enhance their gain stability. We adopted a SiO2 sol-gel method to form PbS QDs doped films on the surface of a fused bi-conical SMFs coupler which is coated with QDs doped films. As Fig. 3 shown, a 1550 nm semiconductor light emitting diode (SLED) and a 980 nm laser diode (LD) are used as the signal and pump respectively. The signal and pump sources are injected into the tapered region simultaneously. The 980 nm pump excites the doped PbS QDs at the tapered region through an evanescent wave. The 1550 nm signal interacts with excited QDs through the evanescent wave and then can be amplified. A 1550nm/980nm wavelength division multiplex (WDM) is connected to an output of the QDFA. The amplified signal is analyzed by an optical spectrum analyzer (QSA).

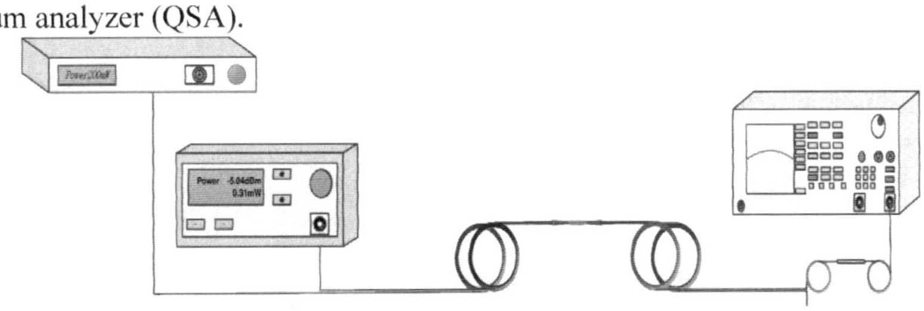

Fig. 3. The characterization system for PbS QDs optical fiber amplifier.

In this paper, we studied temperature-dependent stabilities of QDFA's gains on different fluorine contents. Fig. 4 shows the gain at 1550 nm at different temperature. The gain dropped by 93%, 57.66% and 74.68%, respectively, for QDFA modified with FCPs having 0, 15% and 30% fluorine. The gain declination is caused by vigorous internal electron thermal motion of QDs with rising temperature, leading to easier formation of spontaneous emission.

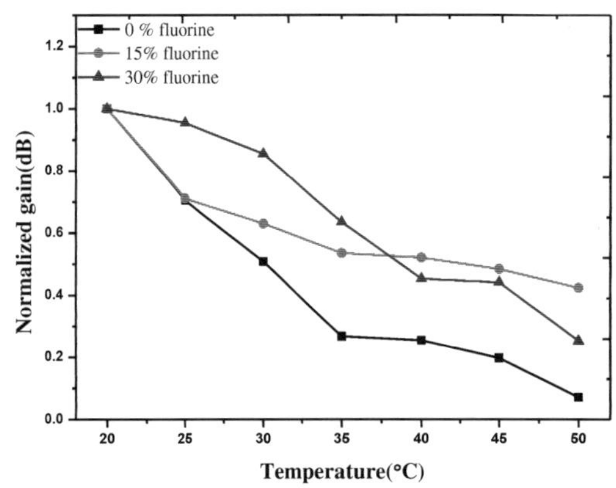

Fig. 4. temperature-dependent stability of gain of optical fiber amplifiers based on PbS QDs

Previously, we used an amphiphilic copolymer to encapsulate hydrophobic PbS QDs that were capped with either oleylamine or oleic acid ligands[6-7]. In this work, the FCPs we designed have the oligo (ethylene glycol) side chain that is hydrophilic and servers to maintain colloidal stability of QDs in silica sol. The carboxylic acid side group is used to ligand-exchange with oleylamine on the surface of PbS QDs that were synthesized in organic solution. Ligand exchange using these multifunctional copolymers not only results in polymer coating of QDs but also renders PbS QDs more stable because carboxylic acid is a stronger ligand than oleylamine.

4. Conclusion

The properties of PbS QDs and QDFAs which were modified by polymers with different fluorine content were investigated. The fluorescence intensity drop rate of QDs with no fluorine-containing polymer was about 3 times than QDs with 30% fluorine-containing polymer. Similarly, thermal stability of QDFA with fluorine-containing polymer exhibited more excellent constancy than QDFA with no fluorine-containing polymer. Due to the good fluorescence performance of the PbS QDs and improved gain stability of QDFA, the proposed fiber amplifier will find potential applications in broadband and high speed fiber-optic communication.

References

[1] Lu Y B, Chu P L, Alphones A, et al. A 105-nm ultrawide-band gain-flattened amplifier combining C-and L-band dual-core EDFAs in a parallel configuration[J]. IEEE Photonics Technology Letters, 2004, 16(7): 1640-1642.

[2] Yi L L, Zhan L, Taung C S, et al. Low noise figure all-optical gain-clamped parallel C+ L band Erbium-doped fiber amplifier using an interleaver[J]. Optics express, 2005, 13(12): 4519-4524.

[3] Akiyama T, Sugawara M, Arakawa Y. Quantum-dot semiconductor optical amplifiers[J]. Proceedings of the IEEE, 2007, 95(9): 1757-1766.

[4] Erneux T, Viktorov E A, Mandel P, et al. The fast recovery dynamics of a quantum dot semiconductor optical amplifier[J]. Applied physics letters, 2009, 94(11): 113501.

[5] Pang F, Sun X, Guo H, et al. A PbS quantum dots fiber amplifier excited by evanescent wave[J]. Optics express, 2010, 18(13): 14024-14030.

[6] Sun X, Xie L, Zhou W, et al. Optical fiber amplifiers based on PbS/CdS QDs modified by polymers[J]. Optics express, 2013, 21(7): 8214-8219.

[7] Sun X, Dai R, Chen J, et al. Enhanced thermal stability of oleic-acid-capped PbS quantum dot optical fiber amplifier[J]. Optics express, 2014, 22(1): 519-524.

[8] Moreels I, Justo Y, De Geyter B, et al. Size-tunable, bright, and stable PbS quantum dots: a surface chemistry study[J]. Acs Nano, 2011, 5(3): 2004-2012

IOP Conf. Series: Journal of Physics: Conf. Series **844** (2017) 012064 doi:10.1088/1742-6596/844/1/012064

Two-photon luminescence and stimulated emission depletion with gold nanorods by a single wavelength

Jianling Chen[1], Minghai You[1], Yiru Peng[2], Hongqin Yang[1,*], Shusen Xie[1]

[1]Key Laboratory of Optoelectronic Science and Technology for Medicine of Ministry of Education, Provincial Key Laboratory for Photonics Technology, Institute of Laser and Optoelectronics Technology, Fujian Normal University, Fuzhou, 350007

[2]College of Chemistry & Engineering, Fujian Provincial Key Laboratory of Polymer Materials, Fujian Normal University, Fuzhou, 350007,

*Corresponding author, e-mail: hqyang@fjnu.edu.cn.

Abstract Single-wavelength two photon STED technique simplifies common STED setup, and decreases the cost of the system. However, this technique limits the fluorescent dye to a few kinds. Because the dye should be excited and depleted by the same single wavelength laser. And the gold nanorods can be excited by near-infrared pulsed laser and produces two photon luminescence, which is brighter than general two-photon autofluorescence by three orders. We found that the gold nanorods can also be depleted by near-infrared pulsed laser. We studied the stimulated emission depletion of gold nanorods with a single wavelength laser in the near infrared. Two photon luminescence was excited with a femtosecond pulse, then depleted by stimulated emission with a stretched pulse. The depletion efficiency with different wavelength was measured to pick out the optimal wavelength. We showed that this method can be applied to super-resolved STED microscopy, which combines the high brightness of the nanorods and the simplicity of the single-wavelength STED system.

Keywords: nanorod, luminescence, STED, super-resolution

1. Introduction

Stimulated emission depletion (STED) microscopy has been developed by Hell and Wichmann two decades ago, providing spatial resolution well beyond the diffraction limit [1]. STED microscopy is generally based on the luorescence switching because of stimulated emission depletion produced by a second red shifted beam, named STED beam. The STED beam is normally shaped to a doughnut-like distribution featuring zero intensity in the center. Earlier STED microscopy used two synchronized pulsed lasers: one for excitation of typically less than 100 ps duration and an another 250 ps pulse for depletion, which requires a relatively complex and expensive setup [1-3]. And later STED microscopy has been developed with continuous wave (CW) lasers, simplifying the mplementation [4]. Moreover, researchers have coupled STED with two-photon microscopy to combine the advantages of two-photon microscopy with the super-esolution ability of STED microscopy [5-7]. All the above-mentioned STED microscopy, the excitation and depletion are implemented by two different wavelengths. And single-wavelength two-photon excitation stimulated emission depletion microscopy has been developed, simplifying the STED optical scheme by using a single wavelength for both two photon excitation and depletion [8-9]. However, this single-wavelength STED limits the fluorescent dye to a few kinds. Because the dye should be excited and depleted by the same single wavelength laser.

On the other hand, gold nanorods (NRs) has attracted much interest for their highly efficient absorption in the near-infrared (NIR) region. Gold nanorods exhibit highly efficient two-photon luminescence, significant enhanced by longitudinal surface plasmon resonance (LSPR) [10]. And Two-photon luminescence (TPL) has been regarded as a serial process, that involves sequential absorption of photons and emission from the recombination of electrons in the sp-band and holes in the d-band [11]. And several progresses have been made in understanding the optical properties of gold nanorods [11-12]. Their LSPR peak can be finely tuned from visible light to NIR region, as a function of aspect ratio [12]. Furthermore, the LSPR are sensitive to the polarization of the incident excitation [13]. And researchers have made efforts to explore the applications of gold nanorods as contrast agents for imaging [13]. And NRs were used for molecularly specific nonlinear imaging, and the results showed that the TPL intensity from NRs labeled cancer cells was three orders of magnitude brighter than the two-photon autofluorescence (TPAF) from the unlabeled cancer cells

IOP Conf. Series: Journal of Physics: Conf. Series **844** (2017) 012064 doi:10.1088/1742-6596/844/1/012064

[15]. We found that the NRs could also be depleted by near-infrared pulsed laser. And we aimed to combine the high brightness of the nanorods and the simplicity of the single-wavelength STED system.

2. Materials and methods

2.1. Experimental setup

The scheme of our experimental setup is shown in Fig. 1. An ultrafast tunable (680-1080 nm) Ti:sapphire laser (Chameleon Ultra II; Coherent) was used as light source for both two-photon excitation and stimulated emission of the gold nanorods. The laser beam is split into two paths by a polarizing beam splitter (PBS), and the relative power of the two paths can be tubed by a half-wave plate in front of the PBS. The first beam is used to excite the gold nanorods and the second beam is used to deplete the excited state by stimulated emission. And the power of the second beam is be further adjusted by another pair of PBS and half-wave plate. The STED beam is expanded with two lenses, and its pulses is stretched with 30-cm glass rod and 100-m polarization-maintaining optical fiber in order to avoid any linear or nonlinear excitation. On this beam path, after the fiber, an optical delay line, made with a pair of mirrors mounted on a linear stage, is used to temporally overlap the pulses of the two beams at the sample. And the final position of the delay line is chosen by optimizing the maximum depletion. And a 0 to 2 π vortex phase plate is used to create doughnut-like pattern at the focus. The two-photon excitation beam path is expanded and recombined with the STED beam by a PBS, then their polarization is made circular by a quarter wave-plate. With a pair of mirrors, the recombined beams are imported into the infrared port of the confocal unit of the microscope (Leica TCS SP8; Leica Microsystems). The two beams are raster scanned into the back aperture of an water-immersion objective lens (N.A. = 1.20 ,63 ×, Leica) with a set of galvanometric scanning mirrors. Emitted light is passed through a laser filter (blocks >680 nm), detected with a photomultiplier tube. And the system collected emission light between 500 and 650 nm.

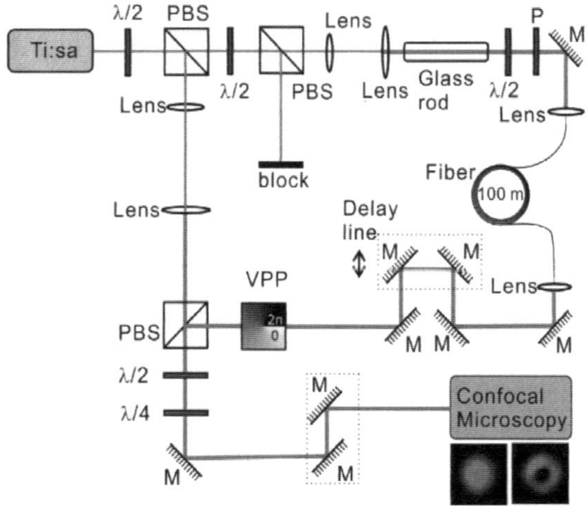

Fig. 1. Scheme of the experimental setup. P is polarizer and VPP represents vortex phase plate.

2.2. Sample preparation

Gold nanorods were synthesized in solutions by a seed-mediated method as reported previously [16, 17]. The gold nanorod samples for imaging were prepared by dispersing 5 μl dilution onto a glass slide and covered by a coverslip.

3. Results

Figure 2 shows a transmission electron microscopy (TEM) image of synthesized gold nanorods. We first examined the depletion of the two-photon luminescence of the gold nanorods with the single wavelength (770 nm). As shown in Figs. 3(a)-(c), we can observe obviously luminescence quenching at the presence of STED beam, and luminescence recovery without STED beam. The two-photon luminescence quenching is about 75%, when the STED pulse is on, which can be observed from Fig. 3(g).

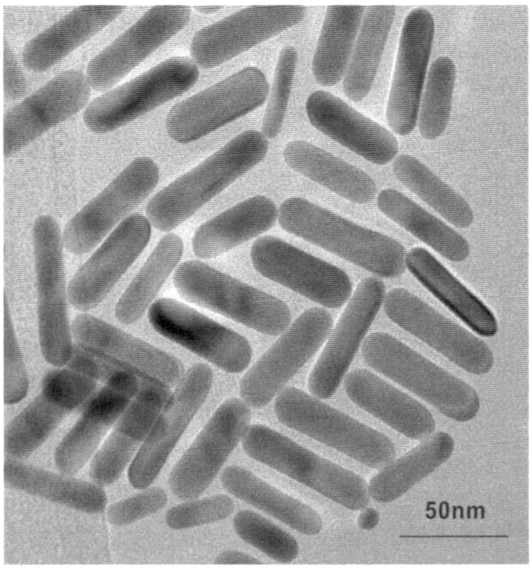

Fig.2 TEM image of gold nanorods.

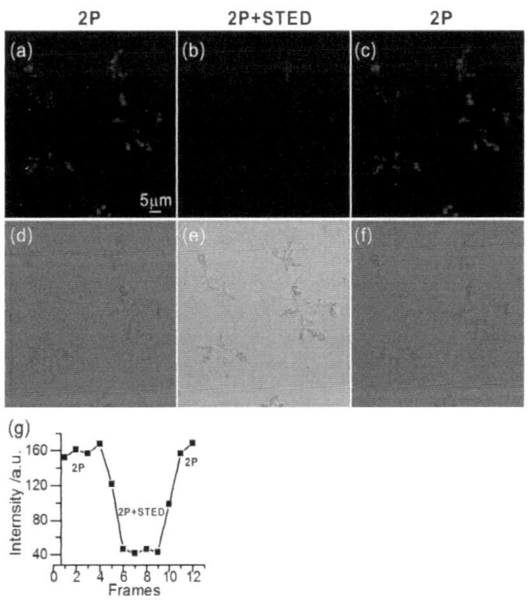

Fig. 3. The depletion of the two-photon luminescence of the gold nanorods. (a)-(c) are the luminescence images, and (d)-(f) are the transmission image with the wavelength equal to 770 nm. (d) Quantification of the signal intensity from the nanorods in (a)-(c).

In order to pick out the optimal wavelength for both excitation and depletion, we evaluated the two-photon luminescence with and without the STED beam, and the depletion efficiency as function of wavelength. Figure 4 shows the wavelength dependencies, remaining the two-photon and STED intensities constant. The black line shows the luminescence induced by the two-photon beam without the presence of the STED beam. Meanwhile we measured the luminescence while superimposing the two-photon and the STED beam (red line, Fig. 4). And then the depletion efficiency was calculated, as shown in Fig. 3 with the green line. As seen from Fig. 4, obvious depletion (depletion efficiency >0.7) can be observed over the measured wavelength from 730 nm to 820 nm. And the maximum excitation was produced at the wavelength of 770 nm, and the maximum depletion was obtained at the wavelength of 760 nm. Because the NRs are bright contrast agent, we chose the wavelength of 760 nm for latter experiment for the better depletion.

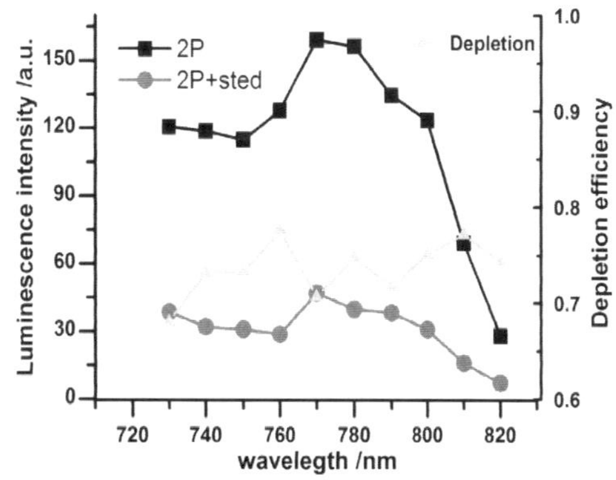

Fig. 4. Two-photon luminescence with (red dots) and without (black squares) the STED beam, and the depletion efficiency (green triangles) as function of wavelength, keeping the two-photon and STED intensities constant.

We analyzed the depletion efficiency by measuring the luminescence depending on the STED beam intensity, while keeping 2P excitation power constant at 6 mw. The results are reported in Fig. 5. The curve shows the luminescence reduction due to the STED effect. As predicted, luminescence quenching increases exponentially with the STED beam intensity.

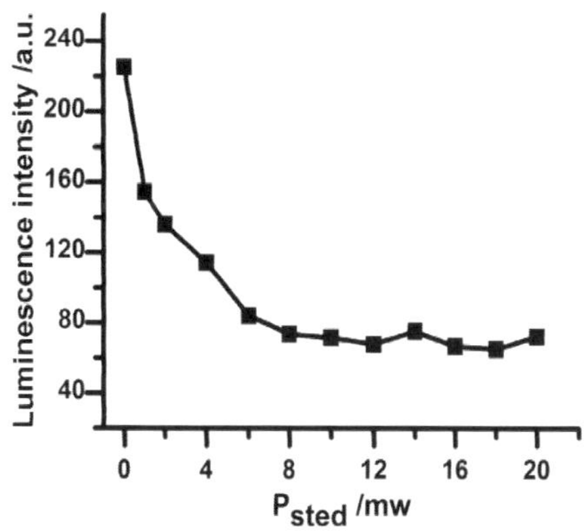

Fig. 5. Depletion of the gold nanorods luminescence, 2P excited and depleted at 760 nm, as a function of the STED beam power P_{STED}. The power of 2P beam is 6 mw.

We used the above optical setup to image gold nanorods to demonstrate the ability of single wavelength 2P STED to image with sub-diffraction limited resolution, with the STED beam on or off. And the results are shown in Fig. 6. Comparison of 2P imaging, single wavelength 2P STED reveals an improvement of lateral resolution when the STED laser is on (Fig. 6 (a) and (b)). And the resolution was quantitatively calculated. Under 2P imaging, images of displayed a full-width at half maximum (FWHM) of 265 nm, shown in Fig. 6(c). In contrast, single wavelength 2P STED images exhibited a FWHM of 190 nm, shown in Fig. 6(d), indicating about 1.4-fold improvement in resolution.

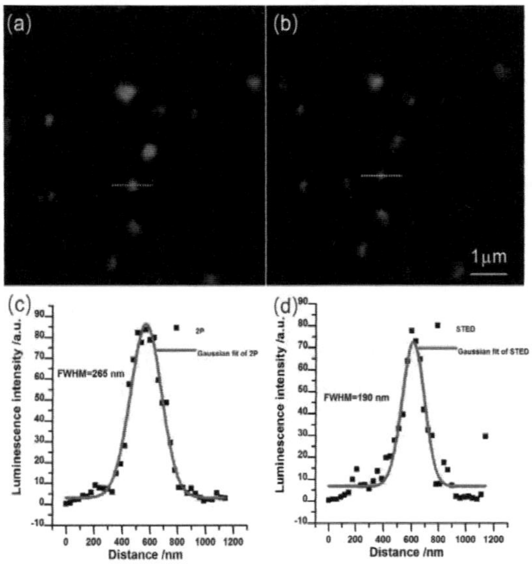

Fig. 6. Two-photon luminescence images of gold nanorods acquired with 2P (a) and single wavelength 2P STED (b). The images were collected by switching the STED laser on and off. (c) and (d) are the profile of intensity along the dashed lines in (a) and (b), respectively.

4. Discussion and conclusion

We showed that gold nanorods can be depleted by near-infrared pulsed laser, which induce two-photon luminescence of the nanorods. With the properties, gold nanorods can be used for single wavelength two-photon STED imaging, which simplifies the optical setup in terms of number of dichroic filters and laser sources of traditional STED microscopy. And The depletion efficiency with different wavelength was measured to pick out the optimal wavelength for single wavelength two-photon STED imaging of gold nanorods. The resolution of single wavelength 2P STED image indicated about 1.4-fold improvement in resolution than that of 2P image. And the possible reason of a little resolution improvement is that the two beam are not well overlapped in the 3D space. And we are making the effort to optimize the arrangement of the beams to improve the resolution. The proposed method combines the high brightness of the nanorods and the simplicity of the single-wavelength STED system, which seems a promising technique to imaging of biological samples at nanoscale resolution.

Acknowledgments

This work is partly supported by the National Key Basic Research Program of China (973 Program) under Grant No. 2015CB352006 and the National Natural Science Foundation of China under Grants Nos. 61335011 and 61405035, and Program for Changjiang Scholars and Innovative Research Team in University under Grant No. IRT 15R10.

References

[1] S. W. Hell and J. Wichmann, "Breaking the diffraction resolution limit by stimulated emission: stimulated-emission-depletion fluorescence microscopy," Opt. Lett. **19**(11), 780-782 (1994).

[2] T. A. Klar, S. Jakobs, M. Dyba, A. Egner, and S. W. Hell, "Fluorescence microscopy with diffraction resolution barrier broken by stimulated emission," Proceedings of the National Academy of Sciences **97**(15), 8206-8210 (2000).

[3] G. Donnert, J. Keller, R. Medda, M. A. Andrei, S. O. Rizzoli, R. Luhrmann, R. Jahn, C. Eggeling, and S. W. Hell, "Macromolecular-scale resolution in biological fluorescence microscopy," Proceedings of the National Academy of Sciences **103**(31), 11440-11445 (2006).

[4] K. I. Willig, B. Harke, R. Medda, and S. W. Hell, "STED microscopy with continuous wave beams," Nat. Methods **4**(11), 915-918 (2007).

[5] G. Moneron and S. W. Hell, "Two-photon excitation STED microscopy," Opt. Express **17**(17), 14567-14573 (2009).

[6] J. B. Ding, K. T. Takasaki, and B. L. Sabatini, "Supraresolution Imaging in Brain Slices using Stimulated-Emission Depletion Two-Photon Laser Scanning Microscopy," Neuron **63**(4), 429-437 (2009).

[7] Q. Li, Y. Wang, D. Chen, and S. S. Wu, "2PE-STED microscopy with a single Ti:sapphire laser for reduced illumination," Plos One **9**(2), e88464 (2014).

[8] T. Scheul, C. D'Amico, I. Wang, and J. C. Vial, "Two-photon excitation and stimulated emission

depletion by a single wavelength," Opt. Express **19**(19), 18036-18048 (2011).

[9] P. Bianchini, B. Harke, S. Galiani, G. Vicidomini, and A. Diaspro, "Single-wavelength two-photon excitation-stimulated emission depletion (SW2PE-STED) superresolution imaging," Proceedings of the National Academy of Sciences **109**(17), 6390-6393 (2012).

[10] M. B. Mohamed, V. Volkov, S. Link, and M. A. El-Sayed, "The `lightning' gold nanorods: fluorescence enhancement of over a million compared to the gold metal," Chem. Phys. Lett. **317**(6), 517-523 (2000).

[11] K. Imura, T. Nagahara, and H. Okamoto, "Near-Field Two-Photon-Induced Photoluminescence from Single Gold Nanorods and Imaging of Plasmon Modes," The Journal of Physical Chemistry B **109**(27), 13214-13220 (2005).

[12] T. Wang, D. Halaney, D. Ho, M. D. Feldman, and T. E. Milner, "Two-photon luminescence properties of gold nanorods," Biomed. Opt. Express **4**(4), 584-595 (2013).

[13] H. Wang, T. B. Huff, D. A. Zweifel, W. He, P. S. Low, A. Wei, and J. X. Cheng, "In vitro and in vivo two-photon luminescence imaging of single gold nanorods," Proceedings of the National Academy of Sciences **102**(44), 15752-15756 (2005).

[14] X. Huang, I. H. El-Sayed, W. Qian, and M. A. El-Sayed, "Cancer Cell Imaging and Photothermal Therapy in the Near-Infrared Region by Using Gold Nanorods," J. Am. Chem. Soc. **128**(6), 2115-2120 (2006).

[15] N. J. Durr, T. Larson, D. K. Smith, B. A. Korgel, K. Sokolov, and A. Ben-Yakar, "Two-Photon Luminescence Imaging of Cancer Cells Using Molecularly Targeted Gold Nanorods," Nano Lett. **7**(4), 941-945 (2007).

[16] B. Nikoobakht and M. A. El-Sayed, "Preparation and Growth Mechanism of Gold Nanorods (NRs) Using Seed-Mediated Growth Method," Chem. Mater. **15**(10), 1957-1962 (2003).

[17] T. K. Sau and C. J. Murphy, "Seeded High Yield Synthesis of Short Au Nanorods in Aqueous Solution," Langmuir **20**(15), 6414-6420 (2004).